ELEMENTS
OF
APPLIED
THERMODYNAMICS

ELEMENTS OF APPLIED THERMODYNAMICS

FOURTH EDITION

by

Robert M. Johnston
Late Professor of Mechanical Engineering
United States Naval Academy

William A. Brockett
Rear Admiral, United States Navy (Retired)
Former Chief, Bureau of Ships, Navy Department

Arthur E. Bock
Professor of Naval Systems Engineering
United States Naval Academy

Eugene L. Keating
Assistant Professor of Mechanical Engineering
United States Naval Academy

Naval Institute Press
Annapolis, Maryland

Copyright © 1978, 1958, 1954, 1951 by the United States Naval Institute, Annapolis, Maryland.

Fourth edition

All rights reserved.

No part of this book may be reproduced without written permission from the publisher.

Library of Congress Catalogue Card No. 77-073341

ISBN 0-87021-169-2

Printed in the United States of America.

PREFACE TO THE FOURTH EDITION

The demonstrated success of earlier versions of this textbook in the teaching of fundamental thermodynamic concepts to the undergraduate student was largely responsible for this new edition. Additional stimuli were the need for updating certain thermodynamic data, the desire to include additional operating component characteristics and the opportunity to introduce the International System of Units (SI). The current edition introduces the thermodynamic properties of air from the Gas Tables, an expanded discussion of incompressible flow and liquid pumps and material on axial and centrifugal flow compressors. New example problems in certain areas and additional chapter-end problems, some in SI units, have been included.

Several minor revisions of the First Edition were made by the authors in later years in order to incorporate new practical applications of the basic concepts, with particular reference to the rapid developments in gas turbine technology. The efforts of Capt. B. P. Miller, USAF, who was responsible for the data on aircraft gas turbines, and of Assistant Professor N. F. Spraggins, who prepared the accompanying illustrations, are particularly appreciated. Throughout the preparation of this new edition the interest and constructive comments of the faculty and staff of the Engineering and Weapons Division have proved most helpful. The authors also express their deep appreciation to Mrs. E. L. Keating for her considerable contribution in typing and checking the many details in manuscript preparation. Following the untimely death of Professor Johnston, the two other original authors were joined by Professor Keating in the current effort.

All data for steam in its various phases have been based on the *Steam Tables (English Units)* by Keenan, Keyes, Hill and Moore, published by John Wiley & Sons, Inc., 1969. Properties of air have been based on the *Gas Tables* by Keenan and Kaye, published by John Wiley, 1948.

San Diego, CA
Annapolis, MD
January 1978

W. A. BROCKETT
A. E. BOCK
E. L. KEATING

PREFACE TO THE FIRST EDITION

The aim of this textbook is to present to the undergraduate the fundamental concepts of thermodynamics in a clear, concise fashion and to give those concepts real physical significance by early and frequent application to actual thermal equipment. The authors feel that a thorough understanding of the subject and the ability to reason logically are developed by continuing emphasis on the relatively few basic principles together with repeated application of these principles to a variety of practical problems. Example problems are used freely throughout and are designed to illustrate, emphasize and broaden the associated descriptive material.

The arrangement of the material is in support of the objectives outlined above. The preliminary chapters are concerned with the First Law of Thermodynamics as embodied in the General Energy Equation and with the properties of the common working substances. A study of the air compressor follows, giving an immediate application of theory to an actual machine. The Second Law of Thermodynamics and other fundamental cyclic concepts are next presented and then applied to the important ideal cycles. Subsequently the operating principles and performance criteria of the major components in the steam power cycle are considered with particular emphasis on the steam turbine. In addition, the book deals with heat transfer, air conditioning, and the combustion of fuel-air mixtures.

The developments employed presume that the student is familiar with general physics and the elementary calculus.

The authors are indebted to Capt. C. S. Seabring, U.S. Navy, Head of the Department of Marine Engineering, United States Naval Academy, for his interest and encouragement in the preparation of this text. Thanks are also due the members of the instructional staff teaching thermodynamics for their constructive criticism of the text material and assistance in the preparation and solution of problems. The efforts of Assistant Professor N. F. Spraggins, who prepared the accompanying illustrations, are particularly appreciated.

Annapolis, Maryland
September 1950

R. M. JOHNSTON
W. A. BROCKETT
A. E. BOCK

CONTENTS

PREFACE TO THE FOURTH EDITION . v
PREFACE TO THE FIRST EDITION . vii
SYMBOLS WITH ASSOCIATED BASIC UNITS IN ENGINEERS' AND (SI) UNIT SYSTEMS .
ABBREVIATIONS .

CHAPTER 1

INTRODUCTION AND THE FIRST LAW OF THERMODYNAMICS 1
(1) General. (2) Dimensions and units. (3) Working substances. (4) Properties. (5) Specific volume, specific weight and density. (6) Pressure. (7) Manometers. (8) Temperature. (9) Energy. (10) Potential energy. (11) Kinetic energy. (12) Internal energy. (13) Heat. (14) Mechanical work. (15) Flow work. (16) The First Law. (17) The Steady Flow Energy Equation. (18) Power.

CHAPTER 2

BASIC CONCEPTS AND THE GENERAL NON-FLOW PROCESS 31
(1) The approach. (2) Properties, states and processes. (3) The reversible process. (4) The energy equation for non-flow processes. (5) Enthalpy. (6) Mechanical work and $\int p\,dv$ for a reversible non-flow process. (7) the p-v diagram. (8) Heat, entropy and the T-s diagram. (9) Specific heat (general). (10) Specific heats of a perfect gas. (11) Specific heats of real gases. (12) Closure.

CHAPTER 3

CHARACTERISTICS OF GASES AND THE NON-FLOW GAS PROCESSES . . . 56
(1) Introduction. (2) The gas laws. (3) The polytropic process equation. (4) Introduction to the non-flow reversible gas processes. (5) Work and the p-v diagram. (6) Internal energy of a gas. (7) Enthalpy of a gas and some thermodynamic relationships. (8) Heat and entropy. (9) The T-s diagram for non-flow gas processes. (10) Applications of the non-flow gas processes. (11) The Air Table.

CHAPTER 4

PROPERTIES AND CHARACTERISTIC BEHAVIOR OF STEAM 94
(1) Introduction. (2) Graphical representation and the saturation lines. (3) General nature of the Steam Tables. (4) The saturated liquid and saturated vapor. (5) The wet vapor. (6) The superheated vapor. (7) The compressed liquid. (8) The p-v diagram for steam. (9) The T-s diagram for steam. (10) The h-s diagram and Mollier chart for steam. (11) Reversible non-flow processes with steam.

CHAPTER 5

ELEMENTARY STEADY FLOW PROCESSES . 124
(1) Steady flow. (2) The energy equation for steady flow. (3) Nozzles. (4) Turbines. (5) Boilers. (6) Incompressible flow and liquid pumps. (7) Compressors. (8) Throttles. (9) The throttling calorimeter. (10) The two-fluid heat balance. (11) The multi-fluid energy balance.

CONTENTS

CHAPTER 6

AIR COMPRESSION 153
(1) Background. (2) The reciprocating compressor. (3) Ideal air compression, single stage. (4) Best compression process and desirability of low value of n. (5) Clearance and effects of clearance. (6) Multi-stage compression. (7) The air compressor as a steady flow machine. (8) Economy resulting from staging and cooling. (9) Performance of an actual reciprocating compressor. (10) Steady flow compressive turbomachinery.

CHAPTER 7

CYCLES, AVAILABLE ENERGY AND THE SECOND LAW OF THERMODYNAMICS .. 182
(1) Cycles and their requirements. (2) Thermal efficiency and The Second Law. (3) Types of irreversibility. (4) Ideal cycle analysis. (5) Pump work and compression work. (6) The reversed Carnot cycle. (7) The Carnot principle. (8) Available energy. (9) Actual thermal and engine efficiencies. (10) Entropy as an index of unavailability and irreversibility. (11) Closure.

CHAPTER 8

POWER CYCLES (GAS) 223
(1) Early development. (2) Methods of analysis. (3) The Stirling and Ericsson cycles. (4) The Otto cycle. (5) The Diesel cycle. (6) The dual combustion cycle. (7) The Joule or Brayton cycle. (8) The open cycle gas turbine. (9) Brayton cycle aircraft gas turbine analysis. (10) Air Table analysis of gas power cycles.

CHAPTER 9

POWER CYCLES (STEAM) 271
(1) Introduction. (2) The Rankine cycle. (3) The Rankine reheat cycle. (4) The Rankine regenerative cycle. (5) The Rankine regenerative reheat cycle. (6) Nuclear reactor steam plants and cycles. (7) The incomplete-expansion and non-expansion cycles.

CHAPTER 10

REFRIGERATION CYCLES 299
(1) Background. (2) Ideal refrigeration—the reversed Carnot cycle. (3) Refrigerating effect, capacity, coefficient of performance. (4) Vapor compression refrigeration. (5) The refrigerant rating cycle. (6) Practical aspects of vapor compression systems. (7) Vapor refrigerants. (8) Air compression system—reversed Joule and Brayton cycles. (9) Vapor absorption systems. (10) Heating with a refrigeration system.

CHAPTER 11

RECIPROCATING STEAM ENGINES 321
(1) Foreword. (2) Apparatus for steam engine cycles. (3) Work of ideal incomplete-expansion steam engine. (4) Work of the ideal non-expansion steam engine. (5) The actual indicator diagram. (6) Indicated horsepower. (7) Engine efficiency and steam rate. (8) Steam

CONTENTS

engine losses. (9) The Uniflow steam engine. (10) Multi-cylinder engines. (11) Effects of superheat and back pressure. (12) Mechanical efficiency.

CHAPTER 12

NOZZLES . 335
(1) Foreword. (2) Definition, function and uses of nozzles. (3) The nozzle energy equation. (4) Velocity of gases in ideal nozzles. (5) Fluid flow and nozzle shape. (6) Nozzle flow and stagnation conditions. (7) Critical flow with compressible fluids. (8) Performance of the actual nozzle, velocity coefficient and nozzle efficiency. (9) Flow of vapors in nozzles. (10) Steam flow in nozzles. (11) Overexpansion and underexpansion in nozzles.

CHAPTER 13

STEAM TURBINES . 358
(1) Introduction. (2) Turbine types. (3) Impulse blading. (4) The simple impulse stage. (5) The reaction stage. (6) Comparison of impulse and reaction stages. (7) The large multi-stage turbine. (8) Internal work and internal engine efficiency. (9) Governing. (10) Reheat and reheat factor. (11) Brake work and brake engine efficiency. (12) Other performance criteria.

CHAPTER 14

HEAT TRANSMISSION . 392
(1) Introduction. (2) Modes of heat transfer. (3) Conduction of heat through a homogeneous material. (4) Fluid films and effect of convection upon heat transfer. (5) Overall coefficient of heat transfer. (6) The resistance concept. (7) Conduction through walls of nonuniform area. (8) Fluid films—desirable and undesirable. (9) Radiation.

CHAPTER 15

HEAT TRANSFER EQUIPMENT . 411
(1) Foreword. (2) Mean temperature difference. (3) Parallel and counter flow. (4) Heat transfer with parallel and counter flow. (5) Actual heat transfer equipment.

CHAPTER 16

PROPERTIES OF MIXTURES . 424
(1) Foreword. (2) Dalton's law and the universal gas constant. (3) Volumetric analysis, partial pressures and mass analysis. (4) Specific heats and thermal properties. (5) Atmospheric air. (6) Partial pressures, specific humidity, relative humidity, and dewpoint. (7) Psychrometry. (8) Enthalpy of air-vapor mixtures. (9) Adiabatic saturation, psychrometry and psychrometric charts.

CHAPTER 17

AIR CONDITIONING . 445
(1) Introduction. (2) Conditioned air. (3) Psychrometric processes for air conditioning. (4) Winter air conditioning. (5) Summer air conditioning.

CONTENTS

CHAPTER 18

BOILERS AND THE CHEMISTRY OF COMBUSTION 461
(1) Introduction. (2) Boiler terminology. (3) Types of fossil fuel. (4) Properties of fuel oil. (5) Combustion equations. (6) Combustion with excess air. (7) Effects of air deficiency on combustion. (8) Analysis of flue gases. (9) The boiler energy balance.

APPENDIX A . 487
Engineering systems of dimensions and units.

APPENDIX B . 494
Thermodynamic properties and pressure-enthalpy diagram of dichlorodifluoromethane (Freon-12); through the courtesy of "Kinetic" Chemicals division, E. I. du Pont de Nemours and Company.

APPENDIX C . 504
Thermal efficiencies of cold air standard Diesel and dual combustion cycles. Max. work related to pressure ratio for Brayton cycle. Regenerative Brayton cycle thermal efficiency.

APPENDIX D . 509
The generalized area and velocity relationship for compressible flow.

APPENDIX E . 511
Thermodynamic properties of air at low pressures.

INDEX . 517

SYMBOLS WITH ASSOCIATED BASIC UNITS IN ENGINEERS' AND (SI) UNIT SYSTEMS

a	linear acceleration, ft/sec^2 (m/sec^2)
A	area, ft^2 (m^2)
c	specific heat, Btu/lb · °R (J/kg · K)
c	velocity of sound, ft/sec (m/sec)
c_p	specific heat at constant pressure
c_v	specific heat at constant volume
C	coefficient or mathematical constant
C_b	blade velocity coefficient
C_n	nozzle velocity coefficient
d	diameter, ft (m)
F	force, lbf (N)
f	subscript applied to symbols h, s, u, v to denote the saturated liquid properties
fg	subscript applied as above to denote change of property during constant pressure-temperature evaporation
g	subscript applied as above to denote the saturated vapor properties
g	acceleration of gravity, ft/sec^2 (m/sec^2)
g_c	dimensional constant, 32.2 lb · ft/lbf · sec^2 (1.0 kg · m/N · sec^2)
h	specific enthalpy (en · thal' pi), Btu/lb (J/kg)
h	surface (film) coefficient, Btu/hr · ft^2 · °F (J/hr · m^2 · °C or W/m^2 · °C)
H	total enthalpy, Btu (J)
H.H.V.	higher heating value of a fuel, Btu/lb (J/kg)
J	mechanical equivalent of heat, 778 ft · lbf/Btu
k	specific heat ratio, c_p/c_v
k	thermal conductivity, Btu/hr · °F · ft (J/hr · °C · m or W/m · °C)
L	length, ft (m)
m	moisture, percent or decimal
M	mass, lb (kg)
\dot{M}	mass rate of flow, lb (kg) per unit time
M	molecular weight, lb/lb mol (kg/kg mol)
n	exponent of polytropic gas process ($pv^n = C$)
N	revolutions per second (rps) or minute (rpm)
N_m	Mach number
p	pressure, lbf/ft^2 (N/m^2 or Pa)
p_r	relative pressure as used in Air Tables
P	power, work per unit time
q	quantity of heat per unit weight of substance, Btu/lb (J/kg)

xiii

SYMBOLS WITH BASIC UNITS

Q	total quantity of heat, Btu (J)
\dot{Q}	rate of heat flow (transfer), Btu (J) per unit time
r	radius, ft (m)
r_p	pressure ratio
r_v	compression ratio
R	specific gas constant, ft · lbf/lb · °R (J/kg · K)
s	specific entropy (ĕn′trō · pĭ) Btu/lb · °R (J/kg · K)
S	total entropy, Btu/°R (J/K)
$S.H.$	superheat, °F (°C)
t	temperature, Fahrenheit, °F (Celsius, °C)
t	time, sec (T and τ are also used)
t	thickness, ft (m)
T	absolute temperature, °R (K)
u	specific internal energy, Btu/lb (J/kg)
U	total internal energy, Btu (J)
U	overall coefficient of heat transfer, Btu/hr · ft² · °F (J/hr · m² · °C)
v	specific volume, ft³/lb (m³/kg)
v_r	relative volume as used in the Air Tables
V	total volume, ft³ (m³)
V	velocity, ft/sec (m/sec)
W	total weight, lbf (N)
wk	specific work, ft · lbf/lb (J/kg)
Wk	total work, ft · lbf (J)
$\dot{W}k$	work per unit time, power
x	quality (of wet vapor), percent or decimal
y	reheat, Btu/lb (J/kg)
Z	height (elevation above a datum), ft (m)

α	alpha	nozzle angle, degrees
β	beta	blade angle, degrees
γ	gamma	specific weight, lbf/ft³ (N/m²)
δ	delta	differential of a path function (heat or work)
Δ	Delta	change or difference, usually of a property during a process or between established values of property
η	eta	efficiency; a subscript is always used. η_m is mechanical efficiency, η_t is thermal efficiency, etc.
θ	theta	log mean temperature difference, °F (°C)
π	pi	(3.1416)
ρ	rho	velocity ratio (blade speed-steam speed ratio)

SYMBOLS WITH BASIC UNITS

ρ	rho	density, lb/ft^3 (kg/m^3)
σ	sigma	Stefan-Boltzmann constant
Σ	Sigma	summation
ϕ	phi	relative humidity, decimal or percent
ϕ	phi	entropy function as used in the Air Tables

Notes: An encircled property symbol adjacent to a process line on any diagram or in conjunction with any term in an equation indicates that the circled property remains constant during the process. Thus ⓢ indicates an isentropic process, ⓟ a constant pressure process.

A dot over a symbol indicates a time rate of change. Thus M signifies mass, and \dot{M} signifies mass rate of flow.

ABBREVIATIONS

abs	absolute
atm	atmosphere
avg	average
bhp or BHP	brake horsepower
Btu	British thermal unit
°C or C	degree(s) Celsius
cos	cosine
deg or °	degree(s)
diam or dia	diameter
°F or F	degree(s) Fahrenheit
ft	foot (feet)
ft · lbf	foot pound force
ft/min or fpm	feet per minute
ft/sec or fps	feet per second
ft^3	cubic foot (feet)
ft^3/min or cfm	cubic feet per minute
ft^3/sec or cfs	cubic feet per second
gal	gallon(s)
gpm	gallons per minute
gps	gallons per second
Hg	mercury
hp	horsepower
hp · hr	horsepower hour
hr	hour
ID	inside diameter
ihp or IHP	indicated (or internal) horsepower
in	inch(es)
J	joule (newton metre)
K	kelvin, SI unit for absolute temperature
kg	kilogram(s) mass
kW	kilowatt(s)
lb	pound mass
lbf	pound force
lbf/in^2 or psi	pounds force per square inch
m	metre(s)
min	minute(s)
mol	mole of substance
N	newton (kilogram metre per second squared)
OD	outside diameter
Pa	pascal (newton per metre squared)

psia	pounds force per square inch absolute
°R or R	degree(s) Rankine (Fahrenheit absolute)
rpm	revolutions per minute
rps	revolutions per second
sec	second(s)
shp or SHP	shaft horsepower
sin	sine
sq ft or ft²	square foot (feet)
W	watt (joules per second)

ELEMENTS
OF
APPLIED
THERMODYNAMICS

CHAPTER 1

INTRODUCTION AND THE FIRST LAW OF THERMODYNAMICS

1-1. General. Thermodynamics, in the broadest sense, is that branch of physics which deals with forms, transformation and evaluation of energy. Thermodynamics emerged from general physics as a science in itself coincident with the beginning of the Machine Age and developed rapidly along logical, mathematical and experimental lines. As might be inferred from the definition above, thermodynamics as such embraces a very wide field. In this text we shall confine ourselves to those aspects of the general science which are of immediate interest to the engineer, together with the basic principles of heat transfer. This branch of the subject is commonly known as **engineering thermodynamics** or **applied thermodynamics.** The principles developed provide an understanding of the fundamental design considerations and performance criteria of equipment such as the turbine, the internal combustion engine, the air compressor, the refrigerating machine, the air-conditioning system, the boiler and various other heat transfer devices.

1-2. Dimensions and Units. Thermodynamics is concerned with both thermal and mechanical properties of substances and their measurement, incident to the development of useful engineering knowledge. Such measurements are expressed in terms of dimensions which signify the character of the quantity under consideration. For example, two basic measurements or dimensions of importance in thermodynamics are length (L) and time (T). In addition, the dimensions of force (F), mass (M) and temperature (t) are fundamental to the development of thermodynamics.

The evaluation of a quantity or dimension can be expressed in terms of arbitrary increments or units. For instance, the dimension of length can be written in units of metres or feet. Other units of larger or smaller size such as kilometre or inch may be derived from the fundamental units.

Force (F) and mass (M) are related with length (L) and time (T) through Newton's second law of motion by:

$$F \sim Ma \qquad (1\text{-}1)$$

or dimensionally

$$F \sim M \cdot L \cdot T^{-2}$$

INTRODUCTION AND THE FIRST LAW OF THERMODYNAMICS

In order to make the above statement of proportionality an equality, a constant must be introduced which is both dimensionally consistent and numerically valid. This necessary constant is assigned the symbol g_c, and introducing it into the above equation yields:

$$F = \frac{Ma}{g_c} \tag{1-2}$$

Rearranging:

$$g_c = \frac{Ma}{F} \tag{1-3}$$

The dimensions of g_c are thus established as $MLT^{-2}F^{-1}$ and are independent of the unit system used. However, the value of g_c is a function of the system in use (see Appendix A), but remains a true constant for any one particular system of units.

The selection and maintenance of standards for use in the physical sciences and engineering are accomplished by international committees such as the General Conference on Weights and Measures. The international standards for length and time are established by precise spectroscopic measurements of radiation characteristics of krypton and cesium. The standard for mass is a platinum/iridium cylinder kept at the International Bureau of Weights and Measures in France and designated as one kilogram. Several different measurement systems have been developed and used in engineering calculations based on such standards. These include the British Gravitational System, the English Engineers' System, the Metric System, the International System of Units, the Metric Gravitational System and the British Absolute System. The basic system used in this text is the Engineers' System. In addition, some exposure to the International System of Units, a modernized metric system, abbreviated SI, will also be provided. A further discussion of dimensions and units for the above systems and some useful conversion factors between Engineers' units and SI units can be found in Appendix A.

In the Engineers' System, there are *four* selected primary dimensions: mass (M), force (F), length (L) and time (T). The corresponding basic units of measurement are pounds (lb), pounds force (lbf), feet (ft) and seconds (sec). From physics we know that one pound mass (lb) will be accelerated at 32.1740 ft/sec² when acted upon by one pound force (lbf). Then:

$$g_c = \frac{Ma}{F} = \frac{(1)\text{lb}(32.1740)\text{ft} \cdot \text{sec}^{-2}}{(1)\text{lbf}} = 32.1740 \frac{\text{lb} \cdot \text{ft}}{\text{lbf} \cdot \text{sec}^2} \tag{1-3A}$$

The above expression is valid *only* for the Engineers' System and the

INTRODUCTION AND THE FIRST LAW OF THERMODYNAMICS

particular units shown. In any other system, the numerical value of g_c and the units it carries will necessarily be different.

In SI the *three* primary dimensions are mass (M), length (L) and time (T). The corresponding base units are the kilogram (kg), metre (m) and second (sec). Force (F) is a secondary dimension, expressed in a derived unit known as the newton (N). The newton is defined as that force which gives a standard kilogram mass an acceleration of one metre per second squared (m/sec²). For the International System, then Equation (1-3) becomes:

$$g_c = \frac{Ma}{F} = \frac{(1)\text{kg}(1)\text{m} \cdot \text{sec}^{-2}}{(1)\text{N}} = 1 \frac{\text{kg} \cdot \text{m}}{\text{N} \cdot \text{sec}^2} \quad (1\text{-}3\text{B})$$

Since the newton is defined as $1\text{N} \equiv 1 \text{ kg} \cdot \text{m/sec}^2$ then

$$g_c = 1 \frac{\text{kg} \cdot \text{m}}{\left(\frac{\text{kg} \cdot \text{m}}{\text{sec}^2}\right) \cdot \text{sec}^2} = 1.0$$

The primary dimensions of M, L, T and derived dimension of F give a magnitude of *unity* and a *dimensionless* character to g_c in SI.

As indicated earlier, we shall confine ourselves to the Engineers' and SI systems in this text. By convention, when pound (lb) with no modifier or kilogram (kg) appears in the text material, the reference is to mass. In SI, newton (N) clearly refers to force and should not be a source of confusion. In the Engineers' System, pound(s) force will always carry the "force" modifier and be abbreviated "lbf." The differentiation between pound (lb) and pound force (lbf) should not present any great difficulty, but does require continuing attention because of the similarity of terms.

As a practical matter in engineering, mass is determined indirectly by the direct measurement or comparison of weight, the force originating from the effect of gravity. Equation (1-2) then takes on this special form:

$$F = W = \frac{Mg}{g_c} \quad (1\text{-}4)$$

where g_c is the constant derived above and g is the local acceleration of gravity. The value of g varies from place to place, but the variation is small over the surface of the earth. The "standard locality" values are 32.1740 ft/sec² or 9.80665 m/sec², but the assumption that g is constant (for Earth's surface problems) and equal to 32.2 ft/sec² or 9.81 m/sec² is sufficiently accurate for most engineering calculations and will be employed in this text. Similarly, the constant g_c in Engineers' units may be rounded off to 32.2 lb · ft/lbf · sec² as a practical matter.

INTRODUCTION AND THE FIRST LAW OF THERMODYNAMICS

TABLE 1-I

Description	Dimensions	Basic units	
		Engineers'	SI
Length	L	ft	m
Time	T	sec	sec
Force	F	lbf	N
Mass	M	lb	kg
Area	L^2	ft^2	m^2
Volume	L^3	ft^3	m^3
Velocity	LT^{-1}	ft/sec	m/sec
Acceleration	LT^{-2}	ft/sec^2	m/sec^2
Weight	F	lbf	N
Work	FL	ft · lbf	N · m (joule)
Power	FLT^{-1}	ft · lbf/sec	N · m/sec (watt)

The other fundamental dimension having direct application in thermodynamics is temperature (t). A discussion of its nature and appropriate symbols is given in Art. 1-8.

Table 1-I indicates some of the quantities commonly encountered in engineering calculations with dimensions and units in the Engineers' and SI systems. The student is assumed to be familiar with the significance of these quantities, and no detailed discussion of them is considered necessary. A complete list of symbols and abbreviations is to be found in the prefatory material of this text.

1-3. Working Substances. The fluid which receives, transports and transfers energy in a thermodynamic system is known as a working substance. The fluid used may be in the liquid, the vapor or the gaseous phase, and it may assume different phases within the system in which it functions. Water (steam) is one of the most common working substances, and its use in a conventional *steam* power plant serves to illustrate the preceding statement. Here the working substance is in the liquid phase from the condenser to the boiler drum, but leaves the boiler drum via the dry pipe as steam or water vapor. If the vapor leaving the drum has its temperature increased considerably with no appreciable pressure change as in a superheater, it will approximate the characteristics of a gas. A vapor may be defined as a substance which can be liquefied by small changes of temperature and pressure, whereas a gas can be liquefied only by large changes of temperature and pressure. The difference is merely one of degree, since all gases can be liquefied and vapors will behave as gases if their temperatures are raised sufficiently without an increase in pressure.

INTRODUCTION AND THE FIRST LAW OF THERMODYNAMICS

Air, classified as a gas, is another common working substance, being used in internal combustion engines and air compressors. Refrigerants, such as ammonia and Freon, are classed as vapors, since liquefaction without excessive changes of pressure and temperature is inherent in the refrigerating systems concerned.

1-4. Properties. The condition or state of a working substance at any point within a thermodynamic system can be established if sufficient information is available as to the properties at that point. The properties of interest in applied thermodynamics include temperature, pressure, specific volume, internal energy, enthalpy and entropy. Temperature, pressure and specific volume are terms with which the student is already familiar, and their magnitudes are readily determined with simple engineering instruments. The remaining properties do not have such obvious physical manifestations; they will be discussed as the necessity for their use arises.

1-5. Specific Volume, Specific Weight and Density. *Specific volume (v)* is defined as volume per unit mass. *Density* is the reciprocal of specific volume, or mass per unit volume. The symbol used for density is the Greek letter ρ (rho). Mathematically:

$$v = \frac{V}{M} \text{ ft}^3/\text{lb of substance (m}^3/\text{kg)} \tag{1-5}$$

and

$$\rho = \frac{1}{v} = \frac{M}{V} \text{ lb/ft}^3 \text{ of substance (kg/m}^3) \tag{1-6}$$

Specific weight is defined as the weight per unit volume of a substance and is symbolized by the Greek letter γ (gamma). Then by definition:

$$\gamma = \frac{W}{V} \text{ lbf/ft}^3 \text{ of substance (N/m}^3)$$

But from Equation (1-4) $W = \dfrac{Mg}{g_c}$. Therefore, substituting for W:

$$\gamma = \frac{M}{V}\frac{g}{g_c}$$

Using Equation (1-6),

$$\gamma = \rho \frac{g}{g_c} \text{ lbf/ft}^3 \text{ of substance (N/m}^3) \tag{1-7}$$

INTRODUCTION AND THE FIRST LAW OF THERMODYNAMICS

Making a unit check for the Engineers' System, the right-hand side yields

$$\left(\frac{\cancel{lb}}{ft^3}\right)\left(\frac{\cancel{ft}}{\cancel{sec^2}}\right)\left(\frac{lbf \cdot \cancel{sec^2}}{\cancel{lb} \cdot \cancel{ft}}\right) = \frac{lbf}{ft^3}$$

Also, since $\dfrac{M}{V} = \rho = \dfrac{1}{v}$,

$$\gamma = \frac{1}{v}\frac{g}{g_c} \text{ lbf/ft}^3 \text{ of substance } (N/m^3) \qquad (1\text{-}8)$$

Considering a simple substance with a fixed molecular weight, a relatively high numerical value of density or specific weight implies closely packed molecules as encountered in the liquid or solid phase. On the other hand, a relatively high value of specific volume implies widely separated molecules, which condition is characteristic of vapors and gases. In thermodynamics, specific volume is used as a matter of convenience in preference to specific weight or density. The reasons for this choice will become obvious to the student as he becomes familiar with the subject. The various tables of thermodynamic properties of working substances normally indicate only specific volumes for given conditions of temperature and pressure. However, specific weight may be determined through Equation (1-7) or (1-8) and density from (1-6).

Example 1-1: The specific volume of steam at atmospheric pressure and 212°F is 26.80 ft³/lb. Find (a) its density and (b) its specific weight.
Solution:

(a) $$\rho = \frac{1}{v} = \frac{1}{26.80 \text{ ft}^3/\text{lb}} = \underline{0.03731 \text{ lb/ft}^3}$$

(b) $$\gamma = \frac{1}{v}\frac{g}{g_c} = \rho \frac{g}{g_c}$$

$$= (0.03731 \text{ lb/ft}^3)(32.2 \text{ ft/sec}^2)\left(\frac{1}{32.2}\frac{lbf \cdot sec^2}{lb \cdot ft}\right)$$

$$= \underline{0.03731 \text{ lbf/ft}^3}$$

Example 1-2: If the density of mercury is 13,600 kg/m³, find (a) its density in lb/ft³ and (b) its specific weight in N/m³.
Solution:

(a) From Appendix A using the appropriate conversion factors

$$\rho = (13,600 \text{ kg/m}^3)(0.4535924 \text{ kg/lb})^{-1}(0.3048 \text{ m/ft})^3$$

or $(13,600 \text{ kg/m}^3)\left(16.01846 \dfrac{\text{kg/m}^3}{\text{lb/ft}^3}\right)^{-1}$

$$= \underline{849.0 \text{ lb/ft}^3}$$

INTRODUCTION AND THE FIRST LAW OF THERMODYNAMICS

(b)
$$\gamma = \rho \frac{g}{g_c}$$

$$\gamma = (13{,}600 \text{ kg/m}^3) \frac{(9.81 \text{ m/sec}^2)}{(1.0 \text{ kg}\cdot\text{m/N}\cdot\text{sec}^2)}$$

$$= \underline{133{,}416 \text{ N/m}^3} = \underline{133.4 \text{ kN/m}^3}$$

The preceding examples raise the question of significant figures carried through into the ultimate answer of a calculation. As a general rule, the indicated accuracy should be consistent with the basic data available including constants introduced in the applicable equations.

1-6. Pressure. Pressure (p) is defined as force per unit area and, in the case of a vapor or a gas, arises from the continual impingement of the moving molecules on the boundaries of the container. Thus the pressure exerted will be proportional to the number, mass, and activity of the molecules. Pressure is normally expressed in pounds force per square inch (psi) or pascals (N/m^2). Since the basic linear unit is the foot, it will usually be necessary to convert to pounds force per square foot when using equations involving a pressure term. Pressure is also expressed in terms of an equivalent column of liquid, such as inches or millimetres of mercury and inches or feet of water. Standard atmospheric pressure at sea level has been established as 760.000 millimetres of mercury, which is equivalent to approximately 29.92 inches of mercury, 14.7 pounds force per square inch, 33.9 feet of water or 101.3 kPa. *In this text, whenever a definite value of atmospheric pressure is not given, it should be assumed that standard atmospheric pressure prevails.*

Most pressure measuring instruments in common use measure pressure relative to atmospheric pressure. Atmospheric pressure varies from place to place with elevation and, in any one locality, with climatic conditions. Therefore, gage pressure is unsuitable for scientific work, since it is referred to a variable datum. A fixed datum is obtained by introducing the concept of absolute pressure, which is pressure referred to a perfect vacuum. All thermodynamic equations and calculations involving pressure are based on the use of *absolute pressure.*

Absolute pressure of the atmosphere is measured by a barometer, usually in inches, millimetres or centimetres of mercury. Pressures above atmospheric are normally measured by pressure gages of the Bourdon tube type, and the readings yield a gage pressure. Absolute pressure in this case is determined from the equation:

Absolute pressure = atmospheric pressure + gage pressure (1-9)

Pressures below atmospheric pressure are normally determined from

vacuum gages calibrated in inches or millimetres of mercury. The conventional vacuum gage will read zero when connected to a vessel in which atmospheric pressure exists. It will read 760 mm of mercury when connected to a vessel in which a perfect vacuum exists, assuming standard atmospheric conditions prevail externally. It should be noted that a high numerical vacuum reading denotes a low absolute pressure. Using consistent units, the following equation yields absolute pressure in this case:

Absolute pressure = atmospheric pressure − vacuum (1-10)

Comparison of Equations (1-9) and (1-10) indicates that *vacuum is a negative gage pressure*. Figure 1-1 provides a graphical representation of the various pressure relationships outlined above.

FIGURE 1-1

Example 1-3: A pressure gage connected to a turbine inlet reads 400 psi. A vacuum gage connected in the exhaust trunk of the same turbine reads 28.0 in. Hg. The barometer reads 30.1 in. Hg. Find: (a) the absolute pressure at the turbine inlet, psia; (b) the absolute pressure in the turbine exhaust trunk, in. Hg abs.

Solution:

(a)
$$1 \text{ in. Hg} = \frac{14.7}{29.92} = 0.491 \text{ psi}$$

$$30.1 \text{ in. Hg} = 30.1 \times 0.491 = 14.8 \text{ psi}$$

$$p = 14.8 + 400 = \underline{414.8 \text{ psia}}$$

INTRODUCTION AND THE FIRST LAW OF THERMODYNAMICS

(b) $p(\text{gage}) = 28.0$ in. Hg
 $p(\text{atm}) = 30.1$ in. Hg
 $p(\text{abs}) = p(\text{atm}) - p(\text{vac})$
 $p(\text{abs}) = 30.1 - 28.0 = \underline{2.1 \text{ in. Hg abs.}}$

1-7. Manometers. The manometer measures pressure by balancing an unknown pressure against a column of liquid. Manometers are used where considerable accuracy is desired in the measurement of small pressures or pressure differences. The manometric fluids commonly used are oil, mercury and water. In the open manometer, Fig. 1-2a, one leg is connected to the pressure point and the other leg is open to the atmosphere. In the differential manometer, Fig. 1-2b, each leg is connected to a separate pressure point, and the difference in pressure at the two points is established by the displacement of the manometric fluid in the two legs. The barometer, Fig. 1-2c, is in essence a manometer for measuring atmospheric pressure. In the barometer, one "leg" is connected to a nearly perfect vacuum and the other "leg" to the atmosphere.

FIGURE 1-2

The conversion of manometer readings to pressures is readily accomplished by: (1) recognizing that equal heights of the same liquid in opposed legs are equivalent and nullify each other; and (2) by use of the equation:

$$\Delta p = \gamma L = \rho \frac{g}{g_c} L \qquad (1\text{-}11)$$

where Δp = pressure difference between top and bottom of a liquid column, lbf/ft² (N/m²)

INTRODUCTION AND THE FIRST LAW OF THERMODYNAMICS

γ = specific weight of manometric fluid, lbf/ft³ (N/m³)
L = length of liquid column measured vertically, ft (m)
ρ = density of manometric fluid, lb/ft³ (kg/m³).

Example 1-4: In the barometer shown in Fig. 1-2(c), the mercury level in the vertical tube is exactly 760 mm above the mercury level in the reservoir ($y = 760.000$ mm). The following data also apply: "water" and "mercury" can be considered to be incompressible fluids having densities of 1,000 kg/m³ and 13,595.1 kg/m³ respectively. The standard local acceleration of gravity is equal to 9.80665 m/sec² or 32.1740 ft/sec². Find the barometric pressure in:

(a) in. Hg
(b) ft H₂O
(c) pascals, Pa (N/m²)
(d) lbf/ft²
(e) psi (lbf/in²)

Solution:

(a)
$$L = y = 760 \text{ mm} = 0.76 \text{ m} = 76 \text{ cm}$$

$$L = \frac{76 \text{ cm}}{(2.54 \text{ cm/in})} = \underline{29.9213 \text{ in. Hg}}$$

(b)
$$\rho_1 = 1,000 \text{ kg/m}^3 \text{ (water)}$$
$$\rho_2 = 13,595.1 \text{ kg/m}^3 \text{ (mercury)}$$

From Equation (1-11):

$$\Delta p = \rho_1 L_1 \frac{g}{g_c} = \rho_2 L_2 \frac{g}{g_c}$$

$$L_1 = \frac{\rho_2}{\rho_1} L_2 = \frac{13,595.1}{1,000} \times 76 \text{ cm}$$

$$L_1 = \left(\frac{13,595.1}{1,000}\right)(76 \text{ cm})\left(\frac{1}{2.54}\frac{\text{in}}{\text{cm}}\right)\left(\frac{1}{12}\frac{\text{ft}}{\text{in}}\right)$$

$$= \underline{33.8985 \text{ ft H}_2\text{O}}$$

(c)
$$\Delta p = \rho \frac{g}{g_c} L$$

$$= (13,595.1 \text{ kg/m}^3)\left(\frac{9.80665 \text{ m/sec}^2}{1.0 \text{ kg} \cdot \text{m/N} \cdot \text{sec}^2}\right)(0.76 \text{ m})$$

$$= \underline{101,325 \text{ N/m}^2 \text{ or Pa} = 101.325 \text{ kPa}}$$

(d)
$$\Delta p = \rho \frac{g}{g_c} L$$

Using water data from (b), $L = 33.8985$ ft H₂O

$$\rho = \frac{1,000 \text{ kg/m}^3}{16.0185 \frac{\text{kg/m}^3}{\text{lb/ft}^3}} = 62.4278 \text{ lb/ft}^3$$

INTRODUCTION AND THE FIRST LAW OF THERMODYNAMICS

$$\Delta p = \left(62.4278 \frac{\text{lb}}{\text{ft}^3}\right)\left(\frac{32.1740 \text{ ft/sec}^2}{32.1740 \text{ lb} \cdot \text{ft/lbf} \cdot \text{sec}^2}\right)(33.8985 \text{ ft})$$
$$= 2116.21 \text{ lbf/ft}^2$$

(e) $$\Delta p = \frac{2116.21 \text{ lbf/ft}^2}{144 \text{ in}^2/\text{ft}^2} = 14.6959 \text{ lbf/in}^2$$

The foregoing example develops the substantially exact values of one standard atmosphere for the several ways in which pressure is commonly expressed. It also indicates that the approximations for standard atmospheric pressure, given in the first paragraph of this article, are very close to the "exact" values. Hence, the rounded-off approximate values will be used throughout the text unless otherwise indicated. Attention is also invited to Table 10 of the Steam Tables (Keenan, Keyes, Hill and Moore), which gives the pressure relationships derived above plus other useful conversion factors.

Example 1-5: If the difference in height of the water legs in the open manometer shown in Fig. 1-2a is 20 in., (dimension y), find the absolute pressure at point A in psia. The specific weight of water is 62.4 lbf/cu ft.
Solution:
Starting from point A, and writing an equation of equilibrium through the manometer:

$$p_A + \Delta p \text{ (gas leg)} - \Delta p \text{ (water leg)} - p(\text{atm}) = 0$$
Δp (gas leg) is a negligible numerical quantity

$$\Delta p \text{ (water leg)} = \gamma L; \gamma = 62.4 \text{ lbf/cu ft}; L = 20 \text{ in.} = \frac{20}{12} \text{ ft}$$

$$\Delta p = 62.4 \frac{\text{lbf}}{\text{ft}^3} \times \frac{20}{12} \text{ ft} = 104 \text{ lbf/ft}^2$$

$$\Delta p = \frac{104}{144} = 0.72 \text{ lbf/in.}^2 \text{ (psi)}$$

$p(\text{atm}) = 14.7$ psi (assumed since atmospheric pressure is not specifically given)
$p_A = \Delta p \text{ (water leg)} + p(\text{atm}) = 0.72 + 14.7 = \underline{15.42 \text{ psia}}$

Example 1-6: In the differential manometer shown in Fig. 1-2b, points A and B are at the same height. The mercury used as a manometric fluid stands 9 inches higher in the B leg than in the A leg ($y = 9$ in.). Find the difference in pressure between points A and B in psf. Specific weight of mercury is 849 lbf per cu ft.
Solution:
Writing the equilibrium equation starting from point A.

$$p_A + \gamma(y \text{ in. water}) - \gamma(y \text{ in. Hg}) - p_B = 0$$
$$p_A - p_B = \gamma(y \text{ in. Hg}) - \gamma(y \text{ in. water})$$

$$\Delta p(\text{water}) = \gamma L = 62.4 \times \frac{9}{12} = 46.8 \text{ lbf/ft}^2$$

$$\Delta p(\text{Hg}) = \gamma L = 849 \times \frac{9}{12} = 636.8 \text{ lbf/ft}^2$$

$$p_A - p_B = 636.8 - 46.8 = \underline{590 \text{ lbf/ft}^2}$$

INTRODUCTION AND THE FIRST LAW OF THERMODYNAMICS

There are many types of manometers in use, but judicious application of the principles outlined above will allow solution for the unknown pressure, or pressure difference, regardless of the manometer type. For instance, in the inclined tube manometer, or "draft gage," the analysis is simple if it is remembered that the *vertical* height of the liquid column must be used in Equation (1-11). Similarly, if several U-tube manometers are connected in series to measure a large pressure difference, the solution is readily obtained by writing the equilibrium equation through the whole system.

1-8. Temperature. As discussed earlier, time, length, mass and force are the fundamental mechanical dimensions of interest in thermodynamics. Similarly, temperature is a fundamental measure or dimension of thermal properties.

Temperature is conventionally measured with thermometers graduated in degrees Fahrenheit or degrees Celsius (Centigrade). The Celsius (Centigrade) scale is based on a 100-degree interval between the ice point and the boiling point of fresh water at standard atmospheric pressure. In this country, Celsius (C) is widely used by physicists, while engineers commonly use the Fahrenheit scale. The interval between the ice point and the boiling point on the Fahrenheit scale is 180 degrees. The Fahrenheit thermometer reads 32°F at the ice point and 212°F at the boiling point. The symbol for temperature as read directly from either the Fahrenheit or Celsius scale is t.

Both the Fahrenheit and Celsius scales are arbitrary in nature. The numerical readings have no fundamental significance, merely indicating the relative "hotness" of a body. From everyday experience it is reasonable to associate temperature with energy content, since we know that the addition of energy to a body as heat will result in a temperature rise. If temperature is to be used as one yardstick for energy measurement, it is necessary to establish a temperature datum at which the energy content is zero. Once such a datum is established, a thermodynamic or energy thermometric scale may be devised. Temperature measured on this scale is known as absolute temperature and designated by the symbol T. Absolute zero, where all bodies are postulated to have zero energy, has been fixed at approximately $(-)273°C$ or $(-)460°F$. Much of the original development of the energy scale concept was accomplished by Lord Kelvin (circa 1848). For this reason, absolute temperatures expressed with reference to the Celsius scale are customarily said to be measured on the Kelvin scale, and are stated in kelvins, or K.

$$T(\text{K}) = t(°\text{C}) + 273 \qquad (1\text{-}12)$$

INTRODUCTION AND THE FIRST LAW OF THERMODYNAMICS

Absolute temperatures referred to the Fahrenheit scale are stated in degrees Rankine (for W. J. M. Rankine, a famous Scottish engineer and contemporary of Kelvin).

$$T(°R) = t(°F) + 460 \qquad (1\text{-}13)$$

1-9. Energy. Energy is defined as the capacity for producing an effect. Since the effect produced is frequently, but not always, mechanical work, a more limited but common definition of energy is "capacity for doing work." The definition of energy in terms of the effect produced is logical, since energy itself is rather intangible in certain forms but the effects produced are tangible and can usually be evaluated.

Energy in all its forms falls into two general classes, namely *stored energy* and *energy in transition*. Energy in transition, as the term implies, is a momentary energy form intermediate between two or more stored forms. All energy in transition begins and eventually ends in one of the several stored forms. Stored energy is associated with systems of bodies. Within any system of bodies, the stored energy is present fundamentally in either the kinetic or the potential form, or both. The bodies concerned may be molar, molecular, atomic or subatomic. In this connection, another method of classifying energy is based upon the size of the particles associated with the energy form. Energy, which stems from the position or motion of relatively *large (molar) bodies,* is called **mechanical energy.** Energy associated with the configuration and motion of *molecules* is known as **thermal energy.** Thermal energy is characterized by its ability to be transferred from one body to another by *temperature difference* alone, through the medium of colliding molecules or electromagnetic waves with an attendant exchange of molecular energy. The transference of mechanical energy is independent of temperature, and is accomplished by physical contact between large masses, either directly or through a machine. The two methods of classification are not contradictory, but serve to complement one another for descriptive purposes.

The general symbol for energy in any form is E. In the Engineers' System, the basic unit of energy is the foot pound force. In the SI System, the basic unit is the newton metre (N · m) commonly called the joule (J). The symbols assigned to the various forms of energy with which we are concerned in engineering thermodynamics will be given in the detailed discussion which follows.

1-10. Potential Energy. Potential energy is the energy stored within a system by virtue of the stresses existing in the system. Thus potential energy may be stored in a spring or may arise from the force of attraction

INTRODUCTION AND THE FIRST LAW OF THERMODYNAMICS

between two bodies. For our purposes the energy associated with the gravitational attraction between the earth and a tangible object is of particular interest. Energy of this type is to be inferred when the term **potential energy** is used in the text without other qualification. This energy is *stored* in the system, composed of the earth and object, and is *mechanical* in nature since the bodies involved are molar. The magnitude of the stored energy associated with the system is conventionally evaluated by the work done against gravity in elevating the object from some arbitrary reference plane, or datum level, to some higher level. Work is defined as force multiplied by distance. Work against gravity is equal to the weight of the body multiplied by the vertical distance through which that body is transported. Then

$$P.E. = \int_0^Z W \, dL$$

Since $W = \dfrac{Mg}{g_c}$ from Equation (1-4) and $dL = dz$, then

$$P.E. = \int_0^Z \frac{Mg}{g_c} \, dz$$

Integrating between limits:

$$P.E. = \frac{Mg}{g_c} Z \qquad (1\text{-}14)$$

where $P.E.$ = potential energy, ft · lbf (N · m or joules)
M = mass of body, lb (kg)
Z = vertical distance (elevation) of body above reference plane, ft (m).

In thermodynamics it is customary to deal with a unit mass of substance, giving:

$$\mathbf{\textit{P.E.} = \frac{g}{g_c} Z \text{ ft} \cdot \text{lbf/lb of substance (J/kg)}} \qquad \underline{(1\text{-}15)}$$

It should be recognized that the selection of an arbitrary datum leads to an evaluation of potential energy which is relative rather than absolute.

Example 1-7: In Fig. 1-3, what is the difference in potential energy per pound mass of fluid at B and at A?
Solution:

$$P.E. = \frac{g}{g_c} Z \text{ ft} \cdot \text{lbf/lb}$$

INTRODUCTION AND THE FIRST LAW OF THERMODYNAMICS

$$P.E._B = \frac{g}{g_c} Z_B = 30 \text{ ft} \cdot \text{lbf/lb}$$

$$P.E._A = \frac{g}{g_c} Z_A = 10 \text{ ft} \cdot \text{lbf/lb}$$

$$P.E._B - P.E._A = 30 - 10 = \underline{20 \text{ ft} \cdot \text{lbf/lb}}$$

FIGURE 1-3

1-11. Kinetic Energy. Kinetic energy is the energy stored in a system by virtue of relative motion between component bodies of the system. As used herein, the term kinetic energy, without other qualification, designates the stored mechanical energy arising from the velocity of a molar body relative to the earth or the work done to bring the body from rest to its observed velocity relative to the earth.

From physics, we know that work is a function of force and the distance through which that force acts. Then:

$$K.E. = Wk = \int_0^L F \, dL$$

From Equation (1-2)

$$F = \frac{Ma}{g_c}$$

INTRODUCTION AND THE FIRST LAW OF THERMODYNAMICS

Then:

$$K.E. = \int_0^L \frac{Ma}{g_c} dL$$

Since $a = \dfrac{dV}{dt}$ and $V = \dfrac{dL}{dt}$, $dL = Vdt$:

$$K.E. = \int_0^L \frac{M}{g_c} \frac{dV}{dt} dL = \int_0^V \frac{M}{g_c} V \, dV$$

Integrating between the limits:

$$K.E. = \frac{MV^2}{2g_c} \tag{1-16}$$

where $K.E.$ = mechanical kinetic energy of a body, ft · lbf (N · m or J)
M = mass of body or substance, lb (kg)
V = velocity of body relative to the earth, ft/sec (m/sec)
g_c = 32.2 lb · ft/lbf · sec² (1 kg · m/N · sec²).

The kinetic energy possessed by a body or substance per unit of mass is:

$$K.E. = \frac{V^2}{2g_c} \text{ ft · lbf/lb of substance (J/kg)} \tag{1-17}$$

Example 1-8: Steam is flowing in a pipe at a velocity of 100 ft per sec. What is the associated kinetic energy per pound of steam flowing?
Solution:

$$K.E. = \frac{V^2}{2g_c} \text{ ft · lbf/lb}$$

$$g_c = 32.2 \text{ lb · ft/lbf · sec}^2;\ V = 100 \text{ ft/sec}$$

$$K.E. = \frac{(100)^2}{2 \times 32.2} = \underline{155.3 \text{ ft · lbf/lb}}$$

1-12. Internal Energy. The two preceding articles have outlined the significance of potential and kinetic energy as associated with the position and velocity of molar bodies. The internal energy of a substance derives from the potential and kinetic energy associated with the individual molecules which make up the substance. **Internal energy is thermal energy stored within the substance itself.**

Internal potential energy is associated with the forces of attraction which exist between molecules. The attractive force pattern is complicated since each molecule in a substance is attracted more or less to every other molecule. However, it is believed that the magnitude of the intermolecular forces is a function of the mass of the molecules and their average

INTRODUCTION AND THE FIRST LAW OF THERMODYNAMICS

separation. For a given substance, then, a change of internal potential energy will result whenever the mean distance between molecules is changed. Such a change is readily evident when a substance changes phase. Thus, a considerable amount of energy is required to vaporize water. This energy is expended in effecting the change from the closely packed molecular configuration of water to the widely separated configuration of steam, and is stored as internal potential energy in the steam.

Internal kinetic energy is associated with the activity of molecules. This activity consists of three forms of motion:

(1) The random translatory motion of the individual molecules.
(2) The rotation of the individual molecules about their respective centers of mass.
(3) The vibratory motion of the atoms composing the molecule relative to one another.

The total internal *kinetic* energy of a substance will be a summation of the translational, rotational and vibratory energies of all the molecules of the substance.

In applied thermodynamics, the evaluation of changes in internal energy is of primary concern. However, some understanding of the nature of internal energy as put forward in classical physics is of more than academic importance. The discussion which follows will be confined to gases and is admittedly oversimplified. As a further preface to that which follows, we must accept the classic definition of a gas as the state in which average molecular separation is so great that the intermolecular forces are extremely small and the associated internal potential energy is effectively zero. This presupposes the gas to be a single-phase substance with the zero energy datum selected accordingly.

Let us first consider a monatomic gas such as helium or argon. According to the kinetic theory of gases, all the molecules of a gas are in random translatory motion, and their average kinetic energy is a function of temperature. Thus the monatomic gas (and every other gas) at any temperature above absolute zero will always possess internal kinetic energy of the translatory type. On the other hand, there is ample evidence that the monatomic gases will not store any measurable quantity of internal kinetic energy of the rotational type. The classic explanation of this phenomenon is based on the structure of the atom. Almost all the mass of an atom is concentrated in the extremely dense nucleus and the monatomic molecule can be considered a point mass. Incident to the random translatory motion of the molecules, collisions will occur. Since the monatomic molecules are substantially point masses, they will bounce off one another without any resultant spin. The mechanism is analogous to

the collision of a moving and a stationary billiard ball where the line of motion coincides with the line of centers. From another standpoint, even if a point mass were to rotate about an axis through its center of gravity, it could store no rotational energy in the absence of a finite radius of gyration. Finally, it is obviously impossible for a monatomic gas to store any of the third type of energy listed above, since at least two atoms are required in the molecule for the internal vibratory type of energy to be present. Thus the internal kinetic energy of a monatomic gas derives entirely from the random translatory type of molecular activity, and the magnitude of this activity is directly reflected by temperature. This leads to the conclusion that the total internal energy of a monatomic gas can be evaluated in terms of temperature and some *constant* factor of proportionality. The foregoing statement is exactly true for the ideal or perfect gas, which, by postulation, is monatomic in character and devoid of the intermolecular forces associated with internal potential energy. It is also true for all practical purposes for *real monatomic* gases as long as the molecular spacing remains large enough to permit neglect of internal potential energy. When the latter becomes a factor, as may result under very high pressures, the "gas" is really a vapor and other methods of attack are necessary.

The case of the diatomic gas differs from the monatomic in that both the vibrational and rotational forms of internal kinetic energy come into play. The diatomic molecule can be thought of as a dumbbell with the heavy nuclei of the two atoms concentrated in the end spheres. Such an arrangement is obviously suited to the storage of rotational energy, since the two principal masses are relatively widely separated and glancing collisions between molecules will invariably cause spin. Secondly, the simple presence of two atoms makes the configuration suitable for the storage of the vibratory form of internal kinetic energy.

Let a diatomic gas, initially at a very low temperature and rigidly contained, be heated at a *constant rate*. Immediately and throughout the process, the temperature of the gas will rise, evidencing an increase in the translational kinetic energy of the molecules. For a time the temperature will increase at a constant rate; then, rather abruptly, the *rate* of temperature rise will fall off. This is physical evidence that the energy being added as heat is no longer being entirely stored in the translational form. Theoretical physics demonstrates that the gas is now storing some of the added energy in the rotational form. Finally, at some still higher temperature, the vibratory motion of the atoms will become significant, an appreciable portion of the added energy will be stored in the vibratory form, and the *rate* of temperature rise will further decrease. Taking the converse side of the proposition, in order to get a given temperature rise

INTRODUCTION AND THE FIRST LAW OF THERMODYNAMICS

more heat must be added when the gas is in a high temperature range than when it is in some lower range. Thus the factor of proportionality relating internal energy and temperature is no longer a constant with temperature as it is for the monatomic gas. Modern molecular theory permits quite accurate prediction of the ultimate division of the three forms of internal kinetic energy when the configuration of the atoms within the molecule is known. However, the temperatures at which the rotational and the vibratory forms in turn become significant and the temperatures at which each finally reaches its theoretical maximum must be determined experimentally for each individual gas. For instance in the table below for hydrogen (H_2), from absolute zero to $-375°F$ only the translatory form of internal kinetic energy is present; above $-375°F$ the rotational form becomes significant and reaches its theoretical (and actual) maximum at approximately 60°F; somewhere above 60°F the vibrational form appears and finally reaches its theoretical maximum at a temperature determined by extrapolation to be about 9000°F. The polyatomic

Approx. temp.	Temp. rise	Heat added	Internal energy stored as:		
			Trans.	Rot.	Vib.
$-375°F$	Small but equal	3 Btu	3	0	0
60°F		5 Btu	3	2	0
9000°F		7 Btu	3	2	2

gases, of which the diatomic is the simplest form, display similar characteristics. The variation is essentially one of degree, in that the proportionate ability to store internal kinetic energy in the rotational and vibrational forms increases as the number of atoms in the molecule increases.

From the discussion to this point, several important conclusions can be drawn:

(1) The internal energy level of any substance is determined by the summation of the molecular potential energy and the three forms of molecular kinetic energy. (It is this aggregate internal energy with which engineering thermodynamics is fundamentally concerned and which is implied where the term "internal energy" is used without qualification.)

(2) Any change in the internal kinetic energy of a substance invariably includes a change in the kinetic energy associated with the translatory type of molecular motion.

(3) Every change of internal *kinetic* energy will be accompanied by a change of temperature.

INTRODUCTION AND THE FIRST LAW OF THERMODYNAMICS

(4) The internal energy of an ideal or perfect gas is made up entirely of the translatory type of kinetic energy, and the factor of proportionality relating internal energy and temperature is a *constant*.

(5) The foregoing statement is true for all practical purposes in the case of real monatomic gases.

(6) In the case of real polyatomic gases, internal energy is also a function of temperature, but the relating factor of proportionality is no longer a constant but varies with the gas temperature.

It will be shown in the following chapters that the factor which relates the temperature and the internal energy of a gas is the specific heat at constant volume of that gas.

Having established some idea of the basic nature of internal energy, we can proceed to identify it further on the basis of a simple physical manifestation. Suppose we have a perfectly insulated cylinder filled with compressed air and a frictionless piston above it locked in place. If the piston is unlocked, it will move through a certain distance with a gradual decrease in the cylinder air pressure. Work will have been performed at the expense of some of the internal energy originally contained in the air. The internal energy will have decreased by an amount exactly equal to the work done. It should be noted that the work potential is provided by a high pressure level within the cylinder, whereas the magnitude of the work performed may be evaluated by determining the change of internal energy. In this process, which is typical, the primary concern is not with the absolute values of internal energy at the beginning and end of the process, but rather with the change which occurs. For that reason, it is customary to select a convenient datum for a substance from which to measure internal energy, rather than attempting to make an absolute evaluation. *The change of internal energy which occurs during a process such as that described above is independent of the datum chosen.* However, the absolute values of internal energy at the beginning and end of such a process are completely dependent on the selected datum.

Internal energy is usually expressed in British thermal units in the Engineers' System and in joules in the SI System. The symbol U will be used to signify total internal energy, Btu or joules; the symbol u will be used for specific internal energy, Btu per pound of substance or joules per kilogram.

1-13. Heat. The term heat, as used in thermodynamics, is defined as **thermal energy in transition.** It includes all energy which is *transferred* from one region to another by reason of a *temperature difference*. The two modes of heat transfer with which we shall be concerned are conduction and radiation. Conduction is predicated to arise from molecular collisions

INTRODUCTION AND THE FIRST LAW OF THERMODYNAMICS

with a resultant transfer of internal kinetic energy from the hotter to the colder substance. Obviously conduction requires physical contact or a tangible connecting medium between the hotter and the colder substance. Radiation, on the other hand, is not dependent on a physical conducting medium but occurs through space and is an electromagnetic wave phenomenon. Thus the heat from the sun is received as radiation through interplanetary space. Both radiation and conduction are dependent on a temperature difference for the transfer of energy.

Convection is sometimes loosely linked with radiation and conduction as a "heat" transfer process. From the thermodynamic standpoint, convection is simply the mechanical transportation of a substance from one region to another in bulk, and with that bulk goes the energy content. Thus convection represents the transportation of stored energy within a substance as opposed to the transfer of energy to or from the substance. Since no energy in transition is involved, convection cannot be concerned with heat within our definition. Examples of convection are steam flowing in a pipe, or the gases of combustion flowing in a boiler. Convection always involves flow of the substance and is not directly dependent on a temperature difference for existence.

It is important to understand the difference between heat (thermal energy in transition) and internal energy (stored thermal energy). Suppose we have an iron bar that has been heated to an elevated temperature and a tank of water at room temperature. If the hot bar is quenched in the water, a decrease in the temperature of the bar and an increase in the temperature of the water immediately occur. After a period of time it will be found that both water and bar are at the same temperature. Since temperature is a measure of internal kinetic energy, the obvious end result is an increase in the internal energy of the water and a decrease in the internal energy of the iron bar. A certain amount of the thermal energy originally stored in the bar is now stored in the water. Since there has been an interchange of thermal energy between the two substances, the conditions of thermal energy in transition are satisfied; the thermal energy during this transitory period is known as heat. Once both substances reach the same temperature, all heat transfer ceases and all thermal energy is again in the stored form. An understanding of the fundamental difference between stored thermal energy (or internal energy), and thermal energy in transition (or heat), will materially assist the student when he passes from the qualitative to the quantitative study of energy.

Heat is usually expressed in British thermal units in the Engineers' System and in joules in the SI System. The symbol Q will be used to signify total heat, Btu or joules. The symbol q will be used for heat per unit mass of substance, Btu per pound or joules per kilogram.

INTRODUCTION AND THE FIRST LAW OF THERMODYNAMICS

1-14. Mechanical Work. Mechanical work is defined as **mechanical energy in transition.** It is a transitory energy form existing between two stored forms. The energy transfer is effected and evidenced by the action of a force through a distance. Consider a long vertical cylinder containing compressed air with a frictionless plunger fitted above it. If the plunger is released, it will rise in the cylinder until equilibrium is attained. A force has acted through a distance, and work has been done. Some energy that originally existed as internal energy stored in the air now is stored as mechanical potential energy in the plunger mass. The mechanical work accomplished is simply the physical manifestation of a transfer of stored energy.

Mechanical work as such is the stock in trade of the engineer. In the usual thermodynamic system, either: (1) the working substance undergoes a series of processes, such that useful work is produced as in a steam power cycle; or (2) work is supplied to the working substance to produce a desired effect, as in an air compressor. Thermodynamics is very much concerned with work, and its development as a science may be attributed to continuing efforts to obtain a maximum of work output for a minimum energy input; or to produce a given effect with a minimum of work input.

In the Engineers' unit system work is usually expressed in foot pounds force (ft · lbf), although in thermodynamics it is frequently useful to use the Btu unit. In the SI System, the basic unit for energy, including both heat and work, is the joule. In this text, the symbol Wk will be used to express total work in ft · lbf or joules (J), and the symbol wk to express specific work in ft · lbf/lb, or J/kg. The term work without qualification refers to mechanical work as discussed in this article.

1-15. Flow Work. Flow work, or displacement energy, is the energy necessary to maintain the continuous steady flow of a stream of fluid. It is mechanical in nature, and is stored in the moving fluid. *Flow work is not present unless there is flow.* It is numerically equal to the product of the force acting on any cross section of the stream, and the distance through which that force must act to cause any selected mass of fluid to pass that cross section.

Figure 1-4 represents a passage filled with fluid that is flowing steadily from left to right. Let us consider a small slice of the fluid passing point (1) as indicated by the shaded area. Let this slice have a cross-sectional area A, a mass M and a length L. The pressure acting on the slice is p, and the specific volume of the fluid is v. In order to get M mass units past point (1), work must be done. The force acting on the slice is pA and the distance through which it acts is L. Then the work done on M units of fluid, in order to maintain its steady flow past point (1), is pAL. The volume of the

INTRODUCTION AND THE FIRST LAW OF THERMODYNAMICS

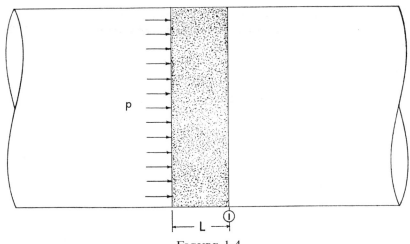

FIGURE 1-4

slice under consideration is equal to the cross-sectional area multiplied by the length or $V = AL$. But the volume is also equal to the mass of the slice multiplied by the specific volume of the fluid, or $V = Mv$.

By an identity then, $AL = Mv$, and the flow work associated with M units of fluid is equal to pAL or Mpv. On a unit mass basis

Flow Work $= pv$ ft · lbf/lb of fluid flowing (J/kg) (1-18)

Thus we find that flow work can be *evaluated* by the product of two properties of the moving fluid, namely pressure and specific volume.

It should be noted that the product pv will always have a numerical value, but represents flow work *only when the fluid is flowing in a steady continuous stream*. The flow work or displacement energy can be thought of as stored in the fluid and is associated with its movement.

1-16. The First Law. A well-known physical law is that of the Conservation of Energy, which states:

(1) *Energy can be neither created nor destroyed but only transformed.*

This general law applies to all forms and types of energy and is still valid under the theory of relativity which holds that mass may be transformed into energy according to a definite conversion factor.

Starting with the broad generalization of the law stated above, we may restrict it to more specialized cases by imposing certain limitations. In the case of a single specified system, for instance, the law may be stated as follows:

(2) *The total amount of all types of energy within any specified system will remain constant unless energy is entering or leaving the system through its*

INTRODUCTION AND THE FIRST LAW OF THERMODYNAMICS

boundaries, and the net increase or decrease of the energy within the system will be exactly equal to the net amount of energy that has passed through the boundaries.

The conversion of mechanical to thermal energy through the phenomenon of friction is a common everyday occurrence. A simple example is the gradual temperature rise of any shaft bearing from the stopped to the steady running condition. The conversion of energy in this case is undesirable but unavoidable. In thermodynamics, we shall be very much concerned with the interconversion of thermal and mechanical energy, but our interest will lie in the production of useful and desirable effects.

The original insight on the conversion of work (mechanical energy) to heat (thermal energy) was provided by Benjamin Thompson (Count Rumford) in the latter part of the eighteenth century. The first accurate experiments in support of Rumford's theories were made by Sir James Prescott Joule beginning in 1843 and continuing with increasing refinement over the next 35 years. It was Joule's experimental work which proved equivalence of the heat and work energy forms and their immutable interrelationship—i.e., that whenever a given quantity of mechanical energy was converted to heat, the same quantity of heat always resulted. And the converse.

In order to deal quantitatively with the energy conversion phenomenon, it is necessary to understand the mathematical relationship between the units used to express mechanical energy and thermal energy. In the SI System, where the newton metre (or joule) has been chosen as the basic unit for both thermal and mechanical energy, the interchange is on a one-for-one basis. Stated in another way, the factor of proportionality is unity. In the Engineers' System, on the other hand, the basic unit for mechanical energy is the foot pound force (ft · lbf) and for thermal energy is the British thermal unit (Btu), the same units which were in vogue in England in Joule's time. The Btu was originally conceived as a unit associated only with heat, in essence the British counterpart of the calorie in the metric system, and defined as the quantity of heat required to raise the temperature of one pound of pure fresh water one degree Fahrenheit. This usage predates the recognition that thermal and mechanical energy are mutually convertible. Joule's efforts, in addition to verifying the basic premise of interchangeability, were also directed toward finding the mathematical relationship between the common English units for work and heat, ft · lbf and Btu. From his work and that of subsequent investigators, it was determined that one Btu is equivalent to approximately 778 ft · lbf. This factor of proportionality is known as Joule's equivalent or the mechanical equivalent of heat and is assigned the symbol J. Then

$$J = 778 \text{ ft} \cdot \text{lbf/Btu} \tag{1-19}$$

INTRODUCTION AND THE FIRST LAW OF THERMODYNAMICS

The foregoing material leads directly to the **First Law of Thermodynamics,** in the parlance of English-speaking engineers:

(3) Mechanical energy and thermal energy are mutually convertible, and when one is converted into the other, the ratio is 778 ft · lbf to 1 Btu.

It should be noted that this law is essentially a restatement of the Law of Conservation of Energy, restricted to the forms of energy of immediate interest in thermodynamics and introducing the numerical value of the mechanical equivalent of heat. There are other ways of stating the First Law, the most usual being similar to (3) above, but further restricting the forms of energy under consideration to heat and work alone. In a broad sense, statement (1) above can be quoted as the First Law of Thermodynamics, and this usage is occasionally encountered. Statement (2) is another acceptable version of the First Law, but in more restricted terms.

1-17. The Steady Flow Energy Equation. The forms and types of energy with which we shall be directly concerned in thermodynamics have now been discussed and are recapitulated in Table 1-II.

TABLE 1-II

Name	Nature	Formula	Units Engineers'	SI
Potential	Stored mechanical	$\dfrac{g}{g_c}Z$	ft · lbf/lb	J/kg
Kinetic	Stored mechanical	$\dfrac{V^2}{2g_c}$	ft · lbf/lb	J/kg
Flow work	Stored mechanical	pv	ft · lbf/lb	J/kg
Internal	Stored thermal	u	Btu/lb	J/kg
Mechanical work	Mechanical—in transition	wk	ft · lbf/lb	J/kg
Heat	Thermal—in transition	q	Btu/lb	J/kg

Let us now consider a generalized thermodynamic *steady flow system* as represented schematically in Fig. 1-5. The *system* may include a single device, a portion of a device, or a group of related devices, depending upon where the system boundaries are established. In the figure the solid lines are intended to represent physical boundaries which constrain the working substance. Location (1) is at the inlet to the system, and location (2) is at the outlet. As a general rule there is complete freedom of choice in fixing these two locations. Thus in a steam plant, (1) and (2) might be at the entrance to the boiler and the exhaust from the turbine; at the entrance to the turbine and outlet from the condenser; at inlet and exhaust of the turbine; or simply across one or more stages of the turbine.

INTRODUCTION AND THE FIRST LAW OF THERMODYNAMICS

FIGURE 1-5

Regardless of where the two points are located, a *specified system* under the definition of statement (2) in Art. 1-16 is delineated by cross-sectional cuts through locations (1) and (2) and the intervening physical boundaries. Stored energy comes into the specified system with the working substance at (1) and leaves the system in the working substance at (2). Between (1) and (2), the working substance is constrained and energy can cross the physical boundaries only in the transition forms of heat and work.

To meet the steady flow specification, the system must be free of all transient effects. In thermodynamic usage, steady flow obtains when the three following conditions are met:

(a) The properties of the working substance at any selected location are constant with time.
(b) The working substance is continuously entering and leaving the system at the same mass rate, so that the mass of the fluid contained in the system is constant.
(c) The rates at which heat and work traverse the boundaries are constant.

While these requirements may appear rigorous at first glance, they represent the conditions under which most real devices operate most of the time. Reduced to essentials, the stated conditions simply establish an energy equilibrium such that the amount of energy stored within the system remains constant with time. With this in mind and referring to Fig. 1-5, it is evident that the energy in the fluid entering the system at (1) plus any energy added between (1) and (2) must equal the energy in the fluid

INTRODUCTION AND THE FIRST LAW OF THERMODYNAMICS

leaving at (2) plus any energy departing between (1) and (2). Taking into consideration all the forms of energy listed in Table 1-II, the previously stated facts can be reduced to equation form. This yields the **Energy Equation for Steady Flow,** essentially a mathematical expression of the First Law of Thermodynamics as set forth in Statement (2), Art. 1-16:

$$\text{Energy in} = \text{Energy out}$$

$$\frac{gZ_1}{g_c(J)} + \frac{V_1^2}{2g_c(J)} + \frac{p_1v_1}{(J)} + u_1 + q_{12} = \frac{gZ_2}{g_c(J)} + \frac{V_2^2}{2g_c(J)} + \frac{p_2v_2}{(J)} + u_2 + \frac{wk_{12}}{(J)}$$

Btu/lb of working substance flowing (J/kg) (1-20)

The foregoing equation is of basic and continuing importance and the student should be thoroughly versed in its significance and use. It establishes a simple system of energy bookkeeping and has a wide variety of applications. When one is using the Steady Flow Energy Equation, all terms must be written using a consistent set of units. As discussed in Art. 1-16 and shown in Table 1-II, the mechanical equivalent of heat (J) is peculiar to the English measurement systems and has no counterpart in SI. Hence the symbol "J" has been placed in parentheses in Equation (1-20) simply to indicate that it is a floating term. *Use J (778 ft · lbf/Btu) when dealing with Engineers' units; omit it when dealing with SI units.*

In interpreting and using Equation (1-20), there are several points to be noted:

(a) In SI units both the mechanical and the thermal energy terms in Equation (1-20) are expressed in J/kg, and J goes to unity.

(b) In the Engineers' units each mechanical energy term in Equation (1-20) requires J in the denominator. Multiplying every term by J will make the units of Equation (1-20) ft · lbf/lb, a form that is occasionally useful.

(c) Each stored energy term has a single subscript, indicating that its value can be definitely fixed at a point at any given instant.

(d) The energy in transition terms (heat and work) have double subscripts, indicating that they are a summation of effects occurring between two points. *Neither is a property of the fluid, and neither can be evaluated at a point.* Each requires the element of time for its realization.

(e) Referring to Fig. 1-5 and Equation (1-20) as written, the positive sign for q_{12} signifies *heat added* to the working substance. A *negative* sign for q_{12} must then signify *heat abstracted* from the working substance.

(f) Similarly a *positive* sign for wk_{12} signifies work *done by* the working

INTRODUCTION AND THE FIRST LAW OF THERMODYNAMICS

substance, and a *negative* sign indicates work *done on* the working substance.

1-18. Power. Power is defined as the rate of doing work. Equation (1-20) becomes a power rather than an energy equation when multiplied by \dot{M}, the mass rate of flow, expressed in units such as lb/hr or kg/hr. The following basic relationships, with which the student should be familiar, are quoted below:

1 hp = 33,000 ft · lbf/min = 550 ft · lbf/sec
1 Btu = 778 ft · lbf
1 kW = 1.341 hp

Using the foregoing relationships, the following approximate equivalents may be derived:

1 hp = 2545 Btu/hr = 42.42 Btu/min
1 hp · hr = 2545 Btu
1 kW = 3413 Btu/hr
1 kW · hr = 3413 Btu

The derivation of these equivalents is left to the student as an exercise. See Appendix A for some additional conversion factors.

Example 1-9: The work delivered by a turbine is 400 Btu per lb of steam supplied when the steam flow is 20,000 lb per hour. Find:

(a) the specific work in ft · lbf/lb
(b) the power delivered in hp
(c) the power delivered in kW

Solution:

(a) $$\frac{wk_{12}}{J} = 400 \text{ Btu/lb}$$

$$wk_{12} = 400 \times 778 = 311{,}200 \text{ ft} \cdot \text{lbf/lb}$$

(b) $$\dot{M} = 20{,}000 \text{ lb/hr} = \frac{20{,}000}{60} \text{ lb/min}$$

1 hp = 2545 Btu/hr = 33,000 ft · lbf/min

$$P = \frac{\dot{W}k}{J} = \dot{M} \times \frac{wk_{12}}{J} = 20{,}000 \times 400 = 8 \times 10^6 \text{ Btu/hr}$$

$$= \frac{8 \times 10^6}{2545} = \underline{3143 \text{ hp}}$$

INTRODUCTION AND THE FIRST LAW OF THERMODYNAMICS

or

$$P = \frac{\dot{W}k}{J} = \dot{M} \times wk_{12} = \frac{20{,}000}{60} \times 311{,}200 = 103.73 \times 10^6 \text{ ft} \cdot \text{lbf/min}$$

$$= \frac{103.73 \times 10^6}{33{,}000} = 3143 \text{ hp}$$

(c) $1 \text{ kW} = 3413 \text{ Btu/hr} = 1.341 \text{ hp}$

$$P = 8 \times 10^6 \text{ Btu/hr} = 3143 \text{ hp}$$

$$P = \frac{8 \times 10^6}{3413} = 2344 \text{ kW} \quad \text{or} \quad P = \frac{3143}{1.341} = 2344 \text{ kW}$$

Problems

1-1. What is a working substance? Name three common working substances.

1-2. List three properties of any working substance.

1-3. Calculate the absolute temperature corresponding to 60°F.

Ans. 520°R

1-4. Calculate the Celsius and Kelvin temperature equivalents for 60°F.

Ans. 15.6°C, 288.6 K

1-5. Name the two transient forms of energy included in the Energy Equation for Steady Flow.

1-6. A steam turbine receives steam at 1379 kPa gage and exhausts it to a condenser where the vacuum is indicated as 635 mm of Hg by a gage. For a barometric pressure of 760 mm of Hg, calculate (a) the absolute pressure at the turbine inlet, kPa, and (b) the absolute pressure corresponding to the vacuum gage reading, kPa and mm of Hg.

Ans. (a) 1480 kPa; (b) 16.7 kPa, 125 mm Hg

1-7. A boiler feed pump takes water at an elevation of 10 feet above a chosen datum and pumps it into a boiler drum at elevation 30 feet. Calculate the change in potential energy for the water from pump to boiler in (a) ft·lbf/lb, and (b) Btu/lb.

Ans. (a) 20 ft·lbf/lb; (b) 0.0257 Btu/lb

1-8. High pressure steam enters a turbine with a velocity of 125 ft/sec and leaves the exhaust trunk of the turbine with a velocity of 1000 ft/second. Calculate the kinetic energy of the steam in ft·lbf/lb and Btu/lb at (a) the entrance to the turbine, and (b) at the exhaust trunk.

Ans. (a) 242.6 ft·lbf/lb, 0.312 Btu/lb; (b) 15,530 ft·lbf/lb, 19.96 Btu/lb

1-9. High pressure steam enters a turbine with a velocity of 38.0 metres/sec and leaves the exhaust trunk with a velocity of 305 metres/sec. Calculate the kinetic energy of the steam in kJ/kg and N·m/kg at (a) the entrance to the turbine, and (b) the exhaust trunk.

Ans. (a) 0.722 kJ/kg, 722 N·m/kg; (b) 46.5 kJ/kg, 46,500 N·m/kg

1-10. Feedwater enters a boiler drum at 700 psia, 300°F and a specific volume of 0.0174 ft³/lb. Steam leaves the superheater of the boiler at 600 psia, 550°F and a specific volume of 0.8749 ft³/lb. Calculate (a) the flow work in Btu/lb associated with the water

INTRODUCTION AND THE FIRST LAW OF THERMODYNAMICS

entering the boiler drum, and (b) the flow work in Btu/lb associated with the steam leaving the superheater.

Ans. (a) 2.25 Btu/lb; (b) 97.2 Btu/lb

1-11. Feedwater enters a boiler drum at a pressure of 4826 kPa abs, 148.9°C and a specific volume of 0.00109 m³/kg. Steam leaves the superheater of the boiler at 4137 kPa abs, 287.8°C and a specific volume of 0.0546 m³/kg. Calculate (a) the flow work in kJ/kg associated with the water entering the boiler drum, and (b) the flow work in kJ/kg associated with the steam leaving the superheater.

Ans. (a) 5.26 kJ/kg; (b) 225.9 kJ/kg

1-12. A small air compressor delivers 1 lb per min of air while supplying 3 horsepower to the air. Calculate the specific work, ft·lbf/lb, supplied to the air.

Ans. 99,000 ft·lbf/lb

1-13. A steam turbine develops 200 horsepower while receiving 400 Btu from each pound of steam passing through the turbine. Calculate the pounds of steam per hour required by the turbine to produce this power.

Ans. 1272.5 lb/hr

1-14. A steam turbine develops 149 kW while receiving 930 kJ from each kilogram of steam passing through the turbine. Calculate the kilograms of steam per hour required to produce this power.

Ans. 576.8 kg/hr

1-15. Many marine steam propulsion plants achieve a fuel rate of 0.45 pound of fuel per shaft horsepower hour. Determine the equivalent fuel rate in kilograms per shaft kilowatt hour.

Ans. 0.27 kg/kW·hr

1-16. At a certain mountain location where the barometer reads 13.5 psia, the intake manifold gage on a cross-country truck reads 20 inches of mercury vacuum. Calculate the absolute pressure in the truck intake manifold, psia and in. Hg.

Ans. 3.7 psia or 7.5 in. Hg

1-17. The power delivered by an auxiliary turbine is 500,000 Btu/hr when the steam flow is 4000 lb/hr. Find (a) the specific work, ft·lbf/lb, and (b) the power delivered, hp.

Ans. (a) 97,250 ft·lbf/lb; (b) 196.5 hp

1-18. A working substance enters a thermodynamic steady flow system with the following conditions: $p_1 = 20$ psia, $v_1 = 11.7$ ft³/lb, $u_1 = 101.6$ Btu/lb, $V_1 = 150$ ft/sec. The working substance leaves the system with the following conditions: $p_2 = 25$ psia, $v_2 = 10.3$ ft³/lb, $u_2 = 149.0$ Btu/lb, and $V_2 = 500$ ft/sec. Changes in elevation through the system are negligible, and 10 Btu/lb transferred heat is added to the fluid as it passes through the system. Determine the work done on or by the fluid, Btu/lb.

Ans. (−)46.3 Btu/lb (on)

CHAPTER 2

BASIC CONCEPTS AND THE GENERAL NON-FLOW PROCESS

2-1. The Approach. In this chapter, some of the fundamental concepts of thermodynamics will be presented. Before embarking on this new material, it is considered most important that the student establish in his own mind a good, sound study philosophy. He should, first of all, realize that thermodynamics requires a considerable amount of inductive reasoning as opposed to great feats of memory. A thorough knowledge of a few basic general equations, coupled with an understanding of how to modify those equations to meet special situations, is a prime requisite to success. Elementary calculus and algebra are also necessary tools with which the student is assumed to be equipped. It is the purpose of this text to supply the basic equations and to indicate the modifications necessary to meet specific cases. In so doing, equations of limited application will naturally be developed. The student is enjoined to follow the reasoning and mathematics leading to the specialized equations rather than attempting to memorize those equations. In this way, the student will be able to develop the special equations when needed and will not encumber himself with a wealth of unrelated detail. In support of this thesis, the text material will emphasize the fundamental equations and note any restrictions that may apply.

2-2. Properties, States and Processes. The properties of a working substance which are of thermodynamic interest have been mentioned in Art. 1-4. The values of the several properties in combination establish the *state* of a working substance. If any of the properties changes in value as referred to some initial condition, a *change of state* has occurred. If the properties are subsequently found to have resumed their initial values, then the substance must again be at its original state. Every state of a working substance is unique in that each property has one and only one value at that state.

In the general case, the state of a simple substance and the values of the various properties can be established if any two of the properties are known. This follows from the fact that the substances with which we are concerned are subject to physical laws and their properties are definitely interrelated. The interrelationship of properties is not surprising if we recognize that the several properties are basically reflections of molecular

CONCEPTS AND THE NON-FLOW PROCESS

characteristics. In the case of a gas, for instance, pressure is a measure of molecular weight, concentration and activity; temperature is a measure of molecular weight and activity; specific volume is a measure of molecular weight and concentration. A characteristic equation relating these three properties exists and will be discussed later. Thus any two of these properties will establish the third, and the remaining properties can be found from other equations.

The two properties usually employed to fix the state of a substance are temperature and pressure, since both are readily determined with common instruments. As will be brought out later, pressure and temperature alone are not always sufficient to fix the state of a vapor, and one other known property may be required.

State changes may be of major or minor magnitude. Any change of state will be accompanied by a general change in the values of the several properties. Conversely, a change in any of the observed properties is indicative of a change of state. The minor fluctuations of a barometer, for instance, reflect the continual changes of state taking place in the air of the atmosphere. In dealing with working substances classed as vapors, the state changes are frequently great enough to include a change of phase. The term *phase* is used to denote the type of molecular agglomeration of a substance such as solid, liquid, vapor or gas. A change of phase is indicative of a major modification in molecular spacing. An observation of phase is often necessary in establishing the state of a substance.

A thermodynamic *process* occurs whenever a working substance undergoes a change of state. Since two properties establish a state point, it is useful to represent thermodynamic properties x and y on cartesian coordinates (see Fig. 2-1). If each instantaneous state through which the

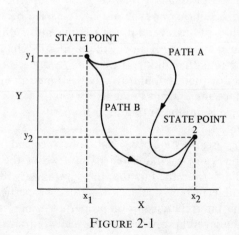

FIGURE 2-1

CONCEPTS AND THE NON-FLOW PROCESS

substance passes, in going from the initial state (*1*) to the final state (*2*), be represented graphically, the line produced is known as a *process line* or *state path*. The solution of many thermodynamic problems is contingent upon a determination of the initial and final states and a knowledge of the type of process followed between those states. The two states fix the two terminal sets of properties (x_1, y_1 and x_2, y_2), and the change of properties associated with the state change can be readily determined by differences. *It is customary to treat property changes as the difference between the* **final** *and* **initial** *value.* Thus, if the internal energy decreases 20 Btu, we write $U_2 - U_1 = -20$ Btu. It should be noted that the change in the value of any property is a function only of the initial and final states and is independent of the process followed between those states. In mathematical parlance, *properties are point functions,* which means that their numerical values are *completely independent of the past history* of the substance. On the other hand, *heat and work are both path functions, and their numerical values are completely dependent upon the process* followed in proceeding from the initial to the final state. Evaluation of the heat and work terms is contingent upon a knowledge of the nature of the process. The conventional subscript system followed in this text should aid the student in keeping point and path functions straight. *All point functions, such as properties, carry only a single subscript, indicating that they exist and are fully defined at a point. Heat and work carry the double subscript, indicating that they are associated with a change of state and will be dependent upon the process followed.* Since a change of state cannot take place instantaneously, we can deduce that the performance of work or the transfer of heat must require time for accomplishment. Therefore, neither heat nor work can possibly exist at a point.

The differentials of point functions, or properties such as *p, v, u,* etc., are exact, and integration is not contingent upon the path followed. Thus:

$$y_2 - y_1 = \int_1^2 dy \qquad x_2 - x_1 = \int_1^2 dx$$

On the other hand, the differentials of path functions such as heat and work are inexact, and evaluation of the inexact differential can be carried out only if the nature of the path followed is known:

$$wk_{12} = \int_1^2 \delta wk \qquad q_{12} = \int_1^2 \delta q$$

The double subscript signifies that the work or heat is dependent upon the particular path followed in proceeding from state 1 to state 2, while the

differential symbol δ is used to indicate that neither heat nor work is an exact differential:

$$wk_{12} \neq wk_2 - wk_1 \qquad q_{12} \neq q_2 - q_1$$

2-3. The Reversible Process. One general approach used in thermodynamics is to establish and investigate an ideal situation and then to make a comparison with real equipment operating in a similar real situation. In this way the thermodynamic effectiveness of the actual machinery can be determined and expressed as some sort of an "efficiency." The processes that occur in an actual thermodynamic device are naturally imperfect, but they can be approximated by certain ideal processes. Fortunately, the ideal processes which are of engineering importance are relatively few in number and fairly simple in nature. A knowledge of these ideal processes enables us to set up and investigate an ideal counterpart for our actual equipment preliminary to making the comparison outlined above. It is logical to begin our studies with a description and analysis of the several important ideal processes.

In thermodynamics, it is customary to refer to the *ideal* processes as *reversible,* and the two terms are interchangeable. A process is reversible if, after completion, it may be made to retrace itself exactly with the working substance returning to the initial state over the identical state path that was originally traversed. A reversible process characteristically brings about the state changes of the working substance in a smooth, orderly fashion, and all the properties follow a definite predictable interrelationship throughout the process. Of particular interest are those reversible processes in which one of the properties remains constant throughout.

The concepts of reversibility will be discussed at length later in the text. For the present discussion, the student need only recognize that "reversible" as referred to a thermodynamic process means "ideal" and is similar in connotation to the term "frictionless" as applied to a machine.

2-4. The Energy Equation for Non-Flow Processes. Most thermodynamic processes of interest to the mechanical engineer fall into two very general classes, namely *steady flow* and *non-flow*. For the time being we will confine ourselves to the non-flow processes. A non-flow process may be defined as one in which the working substance does not flow across its confining boundary during the state change. The molecules present at the end of the process are exactly the same ones which were present at the beginning. Non-flow processes occur in such actual equipment as reciprocating air compressors, reciprocating internal combustion engines, and

CONCEPTS AND THE NON-FLOW PROCESS

reciprocating steam engines during the time both inlet and exhaust valves are closed. Common to all of these machines are pistons and cylinders. Consequently, a piston and cylinder arrangement (Fig. 2-2) has been selected as a typical non-flow device for the discussion and evaluation of non-flow processes. The figure shows an initial condition 1, followed by a

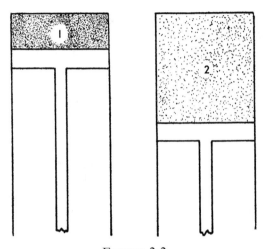

FIGURE 2-2

later condition 2 after a lapse of time. Heat can pass through the boundaries in the general case, and work is physically evidenced by movement of the piston. As the piston moves, a shift in the center of gravity of the captured working substance will occur. Thus the potential energy of the working substance, considered as a single mass, will change slightly. Assuming that one pound of working substance is contained in the pictured cylinder, then the center of gravity must move 778 feet to represent an energy level change of 1 Btu. Since the stroke of all practical devices is measured in inches rather than feet, it is clear that the potential energy change during the depicted non-flow process can be neglected without introducing any measurable errors. From the physical nature of the chosen device, it is evident that the internal energy stored in the working substance will be subject to change. Using the symbolism and energy balance philosophy previously introduced, we can immediately write the non-flow energy equation:

$$u_1 + q_{12} = u_2 + \frac{wk_{12}}{(J)} \text{ Btu/lb (J/kg)} \qquad (2\text{-}1)$$

CONCEPTS AND THE NON-FLOW PROCESS

The student should note that Equation (2-1) *is applicable to any non-flow process, ideal or actual, with any working substance*. This equation can also be inferred from the Steady Flow Energy Equation (1-20), if we consider zero flow as a special limiting case, discard the flow work and kinetic energy terms as not applicable, omit the potential energy terms following the reasoning outlined above and recognize that the subscripts 1 and 2 refer to a separation in time rather than space.

Our immediate interest will be directed toward the reversible non-flow processes. In a typical problem, we can expect to have sufficient data to fix the initial and final states and to know the nature of the process by which the state change is effected. Since all properties of a working substance are fully established at any given state, then the initial and final values of the internal energy will be fixed. Remember that internal energy is a property. In order to complete the solution of Equation (2-1), it is only necessary to establish the value of either the heat term or the work term. With three unknowns determined in the equation, the fourth can readily be found by differences. Since the exact nature of the simple reversible processes can be established, mathematical expressions for heat and work can be developed. The remainder of this chapter will be largely devoted to the development of a general approach to the solution of the reversible non-flow problem. Certain concepts to be presented are applicable to both the flow and non-flow cases, and the student is again enjoined to pay particular attention to the restrictions on the various equations as they are introduced. It should also be noted that our ultimate objective is the determination of the ideal work accomplished and the heat transferred, since heat and work are of primary interest to the engineer.

Example 2-1: In a certain non-flow process, the internal energy decreases 20 Btu/lb, and 77,800 ft·lbf/lb of work is done on the substance. Find the heat added or abstracted.
Solution:

$$u_2 - u_1 = (-)20 \text{ Btu/lb}; \quad wk_{12} = (-)77,800 \text{ ft·lbf/lb}$$

$$u_1 + q_{12} = u_2 + \frac{wk_{12}}{J}$$

$$q_{12} = u_2 - u_1 + \frac{wk_{12}}{J} = -20 - \frac{77,800}{778} = \underline{(-)120 \text{ Btu/lb (abstracted)}}$$

Example 2-2: In another non-flow process involving 2 kg of working substance there is no heat transferred, but the internal energy increases 5000 joules. Find the work done on or by the substance, J/kg.
Solution:

$$q_{12} = 0; \quad U_2 - U_1 = 5000 \text{ J}; \quad M = 2 \text{ kg}$$

$$u_2 - u_1 = \frac{5000}{2} = 2500 \text{ J/kg}$$

$$wk_{12} = q_{12} - (u_2 - u_1) = \underline{(-)2500 \text{ J/kg (on)}}$$

CONCEPTS AND THE NON-FLOW PROCESS

2-5. Enthalpy. As mentioned in Art. 1-4, the properties in which we are interested are pressure, temperature, specific volume, internal energy, entropy and enthalpy. The first four properties mentioned have been discussed at some length in Chapter 1. Before proceeding further, it will be advantageous to explore an additional property, enthalpy.

Inspection of Equation (1-20), the Steady Flow Energy Equation in Engineers' units, reveals that each side of the equation contains the terms u and $pv/(J)$ with the appropriate subscripts. The internal energy term, u, represents a property of the substance. Similarly, the value of the $pv/(J)$ term is fixed by two properties, absolute pressure and specific volume, since J is a constant. No other terms in the equation involve properties. In the interest of simplicity, this suggests combining the two property terms into one. For a given state of a selected working substance, internal energy, absolute pressure and specific volume are all fixed at a definite, unvarying value. That being the case, the sum $u + pv/(J)$ will also have a definite value for the given state. This sum, then, has the characteristics of a property in that it is fixed by the state. Our new property is given the name **enthalpy.** Like internal energy, it is a function of mass and will have both total and specific values. The symbol used for specific enthalpy is h and for total enthalpy is H. Frequently the term enthalpy is used without qualification, and whether specific or total is intended may be determined from the context or units specified. In equation form and by definition:

$$H = U + \frac{pV}{(J)} \text{ Btu (J)} \qquad (2\text{-}2)$$

$$h = u + \frac{pv}{(J)} \text{ Btu/lb (J/kg)} \qquad \underline{(2\text{-}3)}$$

The great practical utility of enthalpy lies in the numerical solution of **steady flow** problems using the Energy Equation. For a given state, we can evaluate both the internal energy and flow work terms by the single determination of an enthalpy rather than by finding the three properties, pressure, specific volume and internal energy, and then carrying out the mathematical operations indicated by Equation (2-3). This practical use of enthalpy to evaluate the internal energy and flow work terms for a steady flow situation occasionally leads to a mistaken definition of enthalpy as the sum of internal energy and flow work. If we were to accept this definition, then the enthalpy of a working substance at rest in a given state would be different from that of the same substance in motion at the same state. This follows from the fact that there can be no flow work without flow. Obviously, if this were true, enthalpy would no longer be a property, since a property is fully defined by the state alone. This leads to the

CONCEPTS AND THE NON-FLOW PROCESS

general definition of enthalpy as simply the sum of internal energy and the $pv/(J)$ *term*. Thus, the enthalpy concept is not limited to flow situations. We certainly can evaluate enthalpy in a non-flow case, and it is possible that it will have some non-flow application.

Example 2-3: At a pressure of 100 psia and 400°F, the specific volume of steam is 4.934 cu ft/lb, and the specific enthalpy is 1227.5 Btu/lb. Find the internal energy at the given state, Btu/lb.

Solution:

$$p = 100 \text{ psia}; \quad v = 4.934 \text{ ft}^3/\text{lb}; \quad h = 1227.5 \text{ Btu/lb}$$

$$h = u + \frac{pv}{J}; \quad u = h - \frac{pv}{J}$$

$$u = 1227.5 - \frac{100 \times 144 \times 4.934}{778} = \underline{1136.2 \text{ Btu/lb}}$$

2-6. Mechanical Work and $\int p\, dv$ for a Reversible Non-Flow Process. The determination of the mechanical work performed during any non-flow reversible process is of primary interest in connection with the study of such processes. In order to develop a general expression for work under these conditions, it is logical to base the discussion on a simple but typical non-flow device. Such a device is the frictionless piston-cylinder arrangement shown in Fig. 2-3.

FIGURE 2-3

In the figure the piston is shown occupying three positions. The initial position is at (1); the next position is at a differential distance, dL, from (1); the final position is at (2). The cross-sectional area of the piston face is A, and the absolute pressure in the cylinder at any time is p. Then the

force acting on the piston face is pA and the work done while the piston moves through the distance dL is given by the equation:

$$\delta(Wk) = pA\,dL$$
since Work = Force × distance
But: $A\,dL = dV$
Then: $\delta(Wk) = p\,dV$

Integrating and establishing limits for the entire process:

$$Wk_{12} = \int_1^2 \delta Wk = \int_1^2 p\,dV \text{ ft} \cdot \text{lbf (J)} \qquad (2\text{-}4)$$

As previously noted, it is usually convenient to deal with a unit mass of working substance rather than with a total mass, M, in the basic equations, in which case Equation (2-4) becomes

$$wk_{12} = \int_1^2 p\,dv \text{ ft} \cdot \text{lbf/lb (J/kg)} \qquad \underline{(2\text{-}5)}$$

where wk_{12} = the mechanical work done per unit mass of working substance during a non-flow reversible process between an initial state (1) and a final state (2)
p = the absolute pressure in lbf/ft² (Pa)
v = the specific volume in ft³/lb (m³/kg).

Inspection of Equation (2-5) reveals several salient points. First, although the equation is limited to the non-flow reversible type of process, the derivation has placed no restriction on the working substance employed. Second, both pressure and specific volume are variables in the general case, and the integration cannot be carried out mathematically for any process unless an equation relating p and v is known. Two important cases are those where, for the specific process, either p or v is constant. Third, when v_2, the final specific volume, is greater than v_1, the initial specific volume, as in an expansion, then the integral will have a *positive* numerical value, indicating that work has been *done by* the working substance. Conversely, if v_2 is less than v_1 as in a compression, the integral will have a *negative* numerical value, indicating that work has been *done on* the working substance. Note that the direction of pressure variation does not control the sign of the numerical evaluation of the integral.

2-7. The *p-v* Diagram. From elementary calculus, the area between a curve, the x-axis, and the ordinates at $x = a$ and $x = b$ is given by the expression, $\int_a^b y\,dx$. Inspection of Equation (2-5) shows that $\int_1^2 p\,dv$ is of the same form and suggests the use of a *p-v* plot with absolute pressure as

the ordinate and specific volume as the abscissa. Such a plot is known as a *p-v diagram* and will be of considerable use in the study of thermodynamics. If each successive state of the working substance during any process is shown on a *p-v* diagram, the line connecting the state points will constitute a state path or process line. In common with any graphical representation, this diagram will assist in visualizing the behavior of the working substance during the process. Furthermore, the area included between the process line, the *v*-axis, and the ordinates of pressure at the beginning and end of the process is equivalent to $\int_1^2 p\, dv$. This area, in turn, represents mechanical work if the state path is for a reversible, non-flow process. Figure 2-4 shows two non-flow reversible processes that might occur in our piston-cylinder arrangement.

FIGURE 2-4

In Fig. 2-4(a), the process proceeds from left to right, the working substance expands, v_2 is greater than v_1, the area under the curve corresponding to $\int_1^2 p\, dv$ is positive in sense and work has been done by the substance. In Fig. 2-4(b) the process proceeds from right to left, the substance is compressed, v_2 is less than v_1, the area under the curve corresponding to $\int_1^2 p\, dv$ is negative in sense and work has been done on the working substance.

CONCEPTS AND THE NON-FLOW PROCESS

Example 2-4: In a reversible non-flow process with a gas, the initial pressure and specific volume are 270 psia and 4 cu ft/lb respectively. The final pressure and specific volume are 80 psia and 9 cu ft/lb. The process is defined by the equation: $pv^{3/2} = C$, where p and v are simultaneous values of pressure and specific volume at any instant during the process and C is a constant.
 (a) Sketch the process on p-v coordinates.
 (b) Find the work done on or by the gas during the process, Btu/lb.
Solution:

(a)

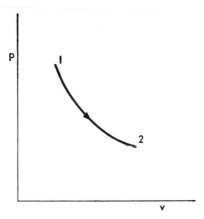

(b) $\quad p_1 = 270$ psia $\quad p_2 = 80$ psia
$\quad\quad\quad v_1 = 4$ ft³/lb $\quad v_2 = 9$ ft³/lb
$\quad\quad\quad p_1 v_1^{3/2} = p_2 v_2^{3/2} = pv^{3/2} = C$

$$wk_{12} = \int_1^2 p\,dv; \quad p = Cv^{-3/2}$$

$$wk_{12} = C\int_1^2 v^{-3/2}\,dv = C\left[\frac{v^{-1/2}}{1-3/2}\right]_1^2 = C[(-)2(v_2^{-1/2} - v_1^{-1/2})]$$

Substituting for C, the values $p_2 v_2^{3/2}$ and $p_1 v_1^{3/2}$:

$wk_{12} = -2(p_2 v_2^{3/2} \cdot v_2^{-1/2} - p_1 v_1^{3/2} \cdot v_1^{-1/2}) = -2(p_2 v_2 - p_1 v_1) = 2(p_1 v_1 - p_2 v_2)$
$wk_{12} = 2(270 \times 144 \times 4 - 80 \times 144 \times 9) = 103,700$ ft·lbf/lb

$$\frac{wk_{12}}{J} = \frac{103,700}{778} = \underline{133.3 \text{ Btu/lb (by)}}$$

Example 2-5: The same gas is brought reversibly from the same initial to the same final state as in Example 2-4, but the state change is effected by a non-flow constant volume, and a non-flow constant pressure process in combination. Draw the p-v diagram and find the work done on or by the gas, if the sequence of processes is:
 (a) constant volume followed by constant pressure.
 (b) constant pressure followed by constant volume.
Solution:

Note: An encircled property symbol adjacent to a property line on any diagram indicates that the property remains constant during the process.

CONCEPTS AND THE NON-FLOW PROCESS

(a)

$p_1 = 270$ psia
$v_1 = 4$ ft^3/lb
$p_2 = 80$ psia
$v_2 = 9$ ft^3/lb

$$wk_{12} = wk_{10} + wk_{02} = \int_1^0 p\,dv + \int_0^2 p\,dv$$

$$p_0 = p_2; \quad v_0 = v_1$$

$$\int_1^0 p\,dv = 0 \quad \text{since} \quad dv = 0 \quad \text{here}$$

$$\int_0^2 p\,dv = p_2 \int_0^2 dv = p_2(v_2 - v_0) \quad \text{since } p \text{ is constant}$$

$$wk_{02} = 80 \times 144(9 - 4) = 57{,}600 \text{ ft} \cdot \text{lbf/lb}$$

$$\frac{wk_{12}}{J} = 0 + \frac{57{,}600}{778} = \underline{74.0 \text{ Btu/lb (by)}}$$

(b)

$p_1 = 270$ psia
$v_1 = 4$ ft^3/lb
$p_2 = 80$ psia
$v_2 = 9$ ft^3/lb

$$wk_{12} = wk_{10} + wk_{02} = \int_1^0 p\,dv + \int_0^2 p\,dv$$

$$p_1 = p_0; \quad v_0 = v_2$$

$$\int_1^0 p\,dv = p_1 \int_1^0 dv = p_1(v_0 - v_1)$$

CONCEPTS AND THE NON-FLOW PROCESS

$$wk_{10} = 270 \times 144(9-4) = 194{,}400 \text{ ft} \cdot \text{lbf/lb}$$

$$wk_{02} = \int_0^2 p\, dv = 0$$

$$\frac{wk_{12}}{J} = \frac{194{,}400}{778} + 0 = \underline{250 \text{ Btu/lb (by)}}$$

2-8. Heat, Entropy, and The *T-s* Diagram. In the preceding article, it has been brought out that the mechanical work during a non-flow process can be represented graphically by an area on the *p-v* diagram. This suggests the possibility of a graphical representation of heat, the other form of energy in transition, as an area on some other diagram. Pressure may be thought of as a mechanical potential causing work to be done, and, for a given substance, it is independent of the mass present. Temperature is also independent of mass and has the characteristics of a thermal potential causing heat to be transferred. By analogy, it is logical to choose absolute temperature as the ordinate of our proposed diagram, just as absolute pressure was the ordinate of the *p-v* diagram.

The selection of the second property for our diagram on which an area will represent heat is not so easy. By inference this property, like volume, should be proportional to mass and have both specific and total values. The property that we want and that will fulfill our specifications is known as **entropy**. The symbol S is used for total entropy, and the symbol s for specific entropy. Then $S = Ms$. The condition to be met by this new property is that, for a reversible process,

$$q_{12} = \int_1^2 \delta q = \int_1^2 T\, ds \text{ Btu/lb (J/kg)} \qquad (2\text{-}6)$$

$$Q_{12} = \int_1^2 \delta Q = \int_1^2 T\, dS \text{ Btu (J)} \qquad (2\text{-}7)$$

Differentiating and rearranging Equation (2-6), we shall arrive at a mathematical definition of entropy:

$$ds = \frac{\delta q}{T} \qquad (2\text{-}8)$$

Integrating between the limits of the initial and final states of any *reversible process*:

$$\int_1^2 ds = \int_1^2 \frac{\delta q}{T}$$

$$s_2 - s_1 = \int_1^2 \frac{\delta q}{T} \text{ Btu/lb} \cdot {}^\circ\text{R (J/kg} \cdot \text{K)} \qquad (2\text{-}9)$$

CONCEPTS AND THE NON-FLOW PROCESS

Equation (2-9) provides a mathematical expression for change of entropy for any *reversible process*. In general, we are more interested in the change of entropy during a process than we are in the particular value at a selected state. It should be noted from the nature of the expressions for entropy that the unit of total entropy, S, is Btu per degree Rankine and the unit of specific entropy, s, is Btu per pound-degree Rankine. The corresponding SI units are joules per kelvin for S and joules per kilogram-kelvin for s. The concept of entropy will be discussed at more length later in the text. For the time being, the student is asked to recognize that it is a property mathematically defined by Equation (2-8) and that it has been introduced because of its utility in representing heat graphically as an area on a coordinate system.

Before discussing the graphical representation of heat further, it will be well to inspect Equation (2-6) in some detail. First, the equation is applicable to any process with any working substance, and the only restriction placed upon it is that the process concerned be *reversible*. Second, both absolute temperature and specific entropy are variables in the general case, and integration cannot be carried out mathematically unless an equation relating T and s is known for the process. Two important cases are those in which either T or s remains constant. Third, when s_2, the final specific entropy, is greater than s_1, the initial specific entropy, then the integral will have a *positive* numerical value, indicating that heat has been *added* to the working substance. Conversely, if s_2 is less than s_1, the integral will have a *negative* numerical value, indicating that heat has been *abstracted* from the working substance. Note that the sign of the numerical value of the integral is determined by the direction of variation of the entropy and is independent of the direction of variation of the absolute temperature.

The coordinate system established by the foregoing discussion employs absolute temperature as the ordinate and specific entropy as the abscissa. Specific entropy rather than total entropy is chosen, since it is generally simpler to deal with one pound of the working substance. The resultant diagram takes the obvious name of the *T-s diagram*, and the area under any reversible process line, either flow or non-flow, represents the heat involved in that process. Figure 2-5 shows the *T-s* diagrams associated with two very simple non-flow processes, namely constant volume heating and cooling of a working substance. In Fig. 2-5(a), the process proceeds from left to right, the entropy increases, the area under the curve corresponding to $\int_1^2 T\,ds$ is *positive* in sense, and heat has been *added* to the working substance. In Fig. 2-5(b), the process proceeds from right to left, the entropy decreases, the area under the curve corresponding to $\int_1^2 T\,ds$ is

CONCEPTS AND THE NON-FLOW PROCESS

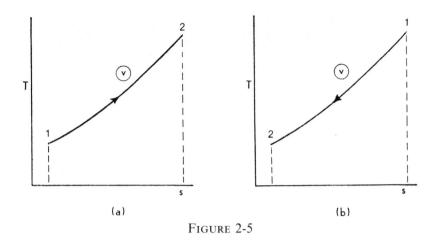

FIGURE 2-5

negative in sense, and heat has been *abstracted* from the working substance.

Example 2-6: Two pounds of steam initially having a specific entropy of 1.6356 Btu/lb · °R and an internal energy of 1100.0 Btu/lb undergo a constant temperature reversible non-flow process at 300°F such that the final entropy is 1.5150 Btu/lb · °R and the internal energy is 1016.4 Btu/lb. Find:
 (a) The transferred heat, Btu
 (b) The work done, ft · lbf
Solution:

(a) $M = 2$ lb $\quad t_1 = t_2 = 300°F$
$s_1 = 1.6356$ Btu/lb · °R $\quad u_1 = 1100.0$ Btu/lb
$s_2 = 1.5150$ Btu/lb · °R $\quad u_2 = 1016.4$ Btu/lb
$T_1 = T_2 = T = 300 + 460 = 760°R$

$$q_{12} = \int_1^2 T\,ds = T\int_1^2 ds \text{ since } T \text{ is constant here}$$

$$Q_{12} = Mq_{12} = MT\int_1^2 ds = MT(s_2 - s_1)$$

$Q_{12} = 2 \times 760(1.5150 - 1.6356) = \underline{(-)\ 183.3 \text{ Btu (abstracted)}}$

(b) $$U_1 + Q_{12} = U_2 + \frac{Wk_{12}}{J}$$

$$\frac{Wk_{12}}{J} = Q_{12} - (U_2 - U_1) = Q_{12} - M(u_2 - u_1)$$

$$\frac{Wk_{12}}{J} = -183.3 - 2(1016.4 - 1100.0)$$

$$= -183.3 + 167.2 = (-)16.1 \text{ Btu}$$

$Wk_{12} = (-)16.1 \times 778 = \underline{(-)12,526 \text{ ft · lbf (on)}}$

CONCEPTS AND THE NON-FLOW PROCESS

2-9. Specific Heat (General). Specific heat capacity, commonly abbreviated to specific heat, is defined as the quantity of heat necessary to raise the temperature of a unit mass of substance one degree. In equation form:

$$c_x = \frac{q_{12}}{(t_2 - t_1)} \frac{\text{Btu/lb} \cdot °\text{F}}{(\text{J/kg} \cdot °\text{C})} = \frac{q_{12}}{(T_2 - T_1)} \frac{\text{Btu/lb} \cdot °\text{R}}{(\text{J/kg} \cdot \text{K})} \quad (2\text{-}10)$$

where
 c_x = the specific heat of the substance
 q_{12} = the heat added, Btu/lb (J/kg)
 $(t_2 - t_1)$ = the observed temperature rise, °F (°C)
 $(T_2 - T_1)$ = the observed temperature rise, °R (K).

Note that the degree intervals for the Fahrenheit and the Rankine temperature scales are equal. Hence $(t_2 - t_1)°\text{F} = (T_2 - T_1)°\text{R}$. Similarly, the one degree Celsius temperature interval equals one kelvin exactly, and $(t_2 - t_1)$ in degrees Celsius equals $(T_2 - T_1)$ in kelvins.

Equation (2-10) represents the experimental method of determining specific heat values by measuring the heat input to a known mass of substance and observing the change of temperature. A value so obtained is the average or mean specific heat for the substance for the conditions under which the heating was carried out. The observed values of specific heat will vary with the temperature range, the pressure range and the amount of restriction imposed on the expansion of the substance during the addition of heat.

The principal utility of specific heat lies in the determination of the quantity of heat added or abstracted from an observation of temperature change. In this context the basic equation becomes:

$$q_{12} = c_x(t_2 - t_1) = c_x(T_2 - T_1) \text{ Btu/lb (J/kg)} \quad \underline{(2\text{-}11)}$$

Practical use of this equation presupposes a known value for c_x. Unfortunately, as pointed out above, specific heat for any one substance is a variable. This variation can be indicated mathematically by writing Equation (2-11) in the form:

$$q_{12} = \int_1^2 c_x \, dt \quad (2\text{-}12)$$

Equation (2-12) can be integrated if c_x can be assumed to be constant or if the relationship between specific heat and temperature is known.

In the case of solids, the value of specific heat is little affected by pressure and expansion but does increase somewhat with temperature. Engineering handbooks customarily quote the specific heat of a particular solid as a mean value over some specified temperature range.

The specific heat of a liquid varies with temperature and pressure but is

CONCEPTS AND THE NON-FLOW PROCESS

little affected by expansion. Over reasonable ranges of temperature and pressure, an average specific heat may be used with little error. As an example, the British thermal unit is frequently defined as $\frac{1}{180}$th of the heat necessary to raise the temperature of 1 pound of fresh water from 32°F to 212°F. This immediately establishes the mean specific heat of fresh water between these temperature limits as 1 Btu/lb · °F. Although not specifically stated, atmospheric pressure may be implied. At higher temperatures, the specific heat of water increases considerably. The effect of pressure is of lesser magnitude, but, in general, an increase in pressure at a given temperature results in a reduction in the value of specific heat. Figures 2 and 3 of the *Steam Tables* (Keenan, Keyes, Hill and Moore) show these effects for water.

The specific heat of a vapor varies considerably with pressure and temperature and expansion characteristics. The effects of temperature and pressure are shown in Fig. 4 of the aforementioned *Steam Tables* for the vapor phase of steam. Chapter 4 provides a more complete discussion of the nature and use of the *Steam Tables*. Suffice it to say at this point that these tables provide another approach such that direct use of the specific heats for the vapor and liquid phases of steam is avoided.

Example 2-7: In a marine propulsion plant, the salt water used to condense the exhaust steam enters the condenser at 60°F and leaves at 80°F. The specific heat of salt water is 0.96 Btu/lb · °F, the density is 64.0 lb/cu ft, the cooling water rate is 5000 gpm and 1 gal = 231 cu in. Find the heat absorbed by the cooling water, Btu/min.
Solution:

$$q_{12} = c_w(t_2 - t_1); \quad \dot{Q}_{12} = \dot{M}q_{12} = \dot{M}c_w(t_2 - t_1)$$
$$c_w = 0.96; \quad \rho = 64.0; \quad \text{cooling water rate} = 5000 \text{ gpm}$$
$$t_1 = 60°F; \quad t_2 = 80°F$$
$$\dot{M} = \frac{5000 \times 231}{1728} \times 64.0 = 42,800 \text{ lb/min}$$
$$\dot{Q}_{12} = 42,800 \times 0.96(80 - 60) = \underline{822,000 \text{ Btu/min}}$$

2-10. Specific Heats of a Perfect Gas. We shall start the discussion of the specific heats of gases with an ideal gas and introduce later any deviations that may be encountered in real gases. For a perfect gas, the value of specific heat is independent of both temperature and pressure. However, the value of specific heat from an experimental determination of the type indicated by Equation (2-10) will very definitely depend upon how the natural tendency for the gas to expand when heated is allowed to proceed. There are two natural extremes to the many possible expansion conditions. One extreme is free expansion and the other is no expansion. With no expansion, the heating process must take place at constant volume. For free expansion, the heating process will take place at constant

pressure. These two heating processes can easily be visualized as occurring in a cylinder fitted with a frictionless piston. For the constant volume process, the piston must be locked in a fixed position. For the constant pressure process, the piston must be free to move against a constant external restraining force. Figure 2-6 shows the necessary mechanical arrangements together with the corresponding T-s diagrams. Let us

FIGURE 2-6

assume that each cylinder contains one pound of gas and that the initial temperature (T_1) is the same for both. Now, let heat be added until the final temperature (T_2) is attained in both cylinders. It will be found that the heat required to bring about the temperature change in the constant pressure device of Fig. 2-6(b) is appreciably greater than that required for the same temperature change in the constant volume device of Fig. 2-6(a). All the heat added in the constant volume process ends up as stored internal energy and is reflected in its entirety by the temperature rise. But the heat added in the constant pressure process is split between internal energy, as evidenced by the temperature increase, and the performance of external mechanical work, as represented by the piston movement against a restraining force. Therefore, for the same temperature change, more heat must be supplied for the constant pressure process than for the constant volume. This fact is reflected in the size of the areas under the process curves on the accompanying T-s diagrams. The foregoing discussion leads to the simple but important conclusion that Equation (2-11)

cannot be used in dealing with gases unless information is available as to the nature of the process followed during heating (or cooling), and the corresponding specific heat is known. In the case of solids and liquids, the expansion effects are neglected in normal calculations, and no process differentiation is necessary.

It is customary to indicate the process applicability of specific heats by an appropriate subscript. Thus "c_x" is intended to signify the specific heat for some process "x." The specific heat at constant volume is given the symbol c_v and the specific heat at constant pressure, c_p. Referring to Fig. 2-6, it is conceivable that there are many processes other than constant volume and constant pressure whereby a gas might be brought from T_1 to T_2. Each such process would have its own particular process line, included area, and specific heat. With this multiplicity of specific heat values for any one gas, the situation would rapidly get out of hand. Fortunately, c_p and c_v are sufficient to handle all the processes with which we are concerned.

2-11. Specific Heats of Real Gases. Real gases differ from the ideal gas in that their specific heat values are affected by both temperature and pressure. The effect of pressure is minor and may be neglected as long as the gas is not so highly compressed as to take on the character of a vapor. On the other hand, the variation with temperature may be appreciable. The magnitude of the variation for a given temperature range will depend upon how much the real gas diverges from the ideal gas. For monatomic real gases such as helium and argon which closely approach the ideal, the specific heats are substantially constant with temperature. All polyatomic real gases show a definite increase in the values of both c_p and c_v with an increase in temperature. The reasons for this variation of specific heat with temperature may be deduced from the discussion of internal kinetic energy in Art. 1-12. Figure 2-7, based on data from the *International Critical Tables,* shows the change in specific heat with temperature for several representative gases.

There are several ways to deal with the variation in the specific heats of real gases with temperature. For highly accurate work, it is necessary to resort to specialized publications such as the *Gas Tables* by Keenan and Kaye. Alternatively, one can start with the equation relating temperature and specific heat that best fits a curve similar to those shown in Fig. 2-7. Equation (2-12) can then be used and the integration carried out. A simpler though less accurate method generally will be used in this text. We shall consider the specific heats of the several gases which are of interest to be constant at some average value over the range of temperatures normally encountered. Using this approach, Equation (2-11) can be modified

FIGURE 2-7 Variation of specific heats of representative gases with temperature.

CONCEPTS AND THE NON-FLOW PROCESS

to give the two fundamental specific heat equations for a gas:

$$q_{12} = c_v(T_2 - T_1) \text{ Btu/lb (J/kg)} \qquad (2\text{-}13)$$
$$q_{12} = c_p(T_2 - T_1) \text{ Btu/lb (J/kg)} \qquad (2\text{-}14)$$

Equation (2-13) is applicable to the heating or cooling at *constant volume* of:
 (a) ideal and monatomic real gases
 (b) other real gases over a reasonable range of temperature provided that the proper average value of c_v is selected

Equation (2-14) is applicable to the heating or cooling at *constant pressure* of:
 (a) ideal and monatomic real gases
 (b) other real gases over a reasonable range of temperature provided that the proper average value of c_p is selected

Example 2-8: During a constant pressure process, the temperature of a certain mass of air is raised from 40°F to 540°F. The specific heat at constant pressure is given by the equation $c_p = \alpha + \beta T - \delta T^2$ where $\alpha = 0.219$, $\beta = 0.342 \times 10^{-4}$ and $\delta = 0.293 \times 10^{-8}$. The average value of c_p which will be employed in this text for air is 0.24 Btu/lb·°F.

Find:
 (a) the heat transferred in Btu/lb, using the variable specific heat expression
 (b) the heat transferred in Btu/lb using the average specific heat

Solution:

(a)
$$q_{12} = \int_1^2 c_p \, dT = \int_1^2 (\alpha + \beta T - \delta T^2) \, dT.$$

Integrating:
$$q_{12} = \left[\alpha T + \frac{\beta T^2}{2} - \frac{\delta T^3}{3}\right]_1^2$$

$$= \alpha(T_2 - T_1) + \frac{\beta}{2}(T_2^2 - T_1^2) - \frac{\delta}{3}(T_2^3 - T_1^3)$$

$T_1 = 40 + 460 = 500°R$; $T_2 = 540 + 460 = 1000°R$

$$q_{12} = 0.219(1000 - 500) + \frac{0.342 \times 10^{-4}}{2}[(1000)^2 - (500)^2]$$

$$- \frac{0.293 \times 10^{-8}}{3}[(1000)^3 - (500)^3]$$

$q_{12} = 109.5 + 12.8 - 0.9 = 121.4$ Btu/lb (added)

(b) $q_{12} = c_p(T_2 - T_1) = 0.24(1000 - 500) = 120$ Btu/lb (added)

Example 2-9: A cylinder contains 0.2 kg of air initially at a temperature of 25°C and a pressure of 140 kPa. After a reversible constant pressure non-flow process, the temperature of the air has risen to 175°C. The initial volume was 0.03 m³ and the final volume is

CONCEPTS AND THE NON-FLOW PROCESS

0.045 m³. The specific heat of air at constant pressure may be taken as 1.0048 kJ/kg·K. Find:

(a) the heat added, kJ
(b) the work done, kJ
(c) the change in internal energy, kJ

Solution:

$$M = 0.2 \text{ kg} \qquad c_p = 1.0048 \text{ kJ/kg·K}$$
$$T_1 = 25 + 273 = 298 \text{ K} \qquad V_1 = 0.03 \text{ m}^3$$
$$T_2 = 175 + 273 = 448 \text{ K} \qquad V_2 = 0.045 \text{ m}^3$$
$$p = p_1 = p_2 = 140 \text{ kPa} = 140 \text{ kN/m}^2$$

(a)
$$Q_{12} = Mq_{12} = Mc_p(T_2 - T_1)$$
$$Q_{12} = (0.2)(1.0048)(448 - 298)$$
$$= \underline{30.1 \text{ kJ (added)}}$$

(b)
$$wk_{12} = \int_1^2 p \, dv = p \int_1^2 dv \quad \text{where } p \text{ is constant}$$
$$wk_{12} = p(v_2 - v_1) \text{ and } Wk_{12} = Mp(v_2 - v_1) = p(V_2 - V_1)$$

Since $V = Mv$,

$$Wk_{12} = (140)(0.045 - 0.03) = \underline{2.10 \text{ kJ (by)}}$$

(c)
$$U_1 + Q_{12} = U_2 + Wk_{12}$$
$$U_2 - U_1 = Q_{12} - Wk_{12}$$
$$= 30.1 - 2.10$$
$$= \underline{28.0 \text{ kJ (increase)}}$$

2-12. Closure. In order to emphasize the more important equations developed in this chapter, they are repeated here, with some comments made as to their applicability:

$$u_1 + q_{12} = u_2 + \frac{wk_{12}}{(J)} \text{ Btu/lb (J/kg)} \qquad (2\text{-}1)$$

Applicable to any non-flow process with any working substance.

$$h = u + \frac{pv}{(J)} \text{ Btu/lb (J/kg)} \qquad (2\text{-}3)$$

Applicable to any working substance as a definition of the property enthalpy.

$$wk_{12} = \int_1^2 p \, dv \text{ ft·lbf/lb (J/kg)} \qquad (2\text{-}5)$$

Applicable to any reversible non-flow process with any working substance.

CONCEPTS AND THE NON-FLOW PROCESS

$$q_{12} = \int_1^2 T\,ds \text{ Btu/lb (J/kg)} \tag{2-6}$$

Applicable to any reversible process, flow or non-flow, with any working substance.

$$q_{12} = c_v(T_2 - T_1) \text{ Btu/lb (J/kg)} \tag{2-13}$$

Applicable as indicated in Article 2-11.

$$q_{12} = c_p(T_2 - T_1) \text{ Btu/lb (J/kg)} \tag{2-14}$$

Applicable as indicated in Article 2-11.

Equations (2-1), (2-5) and (2-6) are fundamental to the solution of the typical non-flow reversible process. In general, the initial and final states can be determined, and the type of process followed is known. The values of u_1 and u_2 are immediately fixed by the initial and final states. If the value of the work term can be determined from Equation (2-5), then q_{12} can be established by solving Equation (2-1). Conversely, if q_{12} can be found from Equation (2-6), then Equation (2-1) will yield the value of wk_{12}. In dealing with a gas, Equations (2-13) and (2-14) may allow determination of q_{12} for *particular processes*. The succeeding two chapters will cover use of the equations listed above in the solution of non-flow reversible problems.

Problems

2-1. Define the following: (a) state of a working substance; (b) thermodynamic process; (c) point function; (d) path function.

2-2. Distinguish between enthalpy and entropy. Are they both point functions or path functions?

2-3. During a non-flow process 120 Btu are removed as heat from each pound of working substance while the internal energy decreases 85.5 Btu/lb. Calculate the work involved in the process in ft·lbf/lb and indicate whether work is done *on* or *by* the working substance.

Ans. ($-$) 26,840 ft·lbf/lb (on)

2-4. At standard atmospheric pressure, saturated steam has a specific volume of 26.80 ft^3/lb. If the enthalpy of that same vapor is 1150.5 Btu/lb, calculate the internal energy of the steam, Btu/lb.

Ans. 1077.6 Btu/lb

2-5. One pound of saturated steam at 100 psia is expanded in a reversible non-flow process from a specific volume of 0.017736 ft^3/lb to a specific volume of 4.434 ft^3/lb. If the pressure remains constant and the internal energy increases 807.5 Btu/lb during the process,

CONCEPTS AND THE NON-FLOW PROCESS

determine how much heat is transferred, Btu/lb, and whether it will be *added to* or *abstracted* from the working substance.

Ans. 889.2 Btu/lb (added)

2-6. If Problem 2-5 is carried out as indicated, but with the additional provision that the process is also performed at a constant temperature of 327.9°F, calculate the change of entropy for the steam during the process, Btu/lb · °R.

Ans. 1.129 Btu/lb · °R (increase)

2-7. One pound of saturated steam at 689.5 kPa is compressed in a non-flow reversible process from a specific volume of 0.2768 m³/kg to a specific volume of 0.001107 m³/kg. If the pressure remains constant at 689.5 kPa and the internal energy decreases 1878.2 kJ/kg during the process, determine how much heat will be transferred and whether it will be added to or abstracted from the working substance.

Ans. (−) 2068.3 kJ/kg (abstracted)

2-8. Steam at a pressure of 50 psia and a temperature of 281°F has an internal energy of 1095.6 Btu/lb. At these conditions the specific volume of the steam is 8.518 ft³/lb. Determine the enthalpy of the steam in Btu/lb.

Ans. 1174.4 Btu/lb

2-9. A reversible, non-flow process of a perfect gas proceeds from a pressure of 400 psia to a pressure of 100 psia with a corresponding increase in specific volume of the gas from 0.518 ft³/lb to 2.072 ft³/lb. During the process the internal energy remains constant at 95.76 Btu/lb. Calculate (a) the enthalpy of the gas at 400 psia and 0.518 ft³/lb in Btu/lb, and (b) the change of enthalpy between the initial and final states.

Ans. (a) 134.1 Btu/lb; (b) 0 Btu/lb

2-10. A reversible non-flow process with a gas is defined by the equation $pv = C$. Show that the work done is given by the equation $wk_{12} = p_1 v_1 \ln \dfrac{v_2}{v_1}$.

2-11. Half a pound of gas undergoes a reversible non-flow process with the pressure remaining constant at 50 psia while the volume increases from 4 ft³ to 8 ft³. Find the work done, Btu/lb.

Ans. 74.0 Btu/lb (by)

2-12. Two kilograms of a gas undergo a reversible non-flow process at 350 kPa while the volume increases from 0.4734 m³ to 0.9468 m³. Find the work done, kJ/kg.

Ans. 82.8 kJ/kg (by)

2-13. During a reversible process with a certain working substance, the entropy remains constant while the temperature increases from 250°F to 450°F. Find the heat transferred.

Ans. 0 Btu/lb

2-14. One pound of water in changing to steam receives 970.3 Btu at 212°F. Determine the change in specific entropy for the process, Btu/lb · °R.

Ans. 1.444 Btu/lb · °R

2-15. Fifty Btu are added to one pound of air during a constant pressure process starting at a temperature of 100°F. Calculate the final temperature of the air, degrees F, for a

constant pressure specific heat of 0.24 Btu/lb · °F. Then calculate the final temperature of the air, starting at 100°F, for the same quantity of heat added during a constant volume process for which the average specific heat is 0.171 Btu/lb · °F.

Ans. $t_2 = 308°F$, $t_2 = 392°F$

2-16. A fuel oil heater receives 23,200 lb of oil per hour at 80°F and discharges it at 160°F. The specific heat of the oil is 0.48 Btu/lb · °F. Find the rate of heat transfer to the oil, Btu/hr.

Ans. 891,000 Btu/hr

CHAPTER 3

CHARACTERISTICS OF GASES AND THE NON-FLOW GAS PROCESSES

3-1. Introduction. Air, classified as a gas, is one of the most common thermodynamic working substances. Certain other gases are occasionally encountered in thermodynamics, and it is important to investigate the properties and characteristic behavior of gases in general. Later in the text the principles developed in this chapter will be used in connection with the study of air compression and the air standard combustion engine cycles.

The student should be familiar with the perfect gas concept as presented in physics. All real gases, such as air, deviate from the perfect gas. This deviation is not great for normal conditions, and the assumption that the real gases follow the perfect gas laws does not introduce serious errors while permitting a relatively simple treatment of principles. Once the fundamental theory has been established, it is possible to introduce corrections on the basis of real gas data when greater accuracy is necessary.

3-2. The Gas Laws. The physicists Boyle and Charles discovered the laws which bear their names as a result of experiment with air and other real gases. These laws may be combined and summarized as follows: *"For a given mass of any gas, the product of the absolute pressure and the volume occupied, divided by the absolute temperature, is a constant."* Mathematically:

$$\frac{p_1 V_1}{T_1} = \frac{p_2 V_2}{T_2} = \frac{pV}{T} \tag{3-1}$$

or

$$\frac{p_1 M v_1}{T_1} = \frac{p_2 M v_2}{T_2} = \frac{pMv}{T}$$

For unit mass of gas,

$$\frac{p_1 v_1}{T_1} = \frac{p_2 v_2}{T_2} = \frac{pv}{T} = R \text{ (a constant)} \tag{3-2}$$

Also $\qquad pv = RT \text{ ft} \cdot \text{lbf/lb (J/kg)} \tag{3-3}$

And $\qquad pV = MRT \text{ ft} \cdot \text{lbf (J)} \tag{3-4}$

p = absolute pressure, lbf/ft² (Pa)
V = total volume, ft³ (m³)

GASES AND THE NON-FLOW PROCESSES

v = specific volume, ft³/lb (m³/kg)
M = total mass, lb (kg)
T = absolute temperature, °R (K)
R = the gas constant, ft · lbf/lb · °R (J/kg · K)

and the subscripts 1 and 2 refer to two different states of the same gas.

Equation (3-3) is the characteristic *equation of state* for a perfect gas; the interrelationship of pressure, temperature and specific volume indicated will always hold for a perfect gas at any selected state. Note that there is no restriction whatsoever as to the process followed in arriving at the selected state. Given the value of R for a particular gas, this equation enables us to determine any one of the properties—pressure, specific volume or absolute temperature—if the other two are known. The value of the gas constant, R, is a function of the molecular weight of the gas under consideration and can be evaluated from the equations:

$$R = \frac{1545}{\text{Molecular weight}} \quad (\text{ft} \cdot \text{lbf/lb} \cdot °R)$$

$$R = \frac{8314}{\text{Molecular weight}} \quad (\text{J/kg} \cdot \text{K})$$

Approximate values of R for some common gases can be found in Table 3-I.

TABLE 3-I

Gas	Molecular weight	R ft · lbf/lb · °R	R J/kg · K
Air	28.97	53.3	287
Oxygen	32.00	48.3	260
Nitrogen	28.02	55.1	297
Hydrogen	2.016	766.0	4120
Helium	4.003	386.0	2080
Steam	18.02	85.7	462

The discussion in this article is strictly applicable only to a perfect gas. A perfect gas, in turn, may be defined as a hypothetical substance that exactly fulfills the relationships developed above and has invariant specific heats. Air and other gases may be treated as perfect over a reasonable range of temperature and pressure with only minor errors. For the purposes of this text, it will be assumed that the real gases with which we deal follow the perfect gas equation of state.

GASES AND THE NON-FLOW PROCESSES

Example 3-1: A tank with a total volume of 10 cu ft is filled with air at a pressure of 20 psi gage and at a temperature of 100°F. Find the specific volume and the total mass of the air in the tank.
Solution:

$$V = 10 \text{ ft}^3$$
$$p = 20 + 14.7 = 34.7 \text{ psia}$$
$$T = 460 + 100 = 560°\text{R}$$
$$R = 53.3 \text{ ft} \cdot \text{lbf/lb} \cdot °\text{R}$$
$$pv = RT; \qquad v = \frac{RT}{p}$$
$$v = \frac{53.3 \times 560}{34.7 \times 144} = \underline{5.97 \text{ ft}^3/\text{lb}}$$
$$M = \frac{V}{v} = \frac{10}{5.97} = \underline{1.675 \text{ lb}}$$

Example 3-2: A 0.1-kg sample of a gas of fixed composition is compressed in a cylinder. Before compression the gas occupied a volume of 0.05 m³ and was at a temperature of 30°C and 102.9 kPa. After compression the gas occupied a volume of 0.008 m³ and was at a pressure of 800 kPa. Find:
(a) the temperature of the gas after compression, °C
(b) the specific volume of the gas before compression, m³/kg
(c) the value of the gas constant, R, for this gas, J/kg·K
Solution:

$$M = 0.1 \text{ kg}$$
$$V_1 = 0.05 \text{ m}^3 \qquad V_2 = 0.008 \text{ m}^3$$
$$T_1 = 30 + 273 = 303 \text{ K} \qquad p_2 = 800 \text{ kPa}$$
$$p_1 = 102.9 \text{ kPa (kN/m}^2\text{)}$$

(a)
$$\frac{p_1 V_1}{T_1} = \frac{p_2 V_2}{T_2} \qquad T_2 = T_1 \frac{p_2 V_2}{p_1 V_1}$$
$$T_2 = \frac{303 \times 800 \times 0.008}{102.9 \times 0.05}$$
$$= 376.9, \text{ say } 377 \text{ K}$$
$$t_2 = 377 - 273 = \underline{104°\text{C}}$$

(b)
$$v_1 = \frac{V_1}{M} = \frac{0.05}{0.1} = \underline{0.5 \text{ m}^3/\text{kg}}$$

(c)
$$p_1 v_1 = RT_1 \qquad R = \frac{p_1 v_1}{T_1}$$
$$R = \frac{102.9 \times 10^3 \times 0.5}{303}$$
$$= \underline{169.8 \text{ J/kg} \cdot \text{K}}$$

3-3. The Polytropic Process Equation. The Gas Laws, as presented in the preceding article, are all fundamentally *equations of state*. In essence all they do is enable us to obtain additional properties at a given state once

GASES AND THE NON-FLOW PROCESSES

two properties are known. As previously stated, reversible processes are of fundamental interest and are those in which the transition between states is carried out with the properties following a definite, predictable interrelationship throughout the process. Since two properties are sufficient to fix the state of a gas, it is reasonable to expect that a *process equation* could be written in terms of any two selected properties. Conventionally, the properties selected are pressure and specific volume, and the equation takes the form:

$$p_1v_1^n = p_2v_2^n = pv^n = C \text{ (a constant)} \qquad (3\text{-}5)$$

This is known as the *general polytropic equation for a gas* and applies to any reversible gas process, either flow or non-flow. The relationship of properties given by the equation holds not only for the initial and final states (1 and 2), but also for any intermediate state between the two. The value of n, known as the **polytropic exponent**, is established by the type of process followed. The word "polytropic" means "multiform" or "varied," and the equation can describe a multiplicity of processes as n varies from zero to infinity. The special significance of several particular values of the polytropic exponent and the range of values encountered in normal engineering processes will be discussed later.

Inspection of the polytropic process equation shows that if the initial state of a gas is known along with the nature of the process followed (i.e.—the value of n), then the state of the gas can be definitely fixed at any point during the process simply by knowing a single property at that point. This is in contravention to the equations of state where no particular process is specified and two properties must be known in order to fix the state.

By employing Equations (3-2) and (3-5) in combination, other useful forms of the polytropic process equation may be derived.

$$\frac{p_1v_1}{T_1} = \frac{p_2v_2}{T_2} \qquad (3\text{-}2)$$

Rearranging:
$$\frac{T_2}{T_1} = \frac{p_2v_2}{p_1v_1} = \left(\frac{p_2}{p_1}\right)\left(\frac{v_2}{v_1}\right)$$

$$p_1v_1^n = p_2v_2^n \qquad (3\text{-}5)$$

Rearranging:
$$\frac{p_2}{p_1} = \left(\frac{v_1}{v_2}\right)^n = \left(\frac{v_2}{v_1}\right)^{-n}$$

and
$$\frac{v_2}{v_1} = \left(\frac{p_1}{p_2}\right)^{1/n} = \left(\frac{p_2}{p_1}\right)^{-1/n}$$

GASES AND THE NON-FLOW PROCESSES

Combining rearranged equations:

$$\frac{T_2}{T_1} = \left(\frac{p_2}{p_1}\right)\left(\frac{p_2}{p_1}\right)^{-1/n} = \left(\frac{p_2}{p_1}\right)^{\frac{n-1}{n}}$$

Also:
$$\frac{T_2}{T_1} = \left(\frac{v_2}{v_1}\right)\left(\frac{v_2}{v_1}\right)^{-n} = \left(\frac{v_2}{v_1}\right)^{1-n} = \left(\frac{v_1}{v_2}\right)^{n-1}$$

In final form:

$$\frac{T_2}{T_1} = \left(\frac{p_2}{p_1}\right)^{\frac{n-1}{n}} = \left(\frac{v_1}{v_2}\right)^{n-1} \qquad (3\text{-}6)$$

Furthermore, since $\dfrac{V_1}{V_2} = \dfrac{Mv_1}{Mv_2} = \dfrac{v_1}{v_2}$,

$$p_1 V_1^n = p_2 V_2^n$$

and
$$\frac{T_2}{T_1} = \left(\frac{p_2}{p_1}\right)^{\frac{n-1}{n}} = \left(\frac{V_1}{V_2}\right)^{n-1}$$

Example 3-3: Air at an initial pressure of 15 psia and temperature of 60°F undergoes a polytropic process such that the final temperature and pressure become 240°F and 90 psia. Find:
(a) the polytropic exponent for the process
(b) the specific volume when the pressure reaches 30 psia during the process
(c) the temperature for the conditions of part (b)

Solution:

(a)
$$p_1 = 15 \text{ psia} \qquad T_1 = 60 + 460 = 520°R$$
$$p_2 = 90 \text{ psia} \qquad T_2 = 240 + 460 = 700°R$$

$$\frac{T_2}{T_1} = \left(\frac{p_2}{p_1}\right)^{\frac{n-1}{n}}; \qquad \frac{700}{520} = \left(\frac{90}{15}\right)^{\frac{n-1}{n}}$$

$$1.346 = (6)^{\frac{n-1}{n}}$$

$$\frac{n-1}{n} = \frac{\ln 1.346}{\ln 6}$$

$$\frac{n-1}{n} = 0.1658; \quad n - 1 = 0.1658n$$

$$n = \frac{1}{1 - 0.1658} = \underline{1.2}$$

(b)
$$p_1 v_1^n = p v^n; \qquad p = 30 \text{ psia}$$

$$v = v_1 \left(\frac{p_1}{p}\right)^{1/n}$$

$$v_1 = \frac{RT_1}{p_1} = \frac{53.3 \times 520}{15 \times 144} = 12.83 \text{ ft}^3/\text{lb}$$

$$v = 12.83 \left(\frac{15}{30}\right)^{1/1.2} = \underline{7.20 \text{ ft}^3/\text{lb}}$$

GASES AND THE NON-FLOW PROCESSES

(c)
$$\frac{T}{T_1} = \left(\frac{p}{p_1}\right)^{\frac{n-1}{n}}; \quad T = T_1\left(\frac{p}{p_1}\right)^{\frac{n-1}{n}}$$

$$T = 520\left(\frac{30}{15}\right)^{2/1.2} = 520 \times 1.1225 = \underline{584°R}$$

or
$$pv = RT; \quad T = \frac{pv}{R}$$

$$T = \frac{30 \times 144 \times 7.20}{53.3} = \underline{584°R}$$

3-4. Introduction to the Non-Flow Reversible Gas Processes. The remainder of this chapter will be largely devoted to a detailed investigation of the various *non-flow reversible gas processes,* together with their representation on p-v and T-s coordinates. The processes of this type that are of engineering importance may be broken into two general groups:

(a) *The general polytropics in which none of the properties remains constant during the process.*

(b) *The special polytropics in which one of the properties remains constant throughout the process.*

The special polytropic processes include the following:
(1) constant volume
(2) constant pressure
(3) constant temperature or isothermal
(4) constant entropy or isentropic

The particular value of n which applies to each of the special processes will now be determined by comparing the general polytropic process equation with the special.

Constant Volume

$$v_1 = v_2 \quad p_1 \neq p_2$$

$$p_1 v_1{}^n = p_2 v_2{}^n \quad \text{or} \quad v_1 = v_2\left(\frac{p_2}{p_1}\right)^{\frac{1}{n}}$$

If $v_1 = v_2$, then the pressure term must be equal to unity. Since $p_1 \neq p_2$, this condition can be satisfied only if the exponent of the pressure term is zero. Then, for a constant volume gas process:

$$n = \infty$$

Constant Pressure

$$p_1 = p_2 \quad v_1 \neq v_2$$

$$p_1 v_1{}^n = p_2 v_2{}^n \quad \text{or} \quad p_1 = p_2\left(\frac{v_2}{v_1}\right)^n$$

GASES AND THE NON-FLOW PROCESSES

By reasoning similar to that above, for a constant pressure gas process:
$$n = 0$$

Constant Temperature
$$T_1 = T_2 \quad \text{so} \quad p_1v_1 = p_2v_2$$
$$p_1v_1{}^n = p_2v_2{}^n$$
Obviously, for an isothermal gas process,
$$n = 1$$

Constant Entropy
$$s_1 = s_2$$
As will be developed later, for this process the characteristic equation is:
$$p_1v_1{}^k = p_2v_2{}^k$$
where
$$k = \frac{c_p}{c_v} \tag{3-7}$$
Thus for the isentropic gas process:
$$n = k$$

The ratio of the specific heat at constant pressure to the specific heat at constant volume is known as the **isentropic exponent** and is assigned the symbol k. The numerical value of k will always be greater than unity, since, as indicated in the discussion of specific heats, c_p is always greater than c_v. The following **mean values for air** will be used in this text and should be memorized by the student because of their frequent application.

$$c_p = \begin{cases} 0.240 \text{ Btu/lb} \cdot {}^\circ\text{R} & (\text{Btu/lb} \cdot {}^\circ\text{F}) \\ 1.0048 \text{ kJ/kg} \cdot \text{K} & (\text{kJ/kg} \cdot {}^\circ\text{C}) \end{cases}$$

$$c_v = \begin{cases} 0.171 \text{ Btu/lb} \cdot {}^\circ\text{R} & (\text{Btu/lb} \cdot {}^\circ\text{F}) \\ 0.7159 \text{ kJ/kg} \cdot \text{K} & (\text{kJ/kg} \cdot {}^\circ\text{C}) \end{cases}$$

$$k = 1.4$$

General Polytropic. For the general polytropic processes wherein none of the properties remains constant, we might infer that n can take any value between zero and infinity. This is theoretically true, but our concern is primarily with processes that approximate those actually encountered in engineering equipment. If we limit ourselves to such processes, the range of values is drastically reduced. For practical purposes it can be stated that

GASES AND THE NON-FLOW PROCESSES

the exponent for most of the general polytropic processes will be positive, greater than unity and less than two. In mathematical notation:

$$1 < n < 2$$

3-5. Work and the *p-v* Diagram. In dealing with the reversible non-flow gas processes, the work done is of particular interest. If the nature of the process is known, and this is the usual case, the relationship between p and v is immediately established. The mechanical work can then be determined from $\int p\,dv$, following a procedure to be outlined later. For any particular process, a sketch of the state path on p-v coordinates is useful in problem solution. The general nature of such a state path may be established in each individual case by considering the physical characteristics of a gas and the conditions imposed on the gas within the cylinder. Figure 3-1 is presented as an aid to the student in representing the several special polytropic processes, pending the development of the ability to analyze each individual process on its own merits. In the diagram all the process

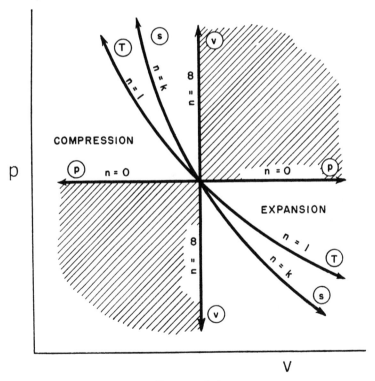

FIGURE 3-1

GASES AND THE NON-FLOW PROCESSES

lines are shown originating from one common initial state point, as a matter of convenience, and proceeding to some final state as represented by the arrowheads. Customarily, where a single process is shown, the initial state is numbered 1 and the final state is numbered 2.

All of the process lines lie in the areas shown within the limits established by the constant pressure ($n = 0$) and the constant volume ($n = \infty$) lines. Except for these limiting processes, *expansion* is characterized by an increase in specific volume, outward movement of the piston, positive work and a drop in pressure; whereas *compression* entails a decrease in specific volume, inward movement of the piston, negative work and an increase in pressure. Process lines that lie in the other two quadrants involve a negative value of n, are highly specialized in nature and will not be considered in this text. Therefore, the corresponding areas on the diagram have been shaded out.

The mechanical work expression for the various non-flow reversible processes will now be developed from $\int p \, dv$ and the applicable process equation.

General Polytropic

$$wk_{12} = \int_1^2 p \, dv \qquad (2\text{-}5)$$

$$p_1 v_1^n = p_2 v_2^n = pv^n = C$$
$$p = Cv^{-n}$$

$$wk_{12} = \int_1^2 Cv^{-n} \, dv = C \int_1^2 v^{-n} \, dv = C \left[\frac{v^{1-n}}{1-n} \right]_1^2$$

$$= \frac{C}{1-n} [v_2^{1-n} - v_1^{1-n}]$$

Substituting either $p_1 v_1^n$ or $p_2 v_2^n$ for C:

$$wk_{12} = \frac{(p_2 v_2^n) v_2^{1-n} - (p_1 v_1^n) v_1^{1-n}}{1-n}$$

Clearing

$$wk_{12} = \frac{p_2 v_2 - p_1 v_1}{1-n} \text{ ft} \cdot \text{lbf/lb (J/kg)}$$

Also since $pv = RT$

$$wk_{12} = \frac{R(T_2 - T_1)}{1-n} \text{ ft} \cdot \text{lbf/lb (J/kg)}$$

GASES AND THE NON-FLOW PROCESSES

Constant Volume

$$wk_{12} = \int_1^2 p\, dv$$

$$v_1 = v_2; \quad dv = 0; \quad n = \infty$$

Since $dv = 0$, $\int_1^2 p\, dv = 0$ and $wk_{12} = 0$

The fact that the mechanical work is zero in this case can also be established by substituting $n = \infty$ in the general polytropic work equation developed above; by recognizing that the piston physically does not move; or by realizing that there is no included area under the constant volume line on p-v coordinates.

Constant Pressure

$$wk_{12} = \int_1^2 p\, dv$$

$$p_1 = p_2 = p = C; \quad n = 0$$

$$wk_{12} = p \int_1^2 dv = p(v_2 - v_1)$$

$$wk_{12} = p_2 v_2 - p_1 v_1 \text{ ft} \cdot \text{lbf/lb (J/kg)}$$

or

$$wk_{12} = R(T_2 - T_1) \text{ ft} \cdot \text{lbf/lb (J/kg)}$$

Similar results can be obtained by substituting $n = 0$ in the general polytropic work equation; by recognizing that the area on p-v coordinates is a rectangle of height p and length $v_2 - v_1$; or by figuring out the work done on or by a piston subjected to a constant pressure force, pA, moving through some distance L in a cylinder containing unit mass of working substance.

Constant Entropy

$$wk_{12} = \int_1^2 p\, dv$$

$$p_1 v_1^k = p_2 v_2^k = p v^k = C$$

This follows the general polytropic exactly with $n = k$, yielding:

$$wk_{12} = \frac{p_2 v_2 - p_1 v_1}{1 - k} \text{ ft} \cdot \text{lbf/lb (J/kg)}$$

or

$$wk_{12} = \frac{R(T_2 - T_1)}{1 - k} \text{ ft} \cdot \text{lbf/lb (J/kg)}$$

GASES AND THE NON-FLOW PROCESSES

Constant Temperature

$$wk_{12} = \int_1^2 p\, dv$$

$$p_1v_1 = p_2v_2 = pv = C; \qquad T_1 = T_2 = T$$

Also $\quad p_1v_1 = p_2v_2 = RT = C$

Since $\quad p = \dfrac{C}{v}, \quad \int_1^2 p\, dv = C\int_1^2 \dfrac{dv}{v} = C[\ln v_2 - \ln v_1]$

$$wk_{12} = p_1v_1 \ln \frac{v_2}{v_1} = p_2v_2 \ln \frac{v_2}{v_1} = RT \ln \frac{v_2}{v_1}$$

But $\quad \dfrac{v_2}{v_1} = \dfrac{p_1}{p_2}$ for this process

and $\quad wk_{12} = RT \ln \dfrac{p_1}{p_2}$ ft · lbf/lb (J/kg)

Because of the special nature of the integral, substitution of $n = 1$ in the general polytropic work equation fails here, as it yields zero over zero, which is indeterminate.

Example 3-4: Air is compressed reversibly in a cylinder from an initial pressure of 15 psia to a final pressure of 60 psia. The initial temperature is 60°F and the initial volume is 1 cu ft. Find:
(a) the mass of the air contained in the cylinder, lb.
(b) the work required if the process is isentropic, ft · lbf
(c) the work required if the process is isothermal, ft · lbf
Solution:

$$p_1 = 15 \text{ psia} \qquad T_1 = 460 + 60 = 520°R$$
$$p_2 = 60 \text{ psia} \qquad V_1 = 1 \text{ ft}^3$$

(a) $\quad p_1V_1 = MRT_1$

$$M = \frac{p_1V_1}{RT_1} = \frac{15 \times 144 \times 1}{53.3 \times 520} = \underline{0.078 \text{ lb}}$$

(b) $$wk_{12} = \frac{R(T_2 - T_1)}{1 - k}$$

$$Wk_{12} = \frac{MR(T_2 - T_1)}{1 - k}$$

$$\frac{T_2}{T_1} = \left(\frac{p_2}{p_1}\right)^{\frac{k-1}{k}}; \quad T_2 = T_1\left(\frac{p_2}{p_1}\right)^{\frac{k-1}{k}}$$

$$T_2 = 520\left(\frac{60}{15}\right)^{\frac{1.4-1}{1.4}} = 520(4)^{.4/1.4} = 520(1.486) = 773°R$$

$$Wk_{12} = \frac{0.078 \times 53.3(773 - 520)}{1 - 1.4} = \underline{(-)2630 \text{ ft} \cdot \text{lbf (on)}}$$

(c) $p_1 = 15$ psia $T_1 = T_2 = 520°R$
 $p_2 = 60$ psia $M = 0.078$ lb

$$wk_{12} = \int_1^2 p\, dv = RT \ln \frac{v_2}{v_1} = RT \ln \frac{p_1}{p_2}$$

$$Wk_{12} = MRT \ln \frac{p_1}{p_2} = -MRT \ln \frac{p_2}{p_1}$$

$$Wk_{12} = -0.078 \times 53.3 \times 520 \ln \frac{60}{15}$$

$$Wk_{12} = -0.078 \times 53.3 \times 520 \times 1.3863 = \underline{(-)2997 \text{ ft} \cdot \text{lbf (on)}}$$

3-6. Internal Energy of a Gas. As discussed in Art. 1-12, the internal energy of a gas is essentially all of the internal kinetic type. Furthermore, any change in total internal kinetic energy always includes a change in the random translatory type of molecular motion, and this, in turn, means a change in temperature. Since temperature change is measurable, a practical method of evaluating internal energy changes in a gas suggests itself, provided some factor relating internal energy and temperature can be established. The constant volume non-flow process presents a simple method of determining that factor. It is emphasized that the choice of the constant volume non-flow process is not significant in itself but is used only because it leads most directly to the physical relationship we are seeking. No work is done during such a process, and any heat added or removed is directly reflected by an equivalent change in the internal energy of the gas. If the gas temperature change and the heat input are simultaneously recorded, then the factor of proportionality between internal energy and temperature is immediately fixed for that particular gas over the range of temperature involved. The student should recognize that the method is exactly that described at the beginning of Art. 2-9 for the experimental determination of specific heats. Further, the factor of proportionality is obviously the specific heat at constant volume. Reducing the foregoing words to equation form:

$$q_{12} = c_v(T_2 - T_1) \tag{2-13}$$

Rearranging the non-flow energy equation and dropping the work term since $wk_{12}/(J) = 0$ gives:

$$q_{12} = u_2 - u_1$$

Equating through q_{12} gives the identity:

$$u_2 - u_1 = c_v(T_2 - T_1) \text{ Btu/lb (J/kg)} \tag{3-8}$$

This equation is extremely important, since it enables a determination of the *change of internal energy* in terms of a temperature change *for a gas for*

GASES AND THE NON-FLOW PROCESSES

any *process* if the mean specific heat at constant volume is known. The equation is similar in nature to the equation of state for gases, since it truly *involves only properties and is independent of the process followed* in proceeding from the initial to the final state.

3-7. Enthalpy of a Gas and Some Thermodynamic Relationships. For a constant pressure non-flow process:

$$wk_{12} = p_2 v_2 - p_1 v_1$$

Rearranging the non-flow energy equation:

$$q_{12} = (u_2 - u_1) + \frac{wk_{12}}{(J)}$$

Substituting for the work term, the non-flow equation for a constant pressure process becomes:

$$q_{12} = (u_2 - u_1) + \frac{p_2 v_2 - p_1 v_1}{(J)}$$

Rearranging:

$$q_{12} = \left[u_2 + \frac{p_2 v_2}{(J)} \right] - \left[u_1 + \frac{p_1 v_1}{(J)} \right]$$

But by definition:

$$h = u + \frac{pv}{(J)} \qquad (2\text{-}3)$$

Finally, for a constant pressure non-flow process:

$$q_{12} = h_2 - h_1$$

But in the specific heat discussion, it was shown that the heat transferred during a constant pressure process with a gas is given by:

$$q_{12} = c_p (T_2 - T_1) \qquad (2\text{-}14)$$

Setting up the identity established through q_{12} in the last two equations, we get:

$$\boldsymbol{h_2 - h_1 = c_p (T_2 - T_1)} \text{ Btu/lb (J/kg)} \qquad \underline{(3\text{-}9)}$$

*This equation provides a means for evaluating the change of enthalpy for any gas for **any** process on the basis of temperature change if the mean value of c_p is known.* This equation, like that for the internal energy of a gas, equates properties only and is independent of the process followed between the initial and final states. On the other hand, note that those equations which contain a heat or work term are generally limited to a

GASES AND THE NON-FLOW PROCESSES

particular process, since neither heat nor work is a property. As a further guide to applicability, the presence of c_v, c_p, or R within an equation indicates that the equation is restricted to gases.

Some useful relationships for gases may now be derived. From the definition of enthalpy we can write:

$$h_1 = u_1 + \frac{p_1 v_1}{(J)}; \qquad h_2 = u_2 + \frac{p_2 v_2}{(J)}$$

or

$$h_2 - h_1 = (u_2 - u_1) + \frac{p_2 v_2 - p_1 v_1}{(J)}$$

But

$$h_2 - h_1 = c_p(T_2 - T_1); \qquad u_2 - u_1 = c_v(T_2 - T_1)$$
$$p_1 v_1 = RT_1; \qquad p_2 v_2 = RT_2$$

Substituting: $c_p(T_2 - T_1) - c_v(T_2 - T_1) + \frac{R}{(J)}(T_2 - T_1)$

Eliminating the common temperature factor:

$$c_p = c_v + \frac{R}{(J)} \text{ Btu/lb} \cdot °R \text{ (J/kg} \cdot \text{K)} \qquad (3\text{-}10)$$

By introducing $k = c_p/c_v$ into this equation, other relationships can be developed as indicated in the example below.

Example 3-5: Helium, for which $R = 386$ ft·lbf/lb·°R and $k = 1.66$, is heated from an initial temperature of 80°F to a final temperature of 180°F in an unknown process. Find:
 (a) the value of c_v
 (b) the value of c_p
 (c) the change of internal energy, Btu/lb
 (d) the change of enthalpy, Btu/lb

Solution:

(a) $\qquad R = 386$ ft·lbf/lb·°R $\qquad k = 1.66$

$$c_p = c_v + \frac{R}{J}; \qquad k = \frac{c_p}{c_v}$$

$$\frac{c_p}{c_v} = \frac{c_v}{c_v} + \frac{R}{Jc_v} \qquad k = 1 + \frac{R}{Jc_v}$$

$$c_v = \frac{R}{J(k-1)} = \frac{386}{778(1.66-1)} = \underline{0.752 \text{ Btu/lb} \cdot °R}$$

(b) $\qquad c_p = kc_v = 1.66 \times 0.752 = \underline{1.25 \text{ Btu/lb} \cdot °R}$

(c) $\qquad T_1 = 460 + 80 = 540°R$
$\qquad T_2 = 460 + 180 = 640°R$
$\qquad u_2 - u_1 = c_v(T_2 - T_1) = 0.752(640 - 540) = \underline{75.2 \text{ Btu/lb}}$

(d) $\quad h_2 - h_1 = c_p(T_2 - T_1) = 1.25(640 - 540) = \underline{125 \text{ Btu/lb}}$

3-8. Heat and Entropy. Having established the equation for the change of internal energy for any process and the special work equations for the several non-flow processes in terms of the initial and final properties, we can now derive the corresponding equations for transferred heat from the rearranged non-flow energy equation:

$$q_{12} = u_2 - u_1 + \frac{wk_{12}}{(J)}$$

Since we do not know the relationships between T and s for the various gas processes, the use of $\int T\,ds$ to evaluate the heat is denied us. For the same reason we are not yet in a position to represent the various processes on T-s coordinates. Because we are dealing with heat, the equations developed will be kept in terms of temperature insofar as possible.

General Polytropic

$$q_{12} = u_2 - u_1 + \frac{wk_{12}}{(J)}$$

$$\frac{wk_{12}}{(J)} = \frac{R(T_2 - T_1)}{(J)(1 - n)}$$

$$u_2 - u_1 = c_v(T_2 - T_1)$$

$$q_{12} = c_v(T_2 - T_1) + \frac{R}{(J)(1 - n)}(T_2 - T_1)$$

$$q_{12} = \left[c_v + \frac{R}{(J)(1 - n)}\right][T_2 - T_1] \text{ Btu/lb (J/kg)}$$

The first bracketed term is sometimes referred to as the polytropic specific heat and its sign will depend upon the numerical value of n.

The change of entropy for this process may be determined from:

$$s_2 - s_1 = \int_1^2 \frac{\delta q}{T} \text{ Btu/lb} \cdot {}^\circ\text{R (J/kg} \cdot \text{K)} \tag{2-9}$$

$$\delta q = \left[c_v + \frac{R}{(J)(1 - n)}\right]dT$$

$$s_2 - s_1 = \left[c_v + \frac{R}{(J)(1 - n)}\right]\int_1^2 \frac{dT}{T}$$

$$s_2 - s_1 = \left[c_v + \frac{R}{(J)(1 - n)}\right]\ln\frac{T_2}{T_1} \text{ Btu/lb} \cdot {}^\circ\text{R (J/kg} \cdot \text{K)}$$

GASES AND THE NON-FLOW PROCESSES

Constant Volume. As previously developed for this process:
$$q_{12} = u_2 - u_1 = c_v(T_2 - T_1) \text{ Btu/lb (J/kg)}$$
For the change of entropy:
$$s_2 - s_1 = \int_1^2 \frac{\delta q}{T}$$
$$\delta q = c_v \, dT$$
$$s_2 - s_1 = c_v \int_1^2 \frac{dT}{T} = c_v \ln \frac{T_2}{T_1} \text{ Btu/lb} \cdot {}^\circ\text{R (J/kg} \cdot \text{K)}$$

Constant Pressure. As previously developed for this process:
$$q_{12} = h_2 - h_1 = c_p(T_2 - T_1) \text{ Btu/lb (J/kg)}$$
$$\delta q = c_p \, dT$$
$$s_2 - s_1 = \int_1^2 \frac{\delta q}{T} = c_p \int_1^2 \frac{dT}{T} = c_p \ln \frac{T_2}{T_1} \text{ Btu/lb} \cdot {}^\circ\text{R (J/kg} \cdot \text{K)}$$

Constant Temperature

$$q_{12} = (u_2 - u_1) + \frac{wk_{12}}{(J)}$$
$$u_2 - u_1 = c_v(T_2 - T_1)$$
$$\frac{wk_{12}}{(J)} = \frac{RT}{(J)} \ln \frac{p_1}{p_2}$$

But $\quad T_1 = T_2 = T, \text{ so } u_2 - u_1 = 0$

Then $\quad q_{12} = \frac{wk_{12}}{(J)} = \frac{RT}{(J)} \ln \frac{p_1}{p_2} \text{ Btu/lb (J/kg)}$

$$s_2 - s_1 = \int_1^2 \frac{\delta q}{T} = \frac{1}{T}\int_1^2 \delta q \text{ since } T \text{ is constant}$$

So $\quad s_2 - s_1 = \frac{q_{12}}{T} \text{ Btu/lb} \cdot {}^\circ\text{R (J/kg} \cdot \text{K)}$

Also $\quad s_2 - s_1 = \frac{RT}{(J)T} \ln \frac{p_1}{p_2} = \frac{R}{(J)} \ln \frac{p_1}{p_2}$

Constant Entropy. This process can now be explored more fully. *Any process during which there is no transfer of heat between an external source and the working substance, either to or from the working substance, is said to be* **adiabatic.** An adiabatic process may be either reversible or irreversible.

71

GASES AND THE NON-FLOW PROCESSES

The only requirement to be met by an adiabatic process is that:

$$q_{12} = 0$$

We also know that, for any *reversible* process:

$$q_{12} = \int_1^2 T\,ds$$

In any real case, T will be finite. Therefore, in the case of an adiabatic process which is also reversible, ds must be equal to zero. As a corollary, the entropy must remain constant throughout the process and s_2 must equal s_1. This leads to the definition of the *isentropic process* as a **reversible adiabatic** characterized by constant entropy and no heat transfer.

It remains for us to verify the value of the isentropic exponent, k, which has been taken on faith as equal to c_p/c_v. In the absence of other information, it is reasonable to assume that the isentropic process will be of polytropic nature. In that case, from the general polytropic work equation with $n = k$ we have:

$$\frac{wk_{12}}{(J)} = \frac{R}{(J)(1-k)}(T_2 - T_1) = \frac{R}{(J)(k-1)}(T_1 - T_2)$$

But if $q_{12} = 0$, the non-flow energy equation becomes:

$$\frac{wk_{12}}{(J)} = u_1 - u_2 = c_v(T_1 - T_2)$$

Equating through the work term:

$$\frac{R}{(J)(k-1)}(T_1 - T_2) = c_v(T_1 - T_2)$$

and

$$\frac{R}{(J)(k-1)} = c_v$$

But

$$\frac{R}{(J)} = c_p - c_v \quad \text{from Equation (3-10)}$$

Then

$$\frac{c_p - c_v}{k-1} = c_v$$

$$k - 1 = \frac{c_p - c_v}{c_v} = \frac{c_p}{c_v} - 1$$

or

$$k = \frac{c_p}{c_v} \qquad \text{Q.E.D.}$$

The relationship between properties during an isentropic process with

GASES AND THE NON-FLOW PROCESSES

a gas can be written simply by substituting k for n in Equations (3-5) and (3-6); thus:

$$p_1 v_1^k = p_2 v_2^k = p v^k \qquad (3\text{-}11)$$

$$\frac{T_2}{T_1} = \left(\frac{p_2}{p_1}\right)^{\frac{k-1}{k}} = \left(\frac{v_1}{v_2}\right)^{k-1} \qquad (3\text{-}12)$$

Example 3-6: A quantity of air undergoes a reversible non-flow constant pressure process from an initial temperature of 400°F to a final temperature of 50°F. Find:
(a) the work done, Btu/lb
(b) the change of internal energy, Btu/lb
(c) the heat transferred, Btu/lb
(d) the change of specific entropy

Solution:
$$T_1 = 400 + 460 = 860°R$$
$$T_2 = 50 + 460 = 510°R$$
$$p_1 = p_2 = p$$

(a)
$$\frac{wk_{12}}{J} = \frac{p(v_2 - v_1)}{J} = \frac{R}{J}(T_2 - T_1)$$

$$= \frac{53.3(510 - 860)}{778} = (-)24 \text{ Btu/lb (on)}$$

(b)
$$u_2 - u_1 = c_v(T_2 - T_1) = 0.171(510 - 860)$$
$$= (-)59.8 \text{ Btu/lb (decrease)}$$

(c)
$$q_{12} = u_2 - u_1 + \frac{wk_{12}}{J} = -59.8 - 24$$

$$= (-)83.8 \text{ Btu/lb (abstracted)}$$

or
$$q_{12} = c_p(T_2 - T_1) = 0.24(510 - 860) = (-)84 \text{ Btu/lb}$$

(d)
$$s_2 - s_1 = \int_1^2 \frac{\delta q}{T} = c_p \ln\frac{T_2}{T_1} = 0.24 \ln\frac{510}{860} = -0.24 \ln\frac{860}{510}$$

$$= -0.24 \times 0.523 = (-)0.1255 \text{ Btu/lb} \cdot °R$$

Example 3-7: During an isentropic non-flow process with air, the pressure drops from 60 psia to 20 psia. The air has a mass of 0.1 lb and the initial temperature was 280°F. Find:
(a) the final temperature, °F
(b) the heat transferred, Btu
(c) the change of internal energy, Btu
(d) the work done, Btu

Solution:
$$p_1 = 60 \text{ psia} \qquad p_2 = 20 \text{ psia}$$
$$T_1 = 460 + 280 = 740°R; \qquad M = 0.1 \text{ lb}$$

(a)
$$\frac{T_2}{T_1} = \left(\frac{p_2}{p_1}\right)^{\frac{k-1}{k}}; \qquad T_2 = T_1\left(\frac{p_2}{p_1}\right)^{\frac{k-1}{k}}$$

$$T_2 = 740\left(\frac{20}{60}\right)^{\frac{1.4-1}{1.4}} = \frac{740}{1.369} = 541°R$$

$$t_2 = 541 - 460 = \underline{81°F}$$

(b) $\quad s_1 = s_2; \quad Q_{12} = Mq_{12} = \underline{0}$

(c) $\quad u_2 - u_1 = c_v(T_2 - T_1)$
$\quad U_2 - U_1 = Mc_v(T_2 - T_1) = 0.1 \times 0.171(541 - 740) = \underline{(-)3.4 \text{ Btu}}$

(d) $\quad U_1 + Q_{12} = U_2 + \dfrac{Wk_{12}}{J}$

$$\dfrac{Wk_{12}}{J} = -(U_2 - U_1) = \underline{+3.4 \text{ Btu (by)}}$$

or $\quad \dfrac{Wk_{12}}{J} = \dfrac{MR(T_2 - T_1)}{J(1-k)} = \dfrac{0.1 \times 53.3(541 - 740)}{778(1 - 1.4)} = \underline{3.41 \text{ Btu (by)}}$

3-9. The T-s Diagram for Non-Flow Gas Processes. In Fig. 3-2 the several special polytropic gas processes are shown on T-s coordinates originating from a common initial state point selected at random. The isentropic and isothermal lines appear as simple straight lines. The relative slopes of the constant volume and the constant pressure lines can be deduced when they originate from the same initial state, by recognizing that more heat is required to bring about a given temperature change at constant pressure than at constant volume, i.e. $c_p > c_v$. Therefore the constant volume line must be steeper with less included area under it. The p-v diagram of Fig. 3-1 is repeated for comparison, and again both the compression and expansion processes are illustrated. A polytropic exponent sequence common to the two diagrams is evident, with a clockwise pattern of 0, 1, k and ∞. By analogy with the p-v diagram, the shaded areas, and the compression and expansion zones on the T-s diagram can be deduced. Similarly, by using the T-s diagram as a reference, we can conclude that the process lines to the left of the isentropic lines on p-v coordinates represent heat abstraction processes, while those to the right represent heat addition processes. All of these inferences can be verified by recourse to the detailed equations previously developed.

For the general polytropic case, the process lines can be spotted on the diagrams by locating them between special lines which bracket the applicable value of n. In most applications it is not necessary to know the exact appearance of any process line as long as its trend is correct. The trends can usually be figured out by laying down the straight line processes on both sets of coordinates and remembering the clockwise sequence of exponents.

Figure 3-2 is intended primarily as an aid in sketching any one gas process on both coordinate systems prior to making a detailed problem solution. If the diagrams are properly drawn between the initial state (1) and the final state (2), it is possible to immediately predict the signs of the heat and work terms and the direction of variation of the several proper-

GASES AND THE NON-FLOW PROCESSES

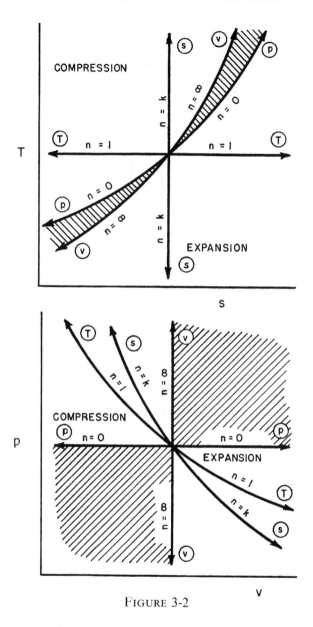

FIGURE 3-2

ties. These predictions then serve as a check on subsequent numerical answers as they arise. The same results can be attained by a thoughtful analysis of the natural tendencies of a gas confined in a piston-cylinder assembly under various imposed conditions. In the next article the line of

GASES AND THE NON-FLOW PROCESSES

reasoning followed for several typical non-flow gas processes will be presented.

Example 3-8: Sketch a reversible non-flow polytropic compression process with air on p-v and T-s coordinates where $n = 1.25$, and predict the direction of variation of the several properties and the signs of the heat and work terms.
Solution:

p increases, v decreases, work is negative (done on)

T increases, u increases, h increases

s decreases, heat is negative (removed)

3-10. Applications of the Non-Flow Gas Processes. The idealized non-flow processes covered in this chapter are basic to the various internal combustion engine cycles and to the reciprocating air compressor. Some of the idealized processes which approximate those encountered in actual equipment will now be analyzed. In every case, the first step is to write the applicable process equation and sketch the process line on p-v and T-s coordinates. Then the general equations listed below with their applicability are used in sequence to yield the particular equations for the process under consideration.

$$wk_{12} = \int_1^2 p\, dv \quad \text{(any non-flow reversible process)}$$

$$u_2 - u_1 = c_v(T_2 - T_1) \quad \text{(any gas process)}$$

$$q_{12} = u_2 - u_1 + \frac{wk_{12}}{(J)} \quad \text{(any non-flow process)}$$

$$s_2 - s_1 = \int_1^2 \frac{\delta q}{T} \quad \text{(any reversible process)}$$

GASES AND THE NON-FLOW PROCESSES

The developments below will follow this procedure but will be skeletonized, since the detailed derivations have been covered previously.

The Otto cycle, which is the prototype for the spark ignition engine cycle, includes a constant volume heating process. This process is shown on p-v and T-s coordinates in Fig. 3-3, along with the piston-cylinder arrangement in which it might occur.

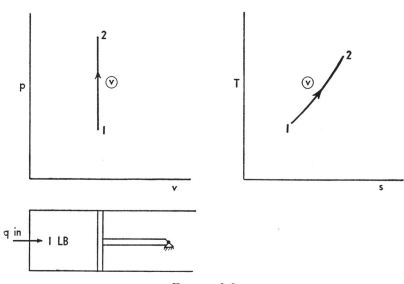

FIGURE 3-3

$$v_1 = v_2 = v; \quad dv = 0$$

$$wk_{12} = \int_1^2 p\, dv = 0$$

$$u_2 - u_1 = c_v(T_2 - T_1)$$

$$q_{12} = u_2 - u_1 = c_v(T_2 - T_1)$$

$$s_2 - s_1 = \int_1^2 \frac{\delta q}{T} = c_v \int_1^2 \frac{dT}{T} = c_v \ln \frac{T_2}{T_1}$$

In this process, heat is added and the entropy must increase. Since the piston is locked, no work can be done. Then the heat added must all be stored as internal energy, and the temperature must rise. From the equation of state ($pv = RT$), if v is constant and T increases, the pressure, p, must also increase.

A constant pressure heating process is included in the ideal Diesel cycle and such a process is shown in Fig. 3-4.

GASES AND THE NON-FLOW PROCESSES

FIGURE 3-4

$$p_1 = p_2 = p$$
$$wk_{12} = \int_1^2 p\, dv = p \int_1^2 dv = p(v_2 - v_1)$$
$$= p_2 v_2 - p_1 v_1 = R(T_2 - T_1)$$
$$u_2 - u_1 = c_v(T_2 - T_1)$$
$$q_{12} = u_2 - u_1 + \frac{wk_{12}}{(J)}$$
$$q_{12} = c_v(T_2 - T_1) + \frac{R}{(J)}(T_2 - T_1) = c_p(T_2 - T_1)$$
$$s_2 - s_1 = \int_1^2 \frac{\delta q}{T} = c_p \int_1^2 \frac{dT}{T} = c_p \ln \frac{T_2}{T_1}$$

In this process, heat is added and the entropy must increase. If the pressure is to be kept constant, the piston must be allowed to move outward as heat is added. The specific volume increases and work is done by the gas. From the equation of state, if p is constant and v increases, then the temperature, T, must also increase. The temperature rise indicates that part of the heat added is stored as internal energy.

The Carnot cycle, which is of considerable theoretical interest, includes an isothermal compression such as is shown in Fig. 3-5.

GASES AND THE NON-FLOW PROCESSES

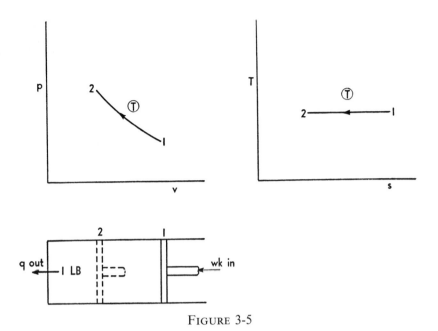

FIGURE 3-5

$$T_1 = T_2 = T; \quad p_1 v_1 = p_2 v_2 = RT$$

$$wk_{12} = \int_1^2 p \, dv = RT \int_1^2 \frac{dv}{v} = RT \ln \frac{v_2}{v_1}$$

$$u_2 - u_1 = c_v(T_2 - T_1) = 0$$

$$q_{12} = u_2 - u_1 + \frac{wk_{12}}{(J)} = \frac{wk_{12}}{(J)}$$

$$q_{12} = \frac{RT}{(J)} \ln \frac{v_2}{v_1} = \frac{RT}{(J)} \ln \frac{p_1}{p_2}$$

$$s_2 - s_1 = \int_1^2 \frac{\delta q}{T} = \frac{q_{12}}{T} = \frac{R}{(J)} \ln \frac{p_1}{p_2}$$

Since this is a compression process, the specific volume must decrease and work must be done on the gas. Since T is constant and v decreases, the equation of state indicates that the pressure, p, must increase. For a gas the internal energy must remain unchanged if the temperature is constant. Therefore the work input cannot be stored in the substance and must all be removed as heat, with a corresponding decrease in entropy.

All of the common internal combustion engine cycles include an isentropic non-flow compression process such as is shown in Fig. 3-6.

GASES AND THE NON-FLOW PROCESSES

FIGURE 3-6

$$p_1v_1^k = p_2v_2^k = pv^k; \quad s_2 - s_1 = 0$$

$$wk_{12} = \int_1^2 p\,dv = \frac{p_2v_2 - p_1v_1}{1-k} = \frac{R(T_2 - T_1)}{1-k}$$

$$u_2 - u_1 = c_v(T_2 - T_1)$$

$$q_{12} = \int_1^2 T\,ds; \quad ds = 0; \quad q_{12} = 0$$

Then
$$0 = u_2 - u_1 + \frac{wk_{12}}{(J)}$$

$$\frac{wk_{12}}{(J)} = u_1 - u_2 = c_v(T_1 - T_2)$$

Since a compression is specified, the specific volume must decrease, work is done on the gas, and from the process equation, $pv^k = C$, the pressure must increase. Since the process is isentropic, the entropy remains constant, no heat is transferred, and the work input must all be stored as internal energy with an accompanying rise in temperature.

In a reciprocating air compressor, one of the processes encountered is a non-flow polytropic compression. Such a process is shown in Fig. 3-7 for an assumed value of n of about 1.30.

GASES AND THE NON-FLOW PROCESSES

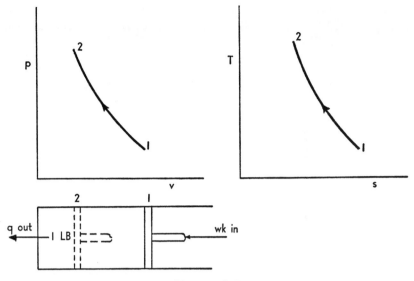

FIGURE 3-7

$$p_1v_1^n = p_2v_2^n = pv^n; \quad wk_{12} = \int_1^2 p \, dv = \frac{p_2v_2 - p_1v_1}{1-n} = \frac{R(T_2 - T_1)}{1-n}$$

$$u_2 - u_1 = c_v(T_2 - T_1)$$

$$q_{12} = u_2 - u_1 + \frac{wk_{12}}{(J)} = c_v(T_2 - T_1) + \frac{R}{(J)(1-n)}(T_2 - T_1)$$

$$s_2 - s_1 = \int_1^2 \frac{\delta q}{T} = \left[c_v + \frac{R}{(J)(1-n)}\right] \ln \frac{T_2}{T_1}$$

Since this is a compression process, the specific volume will decrease, and work is done on the air. The value of n given indicates that the process lies between the isothermal ($n = 1$) and the isentropic ($n = k$). If the process were isothermal, all the energy added as work would have to be removed as heat, with no change in internal energy. If the process were isentropic, no heat would be transferred, and all the energy added as work would be stored as internal energy. By implication, the chosen process will lie between the two extremes, with part of the work input being stored as internal energy and part being removed as heat. The entropy must decrease and the temperature and pressure increase.

Example 3-9: During the constant volume non-flow reversible process which occurs in the Otto cycle, 4.0 Btu of heat are added. The cylinder contains 0.01 lb of air, and the initial temperature and pressure are 650°F and 210 psia respectively. Sketch the process on p-v and T-s coordinates and find:

GASES AND THE NON-FLOW PROCESSES

(a) the final temperature, °F
(b) the final pressure, psia
(c) the work done, Btu
(d) the change of internal energy, Btu/lb

Solution:

$$v_1 = v_2$$
$$Q_{12} = 4 \text{ Btu} \quad M = 0.01 \text{ lb}$$
$$p_1 = 210 \text{ psia} \quad T_1 = 650 + 460 = 1110°R$$

(a) $$Q_{12} = M q_{12} = M c_v(T_2 - T_1)$$

$$T_2 - T_1 = \frac{Q_{12}}{M c_v}; \quad T_2 = T_1 + \frac{Q_{12}}{M c_v}$$

$$T_2 = 1110 + \frac{4}{0.01 \times 0.171} = 1110 + 2340 = 3450°R$$

$$t_2 = 3450 - 460 = \underline{2990°F}$$

(b) $$\frac{p_1 v_1}{T_1} = \frac{p_2 v_2}{T_2}; \quad v_1 = v_2$$

$$p_2 = p_1 \frac{T_2}{T_1} = \frac{210 \times 3450}{1110} = \underline{653 \text{ psia}}$$

(c) $$wk_{12} = \int_1^2 p\, dv = \underline{0}$$

(d) $$u_2 - u_1 = c_v(T_2 - T_1) = 0.171(3450 - 1110)$$
$$= \underline{400 \text{ Btu/lb}}$$

or $$q_{12} = u_2 - u_1 = \frac{Q_{12}}{M} = \frac{4}{0.01} = \underline{400 \text{ Btu/lb}}$$

Example 3-10: Air is compressed in a cylinder during a non-flow reversible polytropic process from an initial temperature and pressure of 80°F and 15 psia to a final temperature and pressure of 285°F and 75 psia. The cylinder contains 0.01 lb of air, and the area of the

GASES AND THE NON-FLOW PROCESSES

piston is 0.2 sq ft. Sketch the process on p-v and T-s coordinates and find:
 (a) the value of the polytropic exponent, n
 (b) the work done, ft·lbf
 (c) the change of internal energy, Btu
 (d) the heat transferred, Btu
 (e) the distance the piston moves during the process, inches
Solution:

$p_1 = 15$ psia $T_1 = 80 + 460 = 540°$R
$p_2 = 75$ psia $T_2 = 285 + 460 = 745°$R
$M = 0.01$ lb $A = 0.2$ ft^2

(a) $$\frac{T_2}{T_1} = \left(\frac{p_2}{p_1}\right)^{\frac{n-1}{n}}; \qquad \frac{745}{540} = \left(\frac{75}{15}\right)^{\frac{n-1}{n}}$$

$$1.380 = (5)^{\frac{n-1}{n}}; \qquad \frac{n-1}{n} = 0.2; \qquad n = \frac{1}{1-0.2} = \underline{1.25}$$

(b) $$Wk_{12} = \frac{MR(T_2 - T_1)}{1-n} = \frac{0.01 \times 53.3(745 - 540)}{1 - 1.25}$$

$$= -\frac{0.01 \times 53.3 \times 205}{0.25} = \underline{(-)437 \text{ ft·lbf (on)}}$$

(c) $$U_2 - U_1 = Mc_v(T_2 - T_1) = 0.01 \times 0.171(745 - 540) = \underline{0.351 \text{ Btu}}$$

(d) $$Q_{12} = U_2 - U_1 + \frac{Wk_{12}}{J} = 0.351 - \frac{437}{778} = \underline{(-)0.211 \text{ Btu (abstracted)}}$$

GASES AND THE NON-FLOW PROCESSES

(e) $\quad V_1 - V_2 = M(v_1 - v_2) = AL; \quad L = \dfrac{M(v_1 - v_2)}{A}; \quad v = \dfrac{RT}{p}$

$$v_1 = \frac{RT_1}{p_1} = \frac{53.3 \times 540}{15 \times 144} = 13.33 \text{ ft}^3/\text{lb}$$

$$v_2 = \frac{RT_2}{p_2} = \frac{53.3 \times 745}{75 \times 144} = 3.68 \text{ ft}^3/\text{lb}$$

$$L = \frac{0.01(13.33 - 3.68)}{0.2} = 0.482 \text{ ft} = \underline{5.78 \text{ in}}$$

Example 3-11: The following sequence of connected processes makes up the ideal Diesel cycle:

> 1-2 Isentropic compression
> 2-3 Constant pressure addition of heat
> 3-4 Isentropic expansion
> 4-1 Constant volume rejection of heat

Sketch the processes on p-v and T-s coordinates, numbering as above and indicating the nature of each process.

Solution:

3-11. The Air Table.
The specific heats of air, as indicated in Fig. 2-7, vary considerably with temperature. Accordingly, the assumption of constant values for c_p and c_v may introduce significant errors in certain calculations. The availability of constant pressure specific heat data for air allows the observed temperature dependency of Fig. 2-7 to be expressed mathematically in the generalized form

$$c_p \langle T \rangle = a + bT + cT^2 + dT^3 + \cdots \qquad (3\text{-}13)$$

where $\langle T \rangle$ is a convention used to indicate that the variation in specific heat with temperature is taken into account in relation to the associated quantity

GASES AND THE NON-FLOW PROCESSES

T is the absolute temperature in °R (K)
a, b, c, \ldots are fitted coefficients.

For temperatures up to approximately 400°F (204°C), the difference in c_p between the precise value using Equation (3-13) and the previously used average value is minimal. Knowing the specific heat data mathematically as a function of temperature, though, allows changes in enthalpy to be evaluated (see Ex. 2-8) using Equations (3-9) and (3-13) as

$$h_2 - h_1 = \int_{T_1}^{T_2} c_p \langle T \rangle \, dT \text{ Btu/lb (J/kg)} \tag{3-14}$$

or

$$= a(T_2 - T_1) + \frac{b}{2}(T_2^2 - T_1^2) + \frac{c}{3}(T_2^3 - T_1^3) + \cdots$$

The results of the above calculations are valid, but the computational work is burdensome. The Air Tables found in Appendix E were developed to ease this complexity and provide values of enthalpy and additional thermodynamic properties for air as a function of temperature. These tables are condensed and presented by permission of the authors and publisher from *Gas Tables*, by J. H. Keenan and J. Kaye, published by John Wiley and Sons, 1948. In these tables a datum enthalpy is taken as equal to zero at zero degrees Rankine, with finite values of h determined as functions of temperature.

The Air Tables presented in Appendix E in both Engineers' units and SI are designed to be entered with a single argument, absolute temperature. A knowledge of two temperatures allows $h_2 - h_1$ to be determined.

As indicated in Art. 1-12, internal energy is a function of temperature only. From the definition of enthalpy the internal energy of air can be expressed as

$$u = h - \frac{pv}{(J)} \tag{2-3}$$

A basic assumption to the formulation of the Air Tables is that the ideal gas equation of state holds:

$$pv = RT \tag{3-3}$$

For the range of temperatures covered by the Air Tables and at the relatively low pressures encountered in engineering practice, the true equation of state varies very little from Equation (3-3). The internal energy entries for the Air Tables can thus be determined quite simply from the following equation:

GASES AND THE NON-FLOW PROCESSES

$$u = h - \frac{RT}{(J)} \text{ Btu/lb (J/kg)}$$

Since both terms on the right-hand side are functions only of temperature, tabulation of internal energy as a function of temperature naturally follows. The datum for internal energy, as for enthalpy, is zero at zero degrees absolute.

As outlined in Art. 1-4, the properties of interest in thermodynamics include temperature, pressure, specific volume, internal energy, enthalpy and entropy. The preceding material outlines the methods used to evaluate internal energy and enthalpy versus absolute temperature, taking into account variable specific heat. Since both properties are functions of temperature alone, the procedures are relatively straightforward. The approach used in the Air Tables for the evaluation of entropy is more complicated, since entropy is a function of both temperature and pressure, as will be shown.

Writing certain equations previously presented, but in differential and rearranged form as appropriate, gives:

$$s_2 - s_1 = \int_1^2 \frac{\delta q}{T} \tag{2-9}$$

$$\delta q = du + \frac{\delta wk}{(J)} \tag{2-1}$$

$$du = c_v \langle T \rangle \, dT \tag{3-8}$$

$$c_p - c_v = \frac{R}{(J)} \tag{3-10}$$

or

$$c_v \langle T \rangle = c_p \langle T \rangle - \frac{R}{(J)}$$

since the difference (R/J) between c_p and c_v remains constant even though the individual values of the two specific heats will vary with temperature. Substituting in Equation (3-8):

$$du = \left[c_p \langle T \rangle - \frac{R}{(J)}\right] dT$$

But $\quad \dfrac{\delta wk}{(J)} = \dfrac{pdv}{(J)} \quad$ (2-5) and $\quad p = \dfrac{RT}{v} \quad$ (3-3)

So $\quad \dfrac{\delta wk}{(J)} = \dfrac{RT}{(J)} \dfrac{dv}{v}$

And $\quad \dfrac{\delta q}{T} = \left[c_p \langle T \rangle - \dfrac{R}{(J)}\right] \dfrac{dT}{T} + \dfrac{RT}{(J)T} \dfrac{dv}{v}$

GASES AND THE NON-FLOW PROCESSES

Substituting the above expression for $\delta q/T$ in Equation (2-9):

$$s_2 - s_1 = \int_1^2 c_p\langle T\rangle \frac{dT}{T} - \frac{R}{(J)}\int_1^2 \frac{dT}{T} + \frac{R}{(J)}\int_1^2 \frac{dv}{v}$$

$$= \int_1^2 c_p\langle T\rangle \frac{dT}{T} - \frac{R}{(J)}\left[\ln\frac{T_2}{T_1} - \ln\frac{v_2}{v_1}\right]$$

But: $\ln\dfrac{T_2}{T_1} - \ln\dfrac{v_2}{v_1} = \ln\dfrac{T_2 v_1}{T_1 v_2} = \ln\dfrac{p_2}{p_1}$, since $\dfrac{p_1 v_1}{T_1} = \dfrac{p_2 v_2}{T_2}$

Then: $$s_2 - s_1 = \int_1^2 c_p\langle T\rangle \frac{dT}{T} - \frac{R}{(J)}\ln\frac{p_2}{p_1}$$

As indicated in the above equation, entropy is a function of both temperature and pressure. The integral on the right involves absolute temperature and temperature-dependent $c_p\langle T\rangle$ only. It can be evaluated by introducing the expression for $c_p\langle T\rangle$ from Equation (3-13) and carrying out the indicated multiple integration. Again this laborious chore is done for us by the Air Tables through the entropy function, ϕ, precomputed and tabulated in the tables against absolute temperature. Thus:

$$\phi_2 - \phi_1 = \int_1^2 c_p\langle T\rangle \frac{dT}{T} \qquad (3\text{-}15)$$

So the change in entropy between any two states, allowing for variable specific heats, can be found from the following equation using the Air Tables to determine the values of ϕ:

$$s_2 - s_1 = (\phi_2 - \phi_1) - \frac{R}{(J)}\ln\frac{p_2}{p_1} \quad \text{Btu/lb} \cdot {}^\circ\text{R} \ (\text{J/kg} \cdot \text{K}) \qquad (3\text{-}16)$$

The zero point for the value of entropy is arbitrarily set at zero degrees absolute and unit pressure. Note that ϕ, like enthalpy and internal energy, is a function of temperature alone, since $c_p\langle T\rangle$, though a variable, is itself dependent only on temperature.

The single entering argument for the Air Tables is absolute temperature, a logical choice since internal energy, u, enthalpy, h, and the entropy function, ϕ, are all functions of temperature only. As mentioned earlier, the two remaining properties of interest are pressure and specific volume. Recalling the equation of state, it should be clear that for any one temperature there can be an infinite number of specific volume and pressure combinations, the only restriction being that the pv product must remain constant and equal to RT. So p and v entries in the table, as such, would be meaningless.

GASES AND THE NON-FLOW PROCESSES

In later chapters the student will find that many of the ideal cycles of practical interest involve an isentropic compression and an isentropic expansion process. With this in mind, the Air Tables have incorporated data that permit the accurate evaluation of isentropic processes under variable specific heat conditions. The device used is relative pressure, p_r, and its concomitant, relative volume, v_r. Starting with an arbitrary datum pressure, p_0, at a datum temperature, T_0, the p_r column in the tables is computed on the basis of an intervening isentropic process. Proceeding from the datum state to some pressure, p, at the corresponding temperature, T, isentropically, $\Delta s = 0$ and Equations (3-15) and (3-16) in combination then give:

$$\frac{R}{(J)} \ln \frac{p}{p_0} = \int_{T_0}^{T} c_p \langle T \rangle \frac{dT}{T} = \phi - \phi_0$$

In consonance with the base line for h, u, and s, T_0 is established as zero degrees absolute and ϕ_0 also goes to zero. Relative pressure can now be defined mathematically:

$$p_r = \left[\frac{p}{p_0}\right]_{(s)}$$

Substituting in the previous equation and rearranging:

$$\ln p_r = \frac{\phi}{R/(J)} \text{ or } p_r = \exp\left(\frac{\phi}{R/(J)}\right) \quad (3\text{-}17)$$

Since ϕ is a function of temperature only, it follows that p_r is also dependent on temperature alone and can be tabulated in the tables against temperature.

For any two specific states, 1 and 2, connected by an isentropic process, we can write:

$$p_{r_1} = \left[\frac{p_1}{p_0}\right]_{(s)} \text{ and } p_{r_2} = \left[\frac{p_2}{p_0}\right]_{(s)}$$

So

$$\frac{p_{r_2}}{p_{r_1}} = \left[\frac{p_2}{p_1}\right]_{(s)} \quad \mathbf{(3\text{-}18)}$$

where p_{r_1} and p_{r_2} are the values tabulated against T_1 and T_2 in the Air Tables. Remember that an isentropic process between the initial and final states is the fundamental hypothesis. The tabular entries for p_r are determined through the use of Equation (3-17) and the values for ϕ previously established.

GASES AND THE NON-FLOW PROCESSES

Relative volume, v_r, is also based on the isentropic hypothesis and is computed from the p_r values through the equation of state:

$$v_r = \frac{RT}{p_r} \tag{3-19}$$

Further, by analogy:

$$\frac{v_{r_2}}{v_{r_1}} = \left[\frac{v_2}{v_1}\right]_{(S)} \tag{3-20}$$

The utility of relative volume and relative pressure lies in the refinements they provide for work with isentropic processes over temperature ranges where the variation in specific heats becomes significant. For an ideal gas and for real gases at moderate temperatures, the isentropic equation is:

$$\frac{T_2}{T_1} = \left(\frac{p_2}{p_1}\right)^{\frac{k-1}{k}} = \left(\frac{v_1}{v_2}\right)^{k-1} \tag{3-12}$$

The tabulated values for p_r and v_r in the Air Tables correct for the fact that k, (c_p/c_v), the isentropic exponent, is not a constant but varies with temperature, just as specific heats do. Because of the exponential relationship, the above equation is quite sensitive to small changes in the value of k.

In sum the Air Tables provide tabulated data for u, h, ϕ, p_r and v_r against a single entering argument, absolute temperature, T. Variable specific heats are accurately taken into account. The data are dependent on reasonable conformance to the equation of state ($pv = RT$), a condition that will be met except where the temperature is very low, approaching absolute zero, or the pressure is exceptionally high, in the thousands of pounds per square inch. The use of the Air Tables is demonstrated in the following example.

Example 3-12: Air is heated in a non-flow process from 540°R to 1500°R at a constant pressure of 90 psia. The air then expands isentropically until the pressure is 15 psia. Assuming constant specific heats, determine:
 (a) the heat input, Btu/lb
 (b) the work output, Btu/lb
 (c) the change in entropy, Btu/lb · °R
 (d) (e) (f) Repeat (a) through (c) using the Air Tables, Appendix E.
 Solution:

$$p_1 = p_2 = 90 \text{ psia} \qquad T_1 = 540°R$$
$$p_3 = 15 \text{ psia} \qquad T_2 = 1500°R$$

(a) $\quad q_{12} = c_p(T_2 - T_1) = 0.24(1500 - 540) = \underline{230.4 \text{ Btu/lb}}$
$\quad q_{23} = 0$ since that process is isentropic

GASES AND THE NON-FLOW PROCESSES

(b) $\quad u_1 + q_{12} = u_2 + \dfrac{wk_{12}}{J}; \quad \dfrac{wk_{12}}{J} = q_{12} - (u_2 - u_1)$

$\qquad\qquad = q_{12} - c_v(T_2 - T_1) = 230.4 - 0.171(1500 - 540) = 66.2$ Btu/lb

$$T_3 = T_2 \left(\dfrac{p_3}{p_2}\right)^{\frac{k-1}{k}} = 1500 \left(\dfrac{15}{90}\right)^{\frac{0.4}{1.4}} = 899°R$$

$$\dfrac{wk_{23}}{J} = \dfrac{R(T_3 - T_2)}{J(1-k)} = \dfrac{53.3(899 - 1500)}{778 \times (-)0.4} = 102.9 \text{ Btu/lb}$$

$$\dfrac{wk_{12}}{J} + \dfrac{wk_{23}}{J} = 66.2 + 102.9 = \underline{169.1 \text{ Btu/lb}}$$

(c) $\quad s_2 - s_1 = c_p \ln \dfrac{T_2}{T_1} = 0.24 \ln \dfrac{1500}{540} = \underline{0.2452 \text{ Btu/lb} \cdot °R}$

(d) $\qquad\qquad\qquad q_{12} = h_2 - h_1$

Entering Appendix E with T_1 and T_2, $h_1 = 129.06$, $h_2 = 369.17$.

$$q_{12} = 369.17 - 129.06 = \underline{240.1 \text{ Btu/lb}}$$

(e) Total work $= wk_{12} + wk_{23}$. From the non-flow energy equation:

$$\dfrac{wk_{12}}{J} = q_{12} - (u_2 - u_1)$$

From Appendix E: $u_1 = 92.04$, $u_2 = 266.34$.

$$\dfrac{wk_{12}}{J} = 240.1 - (266.34 - 92.04) = 65.8 \text{ Btu/lb}$$

$$\dfrac{wk_{23}}{J} = q_{23} - (u_3 - u_2) = u_2 - u_3 \text{ since } q_{23} = 0$$

Since the process is isentropic, p_r can be used.

$$\dfrac{p_{r_2}}{p_{r_3}} = \left(\dfrac{p_2}{p_3}\right)_S; \quad p_{r_3} = p_{r_2}\left(\dfrac{p_3}{p_2}\right)_S = \dfrac{55.86 \times 15}{90} = 9.310$$

Entering Appendix E with $p_r = 9.310$, it can be seen that T_3 lies between 920° and 940°R. Interpolating in the p_r column, we get:

$$T_3 = 925.7°R$$
$$u_3 = 159.13$$
$$\phi_3 = 0.73130$$

GASES AND THE NON-FLOW PROCESSES

$$\frac{wk_{23}}{J} = u_2 - u_3 = 266.34 - 159.13 = 107.21 \text{ Btu/lb}$$

$$\frac{wk_{12}}{J} + \frac{wk_{23}}{J} = 65.8 + 107.2 = \underline{173.0 \text{ Btu/lb}}$$

(f) $$s_2 - s_1 = \phi_2 - \phi_1 - \frac{R}{J}\ln\frac{p_2}{p_1}$$

From Appendix F: $\phi_1 = 0.60078$, $\phi_2 = 0.85416$.

$s_2 - s_1 = 0.85416 - 0.60078 - 0 = \underline{0.25338 \text{ Btu/lb} \cdot {}^\circ\text{R}}$

As a cross check, since $s_3 = s_2$ let us compute $s_3 - s_1$.

$$s_3 - s_1 = \phi_3 - \phi_1 - \frac{R}{J}\ln\frac{p_3}{p_1}$$

$$= 0.73130 - 0.60078 - \frac{53.3}{778}\ln\frac{15}{90}$$

$$= \underline{0.25327 \text{ Btu/lb} \cdot {}^\circ\text{R}}$$

Problems

3-1. During a gas process, the temperature remains constant while the pressure is doubled. How will the specific volume be affected?

Ans. Halved

3-2. A closed tank contains 100 lb of air at a pressure of 200 psia and a temperature of 180°F. The air is subsequently cooled to 80°F. Find (a) the initial specific volume, ft³/lb; (b) the volume of the tank, ft³; (c) the final specific volume, ft³/lb; (d) the final pressure, psia.

Ans. (a) 1.184 ft³/lb; (b) 118.4 ft³; (c) 1.184 ft³/lb; (d) 168.8 psia

3-3. Air initially at 25°C and 150 kPa is heated in a constant volume non-flow process until the pressure reaches 750 kPa. Find the required heat transfer, kJ/kg.

Ans. 853.4 kJ/kg (added)

3-4. Air initially at 75 psia and 65°F is compressed to a final pressure of 300 psia and temperature of 320°F. Find the polytropic exponent for the process.

Ans. $n = 1.4$

3-5. Air initially at 50 psia and 140°F undergoes a polytropic process such that the temperature becomes 40°F. The polytropic exponent for the process is equal to 1.3. Find the final pressure and specific volume.

Ans. 22.7 psia, 8.15 ft³/lb

3-6. Air initially at 400 K expands in a constant pressure non-flow process until the initial volume is doubled. Find (a) the heat transfer, kJ/kg; (b) the work, kJ/kg; and (c) change of entropy, kJ/kg·K, for the process.

Ans. (a) 401.9 kJ/kg (added); (b) 114.8 kJ/kg (by); (c) 0.6965 kJ/kg·K

3-7. Air initially at 15 psia and 60°F is brought to a final temperature of 200°F by a

GASES AND THE NON-FLOW PROCESSES

reversible non-flow process. Identify the process in each case and find the work required in Btu/lb if: (a) $n = 0$, (b) $n = \infty$, (c) $n = k$, (d) $n = 1.2$.

Ans. (a) constant pressure, 9.6 Btu/lb (by); (b) constant volume, 0 Btu/lb; (c) isentropic, $(-)24$ Btu/lb (on); (d) polytropic, $(-)48$ Btu/lb (on)

3-8. Air initially at 15 psia and 60°F is brought to a final temperature of 200°F by a reversible non-flow process. Correlate with problem 3-7, show the processes on p-v and T-s coordinates and find the heat transferred in Btu/lb if (a) $n = 0$, (b) $n = \infty$, (c) $n = k$, (d) $n = 1.2$.

Ans. (a) 33.6 Btu/lb (added); (b) 24 Btu/lb (added); (c) 0 Btu/lb; (d) $(-)24$ Btu/lb (abstracted)

3-9. Air initially at 100°F and 100 psia and occupying a volume of 0.5 ft³ undergoes a reversible non-flow constant temperature process such that the final pressure becomes 20 psia. Find the work done, ft·lbf.

Ans. 11,590 ft·lbf (by)

3-10. The value of the Mechanical Equivalent of Heat, J, from the International Tables is 778.169 ft·lbf/Btu. The value of the Gas Constant, R, for air as used in the *Gas Tables* (Keenan & Kaye) is 53.342 ft·lbf/lb·°R. And from the same source, we find that at 90°F the value of the isentropic exponent, k, for air is 1.400. Using these data, find the values of c_v and c_p for air at 90°F, to four significant figures.

Ans. $c_v = 0.1714$ Btu/lb·°R, $c_p = 0.2399$ Btu/lb·°R

3-11. During an unidentified process with air, the temperature decreases from 600°F to 200°F. Find (a) the change of internal energy and (b) the change of enthalpy.

Ans. (a) $(-)68.4$ Btu/lb (decrease); (b) $(-)96.0$ Btu/lb (decrease)

3-12. Four pounds of a certain gas receive 50 Btu of heat during a constant temperature reversible non-flow process at 165°F. Find (a) the change of specific entropy, Btu/lb·°R, and (b) the work done, Btu. Show the process on p-v and T-s coordinates.

Ans. (a) 0.02 Btu/lb·°R (increase); (b) 50 Btu (by)

3-13. Air initially at 3000°F and 600 psia expands isentropically under non-flow conditions. The final volume is six times the initial volume. Show the process on p-v and T-s coordinates and find (a) the final pressure, psia; (b) the final temperature, °F; (c) the heat transferred, Btu/lb; (d) the work done, Btu/lb; (e) the change of entropy, Btu/lb·°R.

Ans. (a) 48.8 psia; (b) 1230°F; (c) 0 Btu/lb; (d) 302.7 Btu/lb (by); (e) 0 Btu/lb·°R.

3-14. Air undergoes a cycle consisting of a series of non-flow processes listed below:
1-2 constant volume heat addition
2-3 constant pressure heat addition
3-4 constant volume heat rejection
4-1 constant pressure heat rejection

The maximum and minimum values for pressure and specific volume for the cycle are 30 psia, 15 psia, 25 ft³/lb and 12.5 ft³/lb. Show the processes on p-v and T-s coordinates and find (a) heat added, Btu/lb; (b) heat rejected, Btu/lb.

Ans. (a) 329.8 Btu/lb (added); (b) $(-)294.9$ Btu/lb (rejected)

3-15. During a constant pressure reversible non-flow process with air, the temperature increases from 400°F to 1600°F. Show the process on p-v and T-s coordinates and find the heat transferred in Btu/lb (a) using constant specific heat and (b) using the Air Tables.

GASES AND THE NON-FLOW PROCESSES

Ans. (a) 288.0 Btu/lb (added); (b) 314.9 Btu/lb (added)

3-16. During a constant volume reversible non-flow process with air the temperature drops from 1400 K to 320 K. Show the process on p-v and T-s coordinates and find the heat transfer in kJ/kg (a) using constant specific heat and (b) using the Air Tables.

Ans. (a) $(-)773.17$ kJ/kg (rejected); (b) $(-)885.17$ kJ/kg (rejected)

3-17. Rework problem 3-13 using the Air Tables.

Ans. (a) 56.8 psia; (b) 1504°F; (c) 0 Btu/lb; (d) 328.5 Btu/lb (by); (e) 0 Btu/lb·°R

3-18. Air initially at 300 K and 130 kPa undergoes an isentropic non-flow expansion process in which the pressure is reduced to one-fourth the original value. Show the process on p-v and T-s coordinates, and using the Air Tables find (a) the final temperature, K; (b) the final pressure, kPa; (c) the change in internal energy, kJ/kg; (d) the change in enthalpy, kJ/kg; and (e) the change in entropy.

Ans. (a) 201.6 K; (b) 32.5 kPa; (c) -70.4 kJ/kg; (d) -98.6 kJ/kg; (e) 0

3-19. Air initially at 14.7 psia and 80°F is heated in a constant volume non-flow process until the temperature reaches 1500°F. The air is then expanded isentropically until the original volume is tripled. Show the process on p-v and T-s coordinates, and using the Air Tables find (a) the heat added at constant volume, Btu/lb, and (b) the work done, Btu/lb.

Ans. (a) 267.24 Btu/lb (added); (b) 123.23 Btu/lb (by)

CHAPTER 4

PROPERTIES AND CHARACTERISTIC BEHAVIOR OF STEAM

4-1. Introduction. As indicated in Chapter 3, when we are dealing with real gases more than one approach is available. Where precise results are required, the Gas Tables, with useful properties tabulated against temperatures, are used. But for many calculations, use of the perfect gas laws and the assumption of constant specific heats is acceptable. Furthermore certain process equations relating properties can be written for gases. In treating working substances classed as vapors, no such simple approach is applicable. The equations relating properties are involved, and the specific heats are subject to wide variation with both temperature and pressure. A further complication arises from the fact that these working substances usually undergo phase changes in the course of a thermodynamic cycle. This results in a nonuniform variation in properties as the phase changes occur. Figure 4-1 is a plot of heat added versus temperature for several constant pressure processes starting with water at 32°F. The figure indicates the discontinuities associated with phase changes and the difficulties attending any effort to find heat transferred from the specific heat-temperature change product.

All vapors are similar in general behavior, and the trends derived from a study of one vapor will be indicative of how any other vapor will behave. In this chapter our attention will be directed to steam, since it is a representative vapor and a very common working substance. The term *steam* is frequently used to include the liquid (water) phase when we are speaking of the working substance in general. In a more restricted sense, where differentiation from the liquid phase is necessary, the term steam implies the presence of at least a part of the substance in the vapor phase.

The complexity of the equations for establishing the various properties of steam at any known state precludes their repeated solution from a practical standpoint. This seeming impasse is overcome by the use of tables that give the several properties at a particular state as fixed by the easily measurable properties, temperature and pressure. These tables are essentially a series of solutions of the property equations as determined and verified experimentally. The **Steam Tables** (*Thermodynamic Properties of Water Including Vapor, Liquid, and Solid Phases*) by Keenan, Keyes, Hill and Moore and the ASME Steam Tables (*Thermodynamic and Transport Properties of Steam*) published by the American Society of

PROPERTIES OF STEAM

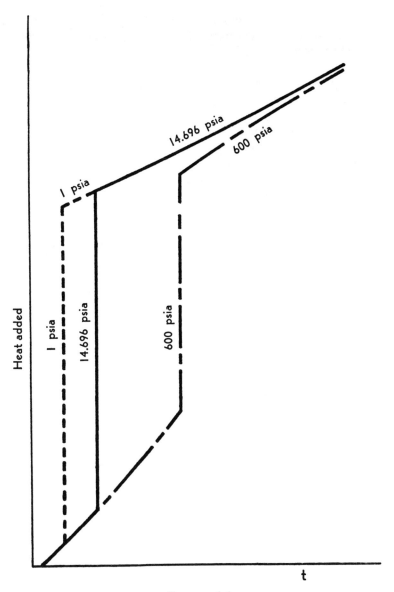

FIGURE 4-1

Mechanical Engineers, are preeminent in this country. The choice between the two sets of tables is a matter of personal preference, since the values extracted are substantially coincident, although the tabular arrangement is considerably different. The former publication has been

PROPERTIES OF STEAM

chosen by the authors for the purposes of this text. Hence, whenever the term "Steam Tables" appears, the work of Keenan, Keyes, Hill and Moore (English Units) is to be inferred, and all work with steam will be based on values taken from these tables. For other thermodynamically important vapors, such as the refrigerants Freon and ammonia, tables of similar nature are also available.

4-2. Graphical Representation and the Saturation Lines. A knowledge of the graphical representation of the properties and processes of steam on conventional coordinate systems is essential, and the saturation lines provide a good point of departure. To demonstrate the saturation concept, suppose that water at room temperature is confined in a cylinder and that standard atmospheric pressure is always imposed on the contents. If heat is added, the temperature of the water will rise until it reaches 212°F. Any further addition of heat at this point will result in vaporization of some of the water. The liquid at 212°F and 14.696 psia is *saturated water*. As more heat is added, the vaporization of the liquid will progress at 212°F until all of the liquid in the cylinder has become vapor. The vapor at 212°F and 14.696 psia, in the absence of any liquid whatsoever, is *saturated steam or dry vapor*. Any further addition of heat will be accompanied by a rise in temperature, and the steam is said to be *superheated*. For any other imposed pressure the vaporization will start and be completed at some other fixed temperature. As indicated in Fig. 4-1, this vaporization temperature will increase with increasing pressure. For any pressure there is a corresponding temperature of vaporization known as the *saturation temperature*. Conversely, for any particular temperature, there is a corresponding *saturation pressure*. As an example the saturation pressure for steam at 212°F is 14.696 psia, and the saturation temperature for steam at 14.696 psia is 212°F. These normally accepted values are at some variance with those found in the Steam Tables we shall use. Referring to Table 1, the saturation pressure for 212°F is given as 14.698 psia. Similarly, from Table 2, the saturation temperature for 14.696 psia is 211.99°F. For practical purposes, the differences are of little import.

Suppose a series of pressures (or temperatures) is selected and some source of information such as the Steam Tables is used to fix all the properties of the steam at the beginning and end of vaporization. This will yield a fully defined family of state points for the saturated liquid and the saturated vapor. A connected plot of these state points on any selected coordinate system will give the **saturated liquid line** and the **saturated vapor line**. The coordinate systems in which we shall be immediately interested are the p-v, the T-s, and a newcomer, the h-s. The saturation lines on these three systems are shown in Fig. 4-2. In the interests of

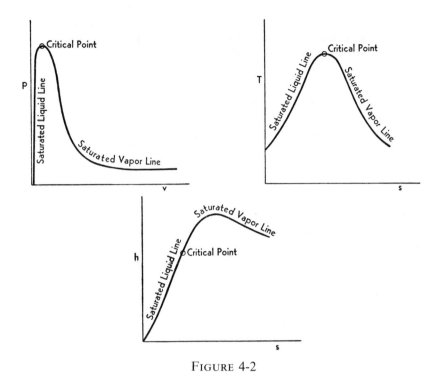

FIGURE 4-2

clarity, the specific volume scale in way of the saturated liquid line on the p-v diagram has been greatly exaggerated. Actually the specific volume of the liquid varies very little with temperature and pressure. Similarly, the ordinate (absolute temperature) on the T-s diagram has been foreshortened at the lower end for simplicity. The value of the diagrams lies more in trends than in details, but the general shape of the saturation lines is important. *No diagram for a vapor has any real significance unless the saturation lines are shown. Their appearance on each coordinate system should be learned and is fundamental required knowledge.*

As shown in Fig. 4-2, the saturated liquid and saturated vapor lines approach one another and finally join smoothly at the **critical point**. At this point the properties of the saturated liquid and the saturated vapor are the same. The critical point has other definite physical significance. Our basic definition of a vapor, as opposed to a gas, stated that a vapor can be made to liquefy by relatively minor changes of pressure or temperature. If the temperature of a vapor is and remains above the critical temperature, the vapor cannot be liquefied regardless of the pressure imposed. If the pressure is greater than the critical value, the vapor and

PROPERTIES OF STEAM

liquid are indistinguishable, and heating or cooling at constant pressure will not result in any recognizable phase change. For steam the critical pressure and temperature are 3203.6 psia and 705.44°F, again from the Steam Tables.

4-3. General Nature of the Steam Tables. The general arrangement and nature of the material contained in the Steam Tables is best determined by a careful perusal of the tables. Particular attention is directed to the pages entitled "Contents" and "Symbols Used In Tables." Note that useful tables other than those having to do with the properties of steam are included.

The entering arguments for the tables having to do with steam are pressure or temperature or both. The tabulated properties within the several tables include specific volume (v), specific internal energy (u), specific enthalpy (h) and specific entropy (s). Throughout the tables, pressure (p) is in pounds force per square inch absolute, psia. Temperature (t) is in degrees Fahrenheit and specific volume in cubic feet per pound. Internal energy and enthalpy are both given in Btu per pound. Entropy is shown in Btu per pound degree Rankine (Fahrenheit absolute). As indicated in Art. 2-8, the use of absolute temperature is fundamental to the concept and definition of entropy. By convention, the values of specific entropy and specific internal energy are both set as equal to zero for the saturated liquid at 32.018°F and 0.08866 psia, the so-called triple point. It is only under these particular conditions of temperature and pressure that the solid (ice), liquid (water) and vapor (steam) phases can coexist in equilibrium. The student should verify the values quoted above for pressure, temperature, entropy and internal energy at the triple point by reference to Tables 1 and 2 of the Steam Tables.

The coverage provided by the more frequently used tables is indicated by the numerals on the T-s diagram shown in Fig. 4-3.

The area to the left of the saturated liquid line is a *liquid region*. Here the liquid is at a temperature below the saturation temperature corresponding to its pressure. Data for this area are normally derived from Table 4.

The area included between the two saturation lines is the *wet region*. Here the liquid and vapor phases coexist at the saturation pressure and temperature and are usually in an intimate mixture. Properties of the substance in this area and on the saturation lines are derived from Table 1 or Table 2 as convenient.

The area above and to the right of the saturated vapor line is the *superheat region*, and the properties are covered by Table 3. Here the actual temperature is greater than the saturation temperature corre-

PROPERTIES OF STEAM

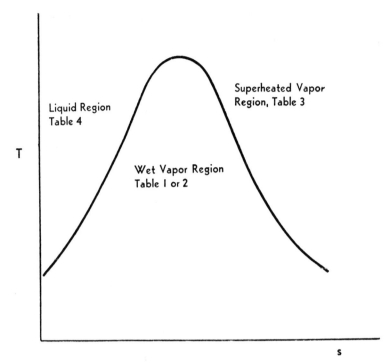

FIGURE 4-3

sponding to the existing pressure. The succeeding articles and accompanying examples will indicate the use of the several tables in some detail.

4-4. The Saturated Liquid and Saturated Vapor. Tables 1 and 2 are both saturation tables and give the properties of the saturated liquid, designated by the subscript f, and the saturated vapor, designated by the subscript g. The basic difference between the two tables is that the entering argument for Table 1 is *temperature* and for Table 2 is *pressure*. From these tables the properties for the saturated condition can be extracted directly. When both temperature and pressure at saturation are known, either table may be used, and the choice will be dictated by the common sense criterion of minimum interpolation.

The gages used to read vacuum, as in a steam turbine condenser, are commonly calibrated in inches of mercury. Unfortunately, Table 2 of the Steam Tables is keyed to pressure in pounds per square inch absolute only. As an ultimate timesaver, the student should derive all the Table 2 values for pressures of 1, $1\frac{1}{2}$ and 2 inches of mercury absolute and record them for future reference.

PROPERTIES OF STEAM

Example 4-1: Find the pressure, specific volume, internal energy, enthalpy and entropy of saturated water at 300°F.
Solution: From Table 1 for 300°F:

$$p = \underline{66.98 \text{ psia}}$$
$$v_f = \underline{0.017448 \text{ ft}^3/\text{lb}}$$
$$u_f = \underline{269.52 \text{ Btu/lb}}$$
$$h_f = \underline{269.73 \text{ Btu/lb}}$$
$$s_f = \underline{0.43720 \text{ Btu/lb} \cdot {}^\circ\text{R}}$$

Example 4-2: Find the temperature, specific volume, internal energy, enthalpy and entropy of saturated steam at a pressure of 1 in. Hg absolute.
Solution:

Saturated steam $p = 1$ in. Hg abs = 0.491 psia

Entering Table 2, an interpolation between 0.48 and 0.50 psia is indicated.

$$t = 78.32 + \frac{11}{20}(79.56 - 78.32) = \underline{79.00\,^\circ\text{F}}$$

Interpolation could be continued in Table 2 in the same manner for the remaining quantities. But there is a convenient shortcut thanks to the saturation temperature turning out to be a whole number. Simply enter Table 1 with 79°F. Then:

$$v_g = \underline{652.7 \text{ ft}^3/\text{lb}}$$
$$u_g = \underline{1036.7 \text{ Btu/lb}}$$
$$h_g = \underline{1096.0 \text{ Btu/lb}}$$
$$s_g = \underline{2.0384 \text{ Btu/lb} \cdot {}^\circ\text{R}}$$

Example 4-3: Find the temperature, specific volume, internal energy, enthalpy and entropy of saturated steam at 850 psia.
Solution:

Saturated steam 850 psia

Using Table 2 and interpolating between 840 and 860 psia:

$$t = 524.01 + \frac{1}{2}(526.76 - 524.01) = \underline{525.39\,^\circ\text{F}}$$
$$v_g = 0.5400 - \frac{1}{2}(0.5400 - 0.5264) = \underline{0.5332 \text{ ft}^3/\text{lb}}$$

or

$$v_g = \frac{1}{2}(0.5400 + 0.5264) = \underline{0.5332 \text{ ft}^3/\text{lb}}$$
$$u_g = 1114.1 - \frac{0.5}{2} = \underline{1113.9 \text{ Btu/lb}}$$
$$h_g = 1198.0 - 0.3 = \underline{1197.7 \text{ Btu/lb}}$$

or

$$h_g = 1197.4 + 0.3 = \underline{1197.7 \text{ Btu/lb}}$$
$$s_g = 1.4106 - \frac{1}{2}(1.4106 - 1.4080) = \underline{1.4093 \text{ Btu/lb} \cdot {}^\circ\text{R}}$$

Simple interpolation can frequently be accomplished by eye or mental arithmetic as indicated above, particularly where the difference between tabulated values is small. Example 4-2 is a definite special case, but

suggests that the concurrent use of the two tables may minimize or simplify interpolation.

4-5. The Wet Vapor. Determination of the properties of the wet vapor, in the area between the two saturation lines, presents a special problem. In this area the pressure and temperature are interdependent and are insufficient to define the state. *Thus steam at saturation pressure and temperature may be saturated water, wet vapor or saturated vapor.* Figure 4-4 is a *T-s* diagram for steam and shows line *fg*, which is defined by saturation temperature and pressure. In order to establish a state point such as (*y*) on that line, an additional property must be known. In the preceding article Tables 1 and 2 were used to determine the properties at (*f*) and at (*g*). The states were definitely fixed by specifying the saturated liquid or saturated vapor condition, and this specification of phase established the additional property necessary to fix the state. At some intermediate point such as (*y*), the substance is a mixture of saturated liquid and saturated vapor. The presence of water, usually as droplets, leads to the

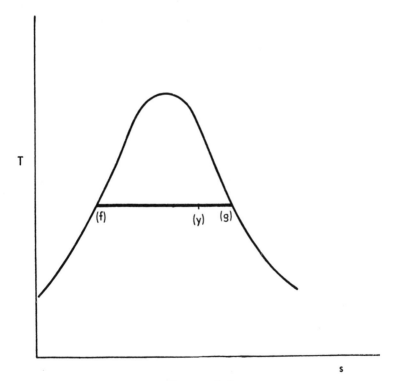

FIGURE 4-4

PROPERTIES OF STEAM

term "wet" in referring to the vapor in this region. The water phase present is spoken of as *moisture*. The **moisture content** of the mixture by mass is assigned the symbol m and is expressed as a percentage or a decimal. Thus at (f), $m = 100$ percent $= 1.0$ and at (g), $m = 0$. On the basis that the best steam contains the least moisture, the term **"quality"** is introduced and assigned the symbol, x. *Quality is the proportion of dry vapor by mass present in the mixture* and is related to moisture content by the following equation:

$$x = (1 - m) = (100 - m)\% \qquad (4\text{-}1)$$

For equation purposes, it is preferable to express both quality and moisture content in decimal form, and this convention will be followed in subsequent developments. Within the wet region, quality and moisture content can be regarded as properties, since both are true state functions. Outside the wet region, they lose all significance and are no longer applicable.

All of the continuous properties, other than pressure and temperature, are associated with mass. This suggests the determination of the properties in the wet region by combining the properties at saturated conditions in proportion to the respective mass of saturated liquid and saturated vapor present. Taking enthalpy as a typical property, its value at some state such as (y) will be given by:

$$h = mh_f + xh_g \qquad (4\text{-}2)$$

Substituting for m:

$$h = mh_f + xh_g = (1 - x)h_f + xh_g$$
$$= h_f - xh_f + xh_g = h_f + x(h_g - h_f)$$

But $(h_g - h_f)$ represents the change of enthalpy that occurs during the complete evaporation process. This value is given the symbol h_{fg} and is tabulated in Tables 1 and 2. Then, in final form:

$$h = h_f + xh_{fg} \qquad (4\text{-}3)$$

Similarly, $h = mh_f + (1 - m)h_g = h_g - m(h_g - h_f)$.

And
$$h = h_g - mh_{fg} \qquad (4\text{-}4)$$

The discussion of the determination of enthalpy in the wet region applies equally well to **specific volume, entropy** *and* **internal energy.** Equations for each of these properties may be written by substituting the proper symbol in place of h in (4-3) and (4-4) above. The change of the property during evaporation, as designated by the subscript *fg*, is not tabulated for specific volume, but v_{fg} can be determined by subtracting the saturated liquid

PROPERTIES OF STEAM

FIGURE 4-5

value (v_f) from the saturated vapor value (v_g). In this particular case, however, use of Equation (4-2) has merit.

Figure 4-5 is a simple sketch showing the physical significance of Equations (4-3) and (4-4). The magnitude of x and m indicates how far the evaporation process has proceeded. *The first equation starts with the saturated liquid properties and adds the proportionate change of properties associated with partial evaporation. The second equation starts with the saturated vapor properties and deducts the proportionate change of properties associated with partial condensation.* Either equation can be used for a particular problem, but it is good practice to choose the one that entails the smaller last term. *Use Equation (4-3) if x is less than 50 percent and (4-4) if m is less than 50 percent.* In most practical cases, the latter condition will obtain.

Whenever the observed values of steam pressure and temperature are coincident with the saturation values of these properties within instrument accuracy, caution is indicated. The unfortunate tendency of the unsophisticated in this situation is to assume that they are dealing with the saturated vapor. Actually *further instrumentation is required in order to determine another property and thereby definitely fix the state.* For this purpose there are various types of calorimeters on the market. The throttling calorimeter, which will be discussed in the next chapter, enables determination of the enthalpy of a steam sample. The separating calorimeter is a simple instrument for determining directly the moisture content and quality of a sample. In this device, the water is separated mechanically from the dry vapor by imposing a radical change of direction on the entering steam sample. The higher inertia of the water droplets causes separation from the lighter dry vapor, and this water is trapped and weighed. The dry vapor continues through a calibrated orifice from which its mass rate of flow can be determined by conventional methods.

Example 4-4: The pressure and temperature of steam in a line are determined to be 65 psia and 298°F. Since these data are inconclusive, a sample of this steam is passed through a separating calorimeter and 0.5 lb of water is collected in 5 minutes. From orifice data the rate of flow of the dry vapor is found to be 0.4 lb per min. Find (a) the quality, (b) the specific volume, (c) the entropy, (d) the enthalpy and (e) the internal energy of the steam in the line.

PROPERTIES OF STEAM

Solution:

(a) $$x = \frac{\text{mass of dry vapor}}{\text{mass of mixture}} = \frac{M_{DV}}{M_M} = \frac{M_{DV}}{M_{DV} + M_{H_2O}}$$

$$M_{H_2O} = 0.5 \text{ lb}$$
$$M_{DV} = 0.4 \text{ lb/min} \times 5 \text{ min.} = 2.0 \text{ lb}$$
$$x = \frac{2.0}{2.0 + 0.5} = 0.8 = \underline{80\%}$$

(b) $$m = (1 - x) = 1 - 0.8 = 0.2$$
$$v = mv_f + xh_g$$

Extracting values for 298°F from Table 1 or for 65 psia from Table 2:

$$v = 0.2 \times 0.017 + 0.8 \times 6.657 = \underline{5.329 \text{ ft}^3/\text{lb}}$$

(c) $$s = s_g - ms_{fg}$$
$$s = 1.6380 - 0.2 \times 1.2035 = \underline{1.3973 \text{ Btu/lb} \cdot {}^\circ\text{R}}$$

(d) $$h = h_g - mh_{fg}$$
$$h = 1179.6 - 0.2 \times 911.9 = \underline{997.2 \text{ Btu/lb}}$$

(e) $$u = u_g - mu_{fg}$$
$$u = 1099.5 - 0.2 \times 832.1 = \underline{933.1 \text{ Btu/lb}}$$

Note that the equations involving moisture are preferable in this example because m is less than 50 percent and that the number of decimal places carried into the final answers is controlled by the basic data extracted from the tables.

Example 4-5: Steam at 210 psia and 386°F has an enthalpy of 1173 Btu per lb as determined by a throttling calorimeter. Find (a) the quality and (b) the specific volume of this steam.

Solution:

$$p = 210 \text{ psia}; \quad t = 386°\text{F}; \quad h = 1173 \text{ Btu/lb}$$

From Table 1 for $t = 386°\text{F}$:

$$p = 210.06 \text{ psia} \quad h_g = 1200.0 \text{ Btu/lb}$$

From Table 2 for $p = 210$ psia:

$$t = 385.97°\text{F} \quad h_g = 1200.0 \text{ Btu/lb}$$

Recognizing the limitations of normal instrumentation, the given data are consistent, and either table can be used. Further, since h is less then h_g, we are clearly dealing with wet steam, but the moisture content is not high since the two enthalpy values are fairly close together.

(a) $$h = h_g - mh_{fg}; \quad m = \frac{h_g - h}{h_{fg}}$$

Using either table, $h_{fg} = 839.9$ and $h_g = 1200.0$.

Then

$$m = \frac{1200.0 - 1173}{839.9} = 0.032$$

$$x = 1 - m = 1.000 - 0.032 = \underline{0.968 = 96.8\%}$$

PROPERTIES OF STEAM

(b) $\qquad v = v_g - m(v_g - v_f)$

From Table 1 for 386°F:

$v_g = 2.183 \qquad v_f = 0.018443$
$v = 2.183 - 0.032(2.183 - 0.018) = \underline{2.114 \text{ ft}^3/\text{lb}}$

From Table 2 for 210 psia:

$v_g = 2.184 \qquad v_f = 0.018443$
$v = 2.184 - 0.032(2.184 - 0.018) = \underline{2.115 \text{ ft}^3/\text{lb}}$

Both answers are right for the given data.

4-6. The Superheated Vapor. *A vapor is said to be **superheated** when its temperature exceeds the saturation temperature for the existing pressure.* The difference between the actual (total) temperature and the saturation temperature is called superheat ($S.H.$) and is stated in degrees Fahrenheit. Superheat is a property peculiar to the superheat region just as quality is to the wet region. The properties specific volume, internal energy, enthalpy and entropy are listed in Table 3 of the Steam Tables for the saturated vapor at the given absolute pressure (across the top of the page) and for the superheated vapor. The entering arguments are temperature and pressure, since these two properties fully define the state in the superheat region. The superheat for a given pressure and temperature can be obtained by subtracting the saturation temperature for that pressure from the actual temperature. The saturation temperature for each tabulated pressure is included in parentheses in the pressure line at the top of the page. Table 3 also shows (in italics) calculated values of v, u, h and s for the supersaturated vapor, a condition where t is below the saturation temperature for the existing pressure. Supersaturation is a metastable state of some engineering interest. However, for the work which will be presented in this text, the aforementioned entries have no direct application other than facilitating interpolation in the case of very low superheats. The use of the table is not difficult to understand, although interpolation is frequently necessary.

Metastable states are highly transient in nature and arise because phase changes do not take place instantaneously. Supersaturation, for instance, is frequently encountered in steam turbines where the expansion associated with the pressure drop across a nozzle normally requires only about 0.0002 second. If the steam entering the nozzle is only slightly superheated and the pressure drop takes the steam below the saturated vapor line and into the wet vapor region, water droplets theoretically should appear immediately. However, the expansion takes place so rapidly that there is actually a small delay before moisture particles begin to form and equi-

PROPERTIES OF STEAM

librium is re-established. This phenomenon is real and a matter for consideration in the detailed design of steam turbines.

Example 4-6: Find (a) the superheat and (b) the enthalpy of steam at 900 psia and 535°F.
Solution:

$$p = 900 \text{ psia} \qquad t = 535°\text{F}$$

(a) $\qquad S.H. = t - t_{sat} = 535 - 532.12 = \underline{2.88°\text{F}}$

(b) $\qquad h = 1193.6$ for 900 psia and 530°F
$\qquad h = 1204.8$ for 900 psia and 540°F

Then for $t = 535°\text{F}$

$$h = 1193.6 + \tfrac{1}{2}(1204.8 - 1193.6) = \underline{1199.2 \text{ Btu/lb}}$$

For points that fall midway between tabulated values, as frequently occurs, calculation is simplified by adding the two appropriate entries and dividing by 2, as follows:

$$h = \frac{1193.6 + 1204.8}{2} = \underline{1199.2 \text{ Btu/lb}}$$

Example 4-7: For steam at 600 psia and 850°F, find (a) the superheat, (b) the specific volume, (c) the internal energy, (d) the enthalpy and (e) the entropy.
Solution:

$$p = 600 \text{ psia} \qquad t = 850°\text{F}$$

(a) $\qquad t_{sat} = 486.33$
$\qquad S.H. = t - t_{sat} = 850 - 486.33 = \underline{363.67°\text{F}}$

(b) At 840°F, $v = 1.2353$ and at 860°F, $v = 1.2577$.

At 850°F, $v = \dfrac{1.2353 + 1.2577}{2} = \underline{1.2465 \text{ ft}^3/\text{lb}}$.

(c) $\qquad u = \dfrac{1292.7 + 1301.2}{2} = \underline{1297.0 \text{ Btu/lb}}$

(d) $\qquad h = \dfrac{1429.8 + 1440.9}{2} = \underline{1435.4 \text{ Btu/lb}}$

(e) $\qquad s = \dfrac{1.6517 + 1.6601}{2} = \underline{1.6559 \text{ Btu/lb} \cdot °\text{R}}$

4-7. The Compressed Liquid. As previously stated, steam at the saturation temperature corresponding to its pressure may be saturated liquid, wet vapor or saturated vapor. Furthermore, if the actual temperature is greater than the saturation temperature for the existing pressure, the steam must be in the superheated vapor phase. The substance may also exist *at a temperature below the saturation temperature corresponding to the imposed pressure.* In this case the substance must be in the *liquid* phase and the state is again fully defined by temperature and pressure alone. This nonsaturated liquid is said to be either subcooled or compressed, de-

PROPERTIES OF STEAM

pending upon the viewpoint. As an example, consider water in a river at 65°F and standard atmospheric pressure. From the Steam Tables, the saturation pressure for 65°F is 0.3057 psia, and the saturation temperature for 14.696 psia is 211.99°F. On the basis of the actual pressure, the liquid is subcooled in that it is 147 degrees below the saturation temperature. On the basis of the actual temperature, the liquid is compressed in that the existing pressure is 14.39 psi above the saturation pressure.

The terms **compressed liquid** and **subcooled liquid** are used interchangeably, but the former is preferable. *The properties of the compressed liquid are established primarily by temperature, and the effects of pressure are relatively small,* as can be seen from Table 4 of the Steam Tables. Hence, for the purposes of this text, whenever the imposed pressure is 100 psia or less, the properties of compressed water will be taken from Table 1 as equal to those of the saturated liquid at the given temperature. If the pressure of the compressed liquid is greater than 100 psia, Table 4 should be used.

Table 4 is straightforward, but the pressure and temperature intervals are such that interpolation is frequently necessary. As in the case of Table 3, certain property values are italicized. Again, these are calculated values for a metastable condition, this time for the "superheated" liquid where the given temperature exceeds the saturation temperature corresponding to the imposed pressure. The entries for zero absolute pressure are of this nature, providing a convenient baseline for determining the properties of the compressed liquid when the pressure lies between 100 and 500 psia.

Example 4-8: In a marine steam propulsion plant, feedwater leaves the feed heater and enters the main feed pump at 270°F and a gage pressure of 35 psi. Find for the water: (a) the specific volume, (b) the internal energy, (c) the enthalpy and (d) the entropy.
Solution:

$$p = 35 + 14.7 = 49.7 \text{ psia}; \quad t = 270°F$$

(a) $v_f = 0.017170 \text{ ft}^3/\text{lb}$
(b) $u_f = 238.82 \text{ Btu/lb}$
(c) $h_f = 238.95 \text{ Btu/lb}$
(d) $s_f = 0.39597 \text{ Btu/lb} \cdot °R$

Example 4-9: Water from the feed pump in the earlier example enters the boiler at 1200 psia and 275°F. Find the enthalpy of the feedwater.
Solution:

$$p = 1200 \text{ psia} \quad t = 275°F$$
Compressed liquid—Use Table 4

At 1000 psia:

$t = 250°F \quad h = 220.61$
$t = 300°F \quad h = 271.46$

PROPERTIES OF STEAM

$$t = 275°F; \quad h = \frac{220.61 + 271.46}{2} = 246.04$$

At 1500 psia:

$$t = 250°F \quad h = 221.65$$
$$t = 300°F \quad h = 272.39$$
$$t = 275°F; \quad h = \frac{221.65 + 272.39}{2} = 247.02$$

At 1200 psia:

$$t = 275°F; \quad h = 246.04 + \frac{2}{5}(247.02 - 246.04)$$
$$h = 246.04 + 0.39 = \underline{246.43 \text{ Btu/lb}}$$

4-8. The p-v Diagram for Steam. Figure 4-6 is a p-v diagram for steam. Such a diagram is indicative of the interrelationship between certain properties and is particularly adapted to the representation of non-flow processes, since the area under a non-flow process line on this coordinate system represents mechanical work. In the figure, lines of constant quality (moisture content), of constant entropy and of constant temperature have been shown. Constant pressure and constant volume lines have been omitted to simplify the diagram but can be mentally spotted in, since they will appear as straight lines. Lines of constant enthalpy and constant internal energy have also been omitted, since they are of limited interest in the non-flow case. *Where more than one line is shown for a particular property, subscripts are used, and the subscript sequence denotes increasing values of the property.*

The two constant quality lines shown both originate at the critical point and remain in the wet region as all such lines must. The points which establish any constant quality line are fixed by dividing the horizontal line connecting the saturation lines in proportion to the selected quality. For example, the line for a constant quality of 75 percent will cross all the constant pressure lines in the wet region at 3/4 of the distance from the saturated liquid line to the saturated vapor line.

Two typical isothermal lines or isotherms are shown. The isothermal line originating in the compressed liquid region descends almost vertically until it intersects the saturated liquid line, then proceeds through the wet region at constant pressure and drops off into the superheat region. All isotherms for temperatures less than the critical value of 705.44°F will have the same general appearance. For temperatures above the critical value, the isotherms remain entirely in the superheat region as illustrated in the figure. It should be observed that the isotherms in the superheat region are similar in shape to those for a gas. This is to be expected, since the superheated vapor is akin to a gas.

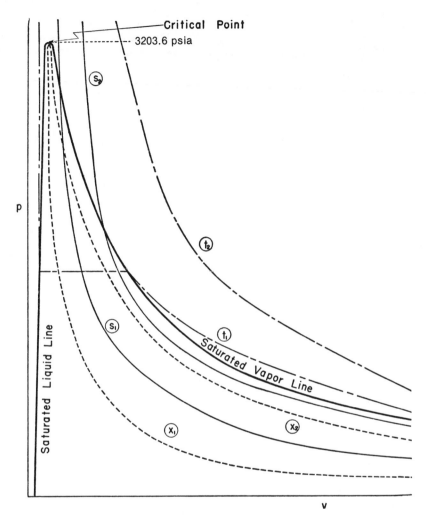

FIGURE 4-6

The isentropic lines shown are typical in that all such lines eventually enter the wet region. The two lines shown both start in the superheat region and are high enough in value to pass to the right of the critical point. For some lower value of entropy such that the line originates in the liquid region to the left of and below the critical point, it will proceed almost vertically to the saturated liquid line, since radical changes in pressure without heat transfer will cause only small changes in the specific volume of water. Upon entering the wet region, the isentropic line will

PROPERTIES OF STEAM

gradually slope to the right, following the general trend of the two lines shown.

The constant property lines discussed cover a rather wide range of pressures and specific volumes. In the usual processes encountered in practice, only a portion of any line would be required, and the process may, of course, proceed in either direction. The *p-v* diagram is seldom used except in the analysis of non-flow steam processes. For the more common flow processes with steam, other diagrams are of more immediate interest.

4-9. The *T-s* Diagram for Steam. The *T-s* diagram has the widest application of any of the several diagrams used in thermodynamics, since the area included under any reversible process line represents heat. It is applicable to both flow and non-flow problems, and heat is always of

FIGURE 4-7

primary interest. Figure 4-7 shows the T-s diagram for steam, including some typical constant property lines. The isentropic and isothermal lines have been omitted for simplicity, since their straight-line nature is obvious. Lines of constant internal energy could be shown but are omitted because of the limited direct application of this property to steam problems. As before, the temperature scale has been foreshortened below 32°F for simplicity.

The construction of the constant quality lines is similar to that followed for the p-v diagram with proportionate intercepts on the horizontal constant temperature lines in the wet region. Again these lines originate at the critical point and remain entirely in the wet region.

The constant superheat lines conform roughly to the shape of the saturated vapor line and move farther into the superheat region with increasing values of superheat.

The constant volume line shown is typical, with a characteristic sharp change of slope on crossing the saturated vapor line.

The constant enthalpy lines on T-s coordinates are of three general types. For values of enthalpy above about 1200 Btu per lb, the lines remain entirely in the superheat region. For values of enthalpy between about 1100 and 1200 Btu per lb, the lines pass from the superheat into the wet region at high temperature and finally re-enter the superheat region at some lower temperature. At values of enthalpy somewhat below 1100 Btu per lb, the lines cross into the wet region from above but do not leave the wet region at low temperatures. Note that the constant enthalpy line for the superheated vapor at low temperature is substantially horizontal. This indicates that the enthalpy is very nearly a function of temperature alone in this region and suggests that the vapor has assumed some of the characteristics of a gas.

The constant pressure lines or isobars are of particular interest, since pressure is a readily observed property and many actual processes occur at approximately constant pressure. A typical constant pressure process line is shown originating in the compressed liquid region. As the liquid is warmed, the temperature rises until the saturation temperature is reached. For clarity, this portion of the line has been distorted on the figure, and the separation from the saturated liquid line is actually much less than indicated. The process continues across the wet region at constant temperature until evaporation is complete. The constant pressure line then breaks sharply upward into the superheat region.

Figure 1 of the Steam Tables is a complete Temperature-Entropy Chart covering a wide range of values and should be studied in connection with the material presented above. Note that it terminates at 32°F and that temperature is in degrees Fahrenheit rather than Rankine.

PROPERTIES OF STEAM

4-10. The *h-s* Diagram and Mollier Chart for Steam. Enthalpy is particularly useful in dealing with flow problems as outlined in the discussion of this property in Art. 2-4. At the same time, the isentropic process is the ideal approximation of many real processes. This suggests the use of an *h-s* diagram. For a perfect gas, the *h-s* diagram will resemble the *T-s* diagram, since enthalpy in this case is a function of temperature alone. The temperature scale of the *T-s* need only be multiplied by c_p to yield a valid *h-s* plot. In the case of steam the similarity between the two diagrams is lost because of the wide variation in specific heat.

Figure 4-8 is an *h-s* diagram for steam. Since enthalpy and entropy are

FIGURE 4-8

PROPERTIES OF STEAM

substantially equal to zero for the saturated liquid at 32°F, the saturated liquid line will start at the origin of coordinates. The critical point marking the junction of the saturated liquid and saturated vapor lines no longer lies at the peak of the "steam dome" as it does on p-v and T-s coordinates. The area included between these lines is still the wet region, although the orientation is askew. Various pertinent constant property lines are shown. The horizontal constant enthalpy lines and the vertical constant entropy lines have been omitted to simplify the construction of the figure. Constant volume and constant internal energy lines could be shown but are not included.

The isobars and isotherms are straight lines in the wet region just as they were on p-v and T-s coordinates. Two such lines for coincident saturation pressure and temperature are shown on the figure. Upon crossing the saturated vapor line, the constant pressure lines show a slight increase in slope, while the constant temperature lines break rather sharply to the right. A third line for a constant temperature greater than the critical value appears in the superheat region. Note that the constant temperature lines become substantially horizontal at high superheat and nearly coincident with lines of constant enthalpy. This indicates that the vapor in this region is approaching the gaseous phase.

The constant quality (moisture content) lines are constructed by dividing the constant temperature (or pressure) lines in the wet region in proportion to the relative amounts of liquid and dry vapor present as specified by the chosen quality. This procedure is exactly the same followed on the p-v and T-s systems, but the pressure-temperature lines are slanted rather than horizontal. Only one constant quality line is shown in the figure. All such lines remain entirely in the wet region and converge at the critical point where the vapor and liquid phases merge.

Two constant superheat lines are shown. The shape of these lines conforms roughly to the saturated vapor line, particularly for the lower values of superheat.

The conventional Mollier Chart for steam includes only a part of the complete h-s diagram. The primary purpose of such a chart is to facilitate the determination of enthalpy at a specified state. The limits of the Mollier Chart, in the flap pocket at the back of the Steam Tables, are delineated in Fig. 4-8. *The chart is a graphical adjunct to the tabulated properties of steam, and the two sources of information should be used jointly.* The tables are basically more accurate, but laborious interpolation can frequently be avoided by using the chart. In the superheat region, Table 3 is about as simple to use as the chart. In the wet region, the chart is definitely preferable unless great accuracy is required. Since the chart does not include specific volume, internal energy or any useful liquid proper-

PROPERTIES OF STEAM

ties, the tables must perforce be used in obtaining these properties. *Careful use of the Mollier Chart will yield values of enthalpy correct to the nearest half Btu per pound, and values of entropy correct to the third decimal place.*

Process lines drawn on *h-s* coordinates are useful in indicating the direction of variation of properties. However, the area under a reversible process line has no quantitative significance as it has on a *T-s* diagram. But a reversible process line proceeding from left to right is accompanied by increased entropy, and we can conclude that heat has been added during the process. Conversely, a reversible process line proceeding from right to left is accompanied by decreased entropy and the rejection of heat. Both *h-s* and *T-s* diagrams are widely used in representing flow processes with steam, either singly or together.

Other charts of vapor properties have been developed for special applications. Among these are the *p-h* diagrams used primarily for refrigerants and the *h-v* or Ellenwood charts for steam which are well suited to turbine design work.

Example 4-10: Steam is admitted to a turbine at 600 psia and 740°F, and exhausts to a condenser at a pressure of 1 psia. Assuming the process to be isentropic, sketch the process on *T-s* and *h-s* coordinates and find the drop in enthalpy, Btu/lb.

Solution:

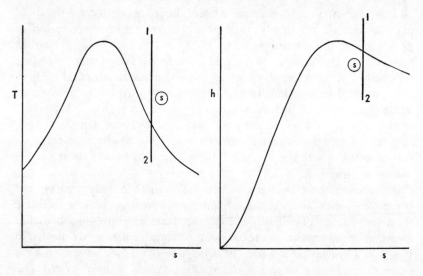

$$p_1 = 600 \text{ psia} \quad p_2 = 1 \text{ psia}$$
$$t_1 = 740°\text{F} \quad s_1 = s_2$$

From Table 3:

$$h_1 = 1373.7$$
$$s_1 = s_2 = 1.6067$$

From Mollier Chart entering with s_2 and p_2:

$$h_2 = 897$$
$$h_1 - h_2 = 1374 - 897 = 477 \text{ Btu/lb}$$

This problem could have been solved using the chart alone or the tables alone. The mixed method used is intended to illustrate how to obtain reasonable accuracy with minimum work.

4-11. Reversible Non-Flow Processes with Steam. The non-flow processes with steam are of some importance since they are encountered in practice in the reciprocating steam engine. Investigation of the corresponding ideal processes is carried out in a manner generally the same as that used for the non-flow gas processes but differing in detail. The basic equations are listed below:

$$wk_{12} = \int_1^2 p\,dv \qquad (2\text{-}5)$$

$$q_{12} = \int_1^2 T\,ds \qquad (2\text{-}6)$$

$$u_1 + q_{12} = u_2 + \frac{wk_{12}}{(J)} \qquad (2\text{-}1)$$

In developing the special equations which apply to a particular process, it is reasonable to assume that sufficient information is available to fix the initial and final states and establish the nature of the process. Knowing the two states, all the properties at the beginning and end of the process can be determined from the Steam Tables. Thus the values of the internal energy terms in the non-flow energy equation are immediately known. Complete solution of the problem will then result if either the heat or work term can be evaluated. There is no simple equation relating p and v for steam such as exists for a perfect gas, and $\int p\,dv$ can be integrated only if p or v remains constant during the process. Similarly, the relationship between T and s is complex, and $\int T\,ds$ can be integrated only if T or s is constant. After the integration for these special cases is carried out, the property values required for evaluation can be extracted from the Steam Tables.

From the discussion above, it is evident that the following reversible non-flow processes with steam are subject to straightforward analysis:
(a) constant volume
(b) constant pressure
(c) constant temperature or isothermal
(d) constant entropy or isentropic

PROPERTIES OF STEAM

There are other non-flow processes with steam that are similar to the general polytropic processes with a gas, but they are of minor importance and major complexity and will not be considered. The four processes listed will be investigated in the succeeding paragraphs. In the accompanying graphical representations, the processes are all shown originating in the wet region and terminating in the superheat region. The equations developed are independent of the location of the initial and final states and of the direction in which the process proceeds. Only numerical answers will be affected by these factors.

Constant Volume

$$v_1 = v_2 \qquad dv = 0$$
$$wk_{12} = \int_1^2 p\, dv = 0$$

Then from the non-flow energy equation:
$$q_{12} = u_2 - u_1 \text{ Btu/lb}$$

There is no work done, and the heat transferred is reflected entirely by a change of internal energy.

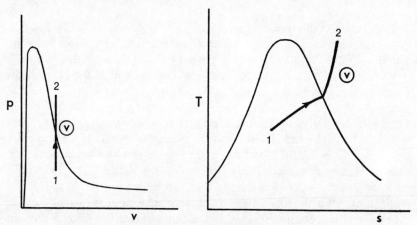

FIGURE 4-9. Non-flow, constant volume heating process.

Constant Pressure

$$p_1 = p_2 = p$$
$$wk_{12} = \int_1^2 p\, dv = p \int_1^2 dv = p(v_2 - v_1)$$

PROPERTIES OF STEAM

or $\quad wk_{12} = p_2v_2 - p_1v_1$ ft · lbf/lb

$$q_{12} = u_2 - u_1 + \frac{wk_{12}}{(J)} = u_2 - u_1 + \frac{p_2v_2 - p_1v_1}{(J)}$$

But by definition, $\quad h = u + \dfrac{pv}{(J)}$

So $\quad q_{12} = h_2 - h_1$ Btu/lb

Also $\quad \dfrac{wk_{12}}{(J)} = q_{12} - (u_2 - u_1) = (h_2 - h_1) - (u_2 - u_1)$

Here the heat transferred is reflected by a change of internal energy and the simultaneous performance of mechanical work. Heat is best evaluated by change of enthalpy in this case, since enthalpy is readily determined from the Steam Tables. Note that the constant pressure process is accompanied by constant temperature in the wet region (Fig. 4-10).

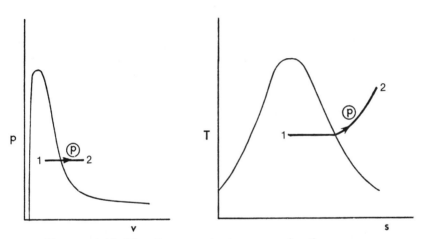

FIGURE 4-10. Non-flow, constant pressure heating process.

Constant Temperature

$$T_1 = T_2 = T$$

$$q_{12} = \int_1^2 T\,ds = T\int_1^2 ds = T(s_2 - s_1) \text{ Btu/lb}$$

$$\frac{wk_{12}}{(J)} = q_{12} - (u_2 - u_1) \text{ Btu/lb}$$

Here the heat transferred is again accompanied by a change of internal energy and the performance of mechanical work. Furthermore, the

PROPERTIES OF STEAM

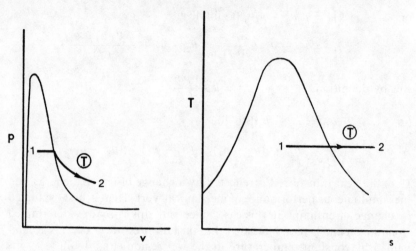

FIGURE 4-11. Non-flow, isothermal heating process.

constant temperature process in the wet region is accompanied by constant pressure (Fig. 4-11). Note in particular the use of absolute temperature in the equation for transferred heat above and the basic necessity for assigning the unit Btu per pound-degree Rankine (Fahrenheit *absolute*) to the entropy values extracted from the Steam Tables.

Constant Entropy

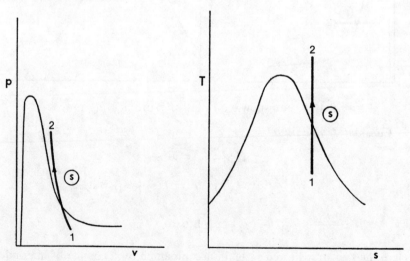

FIGURE 4-12. Non-flow, isentropic compression process.

PROPERTIES OF STEAM

$$s_1 = s_2 \qquad ds = 0$$

$$q_{12} = \int_1^2 T\,ds = 0$$

$$\frac{wk_{12}}{(J)} = u_1 - u_2 = -(u_2 - u_1) \text{ Btu/lb}$$

Here there is no transferred heat, and work is measured entirely by the change of internal energy.

Example 4-11: Steam initially at 10 psia and a quality of 90 percent drops to a pressure of 2.5 psia in a non-flow reversible constant volume process. Show the process on p-v and T-s coordinates and find:
(a) the final quality
(b) the heat transferred, Btu/lb
Solution:

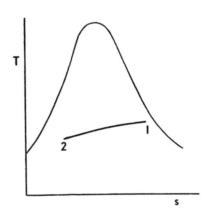

(a)

$p_1 = 10$ psia $\quad x_1 = 0.9 \quad m_1 = 0.1$

$v_1 = v_2$
$v_1 = x_1 v_g + m_1 v_f$

From Table 2:

$v_1 = (0.9 \times 38.42) + (0.1 \times 0.02) = 34.58$
$p_2 = 2.5$ psia $\quad v_2 = v_1 = 34.58$

$$v_g = \frac{146.40 + 135.79}{2} = 141.10 \text{ (interpolating)}$$

$v_f = 0.016267$ (by eye)

$$x_2 = \frac{v_2 - v_f}{v_g - v_f} = \frac{34.58 - 0.02}{141.10 - 0.02} = 0.245 = 24.5\%$$

PROPERTIES OF STEAM

(b) $$u_1 + q_{12} = u_2 + \frac{wk_{12}}{J} \qquad \text{Equation (2-1)}$$
$$\text{Non-flow}$$

$$\frac{wk_{12}}{J} = \int_1^2 p\,dv = 0 \quad \text{since } v_1 = v_2$$

Then
$$q_{12} = u_2 - u_1$$
$$p_1 = 10 \text{ psia} \qquad m_1 = 0.1$$
$$u_1 = u_g - m_1 u_{fg}$$
$$= 1072.2 - 0.1 \times 911.0 = 981.1$$
$$p_2 = 2.5 \text{ psia} \qquad x_2 = 0.245$$
$$u_f = \frac{100.81 + 103.85}{2} = 102.33$$
$$u_{fg} = \frac{953.1 + 951.0}{2} = 952.05$$
$$u_2 = u_f + x_2 u_{fg}$$
$$= 102.33 + (0.245 \times 952.05) = 335.6$$
$$q_{12} = u_2 - u_1 = 335.6 - 981.1$$
$$= (-)645.5 \text{ Btu/lb (abstracted)}$$

Example 4-12: Four pounds of steam initially dry and saturated expand isentropically in a non-flow process from an initial pressure of 275 psia to a final pressure of 125 psia. Show the process on *p-v* and *T-s* coordinates and find:
(a) the final quality
(b) the work done, Btu
Solution:

(a) $\quad p_1 = 275$ psia—saturated vapor; $\quad s_1 = s_2$

From Table 2: $\quad s_1 = 1.5192 = s_2; \quad u_1 = 1117.5$

$\quad p_2 = 125$ psia

PROPERTIES OF STEAM

$$m_2 = \frac{s_g - s_2}{s_{fg}} = \frac{1.5853 - 1.5192}{1.0893} = 0.061$$

$$x_2 = 1 - m_2 = 1 - 0.061 = \underline{0.939 = 93.9\%}$$

(b)
$$u_1 + q_{12} = u_2 + \frac{wk_{12}}{J}$$

$$q_{12} = \int_1^2 T\,ds = 0 \text{ since } s_1 = s_2$$

Then
$$\frac{wk_{12}}{J} = u_1 - u_2$$

$$u_2 = u_g - m_2 u_{fg}$$
$$= 1108.8 - (0.061 \times 793.3) = 1060.4$$

$$\frac{wk_{12}}{J} = 1117.5 - 1060.4 = 57.1 \text{ Btu/lb (by)}$$

$$M = 4 \text{ lb}$$

$$\frac{Wk_{12}}{J} = M \times \frac{wk_{12}}{J} = 4 \times 57.1 = \underline{228.4 \text{ Btu (by)}}$$

Problems

4-1. For steam at 705.44°F and 3203.6 psia (critical conditions) find the specific volume, the internal energy, the enthalpy and the entropy for (a) the saturated liquid using Table 1 and (b) for the saturated vapor using Table 2.

Ans. (a) 0.05053 ft³/lb, 872.6 Btu/lb, 902.5 Btu/lb, 1.0580 Btu/lb · °R; (b) same as (a)

4-2. Find the pressure, specific volume, internal energy, enthalpy and entropy of saturated water at 100°F.

Ans. 0.9503 psia, 0.016130 ft³/lb, 68.04 Btu/lb, 68.05 Btu/lb, 0.12963 Btu/lb · °R

4-3. Find the temperature, specific volume, internal energy, enthalpy and entropy of saturated steam at 306 psia.

Ans. 419.24°F, 1.5149 ft³/lb, 1118.3 Btu/lb, 1204.0 Btu/lb, 1.5098 Btu/lb · °R

4-4. For a pressure of 1½″ Hg absolute, determine the pressure in psia, and compute all the Table 2 entries for that pressure.

Ans. 0.737 psia, 91.64°F, $v_f = 0.016104$, $v_g = 446.5$, $u_f = 59.71$, $u_{fg} = 981.0$, $u_g = 1040.8$, $h_f = 59.71$, $h_{fg} = 1041.7$, $h_g = 1101.4$, $s_f = 0.11461$, $s_{fg} = 1.8893$, $s_g = 2.0039$

4-5. Complete the following table where the given data are shown in boldface type.

p(abs)	t, °F	m	v	u	h	s
1″ Hg	79.0	**0.10**	587.4	937.7	991.1	1.8437
247.1	**400**	0.0629	1.7499	1069.9	**1150.0**	1.4679
350	431.82	0.0256	1.2932	1100.8	1184.5	**1.4750**

121

PROPERTIES OF STEAM

4-6. Complete the following table, given the pressures and temperatures shown in the first two columns in boldface.

p(abs)	t, °F	S.H.	v	u	h	s
570	**850**	369.15	1.3151	1297.9	1436.6	1.6622
320	**425**	1.61	1.4538	1119.4	1205.6	1.5071

4-7. Find the specific volume, enthalpy and entropy of feed water at 1000 psia and 260°F.

Ans. 0.017029 ft³/lb, 230.78 Btu/lb, 0.38013 Btu/lb · °R

4-8. Identify the condition of "steam" at the following pressure and temperature: (a) 1150 psia, 300°F; (b) 1150 psia, 562°F; (c) 1150 psia, 850°F

Ans. (a) compressed liquid; (b) indeterminate—may be saturated liquid, wet vapor or saturated vapor; (c) superheated

4-9. Steam is supplied to a turbine at 600 psia with 200°F superheat and exhausts to the condenser at 0.7 psia. Assume the process to be isentropic and use the Mollier Chart to find (a) the entropy and enthalpy at inlet; (b) the enthalpy and moisture content at exhaust.

Ans. (a) 1.580 Btu/lb · °R, 1342.5 Btu/lb; (b) 865 Btu/lb, 22.6% moisture

4-10. Show on T-s and h-s coordinates the following processes for steam, in continuum and in sequence: (a) constant pressure heating from compressed liquid to superheated vapor such as occurs ideally in a boiler; (b) isentropic expansion from the superheat region to the wet region such as occurs ideally in a turbine; (c) constant pressure cooling from high quality wet vapor to saturated liquid as occurs ideally in a condenser.

4-11. Saturated water at 200 psia is heated at constant pressure in a reversible non-flow process until it becomes saturated vapor. Show the process on p-v and T-s coordinates and find (a) the change of entropy; (b) the change of internal energy; (c) the change of enthalpy; (d) the heat added; (e) the work done.

Ans. (a) 1.0025 Btu/lb · °R; (b) 759.6 Btu/lb; (c) 843.7 Btu/lb; (d) 843.7 Btu/lb; (e) 84.1 Btu/lb (by)

4-12. Steam initially dry and saturated at 150 psia expands isentropically in a non-flow process to a final pressure of 15 psia. Show the process on p-v and T-s coordinates and find (a) the heat transferred; (b) the change of internal energy; (c) the work done.

Ans. (a) 0 Btu/lb; (b) (−)148.1 Btu/lb; (c) 148.1 Btu/lb (by)

4-13. Two pounds of water occupy a container in such state that 0.5 lb is vapor and the remainder is liquid. If the pressure is 30 psia, find (a) the quality and (b) the temperature of the mixture.

Ans. (a) 0.25 or 25%; (b) 250.34°F

4-14. The design pressure and temperature of steam leaving the boiler are shown in bold face type in the table for (a) a U. S. Navy destroyer escort, (b) a 550,000 deadweight tonnage

PROPERTIES OF STEAM

merchant tanker, (c) a public utility central station generating plant—all of recent design. Complete the table.

Service	psia	t, °F	S.H.	v	h	s
Naval Ship	1200	950	382.6	0.6558	1470.3	1.6091
Tanker	900	950	417.9	0.8896	1480.5	1.6464
Central Sta.	3500	1050	**	0.2201	1459.6	1.4956

** Not applicable, since steam conditions are in the supercritical region and there is no saturation temperature corresponding to the imposed pressure.

CHAPTER 5

ELEMENTARY STEADY FLOW PROCESSES

5-1. Steady Flow. In a broad sense steady flow may be defined as continuous flow in which the properties and the velocity of the fluid at any point in the flow channel remain constant with time. More specifically and in direct application to thermodynamic systems, the following conditions must be satisfied when steady flow obtains:

(*a*) *At any selected cross section in the flow system, all the fluid properties and the mean velocity of advance through that cross section are constant with time.*

(*b*) *The total mass of fluid contained in the system at any instant is always the same. For each pound of fluid that enters in a unit time there must be a simultaneous discharge of one pound during the same unit time.*

(*c*) *An equilibrium must exist in that the net rate of heat transfer and work performance as referred to the working fluid must remain constant with time.*

The provisions of (a) and (b) above lead to the Continuity Equation which is an important steady flow corollary. At any cross section, the mass rate of flow is given by:

$$\dot{M} = \frac{AV}{v} \text{ lb/sec (kg/sec)}$$

where A = cross-sectional area, ft²(m²)
V = mean velocity through area A, ft/sec (m/sec)
v = specific volume of the fluid flowing, ft³/lb (m³/kg).

At any particular cross section, the area is constant for a rigid system, and both velocity and specific volume are constant if steady flow exists. Therefore, the mass rate of flow past the cross section is constant with time. Since the flow is continuous and there can be no accumulation or diminution of mass between any two cross sections, the mass rate of flow must be the same at all cross sections. For a single channel the Continuity Equation for steady flow is:

$$\dot{M} = \dot{M}_1 = \dot{M}_2 = \cdots \dot{M}_x$$

or $\quad\quad \dot{M} = \dfrac{A_1 V_1}{v_1} = \dfrac{A_2 V_2}{v_2} = \cdots \dfrac{A_x V_x}{v_x} \text{ lb/sec (kg/sec)} \quad\quad$ (5-1)

ELEMENTARY STEADY FLOW PROCESSES

Also, for cases in which the fluid density, or specific volume, is invariant ($v_1 = v_2 = v_x$), the volume rate of flow is expressed by:

$$A_1 V_1 = A_2 V_2 = \cdots A_x V_x \text{ ft}^3/\text{sec } (\text{m}^3/\text{sec}) \tag{5-1A}$$

For steady flow, the mass rate is constant with time throughout the system as opposed to the properties and velocity which are constant with time at a single location but may vary considerably from one cross section to another.

The provisions of (c) preclude the application of steady flow principles to transient conditions such as those encountered when warming up machinery or when sudden changes of load occur. In practice the conditions of equilibrium may be considered satisfied when the device under observation has "settled out," and the operating criteria such as pressures, temperatures and speed have become substantially constant. By the same token, minor fluctuations in the properties and velocity at a particular location do not prevent the use of steady flow concepts. The usual engineering instruments are either throttled or damped in some fashion to yield average readings which in essence establish equivalent steady flow for the unsteady system that may actually exist.

5-2. The Energy Equation for Steady Flow. The Energy Equation for Steady Flow first presented in Chapter 1 is repeated here:

$$\frac{g Z_1}{g_c(J)} + \frac{V_1^2}{2g_c(J)} + \frac{p_1 v_1}{(J)} + u_1 + q_{12}$$

$$= \frac{g Z_2}{g_c(J)} + \frac{V_2^2}{2g_c(J)} + \frac{p_2 v_2}{(J)} + u_2 + \frac{wk_{12}}{(J)} \text{ Btu/lb (J/kg)} \tag{1-20}$$

This equation has many applications and its modification to suit the non-flow case has already been discussed. There is a particular form of the Steady Flow Energy Equation which is basically applicable to the great majority of thermodynamic flow systems and devices. By reviewing each term in Equation (1-20), this form can be readily established.

The difference in potential energy terms will always be small, since the vertical separation between any two points in a thermodynamic system will not be great. A difference of 778 feet is necessary to produce a potential energy change of 1 Btu per pound. Therefore, the Z terms will be dropped, remembering to recover them if any freak height differences are encountered.

High velocities are frequently established in thermodynamic equipment, and the kinetic energy terms must be retained for possible use.

Note, however, that a velocity of 50 feet per second gives a kinetic energy of only 0.05 Btu per pound.

Since we are dealing with flow, the flow work terms must always be included. Furthermore, the internal energy terms are of continuing importance since our entire study is directed toward thermal considerations. With both u and $pv/(J)$ always present, it is logical to introduce enthalpy in their stead in order to deal with one property rather than three. The succeeding discussions will illustrate the great utility of enthalpy in flow problems.

Heat and work are both of major interest and must be included in our final equation:

$$\frac{V_1^2}{2g_c(J)} + h_1 + q_{12} = \frac{V_2^2}{2g_c(J)} + h_2 + \frac{wk_{12}}{(J)} \text{ Btu/lb (J/kg)} \quad (5\text{-}2)$$

This simplified form of the Energy Equation for Steady Flow is adapted to the majority of thermodynamic applications and will be referred to in the future simply as the **flow energy equation.** The student should observe that this equation is not restricted to reversible processes.

For any particular device, Sections 1 and 2 are usually chosen at entrance and exit respectively. However, the equation may be applied between any two sections if steady flow conditions are satisfied at each section. Machinery such as turbines, centrifugal pumps and heat exchangers have obvious steady flow characteristics. For reciprocating equipment such as air compressors, pumps and steam engines, the steady flow energy equation can still be used if Stations 1 and 2 are chosen in the inlet and discharge pipes far enough from the cylinder so that the periodic interruptions of flow have been largely damped out. The remainder of this chapter will show some practical applications of the flow energy equation to typical equipment.

5-3. Nozzles. The nozzle is essentially a shaped flow channel and has wide application in turbines and ejectors. Figure 5-1 is a sketch of a typical converging-diverging nozzle. The flow energy equation can be applied between any two cross sections, since steady flow occurs throughout. However, the entrance, throat and exit conditions are normally of greatest interest and most easily established. The entrance velocity is usually quite small, and the associated kinetic energy term can be discarded. Any heat transferred will be of small magnitude, since heat transfer takes time and the passage of the working substance through the nozzle occurs very rapidly. There can be no mechanical work performed in the absence of

ELEMENTARY STEADY FLOW PROCESSES

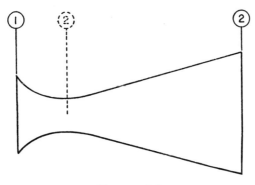

FIGURE 5-1

moving parts. These statements will now be put in mathematical form and the flow energy equation modified accordingly.

$$\frac{V_1^2}{2g_c(J)} + h_1 + q_{12} = \frac{V_2^2}{2g_c(J)} + h_2 + \frac{wk_{12}}{(J)}$$

But
$$\frac{V_1^2}{2g_c(J)} \to 0; \quad q_{12} \to 0; \quad \frac{wk_{12}}{(J)} = 0$$

Then
$$h_1 = \frac{V_2^2}{2g_c(J)} + h_2$$

And
$$\frac{V_2^2}{2g_c(J)} = h_1 - h_2 \quad \text{Btu/lb (J/kg)} \tag{5-3}$$

Inspection of this equation reveals that the nozzle is fundamentally a device for converting energy in the form of enthalpy into kinetic energy. Station 1 is usually chosen at the entrance and Station 2 at either the throat or exit. However, Station 2 can be located anywhere in the nozzle without affecting the final form of the equation. On the other hand, if Station 1 is not located at the entrance, the corresponding kinetic energy term cannot be dropped, since the velocity may well be appreciable. To avoid ambiguity it is always wise to sketch the apparatus under consideration and clearly label the chosen cross sections to which the equation is applied.

Example 5-1: In the first stage of a gas turbine, air enters a group of nozzles at 1200°F and leaves at 950°F. The entering velocity is negligible. Find (a) the kinetic energy, Btu/lb, and (b) the velocity, ft/sec, of the air leaving the nozzle.

ELEMENTARY STEADY FLOW PROCESSES

Solution:

(a)
$$T_1 = 1200 + 460 = 1660°R$$
$$T_2 = 950 + 460 = 1410°R$$

$$KE_2 = \frac{V_2^2}{2g_cJ} = h_1 - h_2; \quad h_1 - h_2 = c_p(T_1 - T_2)$$

$$KE_2 = c_p(T_1 - T_2) = 0.24\,(1660 - 1410) = \underline{60 \text{ Btu/lb}}$$

(b)
$$\frac{V_2^2}{2g_cJ} = h_1 - h_2 = c_p(T_1 - T_2)$$

$$V_2 = \sqrt{2g_cJc_p(T_1 - T_2)}$$

$$V_2 = \sqrt{2 \times 32.2 \times 778 \times 0.24 \times 250} = \underline{1734 \text{ ft/sec}}$$

(When steam is the working substance, values of h_1 and h_2 may be found in Steam Tables. Also, when variable specific heat is invoked for air, the appropriate values of h_1 and h_2 may be found from Appendix E.)

5-4. Turbines. All turbines are essentially steady flow machines, and the energy equation may be applied between any two cross sections. In this discussion only the turbine as a whole will be considered with Station 1 at the entrance and Station 2 at the exhaust, as indicated in Fig. 5-2. In actual turbines the entering velocity is relatively small, although the velocity in the exhaust trunk may become fairly large. For our immediate purposes, it will be assumed that the stations are so selected that the difference between the entering and leaving velocities is negligible. The validity of this assumption will be investigated later when turbines are considered in detail. There are some heat losses from the turbine casing, but these are minimized by insulation and can be neglected. Modification of the flow energy equation on the basis of the foregoing may now be carried out.

$$\frac{V_1^2}{2g_c(J)} + h_1 + q_{12} = \frac{V_2^2}{2g_c(J)} + h_2 + \frac{wk_{12}}{(J)}$$

ELEMENTARY STEADY FLOW PROCESSES

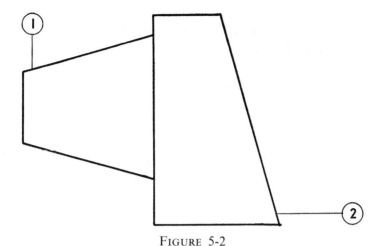

FIGURE 5-2

But $\quad q_{12} \to 0; \quad \dfrac{V_2^2}{2g_c(J)} - \dfrac{V_1^2}{2g_c(J)} \to 0$

Then $\quad h_1 = h_2 + \dfrac{wk_{12}}{(J)}$

And $\quad \dfrac{wk_{12}}{(J)} = h_1 - h_2 \quad \text{Btu/lb (J/kg)} \quad\quad \underline{(5\text{-}4)}$

From the standpoint of thermodynamics, this equation shows that the turbine is basically a device for producing mechanical work at the expense of the enthalpy of the working substance. It should be noted that the final equation as written may not apply at any cross sections other than those chosen because the local velocities may be so great as to preclude omission of the kinetic energy terms. Equation (5-4) and the accompanying discussion also apply to reciprocating steam engines if cross sections 1 and 2 are chosen at locations in the inlet pipe and exhaust trunk well removed from the cylinders.

Example 5-2: A marine propulsion turbine receives steam at the throttle at 875 psia and 940°F at the rate of 100,000 lb/hr. After an irreversible expansion process, the steam exhausts from the turbine at a pressure of 0.6 psia with a moisture content of 10%. Assume the difference between the entrance and exit kinetic energies is negligible and find:
 (a) the work done, Btu/lb
 (b) the power developed, hp

Solution: Refer to Fig. 5-2:

$\quad\quad\quad\quad p_1 = 875 \text{ psia} \quad\quad t_1 = 940°\text{F}$
$\quad\quad\quad\quad p_2 = 0.6 \text{ psia} \quad\quad m_2 = 0.10$

ELEMENTARY STEADY FLOW PROCESSES

(a) $$\frac{wk_{12}}{J} = h_1 - h_2$$

From the Steam Tables:

$$h_1 = 1475.6$$
$$h_2 = 1098.6 - 0.1 \times 1045.4 = 994.1$$
$$\frac{wk_{12}}{J} = 1475.6 - 994.1 = \underline{481.5 \text{ Btu/lb}}$$

(b) $\dot{M} = 100{,}000 \text{ lb/hr}$ \quad 1 hp = 2545 Btu/hr

$$P = \frac{\dot{M}(wk_{12})}{2545\,J} = \frac{100{,}000 \times 481.5}{2545} = \underline{18{,}920 \text{ hp}}$$

5-5. Boilers. The term boiler as used in this text defines an entire steam generating unit and includes the saturated (or generating) section, as well as the superheater, the economizer and the air preheater if used. This is in contradistinction to the practice of restricting the term boiler to only those parts of the unit directly associated with the production of saturated steam.

In applying the flow energy equation to a boiler, we obviously must include the heat term. It will be assumed that the kinetic energy of the feed water entering and of the steam leaving are negligibly different. This is substantially the case in practice with maximum feed velocities limited to about 10 feet per second and exit velocities from the superheater limited to about 125 feet per second. Since a boiler has no moving parts there can be no mechanical work performed. Referring to Fig. 5-3:

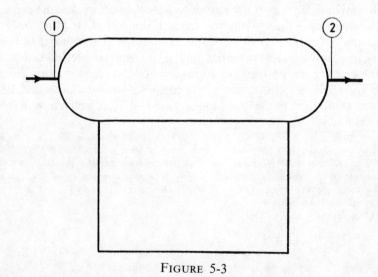

FIGURE 5-3

ELEMENTARY STEADY FLOW PROCESSES

$$\frac{V_1^2}{2g_c(J)} + h_1 + q_{12} = \frac{V_2^2}{2g_c(J)} + h_2 + \frac{wk_{12}}{(J)}$$

But
$$\frac{V_2^2}{2g_c(J)} - \frac{V_1^2}{2g_c(J)} \to 0; \quad \frac{wk_{12}}{(J)} = 0$$

Then
$$h_1 + q_{12} = h_2$$

And
$$q_{12} = h_2 - h_1 \quad \text{Btu/lb (J/kg)} \tag{5-5}$$

This equation has been developed on the basis of the boiler as a whole but is equally applicable to component parts such as the economizer or the superheater. It also has general application to other types of heat exchangers. Since the feed water supplied to a boiler may be rather highly compressed, it frequently will be necessary to use Table 4 of the Steam Tables in determining the enthalpy of the entering feed.

Example 5-3 A boiler receives feed water at 1200 psia and 250°F and delivers steam from the superheater at 900 psia and 950°F. Find the heat added, Btu/lb.
Solution: Refer to Fig. 5-3:

$$p_1 = 1200 \text{ psia} \quad t_1 = 250°F$$
$$p_2 = 900 \text{ psia} \quad t_2 = 950°F$$

$$q_{12} = h_2 - h_1$$

From Table 4: $\quad h_1 = 220.61 + \frac{2}{5}(1.04) = 221.0$

From Table 3: $\quad h_2 = 1480.5$

$$q_{12} = 1480.5 - 221.0 = \underline{1259.5 \text{ Btu/lb (added)}}$$

Example 5-4: If, in the preceding example, the feed water entering has a velocity of 3 m/sec and the steam leaving the superheater has a velocity of 50 m/sec, find:
 (a) the additional heat required to accommodate the change in kinetic energy across the boiler, J/kg
 (b) the percentage error introduced by neglecting the kinetic energy change
Solution:

(a) $\quad V_1 = 3 \text{ m/sec} \quad V_2 = 50 \text{ m/sec}$

$$q_{12} = h_2 - h_1 + \frac{V_2^2 - V_1^2}{2g_c}$$

$$\Delta q_{12} = \frac{V_2^2 - V_1^2}{2g_c} = \frac{(50)^2 - (3)^2}{2}$$

$$= \underline{1245.5 \text{ N} \cdot \text{m/kg or J/kg (added)}}$$

(b) Using Appendix A:

$$q_{12} = (1259.5 \text{ Btu/lb}) \times \left(2{,}326 \frac{\text{J/kg}}{\text{Btu/lb}}\right)$$

$$= 2930 \times 10^3 \text{ J/kg}$$

131

ELEMENTARY STEADY FLOW PROCESSES

$$\text{Error} = \frac{1245.5 \times 100}{2930 \times 10^3} = \underline{0.043\%}$$

5-6. Incompressible Flow and Liquid Pumps. Figure 5-4 shows an elementary steady flow system, a horizontal pipe of uniform diameter carrying an incompressible fluid ($v_1 = v_2 = v_x$). With no work devices present it is immediately evident that no work can be done on or by the fluid. Further, $V_1 = V_2$ from the Continuity Equation and $Z_1 = Z_2$ since the pipe is horizontal. The Steady Flow Energy Equation (1-20) then takes on this form:

$$\frac{p_1 v_1}{(J)} + u_1 + q_{12} = \frac{p_2 v_2}{(J)} + u_2$$

since the work term and the *change* in the kinetic and potential energy terms are all zero. In many cases of incompressible fluid flow, thermal aspects (internal energy and heat transfer) are considered negligible, since no attempt is evident in the system to add or abstract heat. However, if the internal energy and heat transfer terms are dropped from the above equation there results:

$$\frac{p_1 v_1}{(J)} = \frac{p_2 v_2}{(J)} \quad \text{or} \quad p_1 = p_2$$

Experience tells us this cannot be true for an actual flow—there must be a pressure drop. Accordingly, we must return the internal energy and heat transfer terms to the equation, but in a rearranged form to give:

$$\frac{p_1 v_1}{(J)} - \frac{p_2 v_2}{(J)} = (u_2 - u_1) - q_{12}$$

or
$$p_1 - p_2 = \frac{(J)}{v}[(u_2 - u_1) - q_{12}] \; \text{lbf/ft}^2 \; (\text{Pa}) \quad (5\text{-}6)$$

Equation (5-6) indicates that the pressure drop (or head loss) in the system of Fig. 5-4 is equal to the gain in internal energy of the fluid plus any loss of energy in the form of heat by the system (heat transferred from the system would have a negative sign according to the normal sign convention). These latter two effects are associated with friction—between the

FIGURE 5-4

ELEMENTARY STEADY FLOW PROCESSES

fluid and the pipe walls and within the fluid itself due to turbulence.

It is frequently convenient with incompressible fluids to write the energy equation for steady flow in terms of *feet or metres of the fluid flowing*. This may be accomplished by multiplying both sides of Equation (1-20) by $(J)g_c/g$, in which case the resulting Engineers' units will be ft·lbf/lbf, or feet of *head* for each factor in the equation:

$$Z_1 + \frac{V_1^2}{2g} + \frac{(J)g_c u_1}{g} + \frac{p_1 v_1 g_c}{g} + \frac{(J)g_c q_{12}}{g}$$

$$= Z_2 + \frac{V_2^2}{2g} + \frac{(J)g_c u_2}{g} + \frac{p_2 v_2 g_c}{g} + \frac{g_c}{g} wk_{12} \text{ ft (m)} \quad (5\text{-}7)$$

Assuming frictionless, incompressible flow and introducing specific weight, γ, from Equation (1-8), we arrive at Bernoulli's Equation:

$$Z_1 + \frac{V_1^2}{2g} + \frac{p_1}{\gamma} = Z_2 + \frac{V_2^2}{2g} + \frac{p_2}{\gamma} \text{ ft (m)} \quad (5\text{-}8)$$

This equation is sometimes modified to include frictional effects as a *head loss* term, HL, expressed in terms of feet or metres of fluid flowing:

$$Z_1 + \frac{V_1^2}{2g} + \frac{p_1}{\gamma} = Z_2 + \frac{V_2^2}{2g} + \frac{p_2}{\gamma} + HL_{12} \text{ ft (m)} \quad (5\text{-}9)$$

Equation (5-9) is applicable to the steady flow of an *actual, incompressible* fluid with losses expressed as

$$HL_{12} = \frac{(J)g_c}{g}[(u_2 - u_1) - q_{12}]$$

Another modification of Equation (5-7) is useful in dealing with pumps handling incompressible fluids. Such a pump is shown schematically in Fig. 5-5.

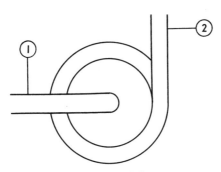

FIGURE 5-5

ELEMENTARY STEADY FLOW PROCESSES

Neglecting the thermal terms in Equation (5-7) yields:

$$Z_1 + \frac{V_1^2}{2g} + \frac{p_1}{\gamma} = Z_2 + \frac{V_2^2}{2g} + \frac{p_2}{\gamma} + \frac{g_c}{g}wk_{12} \text{ ft (m)} \qquad (5\text{-}10)$$

Solving explicitly for the work gives:

$$wk_{12}\left(\frac{g_c}{g}\right) = (Z_1 - Z_2) + \left(\frac{V_1^2}{2g} - \frac{V_2^2}{2g}\right) + \left(\frac{p_1 - p_2}{\gamma}\right) \text{ ft (m)} \qquad (5\text{-}11)$$

The physical configuration and mechanical design of most pump systems are such that all groups of terms in Equation (5-11) are negative in sign. As a matter of convenience, a term *TH*, *total head*, is introduced:

$$TH = (Z_2 - Z_1) + \left(\frac{V_2^2 - V_1^2}{2g}\right) + \left(\frac{p_2 - p_1}{\gamma}\right) \text{ ft (m)} \qquad (5\text{-}12)$$

In this application *TH* represents the total *useful* head developed by the pump and is the sum of the changes in elevation, velocity and pressure heads *through the pump*. Note that the total head equation does not include any provision for internal energy changes in the fluid as it passes through the pump, changes which will occur because of friction effects and other internal losses. When we deal with the pump alone where point 1 is taken at the suction flange and point 2 at the discharge flange, the elevation and velocity head changes are quite small in the typical case, which leads to this common form of the equation:

$$TH = \frac{\Delta p}{\gamma} \text{ ft (m)} \qquad (5\text{-}12\text{A})$$

Pump efficiency using total head (*TH*) as expressed in Equation (5-12) is given by:

$$\eta_{\text{pump}} = \frac{TH}{wk_{\text{brake}}} \text{ in consistent units}$$

or, in terms of power:

$$\eta_{\text{pump}} = \frac{\text{water horsepower}}{\text{brake horsepower}} = \frac{WHP}{BHP} \qquad (5\text{-}13)$$

In Engineers' units:

$$WHP = \frac{\dot{M}(TH)g}{550\, g_c} \text{ hp (useful output)}$$

BHP = brake horsepower (power delivered to the pump by the prime mover)

ELEMENTARY STEADY FLOW PROCESSES

\dot{M} = mass flow rate of fluid, lb/sec
TH = pump total head, ft
g = acceleration of gravity, ft/sec²
$$g_c = 32.2 \frac{\text{lb} \cdot \text{ft}}{\text{lbf} \cdot \text{sec}^2}$$
550 = ft · lbf/sec equivalent of 1 horsepower

The flow energy equation (5-2) can also be used to analyze liquid pumps operating under steady flow conditions. As noted previously, the difference between the inlet and discharge kinetic energy terms is negligible in most pump installations. The velocities are relatively low, although not necessarily equal, since the discharge pipe is frequently made smaller than the suction pipe. The heat transferred will always be small because of the rapidity with which the liquid passes through the pump and the relatively small difference between the temperature of the liquid being pumped and the surroundings. Applying Equation (5-2) to the pump shown in Fig. 5-5:

$$\frac{V_1^2}{2g_c(J)} + h_1 + q_{12} = \frac{V_2^2}{2g_c(J)} + h_2 + \frac{wk_{12}}{(J)} \quad \text{Btu/lb (J/kg)} \quad (5\text{-}2)$$

But $$\frac{V_2^2}{2g_c(J)} - \frac{V_1^2}{2g_c(J)} \to 0; \quad q_{12} \to 0$$

Then $$h_1 = h_2 + \frac{wk_{12}}{(J)}$$

And $$\frac{wk_{12}}{(J)} = h_1 - h_2 \quad \text{Btu/lb (J/kg)} \quad (5\text{-}14)$$

Using this approach, the work done on the fluid by the pump is expressed in terms of the change of enthalpy across the pump. Rewriting Equation (5-14) with enthalpy separated into its component parts gives:

$$\frac{wk_{12}}{(J)} = \frac{p_1 v_1}{(J)} + u_1 - \frac{p_2 v_2}{(J)} - u_2 \quad \text{Btu/lb (J/kg)} \quad (5\text{-}14\text{A})$$

Since the compression process through the pump is essentially adiabatic, the corresponding ideal process is the reversible adiabatic or isentropic. The characteristics of a liquid are such that the increase in internal kinetic energy evidenced by the temperature rise during *isentropic* compression is almost exactly balanced by the decrease in internal potential energy which results from decreased molecular spacing. Therefore, the net change in total internal energy is extremely small and can be neglected for the *isentropic* case. Therefore:

ELEMENTARY STEADY FLOW PROCESSES

$$\frac{wk_{12}}{(J)} = \frac{p_1 v_1 - p_2 v_2}{(J)}$$

Furthermore, the change in specific volume during the isentropic compression of a liquid is always very small, and the exact determination of the final specific volume involves rather laborious calculations. Except in cases where the pressure ratio is extremely high, it is common practice to neglect this change and to assume that the specific volume remains constant at its initial value. In other words, the liquid is truly incompressible. Then:

$$\frac{wk_{12}}{(J)} = \frac{v_1(p_1 - p_2)}{(J)} \text{ Btu/lb (J/kg)} \qquad (5\text{-}15)$$

Note that this equation corresponds to the expression for total head as shown in Equation (5-12A), but that the units are different.

In practice, the isentropic process is used as a standard, and the departures from reversibility are taken into account by introducing pump efficiency, η_{pump}. Then:

$$\frac{wk_{12(actual)}}{(J)} = \frac{wk_{12(isentropic)}}{(J)\eta_{pump}}$$

And referring to Equation (5-14):

$$\Delta h_{(actual)} = \frac{\Delta h_{(isentropic)}}{\eta_{pump}}$$

The difference between the change in enthalpy for the actual and that for the ideal process reflects the change in internal energy, which can be appreciable for the irreversible case.

For pump design work where the primary concern is with the end "useful" work as evidenced by pressure, velocity and elevation heads, neglect of internal energy terms is quite proper. In the final analysis, these terms are buried in the pump efficiency, which to the pump supplier merely reflects the power input that will be required of the prime mover. In thermodynamic analysis, however, the exact condition of the working substance is of continuing interest, and attention must be paid to *all* property changes, including internal energy.

Table 4 of the Steam Tables can be used to determine pump work, provided that the condition of the liquid is known at both suction and discharge. Example problem 5-5 involves many of the relationships just developed.

ELEMENTARY STEADY FLOW PROCESSES

Example 5-5: Saturated water at 250°F enters a centrifugal main feed pump and is discharged at 1200 psia. The pump efficiency is 60% and the delivery rate is 125,000 lb/hr. Find:
- (a) the total head developed by the pump, ft
- (b) the water horsepower, WHP
- (c) the brake horsepower, BHP
- (d) the ideal (isentropic) pump work, Btu/lb
- (e) the actual pump work, Btu/lb
- (f) the estimated temperature of the water at discharge

Solution: Refer to Fig. 5-5:

$t_1 = 250°F$ (sat. water)　　$p_1 = 29.8$ psia
$v_1 = 0.017001$ ft^3/lb　　$p_2 = 1200$ psia
$h_1 = 218.59$ Btu/lb
$\dot{M} = 125,000$ lb/hr　　$\eta_{pump} = 60\%$

(a)　　$V_1 \to V_2;\ Z_1 \to Z_2$

$$TH = \frac{p_1 - p_2}{\gamma};\quad \gamma = \frac{1}{v_1}\frac{g}{g_c} = \frac{32.2}{0.017 \times 32.2} = 58.82\ \text{lbf/ft}^3$$

$$TH = \frac{144(1200 - 29.8)}{58.82} = \underline{2864.8\ \text{ft}}$$

(b)　　$$WHP = \frac{\dot{M}(TH)g}{550\ g_c} = \frac{125,000 \times 2864.8 \times 32.2}{3600 \times 550 \times 32.2} = \underline{180.9\ \text{hp}}$$

(c)　　$$BHP = \frac{WHP}{\eta_{pump}} = \frac{180.9}{0.60} = \underline{301.5\ \text{hp}}$$

(d) Interpolating in Table 4 of the Steam Tables with:

$s_1 = s_2 = 0.36772$　and with　$p_2 = 1200$ psia

we get,　　$h_2 = 222.30$

$$\frac{wk_{12}}{J} = h_1 - h_2 = 218.59 - 222.30 = \underline{(-)3.71\ \text{Btu/lb}}$$

(e)　　$$\frac{wk_{12(actual)}}{J} = \frac{wk_{12(isentropic)}}{J \times \eta_{pump}} = \frac{(-)3.71}{0.60} = \underline{(-)6.18\ \text{Btu/lb}}$$

(f)　　$$h_2 = h_1 - \frac{wk_{12}}{J} = 218.59 + 6.18 = 224.77\ \text{Btu/lb}$$

Entering Table 4 and interpolating between 1000 and 1500 psia gives for 1200 psia:

t	h
250°	221.03
300°	271.83

$$\Delta t_2 = \frac{224.77 - 221.03}{271.83 - 221.03} \times 50° = 3.68°$$

$$t_2 = 250 + 3.7 = \underline{253.7°F}$$

5-7. Compressors. Compressors perform much the same function for gases that pumps do for liquids. There are many different types of compressors such as reciprocating, axial flow and centrifugal. Fans and blowers are also considered compressors, although any compression accomplished is incidental to their primary purpose, which is to produce a large volume rate of flow. In applying the flow energy equation, it is reasonable to assume that the inlet and discharge kinetic energies are substantially the same. The heat term cannot be discarded without investigation of the particular compressor under consideration, since many machines are provided with means for cooling the working substance. Then,

$$\frac{V_1^2}{2g_c(J)} + h_1 + q_{12} = \frac{V_2^2}{2g_c(J)} + h_2 + \frac{wk_{12}}{(J)}$$

Assuming
$$\frac{V_2^2}{2g_c(J)} - \frac{V_1^2}{2g_c(J)} \to 0$$

And
$$\frac{wk_{12}}{(J)} = q_{12} + h_1 - h_2 \quad \text{Btu/lb (J/kg)} \quad (5\text{-}16)$$

In applying this equation, Stations 1 and 2 should be selected in the inlet and discharge sections. Application of the equation to other sections is not feasible because of the complicated internal flow patterns, discontinuities and high local velocities frequently encountered.

Example 5-6: A water cooled reciprocating air compressor takes in air at 15 psia and 60°F and discharges it at 60 psia and 200°F. Heat is removed in the amount of 21.4 Btu/lb. Assume steady flow conditions and find the work done, Btu/lb.

ELEMENTARY STEADY FLOW PROCESSES

Solution:

$$p_1 = 15 \text{ psia}$$
$$T_1 = 460 + 60 = 520°R$$
$$p_2 = 60 \text{ psia}$$
$$T_2 = 460 + 200 = 660°R$$
$$q_{12} = (-)21.4 \text{ Btu/lb}$$
$$\frac{wk_{12}}{J} = q_{12} + h_1 - h_2$$
$$h_1 - h_2 = c_p(T_1 - T_2) = 0.24(520 - 660) = -33.6$$
$$\frac{wk_{12}}{J} = -21.4 - 33.6$$
$$= \underline{(-)55.0 \text{ Btu/lb (on)}}$$

5-8. Throttles. Throttling is a very frequently encountered irreversible process and is characterized by a pressure drop without the development of any mechanical work or other useful form of energy. It is produced simply by a restriction in the flow channel. The orifice and globe valve are common throttling devices. From an energy standpoint, throttling is always accompanied by a reduction in the useful work which the working substance is capable of performing. A typical throttle is shown schematically in Fig. 5-6. As the fluid enters the reduced area offered by the throttle, a sudden increase in velocity will result, with an attendant pressure drop. On the downstream side of the throttle a larger area is immediately presented, and the flow becomes highly confused. The resultant friction and turbulence make for a high degree of irreversibility. The flow eventually smooths out and takes on steady characteristics at some downstream section such as 2, considerably removed from the throttle. The location of Section 2 in relation to the throttle will depend upon the physical dimensions and configuration of the throttle and the exit section. Section 1 is similarly chosen far enough upstream to eliminate

FIGURE 5-6

the possibility of any appreciable flow disturbance at this point. With these provisions, the immediate evidence of the presence of a throttle will be provided by a measurable pressure drop between Sections 1 and 2. In the absence of moving parts, no mechanical work can be performed. Because of the time element, the heat transferred will always be negligible. Since a change in specific volume can be expected between the two sections, there will be a change in velocity even if the two areas are exactly the same. In all practical cases, however, the magnitude of the kinetic energy terms is such that their difference can be neglected. Applying these conditions to the flow energy equation:

$$\frac{V_1^2}{2g_c(J)} + h_1 + q_{12} = \frac{V_2^2}{2g_c(J)} + h_2 + \frac{wk_{12}}{(J)}$$

$$\frac{V_2^2}{2g_c(J)} - \frac{V_1^2}{2g_c(J)} \to 0; \quad q_{12} \to 0; \quad \frac{wk_{12}}{(J)} = 0$$

and $\qquad h_1 = h_2$ **Btu/lb** (J/kg) **(5-17)**

This equation shows that the throttling process is one of **equal enthalpy** for the sections as chosen. For any intermediate sections, this is not true, since the process is in no sense one of constant enthalpy. Immediately after the throttle, for instance, the local velocity will be large, and the associated kinetic energy is attained at the expense of an enthalpy drop as referred to the upstream conditions. In proceeding to Section 2, this velocity is dissipated, and the original enthalpy re-established. In view of the final equality of enthalpies, it is interesting to consider the change, if any, which occurs in the component properties. Rewriting Equation (5-17) gives:

$$u_1 + \frac{p_1 v_1}{(J)} = u_2 + \frac{p_2 v_2}{(J)}$$

From the physical nature of the throttling process, it is already known that p_2 is always less than p_1. Furthermore, it can be shown that v_2 will always be greater than v_1. The internal energy may or may not change, depending upon the substance considered. At any rate the equality of enthalpies is accomplished by compensatory changes in the values of the included properties.

In the interests of maximum useful energy conversion, it is desirable to keep throttling to a minimum. However, throttling is a common method of machinery control, and the losses are accepted in the interests of simplified equipment. Obvious examples are the throttle valves or throttles installed on various types of engines. As will be brought out shortly, the throttling calorimeter is another device that utilizes the throttling principle.

By using the throttling process as an example, some interesting side-

lights on the graphical representation of irreversible processes can be developed. So far, all of the processes shown on the several coordinate systems have been considered reversible. Figure 5-7 shows a throttling process on *T-s* coordinates. A gas has been chosen as the working substance, but the succeeding discussion is equally applicable to vapors. Points 1 and 2 are established on the basis of known conditions, including equality of enthalpies. However, the intermediate nature of the process is unknown and a true process line cannot be drawn. To indicate their indeterminate nature, *irreversible process lines are conventionally drawn dotted* between the two end points as shown. Inspection of the figure reveals an apparent paradox. The area under the dotted line is finite and would represent transferred heat for a reversible process, but the throttling process is essentially adiabatic. Thus the area under an irreversible process line on *T-s* coordinates does not represent heat. Similar reasoning can be applied to the area under an irreversible non-flow process line on *p-v* coordinates in reference to the representation of mechanical work. This leads to the important conclusion that *the area under an irreversible*

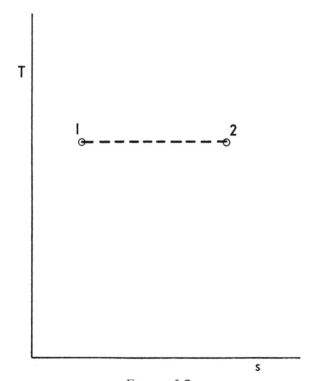

FIGURE 5-7

ELEMENTARY STEADY FLOW PROCESSES

process line on either coordinate system has no significance. However, the diagrams are still useful in showing the initial and final state points. The *h-s* diagram loses nothing, since the area never has any significance, but irreversible processes should be shown dotted on this coordinate system also.

5-9. The Throttling Calorimeter. Whenever wet steam undergoes a throttling process from an initial state 1 to a final state 2, several characteristic property changes occur. These changes are best understood by referring to the Mollier Chart, a portion of which is shown in Fig. 5-8. As already demonstrated, for any substance undergoing throttling:

$$h_1 = h_2 \quad \text{and} \quad p_2 < p_1$$

In the case of steam, it can be seen from the figure that throttling is accompanied by a decrease in moisture content and eventual superheating if the final pressure is sufficiently low. Furthermore a temperature decrease occurs between the initial and final states.

As previously established in the discussion of steam properties, temperature and pressure alone do not define the state of the wet vapor. One other property must be known to fix the state. On the other hand, pressure and temperature alone fully establish the vapor state in the superheat region. Therefore, if a sample of wet steam is throttled to a low pressure region so that it becomes superheated, the final state is established by a measurement of pressure and temperature and the enthalpy can be determined from the Steam Tables. But the throttling process is one of equal enthalpy, so the enthalpy of the wet steam before throttling is also established. This enthalpy together with the original temperature or pressure fully defines the original state.

FIGURE 5-8

All throttling calorimeters operate on the principles outlined above and are essentially instruments for indirectly measuring enthalpy. One type is the Ellison U-path calorimeter shown in Fig. 5-9. Steam at state 1 enters the calorimeter through the sampling pipe, is throttled through an orifice, enters the large cross section, reverses direction, passes around the bulb of the calorimeter thermometer and then is exhausted to atmosphere through a jacketed passage. The pressure in the vicinity of the thermometer, where state 2 is established, may be considered atmospheric without appreciable error. In operation, t_1 or p_1 or both are determined from instruments fitted in the steam line, and for wet steam these readings should be coincident saturation values within instrument accuracy. The temperature, t_2, is determined from the calorimeter thermometer and p_2 from a barometer. It should be noted that this calorimeter is useless if the enthalpy of the sample is less than about 1150 Btu/lb, since atmospheric pressure will be reached within the wet region and no superheating can occur. Such a condition will be reflected by a reading of about 212°F on the calorimeter thermometer, and the use of another type of calorimeter is indicated.

The various properties of the wet steam sample can be computed from the Steam Tables or determined approximately from the Mollier Chart by

FIGURE 5-9. Ellison U-path throttling calorimeter.

ELEMENTARY STEADY FLOW PROCESSES

finding state 2 in the superheat region and proceeding along an equal enthalpy line until the original pressure line is reached.

Example 5-7: A sample of steam is removed from a steam line where the pressure is 215 psia and passed through an Ellison throttling calorimeter. The calorimeter thermometer reads 250°F and barometer is standard. Find for the line steam:
(a) the enthalpy, Btu/lb
(b) the quality
(c) the entropy
(d) the temperature to the nearest whole degree

Solution:

$$p_1 = 215 \text{ psia} \qquad p_2 = 14.696 \text{ psia}$$
$$t_2 = 250°F$$

(a) From Table 3
$$h_2 = 1168.8$$
$$h_1 = h_2 = \underline{1168.8 \text{ Btu/lb}}$$

(b) From Table 2 for 215 psia

$$h_g = 1200.3 \qquad h_{fg} = 838.1$$

$$h_1 = h_g - m_1 h_{fg}; \qquad m_1 = \frac{h_g - h_1}{h_{fg}}$$

$$m_1 = \frac{1200.3 - 1168.8}{838.1} = \frac{31.5}{838.1} = 0.0376$$

$$x_1 = 1 - m_1 = 1 - 0.0376 = 0.9624 = \underline{96.24\%}$$

(c) $\qquad s_1 = s_g - m_1 s_{fg} = 1.5403 - 0.0376 \times 0.9887 = \underline{1.5031 \text{ Btu/lb} \cdot °R}$

(d) $\qquad\qquad\qquad\qquad t_1 = 387.97 = \underline{388°F}$

As an exercise the student should verify the above solution by use of the Mollier chart.

5-10. The Two-Fluid Heat Balance. In the usual heat exchanger, two fluids are involved, with one absorbing the heat rejected by the other. In a boiler, for instance, the combustion gases give up heat to the water and steam contained internally by the boiler. In a marine condenser, the condensing steam transfers heat to the cooling salt water. Similar exchanges of energy in the form of heat occur in such devices as lubricating oil coolers, fuel oil heaters, nonmixing feed water heaters, air heaters, economizers and gas regenerators. Disregarding losses to the surroundings, the heat given up by one fluid during its passage through the device will be numerically equal to the heat absorbed by the second fluid. Thus it should be possible to equate the flow energy equations for the two fluids through their heat terms. In the general case, however, the factor of time must be introduced, since the flow rates of the two fluids are very seldom equal in actual devices. Thus the *rate* of heat transfer is common to both fluids, but the heat rejected by each unit mass of the hotter fluid may be

considerably different from the heat added to each unit mass of the colder fluid. In equation form: (Refer to fig. of Example 5-8 at end of this art.)

$$(-)\dot{Q}_{12} = \dot{Q}_{34}$$

or

$$(-)\dot{M}_A q_{12} = \dot{M}_B q_{34} \text{ Btu/sec (W)} \tag{5-18}$$

where, \dot{M}_A and \dot{M}_B are the mass rates of flow for fluids A and B respectively, lb/sec (kg/sec); subscripts 1 and 2, respectively, refer to the points where fluid A enters and leaves the device; subscripts 3 and 4, respectively, refer to the points where fluid B enters and leaves the device; q_{12} is the heat removed from fluid A, Btu/lb (J/kg) (assuming that A is the hot fluid); q_{34} is the heat added to fluid B, Btu/lb (J/kg) (assuming that B is the cooler fluid). Note that q_{12} and q_{34} will be numerically equal only when \dot{M}_A equals \dot{M}_B. The minus sign in the equation is necessary for algebraic exactness, since q_{12} and q_{34} are of opposite sign in accordance with our previously established convention.

As indicated earlier in this discussion, Equation (5-18) has wide application to heat exchangers. In such devices no mechanical work is performed and the kinetic energy change between entrance and exit is usually so small that it can be safely neglected. The flow energy equations for the two fluid circuits then become:

$$q_{12} = h_2 - h_1 \text{ Btu/lb (J/kg)} \tag{5-5}$$
$$q_{34} = h_4 - h_3$$

and
$$\dot{M}_A q_{12} = \dot{M}_A (h_2 - h_1) \text{ Btu/sec (W)}$$
$$\dot{M}_B q_{34} = \dot{M}_B (h_4 - h_3)$$

but
$$\dot{M}_A q_{12} = (-)\dot{M}_B q_{34}$$

so
$$\dot{M}_A (h_2 - h_1) = (-)\dot{M}_B (h_4 - h_3)$$

or
$$\dot{M}_A (h_1 - h_2) = \dot{M}_B (h_4 - h_3) \text{ Btu/sec (W)}$$

For many of the liquids used in heat exchangers, no tables giving enthalpy values directly are available. It is therefore necessary to make the evaluation through the use of specific heat. For example if fluid B is a liquid from 3 to 4, we can write:

$$q_{34} = h_4 - h_3 = c_B(t_4 - t_3)$$

The values of specific heat for the commonly encountered liquids are contained in most engineering handbooks. No appreciable errors will be introduced if the proper average value over the temperature range is used.

ELEMENTARY STEADY FLOW PROCESSES

A similar approach will suffice for gases (using c_p), but caution must be exercised to select the proper value for the conditions existing. Where enthalpy for the gas can be determined from publications similar to the Air Tables, this method should be used for maximum accuracy. Where a vapor is involved, the wide variations in specific heat with temperature and pressure make the use of an enthalpy table imperative.

Example 5-8: In a lube oil cooler, oil enters at 140°F and leaves at 100°F, at the rate of 400 lb/min. The cooling medium is sea water which enters at 60°F. The average specific heat of the oil is 0.50 Btu/lb · °F and of the salt water is 0.94 Btu/lb · °F. If the flow of the sea water is at the rate of 500 lb/min, find the overboard discharge temperature.

Solution:

$t_1 = 140°F \qquad t_3 = 60°F$
$t_2 = 100°F \qquad \dot{M}_B = 500 \text{ lb/min}$
$\dot{M}_A = 400 \text{ lb/min} \qquad c_B = 0.94$
$c_A = 0.50 \text{ Btu/lb} \cdot °F$

$\dot{M}_A(h_1 - h_2) = \dot{M}_B(h_4 - h_3) \text{ Btu/min}$

$\dot{M}_A c_A(t_1 - t_2) = \dot{M}_B c_B(t_4 - t_3) \text{ Btu/min}$

$$(t_4 - t_3) = \frac{\dot{M}_A c_A(t_1 - t_2)}{\dot{M}_B c_B} = \frac{400 \times 0.50(140 - 100)}{500 \times 0.94} = 17.0$$

$(t_4 - 60) = 17.0; \qquad t_4 = \underline{77.0°F}$

Example 5-9: Steam enters the condenser of a marine propulsion plant at 0.5 psia and a quality of 89% at the rate of 100,000 lb/hr and with a velocity of 1000 ft/sec. It leaves the condenser hotwell as saturated liquid without any change in pressure but at a velocity of 10 ft/sec. The salt water inlet (injection) temperature is 70°F and the discharge (overboard) temperature is 85°F. Sea water has a specific heat of 0.94 Btu/lb · °F and a density of 64 lb/ft³. The injection and overboard velocities are substantially equal. Calculate:

(a) the rate at which energy is extracted from the condensing steam as heat, Btu/min
(b) the flow of sea water required, gallons per minute (gpm)

Solution:

$p_1 = p_2 = 0.5 \text{ psia} \qquad t_3 = 70°F$
$m_1 = 1 - 0.89 = 0.11 \qquad t_4 = 85°F$
$V_1 = 1000 \text{ ft/sec} \qquad c_B = 0.94 \text{ Btu/lb} \cdot °F$
$V_2 = 10 \text{ ft/sec} \qquad \rho_B = 64 \text{ lb/ft}^3$
$\dot{M}_A = 100{,}000 \text{ lb/hr} \qquad 1 \text{ ft}^3 = 7.481 \text{ gal}$

ELEMENTARY STEADY FLOW PROCESSES

(a) $\quad q_{12} = h_2 - h_1 + \dfrac{V_2^2 - V_1^2}{2g_c J}$ (K.E. significant by inspection)

$h_1 = 1096.2 - 0.11 \times 1048.6 = 980.9$ Btu/lb
$h_2 = 47.7$

$q_{12} = 47.7 - 980.9 + \dfrac{(10)^2 - (1000)^2}{2 \times 32.2 \times 778} = (-)953.2$ Btu/lb

$\dot{M}_A q_{12} = \dfrac{100{,}000 \times (-)953.2}{60} = \underline{(-)1{,}588{,}700 \text{ Btu/min (abstracted)}}$

(b) $\quad (-)\dot{Q}_{12} = \dot{Q}_{34} = \dot{M}_B \times c_B(t_4 - t_3)$

$\dot{M}_B = \dfrac{(-)\dot{Q}_{34}}{c_B(t_4 - t_3)} = \dfrac{1{,}588{,}700}{0.94(85 - 70)} = 112{,}670 \text{ lb/min}$

Flow, gpm $= 7.481 \dfrac{\text{gal}}{\text{ft}^3} \times \dfrac{\dot{M}_B}{\rho} = \dfrac{7.481 \times 112{,}670}{64} = \underline{13{,}170 \text{ gpm}}$

5-11. The Multi-Fluid Energy Balance. The two-fluid stream heat exchangers discussed in Art. 5-10 are relatively simple to analyze, since there are two separate and distinct fluid circuits, and the fluids do not physically mix. In other heat exhange devices there is a *mixing* of the fluid streams. Furthermore, it is not uncommon to have more than two inlet streams and one or more outlet streams. This complicates the problem, but the analysis of such systems is similar in principle to that presented in Art. 5-10 for the two-fluid separate systems. The deaerating feed tank (DFT), basically a feed heater which also serves to remove unwanted air from the feed water before it is pumped to the boiler(s), is a good example of a multiple stream, mixing heat exchanger. Hence it has been chosen for exemplary analysis.

ELEMENTARY STEADY FLOW PROCESSES

FIGURE 5-10. Deaerating feed heater (mixing heat exchanger).

Figure 5-10 shows a simple deaerating feedwater heater (DFT) with three inflowing and two outflowing streams. Under conditions of steady flow as specified in Art. 5-1, the thermodynamic properties and the fluid velocities at any selected cross section of the system must be constant with time, the total mass of working substance in the system at any instant must be constant, and the net rate of heat transfer for the system must be constant. Taking the system boundary as the dashed line of Fig. 5-10, Equation (1-20) may be applied as:

$$\Sigma \text{ Energy } rate \text{ in} = \Sigma \text{ Energy } rate \text{ out}$$

$$\Sigma \dot{M}_{in} \left[\frac{gZ}{g_c(J)} + \frac{V^2}{2g_c(J)} + \frac{pv}{(J)} + u \right]_{in} + \dot{Q}$$

$$= \Sigma \dot{M}_{out} \left[\frac{gZ}{g_c(J)} + \frac{V^2}{2g_c(J)} + \frac{pv}{(J)} + u \right]_{out}$$

$$+ \frac{\dot{W}k}{(J)} \text{ Btu/hr (W)} \qquad (5\text{-}19)$$

where \dot{M}, \dot{Q} and $\dfrac{\dot{W}k}{(J)}$ are mass, heat and work *rates*, respectively. For a DFT it is reasonable to disregard heat transfer to the surroundings, no mechanical work is done, and changes in potential and kinetic energies may be considered negligible with respect to other energy changes; Equation (5-19) then may be written:

ELEMENTARY STEADY FLOW PROCESSES

$$\Sigma(\dot{M}h)_{in} = \Sigma(\dot{M}h)_{out} \text{ Btu/hr (W)} \quad (5\text{-}20)$$

To ensure that the mass of fluid in the system is constant with time, in other words that there be no accumulation or diminution of substance in the system with time, the following must also be true:

$$\Sigma \dot{M}_{in} = \Sigma \dot{M}_{out} \text{ lb/hr (kg/sec)} \quad (5\text{-}21)$$

Application of Equations (5-20) and (5-21) to a multi-fluid heat transfer system is demonstrated in Example problem 5-10. In order to determine the enthalpy of each of the streams, knowledge of the properties at entrance or exit, as the case may be, is necessary. The condensate and the fresh water drains going into the DFT are compressed liquid. The pressures obtaining will be somewhat above that in the DFT shell but well below 100 psia. Hence, as pointed out in Art. 4-7, the enthalpy of each can be determined from the observed temperatures at entrance. The effluent from the vent is saturated vapor at the shell pressure. The small amount of entrained air can be neglected for the purposes of the energy balance. The vent vapor efflux results from "boil-off" from the feed water. It follows that the feed water is saturated liquid at the DFT shell pressure. The exhaust steam supplied may be wet, saturated or slightly superheated, dependent upon the operating conditions.

Example 5-10: A deaerating feedwater heater similar to Fig. 5-10 operates at 30 psia shell pressure with the following flow rates and properties:

No.	Item	Flow rate (lb/hr)	Temp. (°F)	Enthalpy (Btu/lb)
1	Condensate (liquid)	100,000	80.0	48.1
2	Vent (vapor)	1,000	250.34	1164.3
3	Drains (liquid)	5,000	100.0	68.1
4	Feed (liquid)	?	250.34	218.9
5	Exhaust steam	?	280.0	1178.7

Determine the exhaust steam flow rate required to operate the heater under these conditions. What quantity of boiler feed will be available from the heater (lb/hr)?

Solution: Refer to Fig. 5-10:

$$\Sigma(\dot{M}h)_{in} = \Sigma(\dot{M}h)_{out} \quad \text{(energy balance)}$$
$$\dot{M}_1 h_1 + \dot{M}_3 h_3 + \dot{M}_5 h_5 = \dot{M}_2 h_2 + \dot{M}_4 h_4 \text{ (Btu/hr)}$$

$$\Sigma(\dot{M})_{in} = \Sigma(\dot{M})_{out} \quad \text{(mass balance)}$$
$$\dot{M}_1 + \dot{M}_3 + \dot{M}_5 = \dot{M}_2 + \dot{M}_4 \text{ (lb/hr)}$$

Since \dot{M}_4 is a function of \dot{M}_5, a direct solution is achieved by solving the mass balance equation explicitly for \dot{M}_4 and then substituting that relationship into the energy balance equation. Because the energy balance equation then has \dot{M}_5 as the only unknown, its magnitude may be determined.

$$\dot{M}_4 = \dot{M}_1 + \dot{M}_3 + \dot{M}_5 - \dot{M}_2$$

and

$$\dot{M}_1 h_1 + \dot{M}_3 h_3 + \dot{M}_5 h_5 = \dot{M}_2 h_2 + (\dot{M}_1 + \dot{M}_3 + \dot{M}_5 - \dot{M}_2) h_4$$

$$\dot{M}_5 (h_5 - h_4) = \dot{M}_2 h_2 + (\dot{M}_1 + \dot{M}_3 - \dot{M}_2) h_4 - \dot{M}_1 h_1 - \dot{M}_3 h_3$$

$$\dot{M}_5 = \frac{\dot{M}_2 h_2 + (\dot{M}_1 + \dot{M}_3 - \dot{M}_2) h_4 - \dot{M}_1 h_1 - \dot{M}_3 h_3}{(h_5 - h_4)}$$

$$\dot{M}_5 = \frac{(1000)(1164.3) + (104,000)(218.9) - (100,000)(48.1) - (5000)(68.1)}{(1178.7 - 218.9)}$$

$$\dot{M}_5 = \frac{18,779,400}{959.8} = \underline{19,566 \text{ lb/hr}}$$

Inserting this result into the mass balance equation gives

$$\dot{M}_4 = \dot{M}_1 + \dot{M}_3 + \dot{M}_5 - \dot{M}_2 = 100,000 + 5000 + 19,566 - 1000$$
$$\dot{M}_4 = \underline{123,566 \text{ lb/hr}}$$

Problems

5-1. Which two specifications for steady flow in a thermodynamic system are also essential to definition of the Continuity Equation?

5-2. Throttling is defined as a process of equal initial and final enthalpies. Why would it be misleading to call the process one of constant enthalpy?

5-3. Determination of the state of wet vapor by use of a throttling calorimeter exhausting to the atmosphere is limited to enthalpies greater than approximately 1150 Btu/lb. What is responsible for this limiting feature?

5-4. Steam leaves a boiler at 600 psia and 750°F at the rate of 75,000 lb/hr through the main steam line, which has a cross-sectional area of 0.322 ft^2. Find the velocity of the steam in the line, ft/sec.

Ans. 73.3 ft/sec

5-5. Steam leaves a boiler at 6550 kPa abs and 510°C at the rate of 45,400 kg/hr through the main steam line, which has a cross-sectional area of 0.030 m^2. Determine the velocity of the steam in the line, m/sec. The specific volume of the steam is 0.0525 m^3/kg.

Ans. 22.1 m/sec

5-6. An air compressor takes in 50 ft^3/min of air at 14.7 psia and 60°F. The air is discharged at 100 psia and 260°F. Find (a) the mass rate of flow of the air, lb/min; (b) the volume rate of flow at discharge, ft^3/min.

Ans. (a) 3.82 lb/min; (b) 10.18 ft^3/min

5-7. Steam enters the first stage nozzles of a large turbine with negligible velocity at a pressure of 540 psia and a temperature of 800°F. The pressure at the nozzle exit is 220 psia. If the process is isentropic, find (a) the final enthalpy, Btu/lb; (b) the kinetic energy at exit, Btu/lb; (c) the velocity at exit, ft/sec.

Ans. (a) 1302.2 Btu/lb; (b) 108.1 Btu/lb; (c) 2327 ft/sec

5-8. An auxiliary turbine receives saturated steam at 300 psia with negligible kinetic energy at the rate of 3000 lb/hr. At exhaust the steam pressure is 1.5 psia, and the moisture content is 10%. The inside diameter of the exhaust trunk is 12 inches. Find (a) the velocity of

ELEMENTARY STEADY FLOW PROCESSES

the steam in the exhaust trunk; (b) the enthalpy of the steam in the exhaust trunk; (c) the work done, Btu/lb; (d) the horsepower developed in the turbine.

Ans. (a) 217.4 ft/sec; (b) 1008.9 Btu/lb; (c) 194.1 Btu/lb; (d) 229 hp

5-9. A boiler receives feed water at 1000 psia and 350°F and delivers superheated steam at 900 psia and 840°F. Neglect any kinetic energy changes and find the heat added, Btu/lb.

Ans. 1093.8 Btu/lb

5-10. Saturated steam is supplied to a fuel oil heater at 150 psia and becomes saturated water at the same pressure. Find the heat transferred from the steam, Btu/lb.

Ans. (−)864.2 Btu/lb

5-11. A main feed pump receives water from a booster pump at 230°F and 55 psia and delivers the water at 1000 psia. Assume the process to be isentropic and find the work done in Btu/lb from (a) Table 4 of the Steam Tables and (b) Equation (5-15).

Ans. (a) (−)3.48 Btu/lb; (b) (−)2.95 Btu/lb

5-12. A condensate pump takes suction from the hotwell of a condenser at an absolute pressure of 6 kPa and discharges the water at 340 kPa absolute. The condensate is at 36°C with a density of 994 kg/m^3. The net change in kinetic energy across the pump is negligible. Calculate the power required to drive the pump, kW, when delivering 1.4 m^3/min of condensate if the pump efficiency is 65%.

Ans. 12.0 kW.

5-13. The total head (*TH*) for a condensate pump handling water at 100°F is 100 ft. Neglecting elevation and velocity changes across the pump, calculate (a) the pressure difference across the pump in psi. If the pump overall efficiency is 60%, determine (b) the power required to drive the pump when pumping one cubic foot per second of condensate.

Ans. (a) 43 lbf/in^2; (b) 18.8 BHP

5-14. Under steady operating conditions a deaerating feed tank maintains its liquid level 18 feet above the centerline of the booster pump, which is pumping 225 gallons per minute of 250°F feed water from the tank through a 6-inch inside diameter pipe. DFT operating pressure is 15 psig, and the friction head loss from DFT to pump suction is 2.0 feet of water. Calculate the pump suction pressure in psig.

Ans. 21.5 psig

5-15. A large water-cooled air compressor takes in air at 15 psia and 50°F and delivers it at 90 psia and 270°F. The compressor delivers 35.6 hp to the air, and heat is rejected by the air at the rate of 452 Btu/min. Find (a) the change of specific enthalpy, Btu/lb; (b) the rate of change of enthalpy, Btu/min; (c) the mass rate of flow, lb/min; (d) the volumetric capacity of the compressor based on the inlet air, ft^3/min.

Ans. (a) 52.8 Btu/lb; (b) 1058 Btu/min; (c) 20 lb/min; (d) 251.7 ft^3/min

5-16 A throttling calorimeter is connected to a steam line, and the pressure and temperature in the line are determined to be 205 psi gage and 390°F **from the connected instruments.** The calorimeter temperature is 240°F. Barometer is standard. Use the Mollier Chart and estimate the enthalpy and quality of the line steam.

Ans. 1164 Btu/lb, 95.6%.

5-17. A deaerating feed tank (DFT) receives 200,000 lb/hr of main and auxiliary condensate at an enthalpy of 110 Btu/lb, 8000 lb/hr of auxiliary exhaust steam at

ELEMENTARY STEADY FLOW PROCESSES

1200 Btu/lb, 2000 lb/hr of low pressure drains (liquid) at 150°F and an undetermined amount of augmenting steam at 1290 Btu/lb. The DFT operates at 13.0 psig and has negligible steam vent flow. Determine the quantity of augmenting steam required, if any, to balance the system, lb/hr.

Ans. 12,180 lb/hr

5-18. The specifications for the fuel oil heater described in problem 5-10 require it to heat 2500 lb/hr of heavy fuel oil from 70°F to 210°F with the steam conditions shown. If the specific heat of the fuel oil is 0.51 Btu/lb · °F, how much steam will be used, lb/hr?

Ans. 206.5 lb/hr.

5-19. Water vapor at 2 psia with a moisture content of 0.15 enters the distilling condenser of a marine evaporator plant at the rate of 1500 lb/hr and leaves at the same pressure as saturated liquid. The coolant is sea water with a specific heat of 0.94 Btu/lb · °F. Operation requires that the temperature of the sea water leaving be limited to a maximum of 100°F. To meet this limit, how much sea water must be pumped through the condenser, lb/min, when the injection temperature is: (a) 60°F; (b) 70°F?

Ans. (a) 578 lb/min; (b) 770 lb/min

5-20. A deaerating feed tank operates at 15 psig with 32.0 psia saturated steam in the auxiliary exhaust line and saturated vapor at DFT pressure at the vent. Total condensate flow entering is 100,000 lb/hr at 110 Btu/lb, and total low pressure drain flow (liquid) is 3000 lb/hr at 180 Btu/lb. There is no evaporator drain flow. Vapor loss through the vent is negligible. Sketch and label the system, and calculate the required flow of auxiliary exhaust steam, lb/hr.

Ans. 11,600 lb/hr

5-21. A marine main propulsion plant delivers 30,000 shaft horsepower at full power. Under these conditions, the deaerating feed tank operates with a shell pressure of 33 psia. The pressure in the auxiliary exhaust steam line is 36 psia. The flow rates and observed temperatures are as shown in the table below. Verify the enthalpy column, and for balanced conditions find (a) the required flow rate of auxiliary exhaust steam and (b) the feed water available, both in lb/hr.

Item	Flow rate (lb/hr)	Temp. (°F)	Enthalpy (Btu/lb)
Condensate	210,000	106	74.0
Vent	200		1166.2
Drains	18,000	210	178.1
Exhaust Steam	?	370	1222.4
Feed Water	?		224.5

Ans. (a) 32,200 lb/hr; (b) 260,000 lb/hr

CHAPTER 6

AIR COMPRESSION

6-1. Background. Compressed air is widely used as the working substance for pneumatic tools such as riveters, hammers and drills, as well as the actuating fluid for a large group of industrial control systems. In addition, compressed air is required in several combustion systems, including the gas turbine. These applications presuppose the availability of an adequate quantity of air at a specific working pressure, and various types of compressors are in general use to this end.

Air compressors are generally classified as either positive displacement machinery or as turbomachinery. Positive displacement machines operate by reducing the volume of a confined amount of air with a corresponding increase in pressure. An example of a positive displacement device is the reciprocating air compressor. The piston cylinder configuration and idealized gas processes developed in Chapter 3 are the basis of analysis for actual reciprocating compressors.

Air compressors categorized as turbomachinery impart dynamic forces to a moving air stream by the rotation of compressive components. The axial and centrifugal flow compressors are basically steady flow machines. As such, the steady flow concepts developed in Chapter 5 are employed for the analysis of turbocompressors.

6-2. The Reciprocating Compressor. The reciprocating compressor consists of a cylinder and piston, the cylinder having suitable valves which permit the admission and the discharge of the air. The indicator card showing the pressure and total volume relationship of the air within such a cylinder throughout the cycle is illustrated in Fig. 6-1. In this chapter, *letter subscripts will be used to designate conditions within the cylinder and numerical subscripts to designate conditions adjacent but external to the cylinder.* When the suction valve opens (point A) air is taken into the cylinder from the atmosphere and continues to enter throughout the remainder of the suction stroke (point B). At this point in the cycle the suction, or inlet, valve closes, the piston direction reverses and the air within the cylinder is compressed as indicated by the process line BC. At C the discharge valve opens, permitting the discharge of the high pressure air from the cylinder to some external tank or receiver, and the discharge process continues for the remainder of the piston stroke to point D. Here the discharge valve closes, the piston again reverses its direction of travel,

AIR COMPRESSION

FIGURE 6-1. Indicator diagram for actual reciprocating compressor.

FIGURE 6-1a. Mechanical engine indicator and reducing mechanism.

AIR COMPRESSION

and the air that remains in the cylinder re-expands to point A for the completion of the cycle. At the end of the compression process the cylinder pressure (point C) is somewhat higher than the receiver pressure (p_2), this difference in pressure being required to overcome the valve inertia and the friction losses that occur as the air flows through the valve. This phenomenon, also observed during the suction process, is inherent in all reciprocating engines and compressors and is known as *valve throttling* or *wire drawing*. Diagrams similar to Fig. 6-1 are obtained from a mechanical instrument known as an *engine indicator*, which is attached to the cylinder as shown in Fig. 6-1a. The actual process lines are traced by the pencil upon a paper wrapped on the oscillating drum, and the completed tracing is known as an *indicator card*. The vertical displacement of the pencil is caused by the cylinder pressure, and the drum positions follow the motion of the piston, showing relative cylinder volumes. Equivalent diagrams also can be obtained electronically using pressure transducer techniques.

6-3. Ideal Air Compression, Single Stage. For the initial analysis of air compressors, a basic cycle is established. Such a cycle assumes a perfect compressor having no clearance, operating without friction or valve throttling with a thermodynamically reversible compression process. This ideal cycle is illustrated in Fig. 6-2. Beginning at point A with zero cylinder volume, atmospheric air enters the cylinder during the suction stroke, A-B,

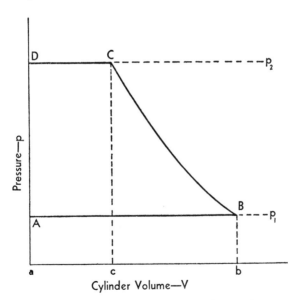

FIGURE 6-2. Indicator diagram of ideal reciprocating compressor.

AIR COMPRESSION

at constant pressure, temperature and specific volume. Actual compression begins at B and continues along the polytropic, B-C. At C the cylinder pressure becomes equal to the receiver pressure, and the discharge valve opens. The remainder of the stroke, C-D, is required for the expulsion of the air from the cylinder, again at constant pressure, temperature and specific volume, until at D the cylinder is completely empty and ready for a new cycle.

For simplicity of analysis the work of the ideal compressor may be divided into three parts: (a) suction work, (b) compression work and (c) discharge work. Adhering to the convention that *positive work is work done by the fluid* and that *negative work is work done on the fluid*, such an analysis is made as follows:

(a) *Suction work*. During the suction stroke air flows into the cylinder and exerts a pressure ($p_A = p_B$) upon the piston. The resulting force upon the piston face is $p_A A$, and the distance this force travels is the length of the suction stroke, L. The work done by the air upon the piston is therefore $p_A AL$. As shown in Chapter 2, the term AL is equivalent to $V_B - V_A$. The suction work is, therefore:

$$Wk_s = p_B V_B - p_A V_A$$
But $\quad V_A = 0$
Hence $\quad Wk_s = p_B V_B \text{ ft} \cdot \text{lbf/cycle (J/cycle)}$

(b) *Compression work*. At the end of the suction stroke the inlet valve closes, entrapping a certain mass of air within the cylinder. As the piston travel is reversed, this mass of air is compressed polytropically from atmospheric to receiver pressure along the process line B-C. Since this process is non-flow, the work equation developed in Chapter 3 can be employed directly, and the work of compression is:

$$Wk_c = \frac{p_c V_c - p_B V_B}{1 - n} \text{ ft} \cdot \text{lbf/cycle (J/cycle)}$$

(c) *Discharge work*. Discharge of the compressed air begins when the cylinder pressure becomes equal to the receiver pressure and continues for the remainder of the stroke. Throughout this part of the stroke $p_C = p_D$, and the air resists expulsion and produces a force $p_C A$ upon the piston. Since suction and discharge are similar in nature, the discharge work is:

$$Wk_d = p_D V_D - p_C V_C$$
However, $\quad V_D = 0$
and $\quad Wk_d = - p_C V_C \text{ ft} \cdot \text{lbf/cycle (J/cycle)}$

The net work required for the ideal cycle is the sum of these individual

AIR COMPRESSION

quantities. Thus, for the general case:

$$Wk(\text{net}) = Wk_s + Wk_c + Wk_d$$

$$= p_B V_B + \frac{p_C V_C - p_B V_B}{1 - n} - p_C V_C \text{ ft} \cdot \text{lbf/cycle (J/cycle)}$$

It should be noted that the terms $p_B V_B$ and $p_C V_C$ represent the areas $ABba$ and $CDac$ respectively, while the term

$$\frac{p_C V_C - p_B V_B}{1 - n}$$

is represented by the area under the polytropic process line, $bBCc$. The above expression for work therefor evaluates the area $ABCD$, and Fig. 6-2 might be considered to be an indicator diagram for an ideal reciprocating compressor. It should also be noted that there are no changes of properties of the air during suction and discharge. The difference between the non-flow polytropic compression process of Chapter 3 and the ideal compressor cycle is simply the change of displacement energy occasioned by the flow of air into and out of the cylinder. During suction,

$$p_A = p_B = p_1, \quad T_A = T_B = T_1, \quad v_A = v_B = v_1$$

and during discharge,

$$p_C = p_D = p_2, \quad T_C = T_D = T_2, \quad v_C = v_D = v_2$$

Therefore, $\quad p_B V_B = M p_B v_B = M p_1 v_1$
and $\quad p_C V_C = M p_C v_C = M p_2 v_2$

The net work of the ideal cycle is:

$$Wk(\text{net}) = M p_1 v_1 + \frac{M p_2 v_2 - M p_1 v_1}{1 - n} - M p_2 v_2$$

and it becomes apparent that in the ideal case the work of a compressor cycle may be calculated on the basis of the change of displacement energy established by the external operating conditions and the polytropic process by which this change is effected.

The work equation can be simplified by stating all terms on the basis of the common denominator, $1 - n$, and reducing. Thus,

$$Wk(\text{net}) = \frac{(1 - n)(M p_1 v_1) + (M p_2 v_2 - M p_1 v_1) - (1 - n)(M p_2 v_2)}{1 - n}$$

$$Wk(\text{net}) = \frac{Mn}{1 - n}(p_2 v_2 - p_1 v_1) \text{ ft} \cdot \text{lbf/cycle (J/cycle)}$$

AIR COMPRESSION

in which, M = mass of air entering and leaving the cylinder, lb/cycle (kg/cycle).

In terms of temperatures,

$$Wk(\text{net}) = \frac{nMR}{1-n}(T_2 - T_1) \text{ ft} \cdot \text{lbf/cycle (J/cycle)}$$

in which, T_1 and T_2 = absolute temperatures at admission to and exhaust from cylinder respectively, in °R (K).

For unit mass of air,

$$wk(\text{net}) = \frac{n}{1-n}(p_2v_2 - p_1v_1) \text{ ft} \cdot \text{lbf/lb (J/kg)} \tag{6-1}$$

and

$$\mathbf{\mathit{wk}(net)} = \frac{\mathbf{\mathit{nR}}}{\mathbf{1-\mathit{n}}}(\mathbf{\mathit{T}_2 - \mathit{T}_1}) \text{ ft} \cdot \textbf{lbf/lb (J/kg)} \tag{\underline{6-2}}$$

For the general case of the ideal cycle the terminal temperatures of the compression process are readily available, and Equation (6-2) is usually the more convenient to use.

Example 6-1: Determine the final temperature and the net work required per pound of air by an ideal compressor in which the air is compressed from an initial pressure of 15 psia to a final pressure of 75 psia. The initial temperature is 60°F, and the value of n during compression is assumed to be 1.30.

Solution: Refer to Fig. 6-2:

$$p_1 = 15 \text{ psia} \quad p_2 = 75 \text{ psia}$$
$$t_1 = 60°F \quad n = 1.30$$
$$T_1 = 60 + 460 = 520°R$$

$$T_2 = T_1\left(\frac{p_2}{p_1}\right)^{\frac{n-1}{n}} = 520\left(\frac{75}{15}\right)^{\frac{1.3-1}{1.3}} = 754°R$$

$$t_2 = 754 - 460 = \underline{294°F}$$

$$wk(\text{net}) = \frac{nR}{1-n}(T_2 - T_1) = \frac{(1.3)(53.3)}{1-1.3}(754-520)$$

$$= \underline{(-)54,000 \text{ ft} \cdot \text{lbf/lb (on)}}$$

Where the final temperature is unknown, or is not required for other considerations, Equation (6-2) may be conveniently expressed in terms of the initial temperature of the air and the pressure ratio, r_p, defined as:

$$r_p = \left(\frac{p_2}{p_1}\right)$$

Then,

$$wk(\text{net}) = \frac{nR}{1-n}(T_2 - T_1) = \frac{nRT_1}{1-n}\left(\frac{T_2 - T_1}{T_1}\right) = \frac{nRT_1}{1-n}\left(\frac{T_2}{T_1} - 1\right)$$

AIR COMPRESSION

But
$$\frac{T_2}{T_1} = \left(\frac{p_2}{p_1}\right)^{\frac{n-1}{n}} = (r_p)^{\frac{n-1}{n}}$$

and
$$wk(\text{net}) = \frac{nRT_1}{1-n}\left[\left(\frac{p_2}{p_1}\right)^{\frac{n-1}{n}} - 1\right]$$

$$= \frac{nRT_1}{1-n}\left[(r_p)^{\frac{n-1}{n}} - 1\right] \text{ ft} \cdot \text{lbf/lb (J/kg)} \quad (6\text{-}3)$$

6-4. Best Compression Process and Desirability of Low Value of *n*. The effect of the compression exponent, n, upon the ideal cycle is illustrated in Fig. 6-3. Between any two pressures the generally desirable compression process is that which will result in the least work for the cycle. The concept

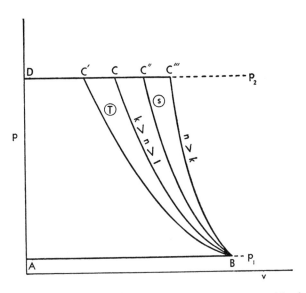

FIGURE 6-3. Effect of compression exponent, n, upon ideal work.

of isothermal compression best fulfills this requirement, and the work cannot be further reduced without maintaining the compression temperature below that of admission, which is not practical. In this respect, the isothermal process constitutes a natural limit for the rejection of heat predicated on perfect cylinder cooling with coolants at or very near the temperature of the inlet air. Additional transfer of heat would require a refrigeration system, the work necessary for such a system being additive to the net compressor work.

AIR COMPRESSION

An expression for minimum work may now be obtained from the general statement

$$Wk(\text{net}) = p_B V_B + \int_{V_B}^{V_{C'}} p\, dV - p_{C'} V_{C'}$$

For the isothermal process,

$$\int_{V_B}^{V_{C'}} p\, dV = p_B V_B \ln\left(\frac{V_{C'}}{V_B}\right)$$

and,

$$p_B V_B = p_{C'} V_{C'}$$

Thence:

$$Wk_{\min} = p_{C'} V_{C'} + p_B V_B \ln\left(\frac{V_{C'}}{V_B}\right) - p_{C'} V_{C'}$$

$$= p_B V_B \ln\left(\frac{V_{C'}}{V_B}\right) = M p_1 v_1 \ln\left(\frac{p_1}{p_2}\right)$$

$$= MRT_1 \ln\left(\frac{p_1}{p_2}\right) \text{ ft} \cdot \text{lbf (J)}$$

and for unit mass

$$wk_{\min} = RT_1 \ln\left(\frac{p_1}{p_2}\right) \text{ ft} \cdot \text{lbf/lb (J/kg)} \quad (6\text{-}4)$$

Example 6-2: Calculate the work required for the isothermal compression of 1 lb of air per cycle from an initial pressure of 15 psia to a final pressure of 75 psia if the initial temperature is 60°F.

Solution: Refer to Fig. 6-3:

$$p_1 = 15 \text{ psia} \qquad p_2 = 75 \text{ psia} \qquad t_1 = 60°F$$

$$wk_{\min} = RT_1 \ln\left(\frac{p_1}{p_2}\right) = (53.3)(60 + 460) \ln\left(\frac{15}{75}\right)$$

$$= (-)44{,}600 \text{ ft} \cdot \text{lbf/lb}$$

Although the isothermal constitutes the ideal compression process, it must be remembered that to accomplish this process heat must be removed at the same rate as the equivalent work is performed. That is, for the isothermal process $Q_{BC'} = [Wk/(J)]_{BC'}$. This condition cannot be fulfilled in practice because sufficient heat cannot be rejected. However, the establishment of an expression for minimum work provides a criterion of performance for all operating compressors.

6-5. Clearance and Effects of Clearance. All operating compressors must have some space remaining between cylinder head and piston at the end of the discharge stroke. This space, including valve ports, is known as

the **clearance volume.** The ratio of this volume to the piston displacement of the unit is, by definition, the **clearance** and is expressed either as a decimal or as a percentage. As illustrated in Fig. 6-4,

$$V_B - V_D = \text{piston displacement}$$
$$V_D = \text{clearance volume}$$

then, by definition,

$$c = \text{clearance} = \frac{V_D}{V_B - V_D}$$

It is also observed from Fig. 6-4 that the discharge process, as compared with Fig. 6-2, is incomplete, since the clearance volume is occupied by air which has been compressed. This amount of air remains within the

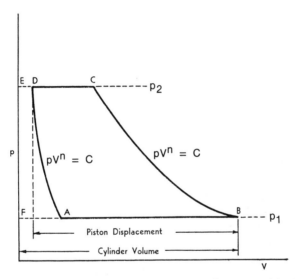

FIGURE 6-4. Ideal reciprocating compressor diagram with clearance.

cylinder throughout the cycle, re-expanding along the path D-A during the first part of the suction stroke, and being compressed with the fresh charge along the compression path B-C. The net result may be analyzed on the basis of *the mass of air discharged* (along C-D). Thus when M is the mass of air discharged, *by definition* equal to $(V_B - V_A)/v_1$, and M' is the mass of air remaining in the clearance volume,

$$M + M' = \text{total mass of air in cylinder at } B$$

The total work required by the cycle *without clearance* is:

AIR COMPRESSION

$$Wk_t = (M + M')\left(\frac{nR}{1-n}\right)(T_C - T_B)$$

which is equivalent to the area *FBCE* of Fig. 6-4. The work recovered during re-expansion is numerically evaluated by the expression,

$$Wk_{exp} = M'\left(\frac{nR}{1-n}\right)(T_D - T_A)$$

the equivalent area being *DAFE*. Thus the work originally performed upon the clearance air is recovered during re-expansion, and the net work required for the cycle (area *ABCD*) is:

$$wk(\text{net}) = (Wk_t - Wk_{exp})/M \text{ ft} \cdot \text{lbf/lb (J/kg)}$$

For the ideal cycle the value of n during re-expansion is assumed to be the same as the value during compression. Further, as in Art. 6-3,

$$T_1 = T_A = T_B, \quad \text{and} \quad T_C = T_D = T_2$$

Thence:

$$wk(\text{net}) = \left(\frac{M+M'}{M}\right)\left(\frac{nR}{1-n}\right)(T_2 - T_1) - \left(\frac{M'}{M}\right)\left(\frac{nR}{1-n}\right)(T_2 - T_1)$$

$$wk(\text{net}) = \left(\frac{nR}{1-n}\right)(T_2 - T_1) \text{ ft} \cdot \text{lbf/lb delivered (J/kg delivered)}$$

Thus, for the ideal compressor, *clearance has no effect upon the work required **per unit mass of air delivered** by the compressor.*

However, further inspection of Fig. 6-4 shows that clearance results in a large reduction in the volume of air entering the cylinder during the suction stroke. The volume of fresh air entering the cylinder is seen to be $V_B - V_A$, and not the equivalent of the piston displacement. This reduction of capacity is conventionally expressed in terms of **volumetric efficiency,** which, by definition, *is the ratio of the volume of air entering the cylinder to the piston displacement of the cylinder.* Thus,

$$\eta_v = \frac{\text{volume entering cylinder}}{\text{piston displacement}} \qquad \underline{(6\text{-}5)}$$

or from Fig. 6-4,

$$\eta_v = \frac{(V_B - V_D) - (V_A - V_D)}{(V_B - V_D)}$$

However,

AIR COMPRESSION

$$V_A = V_D \left(\frac{p_2}{p_1}\right)^{1/n} = V_D(r_p)^{1/n} \quad \text{and} \quad c = \frac{V_D}{V_B - V_D}$$

Then,

$$\eta_v = \frac{(V_B - V_D) - V_D \left(\frac{p_2}{p_1}\right)^{1/n} + V_D}{V_B - V_D}$$

$$= \frac{(V_B - V_D) + V_D \left[1 - \left(\frac{p_2}{p_1}\right)^{1/n}\right]}{(V_B - V_D)}$$

$$= 1 + c\left[1 - \left(\frac{p_2}{p_1}\right)^{1/n}\right] = 1 + c\left[1 - (r_p)^{1/n}\right]$$

$$\eta_v = 1 + c - c\left(\frac{p_2}{p_1}\right)^{1/n} = 1 + c - c(r_p)^{1/n}$$

Because this equation applies only to an ideal compressor, it is usually designated the *apparent volumetric efficiency*. Chiefly because of leakage past valves and piston and some heating of the fresh charge during suction, the actual volumetric efficiency is always less than the apparent volumetric efficiency.

It should be noted that large clearances or high pressure ratios decrease the apparent volumetric efficiency. Thus, for a pressure ratio of 5, with a value of n of 1.30 as in the previous example problem, and a clearance of 2 percent, the apparent volumetric efficiency is 95 percent. If, under the same conditions, the clearance is 10 percent, the apparent volumetric efficiency is 76 percent. Clearance must, therefore, be kept to a minimum. Also, for 2 percent clearance with a value of n of 1.30 but with a pressure ratio of 10, the apparent volumetric efficiency is 90 percent. Thus high pressure ratios result in a decrease in volumetric efficiency, in addition to introducing lubrication difficulties and the possibility of explosions due to the high temperatures involved.

Example 6-3: The piston displacement of the ideal compressor in Example 6-2 is 12.83 cubic feet. If the compressor has 5 percent clearance and discharges 0.7 lb of air per cycle, determine (a) the apparent volumetric efficiency and (b) the actual volumetric efficiency.

Solution: Refer to Fig. 6-4:

$$p_1 = 15 \text{ psia} \quad p_2 = 75 \text{ psia} \quad c = 5\%$$
$$t_1 = 60°F \quad M = 0.7 \text{ lb per cycle}$$

(a) apparent

$$\eta_v = 1 + c - c\left(\frac{p_2}{p_1}\right)^{1/n}$$

$$= 1 + 0.05 - 0.05\left(\frac{75}{15}\right)^{1/1.3} = \underline{87.8\%}$$

volume entering cylinder — V:

$$V = \frac{MRT_1}{p_1} = \frac{(0.7)(53.3)(520)}{144 \times 15} = 8.98 \text{ ft}^3 \text{ per cycle}$$

(b) actual $\quad \eta_v = \dfrac{V}{\text{piston displacement}} = \dfrac{8.98}{12.83} = \underline{70\%}$

6-6. Multi-stage Compression. To reduce lubrication problems and the possibility of explosions, the air temperature leaving a reciprocating compressor stage is limited to about 400°F. This limits the discharge pressure for a single stage unit with adiabatic compression to roughly 75 psig. With cylinder cooling, discharge pressures up to 100 psig are acceptable. Two-stage compressors are used where discharge pressures ranging from 80 psig to 500 psig are required, and additional stages are employed for increasingly higher pressures. The quoted limiting pressures are somewhat flexible, depending upon the effectiveness with which cylinder cooling can be accomplished. The high discharge pressures of multi-stage compressors are made possible by cooling the air between stages in an **intercooler.** Intercooling also reduces the internal work required for compression, as will presently be shown.

The idealized cylinder processes for a two-stage compressor are illus-

FIGURE 6-5. Ideal two-stage reciprocating compressor with clearance.

AIR COMPRESSION

trated in Fig. 6-5. Air enters the low pressure cylinder at pressure p_1 and temperature $T_1 = T_A = T_B$, is polytropically compressed (process B-C) and is discharged at pressure p_i and temperature $T_C = T_D$. The air then enters an intercooler, where heat is removed. In the intercooler the pressure of the air remains constant at p_i, but its temperature is reduced to T_E with an accompanying decrease in specific volume. The second stage of compression is accomplished in the high pressure cylinder with admission from the intercooler at pressure p_i and temperature $T_E = T_F$, polytropic compression along F-G and discharge against pressure p_2 at temperature $T_G = T_H$. Thus the work saved by staging and intercooling (as compared with single stage compression) is represented by the area $CKGFC$.

Since the magnitude of the area $CKGFC$ is dependent upon the interstage pressure, it is necessary to develop an expression for minimum total work in terms of p_i. This is accomplished by equating the first differential of the equation for total work to zero and solving for p_i as follows:

$$wk_t = wk_{LP} + wk_{HP}$$

$$wk_t = \frac{nRT_1}{1-n}\left[\left(\frac{p_i}{p_1}\right)^{\frac{n-1}{n}} - 1\right] + \frac{nRT_E}{1-n}\left[\left(\frac{p_2}{p_i}\right)^{\frac{n-1}{n}} - 1\right]$$

and,

$$\frac{d(wk_t)}{dp_i} = 0 = \frac{nRT_1}{1-n}\left[\frac{n-1}{n}\left(\frac{p_i}{p_1}\right)^{-\frac{1}{n}}\left(\frac{1}{p_1}\right)\right]$$
$$+ \frac{nRT_E}{1-n}\left[\frac{n-1}{n}\left(\frac{p_2}{p_i}\right)^{-\frac{1}{n}}\left(-\frac{p_2}{p_i^2}\right)\right]$$

from which,

$$\frac{nRT_1}{1-n}\left[\frac{n-1}{n}\left(\frac{p_i}{p_1}\right)^{-\frac{1}{n}}\left(\frac{1}{p_1}\right)\right] = \frac{nRT_E}{1-n}\left[\frac{n-1}{n}\left(\frac{p_2}{p_i}\right)^{-\frac{1}{n}}\left(\frac{p_2}{p_i^2}\right)\right]$$

For the ideal case the value of n is assumed the same for both cylinders and, by intercooling, T_E is equal to T_1. Then:

$$\left(\frac{p_i}{p_1}\right)^{-\frac{1}{n}}\left(\frac{1}{p_1}\right) = \left(\frac{p_2}{p_i}\right)^{-\frac{1}{n}}\left(\frac{p_2}{p_i^2}\right)$$

$$\frac{p_i^{-\frac{1}{n}}}{p_1^{\frac{n-1}{n}}} = \frac{p_2^{\frac{n-1}{n}}}{p_i^{\frac{2n-1}{n}}}; \quad p_i^{\frac{2n-2}{n}} = (p_1 \cdot p_2)^{\frac{n-1}{n}}$$

and
$$p_i^2 = p_1 \cdot p_2$$

Thence:
$$p_i = \sqrt{p_1 \times p_2}. \quad \textbf{(6-6)}$$

Thus for the idealized compressor and intercooler, minimum work (for any value of n greater than 1) results when the *pressure ratios* across each

AIR COMPRESSION

stage are equal. Since by intercooling T_E is equal to T_1, the conditions for minimum work result in equal work per stage; and since $T_F = T_E = T_1$, then $T_G = T_C$. When all of these conditions are satisfied, the total work becomes

$$wk_t = 2 \times \text{work per stage}$$

$$wk_t = \frac{2nRT_1}{1-n}\left[\left(\frac{p_i}{p_1}\right)^{\frac{n-1}{n}} - 1\right] = \frac{2nR}{1-n}(T_C - T_1) \text{ ft} \cdot \text{lbf/lb (J/kg)}$$

Similar idealized developments may be made along the same lines for compressors having three or more stages. Well-designed compressors will show fairly close agreement with the ideal specifications. In operation the pressure ratios of the stages will seldom be exactly equal; but if a wide variation from the ideal intermediate pressure exists, some adjustment of valve springs is indicated. Furthermore, the temperature of the air leaving the intercooler will not return exactly to the first stage inlet temperature, and the value of n may well be different in each stage. In this case, the total work must be determined by treating each stage separately and making a summation.

Example 6-4: Determine (a) the best intercooler pressure and (b) the work required per kilogram of air for an ideal two-stage compressor operating between a suction pressure of 105 kPa abs and a discharge pressure of 1260 kPa abs. The suction temperature is 25°C and a value of $n = 1.30$ is assumed for both stages.

Solution: Refer to Fig. 6-5:

$$p_1 = 105 \text{ kPa abs} \qquad p_2 = 1260 \text{ kPa abs} \qquad n = 1.30$$
$$t_1 = t_A = t_B = t_E = 25°\text{C}$$

(a) best intercooler pressure:

$$p_i = \sqrt{p_1 \times p_2} = \sqrt{105 \times 1260} = 363.7 \text{ kPa abs}$$

(b) work of ideal cycle:

$$wk_t = \frac{2nRT_1}{1-n}\left[\left(\frac{p_i}{p_1}\right)^{\frac{n-1}{n}} - 1\right]$$

$$= \frac{(2)(1.3)(287)(25 + 273)}{1 - 1.3}\left[\left(\frac{363.7}{105}\right)^{\frac{1.3-1}{1.3}} - 1\right]$$

$$= \underline{(-)246.1 \text{ kJ/kg}}$$

6-7. The Air Compressor as a Steady Flow Machine. The material heretofore presented has been confined to the cyclic analysis of idealized compressors in terms of unit mass of air. Under steady service conditions a compressor handles air at a constant flow rate, and this flow rate determines the capacity of the unit. By definition, *the **capacity** of an air compressor is the actual volume of air entering the compressor expressed in*

AIR COMPRESSION

cubic feet (metres) per minute at the pressure and temperature of the surrounding atmosphere. Since atmospheric conditions are variable and since volumetric efficiency is affected by discharge pressure, the **rated capacity** *of an air compressor is usually considered to be the volume entering the cylinder under steady flow at rated speed and discharge pressure, expressed in cubic feet (metres) per minute at 14.7 psia and 70°F (101.3 kPa and 21°C).*

As indicated in Chapter 5, the flow energy equation may be applied to compressors if care is exercised in the selection of points. Figure 6-6 is a schematic diagram of a two-stage reciprocating compressor with intercooler and aftercooler, the latter being included because it is frequently utilized to reduce the temperature and volume of the air before storage.

FIGURE 6-6. Flow diagram for two-stage compressor with intercooling and aftercooling.

For an overall energy balance, point 1 is at the compressor inlet while point 2 is at the receiver entrance. The energy equation, based upon the mass flow rate, \dot{M}, is:

$$\dot{Q}_{12} = \dot{M} q_{12} = \dot{M}(h_2 - h_1) + \dot{M}\left[\frac{wk_{12}}{(J)}\right] \text{Btu/min (J/min)}$$

In this equation q_{12} evaluates the entire amount of heat rejected by the unit, including the heat rejected because of cylinder cooling, intercooling and aftercooling; $wk_{12}/(J)$ represents the total work done in

167

AIR COMPRESSION

both stages of the compressor; and $(h_2 - h_1)$ is the total change of enthalpy from compressor inlet to receiver inlet.

By proper selection of the intermediate points, the flow energy equation can be applied to the individual stages, the intercooler or the aftercooler, as desired. Thus, using points 1 and D, Fig. 6-6,

$$\dot{Q}_{1D} = \dot{M} q_{1D} = \dot{M}(h_D - h_1) + \dot{M}\left[\frac{wk_{1D}}{(J)}\right] \text{ Btu/min (J/min)}$$

in which \dot{Q}_{1D} is the heat rejection rate (cylinder cooling) in the first stage, $\dot{M}(h_D - h_1)$ is the rate of change of enthalpy between compressor inlet and intercooler inlet, and $\dot{M}[wk_{1D}/(J)]$ is the power required for the compression of the air in the first stage cylinder, as calculated from the previously developed equations:

$$\dot{M}\left[\frac{wk_{1D}}{(J)}\right] = \frac{\dot{M}nRT_1\left[\left(\frac{p_i}{p_1}\right)^{\frac{n-1}{n}} - 1\right]}{(J)(1-n)} = \frac{\dot{M}nR}{(J)(1-n)}(T_D - T_1)$$

For the intercooler, using points D and E, the energy equation is:

$$\dot{Q}_{DE} = \dot{M} q_{DE} = \dot{M}(h_E - h_D) = \dot{M} c_p (T_E - T_D) \text{ Btu/min (J/min)}$$

As indicated in Chapter 5 and in Art. 6-1, the steady flow analysis applies as well to rotary, centrifugal and axial flow as to reciprocating air compressors. On the general assumptions of Art. 5-7 and employing the polytropic specific heat, $c_n = c_v + R/(J)(1-n)$, as developed in Chapter 3:

$$\frac{wk_{12}}{(J)} = q_{12} + (h_1 - h_2)$$

$$= [c_v + R/(J)(1-n)](T_2 - T_1) + c_p(T_1 - T_2)$$

and, since

$$R/(J) = c_p - c_v$$

$$\frac{wk_{12}}{(J)} = (c_v - nc_v + c_p - c_v)(T_2 - T_1)/(1 - n) - c_p(T_2 - T_1)$$

$$= (c_v - nc_v + c_p - c_v - c_p + nc_p)(T_2 - T_1)/(1 - n)$$

and, finally

$$\frac{wk_{12}}{(J)} = \frac{nR}{(J)(1-n)}(T_2 - T_1) \text{ Btu/lb (J/kg)}$$

It is thus apparent that the net work required for a given compression is independent of the type of compressor employed, and for any particular range of pressures depends solely upon the value of the exponent n.

AIR COMPRESSION

6-8. Economy Resulting From Staging and Cooling. The limit established by isothermal compression has already been presented, together with the desirability of this process on the basis of minimum compressor work. The other natural limit to compressor performance is based on a compressor that is physically impervious to heat, i.e., no cooling can be accomplished and compression is adiabatic. Between any established pressure limits, more work is required in the latter case, and the temperature of the air at discharge is materially increased.

In all cases the high-pressure air delivered by any compressor will ultimately perform work at the expense of the stored energy it then possesses; and an adiabatic compressor would provide high internal energy at discharge. However, compressed air is usually held in an uninsulated receiver pending use. Much of its internal energy is dissipated by heat loss from such a receiver, and any additional internal energy provided by adiabatic compression is seldom utilized. In practice various means of cooling are employed to limit cylinder temperatures, and some reduction of total compressor work results. Well-designed units usually have water-jacketed cylinders which give operating values of n between 1.10 and 1.30. Small, cheaply produced units employ air-cooling with finned cylinder surfaces and are characterized by higher operating values of n. With multi-stage compressors the necessary intercooling also results in a saving of internal power. For either single or multi-stage compressors the economies effected by cooling are based upon the internal power required by an equivalent single stage compressor operating isentropically between the given pressure limits.

The actual saving of work is illustrated in Fig. 6-7. Thus, if the total compression is accomplished isentropically in a single stage, the compression process is BC''. With cylinder cooling and intercooling the actual compression processes are BC and FG and the work saved is represented by the area $BCFGC''$. The actual location of points C and G depends upon the effectiveness of the compressor cooling system, the limiting condition being reached when points C and G coincide with points F and C' respectively to produce single stage isothermal process BC'. Such an "isothermal standard" is occasionally utilized.

Example 6-5: An ideal two-stage reciprocating compressor handles air at a rate of 6 pounds per minute from a suction pressure of 15 psia with a temperature of 60°F, discharging against a pressure of 180 psia. The temperature of the air leaving the intercooler is 60°F and n is 1.30 for both stages. Determine (a) the internal power required, hp; (b) the power saved by cylinder cooling and intercooling, hp; (c) the heat that must be rejected by the compressor between inlet and discharge, Btu/min; (d) the heat rejected in each stage because of cylinder cooling, Btu/min; (e) the heat rejected because of intercooling, Btu/min; (f) the total heat rejected because of cylinder cooling and intercooling, Btu/min.

Solution: Refer to Figs. 6-6 and 6-7:

AIR COMPRESSION

FIGURE 6-7. Comparison of compression processes with and without cylinder and intercooling.

$$p_1 = 15 \text{ psia} \quad t_1 = 60°\text{F} \quad \dot{M} = 6 \text{ lb/min}$$
$$p_2 = 180 \text{ psia} \quad n = 1.30$$
$$T_F = T_B = T_1 = 60 + 460 = 520°\text{R}$$
$$p_i = \sqrt{p_1 \times p_2} = \sqrt{15 \times 180} = 52 \text{ psia}$$
$$T_G = T_C = T_1 \left(\frac{p_i}{p_1}\right)^{\frac{n-1}{n}} = 520 \left(\frac{52}{15}\right)^{\frac{1.3-1}{1.3}} = 693°\text{R}$$

(a)
$$(HP_{LP} + HP_{HP}) = \frac{2\dot{M}nR(T_C - T_B)}{(1 - n)33{,}000}$$
$$= \frac{(2)(6)(1.3)(53.3)(693 - 520)}{(1 - 1.3)(33{,}000)} = \underline{(-)14.5 \text{ hp}}$$

170

AIR COMPRESSION

(b) $$HP_{isentropic} = \frac{\dot{M}kRT_1\left[\left(\frac{p_2}{p_1}\right)^{\frac{k-1}{k}} - 1\right]}{(1-k)33,000}$$

$$= \frac{(6)(1.4)(53.3)(520)\left[\left(\frac{180}{15}\right)^{\frac{1.4-1}{1.4}} - 1\right]}{(1-1.4)(33,000)} = (-)18.2 \text{ hp}$$

$$HP_{SAVED} = 18.2 - 14.5 = \underline{3.7 \text{ hp}}$$

(c) $\dot{M}(h_G - h_1) = \dot{M}c_p(T_G - T_1) = (6)(0.24)(693 - 520)$
$\quad = 249 \text{ Btu/min}$

$$\dot{M}\left(\frac{wk_{12}}{J}\right) = \frac{33,000(HP_{LP} + HP_{HP})}{778} = 42.42(-14.5) = (-)615 \text{ Btu/min}$$

$$\dot{Q}_{rej} = \dot{M}(h_G - h_1) + \dot{M}\left(\frac{wk_{12}}{J}\right)$$

$$= 249 - 615 = (-)366 \text{ Btu/min}$$

(d) $$HP_{LP} = HP_{HP} = HP_{stage} = \frac{\dot{M}nR(T_c - T_B)}{(1-n)33,000}$$

$$= \frac{(6)(1.3)(53.3)(693 - 520)}{(1-1.3)(33,000)} = (-)7.25 \text{ hp}$$

$\dot{M}(\Delta h)_{LP} = \dot{M}(\Delta h)_{HP} = \dot{M}(\Delta h)_{stage} = \dot{M}c_p(T_C - T_B)$
$\quad = (6)(0.24)(693 - 520) = 249 \text{ Btu/min}$

$$\dot{Q}_{LP} = \dot{Q}_{HP} = \dot{Q}_{stage} = \dot{M}(\Delta h)_{stage} + \frac{HP_{stage}}{J}$$

$$= 249 + (42.42)(-7.25) = (-)58.5 \text{ Btu/min}$$

(e) $\dot{Q}_{intercooler} = \dot{M}(h_E - h_D) = \dot{M}c_p(T_E - T_D)$
$\quad = (6)(0.24)(520 - 693) = (-)249 \text{ Btu/min}$

(f) $\dot{Q}_{total} = \dot{Q}_{LP} + \dot{Q}_{intercooler} + \dot{Q}_{HP}$
$\quad = -58.5 - 249 - 58.5 = \underline{(-)366 \text{ Btu/min}}$

6-9. Performance of an Actual Reciprocating Compressor. Throughout the previous discussions attention has been directed toward the energy quantities associated with idealized compressor processes in terms of the changes of properties of the air being compressed. On the basis of stated assumptions means have been provided for calculating the heat rejected *from the air* in passage through a compressor, and the power required to produce the desired *changes of pressure and specific volume in the air being handled.*

In an actual compressor power is supplied to the machine from an external driving unit such as an engine or an electric motor. *The power so received is termed the compressor **brake horsepower*** and is usually the measured output of the driving unit. Due to friction the brake horsepower is always greater than the useful power developed within the compressor.

AIR COMPRESSION

This internal power is designated the **indicated horsepower** *and is calculated,* in Engineers' units, from the equation:

$$IHP = \frac{p_m LAN}{33000}$$

in which,
p_m = mean effective cylinder pressure as determined from indicator cards, psi
L = length of stroke, ft
A = piston area, sq in
N = number of cycles per minute
 = rpm for single-acting compressors

Both indicated horsepower and brake horsepower are used in estimating the performance of reciprocating air compressors. For a frictionless machine these two quantities would be exactly equal. For an actual compressor the difference between the brake and the indicated horsepower is the *friction horsepower,* and the magnitude of this quantity is indicative of the mechanical imperfection of the device. However the generally accepted criterion of mechanical performance is **mechanical efficiency,** which, for a compressor, is the ratio of the indicated to the brake horsepower, or

$$\eta_m = \frac{IHP}{BHP}$$

The internal effectiveness of the compression process is measured by the *compression efficiency,* the ratio of the ideal work (power) to the actual internal work (indicated power), for the same suction pressure and temperature, the same discharge pressure and the same flow rate. In air-cooled reciprocating units and in axial-flow and centrifugal compressors, the heat transfer per unit mass is minimal, and the actual process is substantially adiabatic. So the ideal process is clearly the reversible adiabatic or isentropic. In the case of well-designed, water-cooled reciprocating compressors, the ideal process to be sought is the isothermal. In any case, when an efficiency for an air compressor is quoted, it must be identified to either the isentropic or the isothermal standard for clarity. Using the isentropic standard, the expression for compression efficiency is:

$$\eta_{c_{(S)}} = \frac{wk_{(S)}}{wk_{\text{actual}}} = \frac{HP_{(S)}}{IHP} \qquad (6\text{-}7)$$

AIR COMPRESSION

Combining compression efficiency and mechanical efficiency yields the compressor overall efficiency:

$$\eta_{\text{overall}_{\circledS}} = \eta_m \times \eta_{c_{\circledS}} = \left(\frac{IHP}{BHP}\right)\left(\frac{HP_{\circledS}}{IHP}\right) = \frac{HP_{\circledS}}{BHP} = \frac{wk_{\circledS}}{wk_{\text{brake}}}$$

Should the isothermal standard be used, the above equations apply but with isothermal work (power) substituted for isentropic.

The capacity, brake horsepower, mechanical efficiency, compressor efficiency and overall efficiency of compressors are determined on the basis of actual performance tests. The methods to be used in conducting these tests are prescribed in the Test Codes for such machinery.

Where indicator cards cannot be obtained (in conformity with Code procedure) the indicated horsepower can be approximated by using the ideal equations herein developed in conjunction with observed values of pressure and temperature. In general such an approximation will yield a value slightly less than the power calculated from actual indicator cards. For water-cooled compressors a check on this approximation can be obtained through a compressor heat balance. For this purpose it is assumed that in addition to the heat rejected by the air, the cooling water must also absorb the thermal equivalent of the friction horsepower. The heat actually absorbed by the cooling water can be calculated by measuring the flow rate and the change of temperature of the water in passage through the compressor; thus,

$$\dot{Q}_w = \dot{M}_w q_w = \dot{M}_w c_w (t_{\text{out}} - t_{\text{in}}) \text{ Btu/min}$$

where
\dot{M}_w = water flow rate, lb/min
c_w = specific heat of water, Btu/lb · °F
t_{out} = temp. of water leaving compressor, °F
t_{in} = temp. of water entering compressor, °F

Then,

$$\dot{Q}_w = \dot{Q}_{\text{air}} + \dot{Q}_{\text{frict}}$$

in which,
\dot{Q}_{air} = heat rejected by the air based upon observed temperatures and pressures and a calculated value of n
\dot{Q}_{frict} = thermal equivalent of the friction horsepower

Example 6-6: The following data were obtained from a performance test of a $5\frac{1}{2}$ in. × 4 in. single stage reciprocating air compressor having 3 percent clearance: barometer, 30.26 in. Hg; suction pressure, 0 psig; suction temperature, 72°F; discharge pressure, 58 psig; discharge temperature, 345°F; shaft rpm, 1160; shaft horsepower, 7.45; mass of air

AIR COMPRESSION

delivered, 3.92 lb per min. Determine (a) the actual volumetric efficiency; (b) the approximate indicated horsepower; (c) the isothermal compression efficiency; (d) the mechanical efficiency; (e) the overall efficiency of the unit.

Solution:

$$\dot{M} = 3.92 \text{ lb/min} \quad t_1 = 72°F \quad t_2 = 345°F$$
$$p_1 = 0 + 30.26 \times 0.491 = 14.86 \text{ psia} \quad p_2 = 58 + 14.86 = 72.86 \text{ psia}$$

(a) Air entering cylinder $= \dot{V} = \dot{M}v_1 = \dfrac{\dot{M}RT_1}{p_1} = \dfrac{(3.92)(53.3)(532)}{(144)(14.86)} = 51.9$ cfm

$$\text{P.D.} = \frac{\pi d^2 L N}{4 \times 1728}$$

d = internal cylinder diameter = $5\tfrac{1}{2}$ in.

L = length of stroke = 4 in.

N = number of cycles per min = 1160

$$\text{P.D.} = \frac{\pi (5.5)^2 (4)(1160)}{(4)(1728)} = 63.8 \text{ cfm}$$

$$\eta_v = \frac{\dot{V}}{\text{P.D.}} = \frac{51.9}{63.8} = \underline{81.3 \text{ percent}}$$

(b) $\left(\dfrac{T_2}{T_1}\right) = \left(\dfrac{p_2}{p_1}\right)^{\frac{n-1}{n}} \quad \left(\dfrac{805}{532}\right) = \left(\dfrac{72.86}{14.86}\right)^{\frac{n-1}{n}}$

$$\frac{n-1}{n} = 0.26; \quad n = 1.351$$

$$IHP_{\text{approx}} = \frac{n\dot{M}R(T_2 - T_1)}{(1-n)(33,000)}$$

$$IHP_{\text{approx}} = \frac{(1.351)(3.92)(53.3)(805 - 532)}{(1 - 1.351)(33,000)} = \underline{(-)6.65 \text{ hp}}$$

AIR COMPRESSION

(c) $$HP_{\text{isothermal}} = \frac{\dot{M}RT_1}{33{,}000} \ln\left(\frac{p_1}{p_2}\right)$$

$$= \frac{(3.92)(53.3)(532) \ln\left(\frac{14.86}{72.86}\right)}{33{,}000} = (-)5.36 \text{ hp}$$

$$\eta_{c_{\widehat{T}}} = \frac{\text{isothermal horsepower}}{\text{indicated horsepower}} = \frac{5.36}{6.65} = \underline{80.6\%}$$

(d) $$\eta_{\text{mech}} = \frac{\text{indicated horsepower}}{\text{shaft horsepower}} = \frac{6.65}{7.45} = \underline{89.3\%}$$

(e) $$\eta_{\text{overall}_{\widehat{T}}} = \eta_{c_{\widehat{T}}} \times \eta_{\text{mech}} = 0.806 \times 0.893 = \underline{72.0\%}$$

6-10. Steady Flow Compressive Turbomachinery. Air compressors classified as turbomachinery can be subdivided as being either *axial flow* or *centrifugal* devices. These machines, also termed dynamic compressors, are steady flow devices which increase the kinetic energy of a moving stream of air. In a centrifugal machine an impeller imparts a kinetic energy increase to a radial flow (see Fig. 6-8a), while in an axial flow machine a set of rotating vanes (rotors) and a set of stationary vanes (stators) are used to increase the kinetic energy of the flow (see Fig. 6-8b).

FIGURE 6-8a. Centrifugal compressor.

FIGURE 6-8b. Axial-flow compressor.

AIR COMPRESSION

In both cases, an additional device, termed a diffuser, is required to convert kinetic head into discharge pressure head.

Both the centrifugal and the axial units are high-speed machines. Centrifugal units are essentially constant discharge pressure devices having variable capacity, depending on their operating speed. Axial units conversely are relatively constant-capacity machines with a speed-dependent variable discharge pressure.

The ideal performance analysis of both centrifugal and axial air compressors utilizes the steady flow process as developed in Chapter 5. Potential energy changes between inlet and discharge in the turbomachines are negligible. Also heat loss by the compressed air to the surroundings is minimal because of the short residence time in these high-speed units. In addition, the diffuser reduces the discharge velocity to approximately the inlet value, thereby allowing the kinetic energy change to be neglected. The steady flow energy equation (5-2) then reduces to

$$\frac{wk_{12}}{(J)} = (h_1 - h_2) \text{ Btu/lb of air (J/kg)}$$

and using the constant pressure specific heat for air,

$$\frac{wk_{12}}{(J)} = c_p(T_1 - T_2) \text{ Btu/lb of air (J/kg)} \qquad (6\text{-}8)$$

The ideal axial and centrifugal compression process is isentropic. Consequently, it is convenient to represent the ideal air compression process on a *T-s* diagram (Fig. 6-9).

FIGURE 6-9

AIR COMPRESSION

The processes in real compressive turbomachines will be irreversible because of fluid friction. To analyze an actual compressor, the isentropic compression efficiency is utilized, where

$$\eta_{c_\circledS} = \frac{\text{ideal isentropic compression work (power)}}{\text{actual compression work (power)}} < 1.0 \quad (6\text{-}7)$$

The vertical distance on the T-s diagram is proportional to the air compressor work input [see Equation (6-8)]. Thus the two work terms can be expressed as follows:

ideal compressor work:

$$wk_{12_\circledS} = c_p(T_1 - T_{2_\circledS})$$

where

T_1 = actual suction temperature, °R (K)
T_{2_\circledS} = isentropic discharge temperature, °R (K)

and actual compressor work:

$$wk_{12_a} = c_p(T_1 - T_{2_a})$$

where

T_1 = actual suction temperature, °R (K)
T_{2_a} = actual discharge temperature, °R (K)

The isentropic compression efficiency can be written as:

$$\eta_{c_\circledS} = \frac{wk_{12_\circledS}}{wk_{12_a}} = \frac{c_p(T_{2_\circledS} - T_1)}{c_p(T_{2_a} - T_1)} = \frac{T_{2_\circledS} - T_1}{T_{2_a} - T_1}$$

Since the actual compressor requires a greater energy input than the ideal, the actual discharge state is located on the exhaust pressure line at a higher temperature (see Fig. 6-9).

Axial flow compressors with moderate pressure ratios (8:1 to 10:1) often develop compression efficiencies in the 80% to 86% range. Higher pressure ratios tend to cause a drop in compression efficiency. Tending to offset the advantage of relatively high efficiency for the axial flow compressor are its weight and cost of manufacture.

Centrifugal compressors operate at generally lower pressure ratios than the axial machines and exhibit the same tendency as the axial flow machine in achieving lower compression efficiencies at the higher pressure ratios. Compression efficiencies of 80% or higher are achieved by centrifugal compressors in the 2 to 4 pressure ratio range. Advantages of the centrifugal compressor are its relative simplicity and light weight.

AIR COMPRESSION

Example 6-7: A 2-kW centrifugal compressor operates with suction conditions of 100 kPa absolute and 25°C. The pressure ratio for the unit is 3 and $\eta_{c_{(s)}} = 0.70$. Determine
(a) discharge pressure
(b) discharge temperature (actual)
(c) work input per kg of air

Solution:

(a)
$$r_p = \left(\frac{p_2}{p_1}\right); \quad p_2 = r_p p_1$$
$$p_2 = 3 \times 100 \text{ kPa} = \underline{300 \text{ kPa abs}}$$

(b) discharge temperature:

$$\frac{T_{2_{(s)}}}{T_1} = \left(\frac{P_2}{P_1}\right)^{\frac{k-1}{k}}; \quad T_{2_{(s)}} = (298) \times (3)^{\frac{0.4}{1.4}} = 407.9 \text{ K}$$

$$wk_a = c_p(T_1 - T_{2_a}) = c_p(T_1 - T_{2_{(s)}})/\eta_{c_{(s)}}$$

$$T_{2_a} = \frac{T_1(\eta_{c_{(s)}} - 1) + T_{2_{(s)}}}{\eta_{c_{(s)}}} = \frac{298(0.7 - 1) + 407.9}{0.7}$$

$$= \underline{455 \text{ K}}$$

(c) compressor work:

$$\eta_{c_{(s)}} = \frac{wk_{(s)}}{wk_a} = \frac{c_p(T_1 - T_{2_{(s)}})}{c_p(T_1 - T_{2_a})}$$

$$wk_a = \frac{c_p(T_1 - T_{2_{(s)}})}{\eta_{c_{(s)}}} = \frac{1.0048(298 - 407.9)}{0.70}$$

$$= \underline{(-)157.8 \text{ kJ/kg}}$$

Example 6-8: An axial flow compressor discharges 900 lb/min of air. Inlet conditions are 14.7 psia and 50°F, while the actual discharge conditions are 162 psia and 648°F. For the process (a) sketch the *T-s* diagram; then calculate
(b) the capacity, ft³/min
(c) the isentropic compression efficiency
(d) the isentropic compression power, hp
(e) the actual compression power, hp

Solution:

(a)

AIR COMPRESSION

$p_1 = 14.7$ psia $T_1 = 50 + 460 = 510°R$ $\dot{M} = 900$ lb/min
$p_2 = 162$ psia $T_{2_a} = 648 + 460 = 1108°R$

(b) $$\dot{V} = \frac{\dot{M}RT_1}{p_1} = \frac{(900)(53.3)(510)}{(14.7)(144)} = 11{,}557 \text{ ft}^3/\text{min}$$

(c) $$T_{2_{\circledS}} = T_1\left(\frac{p_2}{p_1}\right)^{\frac{k-1}{k}} = 510\left(\frac{162}{14.7}\right)^{\frac{1.4-1}{1.4}} = 1012.4$$

$$\eta_{c_{\circledS}} = \frac{wk_{\circledS}}{wk_a} = \frac{c_p(T_{2_{\circledS}} - T_1)}{c_p(T_{2_a} - T_1)}$$

$$\eta_{c_{\circledS}} = \frac{1012.4 - 510}{1108 - 510} = 0.84 \text{ or } \underline{84\%}$$

(d) $$HP_{\circledS} = \dot{M}c_p(T_1 - T_{2_{\circledS}})$$
$$= \frac{(900)(0.24)(510 - 1012.4)}{42.42}$$
$$= (-)2558 \text{ hp (on)}$$

(e) $$HP_a = \dot{M}c_p(T_1 - T_{2_a})$$
$$= \frac{(900)(0.24)(510 - 1108)}{42.42}$$
$$= (-)3045 \text{ hp (on)}$$

Problems

6-1. An ideal single stage air compressor without clearance takes in air at 14.5 psia with a temperature of 60°F and delivers it at 60 psia after isentropic compression. Calculate (a) the suction work, ft · lbf/lb; (b) the compression work, ft · lbf/lb; (c) the discharge work, ft · lbf/lb; (d) the total work required for suction, compression and delivery of one pound of air, ft · lbf/lb.

Ans. (a) $+27{,}720$ ft · lbf/lb; (b) $(-)34{,}650$ ft · lbf/lb; (c) $(-)41{,}570$ ft · lbf/lb; (d) $(-)48{,}500$ ft · lbf/lb

6-2. An ideal single stage reciprocating air compressor without clearance has inlet conditions of 21°C and 102 kPa abs. Air is discharged from the compressor at 122°C and 410 kPa abs. Calculate (a) the polytropic exponent, n, for the compression process and (b) the net work required for the compression process, kJ/kg.

Ans. (a) 1.27; (b) $(-)136.3$ kJ/kg

6-3. An ideal compressor without clearance receives air at a pressure of 14.7 psia with a specific volume of 13.10 ft³/lb. The air is discharged against a receiver pressure of 88.2 psia, and the value of n for the compression process is 1.25. Calculate (a) the temperature of the air entering the receiver, °F; (b) the net work required, ft · lbf/lb.

Ans. (a) 284°F; (b) $(-)59{,}700$ ft · lbf/lb

6-4. An ideal single stage air compressor without clearance takes in air at 14.7 psia at 80°F and delivers it at 50 psia with a temperature of 256°F. Calculate (a) the polytropic exponent for the compression process; (b) the net work required per lb of air discharged,

AIR COMPRESSION

ft·lbf/lb; (c) the net work required per lb of air discharged if compression is isothermal, ft·lbf/lb.

Ans. (a) 1.30; (b) (−)40,650 ft·lbf/lb; (c) (−)35,230 ft·lbf/lb.

6-5. A single stage air compressor with a piston displacement of 13.4 ft^3/cycle and 5 percent clearance takes in air at 14.7 psia with a temperature of 72°F and discharges 0.7 lb of air per cycle at a pressure of 73.5 psia. For a value of n of 1.30 calculate (a) the apparent volumetric efficiency; (b) the actual volumetric efficiency; (c) the net work required, ft·lbf/cycle; (d) the net work required per lb of air delivered, ft·lbf/lb.

Ans. (a) 87.8%; (b) 70%; (c) −38,700 ft·lbf/cycle; (d) −55,300 ft·lbf/lb

6-6. An ideal two-stage air compressor with an intercooler takes in air at 14.5 psia at 60°F and delivers it at 500 psia. The temperature of the air leaving the intercooler is 60°F. Calculate (a) the intermediate pressure for minimum total work, psia; (b) the work required per pound of air delivered if compression is isentropic in both cylinders; (c) the work required for isentropic single stage compression between upper and lower pressures specified, ft·lbf/lb.

Ans. (a) 85.2 psia; (b) (−)127,800 ft·lbf/lb; (c) (−)169,800 ft·lbf/lb

6-7. A two-stage air compressor requires an internal power of 12 kW. The compressor receives air at 22°C and 101 kPa abs and delivers it at 1323 kPa abs. For a value of n of 1.3 for the compression process calculate (a) the intermediate pressure for minimum work, kPa abs; (b) the temperature of the air entering the intercooler, K; and (c) the mass flow rate of air, kg/sec.

Ans. (a) 365.5 kPa abs; (b) 397 K; (c) 0.047 kg/sec

6-8. An ideal two-stage compressor with intercooler handles air at the rate of 4.0 lb/min. The suction pressure is 14.7 psia; suction temperature is 70°F; delivery pressure is 600 psia; and the temperature of the air leaving the intercooler is 70°F. If the polytropic exponent, n, is 1.20 for both stages, calculate (a) the intercooler pressure for minimum total work, psia; (b) the temperature of the air entering the intercooler, °F; (c) the heat rejected in the intercooler, Btu/min; (d) the total heat rejected by the compressor, Btu/min; (e) the internal power required, hp.

Ans. (a) 93.9 psia; (b) 262°F; (c) 184.3 Btu/min; (d) 447 Btu/min; (e) 14.9 hp

6-9. A single stage, single-acting air compressor with a bore of 6 in. and stroke of 6 in. operates at 550 rpm with an actual volumetric efficiency of 87 percent. Air is taken in at 14.5 psia and 40°F and is discharged at a pressure of 58 psia. The compressor does work on the air at the rate of 4.65 hp, and the air rejects heat to the cooling system at the rate of 82 Btu/min. Considering the compressor as a steady flow machine, calculate (a) the rate at which air is compressed and discharged, lb/min; (b) the change of enthalpy of the air in passage through the compressor, Btu/lb; (c) the temperature of the air leaving the compressor, °F; (d) the polytropic exponent for the compression process; (e) the compression efficiency based upon isothermal compression, percent.

Ans. (a) 3.68 lb/min; (b) 31.3 Btu/lb; (c) 171°F; (d) 1.20; (e) 88.6%

6-10. A single stage, air-cooled compressor has a capacity of 72 cfm at a suction pressure and temperature of 14.7 psia and 70°F respectively, and is to discharge against a pressure of 80 psia. The isentropic compression efficiency as given by the manufacturer is 94 percent. Assuming a mechanical efficiency of 92 percent, calculate (a) the overall efficiency, percent; (b) the required shaft input, hp; (c) the isothermal compression efficiency under the above conditions, percent.

Ans. (a) 86.5%; (b) 11.6 hp; (c) 73%

AIR COMPRESSION

6-11. A two-stage air compressor has a capacity of 100 cfm free air at a pressure of 14.7 psia with a temperature of 70°F when discharging at a pressure of 235 psia with a shaft input of 21.5 horsepower. Assuming equal work per stage with $n = 1.20$, and ideal intercooling, calculate (a) the internal (approximate indicated) horsepower required under rated conditions, hp; (b) the power saved by cylinder cooling and intercooling, hp; (c) the power saved by cylinder cooling, hp; (d) the compression efficiency based upon isothermal compression, percent; (e) the mechanical efficiency, percent; (f) the overall efficiency, percent; (g) the total heat rejected by the compressor, Btu min.

Ans. (a) 20.0 hp; (b) 7.1 hp; (c) 1.8 hp; (d) 89.0%; (e) 93.0%; (f) 82.8%; (g) 601 Btu/min

6-12. The following data were obtained from a performance test of a 4 in. by $1\frac{1}{2}$ in. by $3\frac{1}{2}$ in. stroke, single acting, water-jacketed two-stage air compressor: pressures—atmospheric, 14.8 psia; leaving first stage, 92 psia; entering second stage, 92 psia; leaving second stage, 609 psia; temperatures—atmospheric, 75°F; leaving first stage, 227°F; entering second stage, 76.5°F; leaving second stage, 260°F; compressor speed, 648 rpm; power input to motor, 5.65 kw. The motor and belt efficiency is 81.5 percent and the following are calculated from test data; air flow rate, 0.853 lb/min; compression exponent for first stage, 1.146, and for second stage, 1.20. Calculate (a) the approximate indicated horsepower; (b) the horsepower for isothermal compression; (c) the shaft horsepower; (d) the first stage piston displacement, cfm; (e) the actual volumetric efficiency, percent; (f) the compression efficiency based on isothermal compression, percent; (g) the mechanical efficiency based on the approximate indicated horsepower, percent; (h) the overall efficiency of the compressor, percent.

Ans. (a) 3.16 hp; (b) 2.74 hp; (c) 6.17 hp; (d) 16.5 cfm; (e) 69.2%; (f) 86.7%; (g) 51.2%; (h) 44.4%

6-13. A centrifugal air compressor has a pressure ratio of 3 and inlet conditions of 27°C and 103 kPa abs. The mass flow rate is 5 kg/sec. measured at inlet. For isentropic compression find (a) the discharge temperature, K; (b) the discharge pressure, kPa abs; and (c) the required internal power, kW.

Ans. (a) 410.6 K; (b) 309 kPa abs; (c) 555 kW

6-14. An axial air compressor with an isentropic compression efficiency of 83% has an internal horsepower of 65 and an overall pressure ratio of 10. Inlet conditions are 14.6 psia and 60°F. Determine (a) the discharge pressure, psia; (b) the isentropic discharge temperature, °R; (c) the actual discharge temperature, °R; (d) the mass flow rate, lb/min; and (e) the capacity, ft³/min.

Ans. (a) 146 psia; (b) 1004°R; (c) 1103°R; (d) 19.7 lb/min; (e) 260 ft³/min

6-15. A centrifugal air compressor having an isentropic compression efficiency of 78% has inlet and discharge pressures of 100 kPa abs and 350 kPa abs, respectively. Inlet temperature to the unit is 25°C and the capacity measured at inlet is 1.0 m³/sec. Determine (a) the pressure ratio; (b) the actual discharge temperature, K; (c) the mass flow rate, kg/sec; and (d) the internal compressor power, kW.

Ans. (a) 3.5; (b) 462 K; (c) 1.17 kg/sec; (d) 193 kW

6-16. A 33,300 internal hp axial air compressor has inlet conditions of 14.5 psia and 65°F and discharge conditions of 160 psia and 647°F. Calculate (a) the overall pressure ratio, r_p; (b) the isentropic discharge temperature, °R; (c) the isentropic compression efficiency, η_c; (d) the mass flow rate of air, lb/min; and (e) the capacity of the unit, ft³/min.

Ans. (a) 11, (b) 1042°R; (c) 89%; (d) 10,110 lb/min; (e) 135,500 ft³/min

CHAPTER 7

CYCLES, AVAILABLE ENERGY AND THE SECOND LAW OF THERMODYNAMICS

7-1. Cycles and Their Requirements. The word cycle is derived from the Greek word meaning circle. Its application in the thermodynamic sense refers to a circle or closed loop of events called processes. A thermodynamic cycle is defined as a *recurring* series of thermodynamic processes used for transformation of energy (either thermal or mechanical) to produce a useful effect. Special emphasis should be placed on the word "recurring" in the previous definition, since it requires the periodic return of the system, made up of the elements of the cycle, to its original state or starting condition. The main distinguishing feature to be developed here between a process and a cycle is the ability of a cycle to produce a desired effect *continuously*. It will be shown later in this chapter that although an ideal process is capable of converting all the energy supplied as heat into work, it cannot do so continuously. It will be shown further that although an ideal cycle cannot convert all the energy supplied as heat into mechanical work, it is capable of converting a portion of that energy *continuously* into mechanical work.

Performance of a cycle of events or processes by a thermodynamic system requires five essential elements:

1. A working substance
2. An engine
3. A source (high-temperature energy reservoir)
4. A sink (low-temperature energy reservoir)
5. A pump

The *working substance* may be a liquid, gas or vapor, and acts as a vehicle for carrying energy throughout the cycle. Steam power plants use water in its liquid and vapor phases as a working substance, while combustion engines use air in combination with suitable fuels.

The *engine* is a device for converting thermal energy of the working substance into useful mechanical work. It may be of a reciprocating nature such as the internal combustion engine, or rotary such as the steam or gas turbine.

The *source* supplies energy (as heat) to the working substance, an example being the combustion products from a boiler furnace in a steam power plant.

THE SECOND LAW OF THERMODYNAMICS

The *sink* absorbs energy (as heat) from the working substance either directly or through an intermediate heat transfer device known as a receiver. Two common examples are the atmosphere for combustion engines and the ocean for marine steam power plants.

The *pump* serves the purpose of moving the working substance from the low pressure region of the cycle to the high pressure region. Whether or not a pump is required as a separate device in a particular cycle is determined solely by the nature of the apparatus used to convert thermal energy into shaft work.

All thermodynamic cycles are subject to classification as "open" or "closed" cycles. "Open" cycles are those in which the atmosphere is employed to complete or close the cycle. An example is a combustion engine cycle in which atmospheric air supplies oxygen for combustion and from which the exhaust products are returned to the atmosphere. "Closed" cycles are those in which the working substance never leaves the system except for inadvertent leakage. An example of a "closed" cycle is the steam power plant cycle shown diagrammatically in Fig. 7-1a and ideally to *h-s* coordinates in Fig. 7-1b. In this cycle water is pumped into

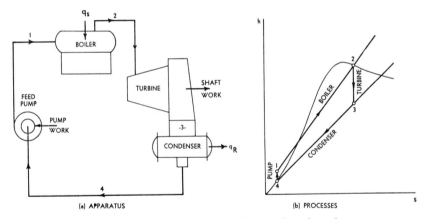

FIGURE 7-1. Steam power plant—closed cycle.

the boiler by the feed pump where energy is added as heat from the combustion of fuel with air. The water is converted to steam, which then flows to the turbine where a part of its thermal energy is converted into mechanical (shaft) work. Leaving the turbine, the steam passes next to the condenser (sink), where the cooling water removes that portion of the thermal energy not capable of being converted into work and necessarily needing to be removed to condense the steam into water. A feed pump,

using a small portion of the mechanical work developed by the turbine, pumps the water back into the boiler.

Rotating steam and gas turbines are used only for expanding a working substance from the upper to the lower pressure limits of a cycle. Consequently, they always have in their cycles a pump or a compressor, respectively, for raising the working substance from the lower to the higher pressures involved. Reciprocating internal combustion engines, where the processes of the cycle are carried out within the engine itself, perform their own compressing or pumping duties and do not have a separate pump included in their apparatus.

7-2. Thermal Efficiency and The Second Law. The First Law of Thermodynamics as expressed in Chapter 1 is essentially an expression of the Law of Conservation of Energy restricted to those energies, heat and mechanical work, of utmost importance in the field of thermodynamics. No restriction is placed on how this conversion from heat to work or vice versa is to take place, nor to what extent it might proceed. The insufficiency of the First Law to adequately and completely express the conditions under which thermal and mechanical energies may be transformed will be shown in the following development.

The steady flow energy equation may be applied to any steady flow cycle without regard for the working substance involved. Furthermore, the very nature of cyclic operations necessarily requires a working substance to return periodically to its original state, while the conditions for steady flow require the properties of a substance to remain fixed at a point in the system with time. Recognizing that a thermodynamic cycle is a means for continuously converting thermal energy into mechanical work, consider the closed cycle of Fig. 7-2 operating under steady flow conditions. Any section across the flow path of the working substance may be chosen as a reference plane. Evaluating the energy equation at point 1 for the beginning and end of *one cycle* gives

$$\frac{gZ_1}{g_c(J)} + \frac{V_1^2}{2g_c(J)} + u_1 + \frac{p_1v_1}{(J)} + q_{1\text{-}1} =$$

$$\frac{gZ_1}{g_c(J)} + \frac{V_1^2}{2g_c(J)} + u_1 + \frac{p_1v_1}{(J)} + \frac{wk_{1\text{-}1}}{(J)}$$

which becomes

$$q_{1\text{-}1} = \frac{wk_{1\text{-}1}}{(J)} \text{ Btu/lb (J/kg)} \tag{7-1}$$

This follows since, at any point in a closed cycle operating under steady-

THE SECOND LAW OF THERMODYNAMICS

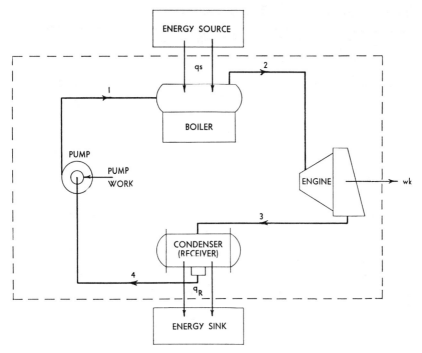

FIGURE 7-2. Basic cycle apparatus.

flow conditions, the kinetic energy, the potential energy, the internal energy and the flow work terms are coincident, leaving only the transient energy terms heat and mechanical work. Since the energy equation specifies no time limitation, the notation of Equation (7-1) represents the conversion of energy taking place in unit mass of any working substance starting at point 1 and traversing the complete cycle until it returns again to point 1.

Investigation of Equation (7-1) in the light of the First Law Statement shows no violation, for it says simply that a quantity of energy as heat was converted into mechanical work. But inspection of Fig. 7-2 shows in reality there were two energy transactions as transferred heat and two conversions in the form of mechanical work. To reveal further the character of these heat and work quantities, it is advantageous to apply the flow energy equation separately to each element of the cycle, neglecting all changes of potential and kinetic energy across individual devices. For the energy received by the working substance from the source (in the boiler):

$$q_{12} = q_S = h_2 - h_1 \text{ Btu/lb (J/kg)}$$

THE SECOND LAW OF THERMODYNAMICS

Energy rejected to the sink by the working substance:

$$q_{34} = q_R = h_4 - h_3 \text{ Btu/lb (J/kg)}$$

Energy converted to work in the turbine:

$$\frac{wk_{23}}{(J)} = h_2 - h_3 \text{ Btu/lb (J/kg)}$$

Turbine work converted to pump work is:

$$\frac{wk_{41}}{(J)} = h_4 - h_1 \text{ Btu/lb (J/kg)}$$

Having evaluated the energy transfers of the cycle by two separate means, it remains for us to develop the relationship existing between the transferred heat and mechanical work terms of each. Starting with the transferred heat terms, $q_{1\text{-}1}$ of Equation (7-1) must necessarily equal the algebraic sum of $q_{1\text{-}2} + q_{3\text{-}4}$, since the heat transfer for the complete cycle must equal that of its components. Likewise, $wk_{1\text{-}1}/(J)$ for the cycle must equal $wk_{2\text{-}3}/(J) + wk_{4\text{-}1}/(J)$. In equation form,

$$\frac{wk_{1\text{-}1}}{(J)} = \frac{wk_{23}}{(J)} + \frac{wk_{41}}{(J)} = q_{1\text{-}1} = q_{12} + q_{34}$$

Reference to an h-s diagram such as Fig. 7-1(b) for the individual enthalpy magnitudes involved in the above equation will show that $wk_{2\text{-}3}/(J)$ and $q_{1\text{-}2}$ are both positive in sign, while $wk_{4\text{-}1}/(J)$ and $q_{3\text{-}4}$ are both negative. Expressed algebraically,

$$\frac{wk_{1\text{-}1}}{(J)} = \frac{wk(\text{turbine})}{(J)} + \frac{(-wk)(\text{pump})}{(J)} = \frac{wk(\text{net})}{(J)}$$

and
$$q_{1-1} = q_S + (-q_R) = q(\text{net})$$

Replacing q_{1-1} and $wk_{1-1}/(J)$ of Equation (7-1) with their equivalents from above, we obtain:

$$q(\text{net}) = \frac{wk(\text{net})}{(J)}$$

$$q_S - q_R = \frac{wk(\text{net})}{(J)} \text{ Btu/lb (J/kg)} \qquad (7\text{-}2)$$

Equation (7-2) shows no violation of the First Law of Thermodynamics, since it merely expresses the fact that, in a cycle, a certain quantity of energy as heat has been transformed into mechanical work. The inadequacy of the First Law mentioned earlier stems from its inability to predict

or evaluate how much of the energy received by a cycle as heat might be converted to useful work. As expressed in Equation (7-2), the net work developed by a cycle is equal to the arithmetic difference (or its equivalent, the algebraic sum) of the cyclic energies externally acquired as heat (q_S), and rejected as heat (q_R). Early efforts to increase the cyclic work produced from a given quantity of energy by reducing the heat rejected suggested the possibility of eliminating that seeming waste altogether. No attempt at so doing has ever proved successful, and from extended consideration of such attempts has come a conclusion referred to as the Second Law of Thermodynamics. Applied specifically to heat engines, it declares: **No engine, actual or ideal, when operating in a cycle can convert all the heat supplied it into mechanical work.** Recognizing then that even the most thermally and mechanically perfect cycles must reject some heat, a thermodynamic power cycle efficiency may be defined as the net cycle work (output) divided by the heat supplied (input), giving

$$\eta_t = \frac{\text{output}}{\text{input}} = \frac{q_S - q_R}{q_S} = \frac{wk(\text{net})}{(J)(q_S)} \tag{7-3}$$

The relationships expressed above are applicable to any thermodynamic power cycle, *actual or ideal*. It is imperative that they be thoroughly understood for much of the succeeding work.

7-3. Types of Irreversibility. Before we continue consideration of Equation (7-3) for actual and ideal cycles, it will be advantageous to extend the discussion of reversible processes as presented in Chapter 2 to differentiate more completely between the types of irreversibilities separating the reversible from the irreversible processes.

Process reversibility requires a thermodynamic system to be able to reverse itself exactly and completely, leaving working substance and surroundings in their original state. Let us classify those features of real processes that prohibit this attainment. First, *irreversibilities* may be classified as *mechanical* or *thermal* in nature. Second, each of these classifications may be subdivided further into *internal* and *external* irreversibilities.

Mechanical irreversibilities in general are associated with friction and turbulence of a system fluid and the sliding or other friction between moving parts of a machine.

Thermal irreversibilities are generally associated with energy transfers as heat between parts of a system at different temperature levels or between the system and its surroundings.

Internal irreversibilities occur within the working substance itself.

External irreversibilities are external to the fluid, if mechanical, or between the system and its surroundings, if thermal.

One example of an internal mechanical irreversibility is the fluid friction and turbulence typical of the throttling process, while the stratified conditions as to temperature at the beginning of combustion in an internal combustion engine cylinder typify internal thermal irreversibility. Relatively speaking, the effects of these internal thermal and mechanical irreversibilities are small in well-designed machines.

To demonstrate external mechanical irreversibility, consider a cylinder with piston connected by a suitable mechanism to a flywheel. To satisfy the requirements of reversibility as closely as possible, assume the mass of the flywheel is large and the angular velocity of the flywheel is small. Under these conditions, the effects of acceleration, both in the piston and the flywheel, might be considered negligible in their effect on equilibrium states of the gas. Such a mechanism, pictured in Fig. 7-3(a), is used to control a non-flow, internally reversible, adiabatic process of a perfect gas.

FIGURE 7-3. Effects of mechanical irreversibilities on an internally reversible process.

With no external friction in the moving parts, all the work of an expansion, as indicated by process line 1-2 of Fig. 7-3(b), is stored in the flywheel as kinetic energy. Reversal of the process would convert the kinetic energy of the flywheel into stored energy in the gas, and both the mechanism and the gas would revert through the same series of state points to the exact state or condition from which it started. This situation typifies external reversibility. However, if there is any mechanical friction present, part of the energy that might have been converted to mechanical work is dissipated as heat and transferred to the surroundings or stored in the metal parts of the mechanism. The kinetic energy stored in the flywheel is less

than before. On reversing such a process, kinetic energy equivalent to, say, area $1'2zy$ rather than $12zx$ of Fig. 7-3b is available for compression of the gas. Even though these *external* irreversibilities do not affect the path of the state point of the gas, they do limit the extent of the return path, and the final state of the system would be represented by some point such as $1'$. Further, the physical apparatus would be incapable of attaining its original position, and would stop at a position equivalent to $1'$. Thus, because of the effects of external friction, the conditions required for reversibility have not been satisfied, with neither the working substance nor the mechanism returning to its initial state.

If the same system is assumed to be externally reversible, but have internal irreversibilities on the reversed path from 2 toward 1, Fig. 7-3c, friction effects would tend to increase the temperature and internal energy of the gas at each point above their value on the expansion path. Consequently, the original specific volume would be reached by the gas at a pressure (point $1''$) higher than the original (assuming enough energy were available from the flywheel) or the original pressure would be reached at a larger specific volume (point $1'$) of the gas. In the first case, the equipment but not the gas would return to its initial state, and in the second, neither the gas nor the equipment would reach its original state. Therefore, both internal and external mechanical irreversibilities reduce the useful effect to be desired and must be minimized by careful design.

External thermal irreversibility occurs in practice whenever heat is transferred between a working substance and an outside agency by means of a temperature difference. Heat transfer is always dependent on a temperature difference, and experience tells us that heat flows naturally from a high temperature region to a low temperature region. Conversely, we know that heat will not of itself flow against a temperature gradient, or from a low to a high temperature region. Thus, if energy is transferred as heat to boiler water at an average temperature of 500°F from a furnace at an average temperature of 3000°F, this process is highly irreversible, since there is no possibility of reversing the process by returning the energy as heat from the water to the furnace. Similarly, the heat rejection process in a condenser is irreversible, since the cooling water temperature always lies below the condensing steam temperature. Here the mean temperature *difference* will be much less than in the boiler, for cooling water temperatures of 60°F and condensing steam temperatures of 100°F are common. Thus, the *degree* of external thermal irreversibility is lessened. The most productive field for cycle improvement lies in the reduction of external thermal irreversibilities. This is practically attained by striving for a high mean temperature of the working substance during heat reception and a low mean temperature during heat rejection, assuming a constant tem-

THE SECOND LAW OF THERMODYNAMICS

perature source and sink. In brief, external thermal irreversibility arises from a temperature difference during heat transfer across thermodynamic system boundaries and can be reduced but never entirely eliminated, since a temperature difference is requisite to the transfer.

7-4. Ideal Cycle Analysis. Article 7-3 developed the fact that any temperature difference involved in transfer of heat necessarily constituted an irreversibility. In order to postulate an *ideal* cycle which receives and rejects energy as heat, these heat transfers must be accomplished at identical temperatures of source and working substance for the heat reception and of working substance and sink for the heat rejection. Furthermore, an ideal cycle must have reversible expansion and compression processes for the working substance if the cycle is to fulfill the requirement of complete reversibility. Several reversible processes are possible for expansion of a working substance from a high to a low pressure, but the advantage of the reversible adiabatic process will be brought out by the following example.

Example 7-1. One pound of air (considered here a perfect gas) with an initial temperature of 200°F is allowed to expand without flow between pressures of 90 and 15 psia. Which of the three processes, $pv = c$, $pv^k = c$, or $pv^{1.5} = c$ will produce the maximum work with minimum heat supplied?

Solution:

$p_1 = 90$ psia $p_2 = 15$ psia
$T_1 = 660°R$ processes: 1-2 isentropic (reversible adiabatic, $n = k$)
 1-2' isothermal ($T = c$, $n = 1.0$)
 1-2" polytropic, $n = 1.5$

THE SECOND LAW OF THERMODYNAMICS

Work:
Isothermal process

$$\frac{wk_{12}'}{J} = \int_1^{2'} \frac{p\,dv}{J} = \frac{pv}{J}\ln\frac{v_2}{v_1} = \frac{RT_1}{J}\ln\frac{p_1}{p_2} = \frac{53.3(660)}{778}\ln\frac{90}{15}$$

$$= \underline{+81.0 \text{ Btu/lb}}$$

Isentropic process

$$T_2 = T_1\left(\frac{p_2}{p_1}\right)^{\frac{k-1}{k}} = 660\left(\frac{15}{90}\right)^{\frac{1.4-1.0}{1.4}} = 396°R$$

$$\frac{wk_{12}}{J} = \int_1^2 \frac{p\,dv}{J} = \frac{p_2v_2 - p_1v_1}{J(1-k)} = \frac{R(T_2 - T_1)}{J(1-k)}$$

$$= \frac{53.3(396 - 660)}{778(1 - 1.4)} = \underline{+45.2 \text{ Btu/lb}}$$

Polytropic process, n = 1.5

$$T_2'' = T_1\left(\frac{p_2}{p_1}\right)^{\frac{n-1}{n}} = 660\left(\frac{15}{90}\right)^{\frac{1.5-1.0}{1.5}} = 363°R$$

$$\frac{wk_{12}''}{J} = \int_1^{2''} \frac{p\,dv}{J} = \frac{p_2''v_2'' - p_1v_1}{J(1-n)} = \frac{R(T_2'' - T_1)}{J(1-n)}$$

$$= \frac{53.3(363 - 660)}{778(1 - 1.5)} = \underline{+40.7 \text{ Btu/lb}}$$

Heat transfer:
Isothermal process

$$q_{12'} = u_{2'} - u_1 + \frac{wk_{12'}}{J} = 0 + 81.0 = \underline{+81.0 \text{ Btu/lb}}$$

Isentropic process

$$q_{12} = \underline{0} \text{ (by definition of process)}$$

Polytropic process

$$q_{12''} = u_{2''} - u_1 + \frac{wk_{12''}}{J} = c_v(T_{2''} - T_1) + \frac{wk_{12''}}{J}$$

$$= 0.171(363 - 660) + 40.7 = -50.8 + 40.7 = \underline{-10.1 \text{ Btu/lb}}$$

	Work, Btu/lb	Transferred heat, Btu/lb	Work from original stored energy, Btu/lb
Isothermal process	+81.0	+81.0	0.0
Isentropic process	+45.2	0.0	45.2
Polytropic process ($n = 1.5$)	+40.7	−10.1	40.7

As might have been inferred from Fig. 3-2, or, for that matter, from the non-flow energy equation, the isothermal process produces work only in the amount of energy transferred to the substance as heat. The isentropic process, producing work at the expense of the stored internal energy of the

substance, produces 45.2 Btu per lb while receiving no energy as heat. The $n = 1.5$ polytropic process actually had to have 10.1 Btu per lb abstracted from it while producing only 40.7 Btu per lb of work. Thus it is logical to deduce that from a given initial state to a fixed final pressure, the isentropic process will produce more work from initially contained energy than any other process.

Having developed the concepts of thermal and mechanical reversibility, thermal efficiency and the production of maximum work from a reversible process with no heat transfer, it will next be advantageous to investigate a *reversible cycle*, that is, a cycle composed entirely of internally and externally reversible processes. An investigation of any one of several such cycles would yield the same fundamental relationships; but because of its ease of analysis and historic background, the Carnot cycle will be employed. It will be used simply as a tool in this case to demonstrate the method of cycle analysis. Later articles in this chapter will develop the philosophy of the Carnot cycle, particularly with respect to the availability of energy and cyclic thermal efficiency.

The required elements of a Carnot cycle include (1) a Carnot *engine*, which consists of a perfectly insulated cylinder fitted with a frictionless piston and a perfectly conducting cylinder head; (2) *a working substance* (perfect gas in the following analysis, although the cycle is just as applicable to a vapor) that fills the space between the perfectly conducting cylinder head and piston: (3) an energy *source*, called the hot body, whose supply of energy at a constant absolute temperature, T_A, is infinite; (4) a receiver (or energy *sink*) called the cold body, whose energy capacity is infinite at a constant absolute temperature, T_B, lower than T_A; (5) a perfect nonconductor of heat for covering the cylinder head. The various mechanical elements of the Carnot cycle are shown, together with the p-v and T-s diagrams of the cycle, in Fig. 7-4.

The cycle is performed as follows: with the working substance at state 1, the nonconducting cover is put on the cylinder head, the piston rod is moved to the left, and isentropic compression takes place until state 2 is reached; then the hot body replaces the nonconducting cover on the cylinder head, and energy is transferred as heat to the working substance isothermally with consequent expansion of the gas and recession of the piston until state 3 is reached; at this point the nonconducting cover is again placed on the cylinder head, and the working substance continues to expand isentropically to state 4; the cold body now replaces the insulating cover and the working substance contracts while rejecting heat isothermally until it has returned to state 1.

Because this Carnot cycle transfers heat only at the constant temperatures of the source and sink and employs reversible adiabatic (isentropic)

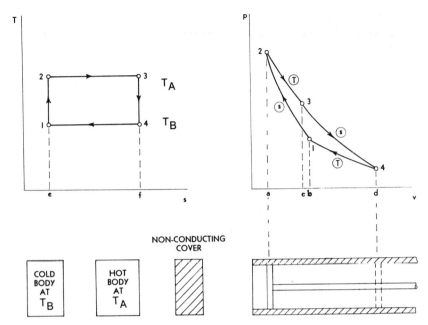

FIGURE 7-4. Carnot engine elements and cycle diagrams.

expansion and compression processes, it has the **highest possible thermal efficiency between the postulated source and sink temperatures.** An indirect proof for this statement will be developed in Art. 7-7.

Questions invariably arise concerning the transfer of heat with the source or sink and the working substance both at the same constant temperature as indicated in the processes 2-3 and 4-1 of the cycle diagrams. Practically, of course, this is impossible. The quantity of heat transferred is a direct function of time, the interconnecting surface area and the temperature difference. By postulating an infinitesimal temperature difference in either direction, however, the conditions of external thermal reversibility are substantially satisfied. In order to transfer a finite quantity of heat under these conditions, either an infinite time or an infinite area must be employed. Herein lies the impracticability of even approaching the theoretical Carnot as an actual power cycle.

A further question arises from the fact that mechanical energy as shaft work is obviously required to complete the isentropic compression of process 1-2. Recalling the actual power plant cycle of Fig. 7-1, a similar compression of the working substance is performed by the pump, using a portion of the work developed by the turbine. In the internal combustion engine, compression is performed by the piston through energy derived

THE SECOND LAW OF THERMODYNAMICS

from the rotating flywheel. Similarly, any suitable ideal (frictionless) mechanism might be assumed to transfer a portion of the positive work of the expansion processes 2-3 and 3-4, to the piston of the Carnot engine for the compression process.

The approach to an analysis of any thermodynamic cycle will be illustrated by a detailed analysis of the Carnot cycle. In this case, the working substance is assumed to be a unit mass of a perfect gas, and Fig. 7-4 shows the Carnot cycle on p-v and T-s coordinates. Using the previously developed general equation for the work of any thermodynamic power cycle in terms of heat supplied and rejected [Equation (7-2)], the terms in this equation will be evaluated first from the standpoint of areas on the T-s diagram. Heat is transferred from the source to the gas isothermally from state 2 to state 3. Evaluating the heat transferred during this non-flow process gives

$$q_S = T_A(s_3 - s_2) \text{ Btu/lb (J/kg)}$$

The area equivalent to this heat added is 23fe.

The heat rejected is

$$q_R = T_B(s_4 - s_1) = T_B(s_3 - s_2) \text{ Btu/lb (J/kg)}$$

and the corresponding area is 14fe.

The last equation deviates from the usual process procedure for evaluating a change in properties, in that it subtracts the final entropy (s_1) from the initial entropy (s_4). Conventional use of $q_S - q_R$ with cycles implies **arithmetic** subtraction of q_R and **discards the negative sign** of the heat rejection process. Therefore, in this and all other thermodynamic power cycles involving q_S and q_R, **the heat rejected term, q_R, will be set up and evaluated as above to give a positive numerical result.**

In terms of q_S and q_R, the net work of the cycle is

$$\frac{wk(\text{net})}{(J)} = q_S - q_R = T_A(s_3 - s_2) - T_B(s_3 - s_2)$$

$$\frac{wk(\text{net})}{(J)} = (T_A - T_B)(s_3 - s_2) \text{ Btu/lb (J/kg)} \qquad (7\text{-}4)$$

and the cycle thermal efficiency is

$$\eta_t = \frac{q_S - q_R}{q_S} = \frac{(T_A - T_B)(s_3 - s_2)}{T_A(s_3 - s_2)} = \frac{T_A - T_B}{T_A} \qquad (7\text{-}5)$$

Using the p-v diagram of the Carnot cycle, the net work of the cycle can be obtained in terms of the non-flow work equations.

The appropriate process relations for the work of each process give:

THE SECOND LAW OF THERMODYNAMICS

$$\frac{wk_{12}}{(J)} = \frac{p_2v_2 - p_1v_1}{(J)(1-k)} = \frac{R(T_2 - T_1)}{(J)(1-k)} \text{ Btu/lb (J/kg)}$$
$$= \text{area } 12ab$$

$$\frac{wk_{23}}{(J)} = \frac{p_2v_2}{(J)} \ln\left(\frac{v_3}{v_2}\right) = \frac{RT_2}{(J)} \ln\left(\frac{v_3}{v_2}\right) \text{ Btu/lb (J/kg)}$$
$$= \text{area } 23ca$$

$$\frac{wk_{34}}{(J)} = \frac{p_4v_4 - p_3v_3}{(J)(1-k)} = \frac{R(T_4 - T_3)}{(J)(1-k)} \text{ Btu/lb (J/kg)}$$
$$= \text{area } 34dc$$

$$\frac{wk_{41}}{(J)} = \frac{p_4v_4}{(J)} \ln\left(\frac{v_1}{v_4}\right) = \frac{RT_4}{(J)} \ln\left(\frac{v_1}{v_4}\right) \text{ Btu/lb (J/kg)}$$
$$= \text{area } 41bd$$

Since each of the above expressions carries the algebraic sign of its particular process, it is necessary to add the expressions algebraically to get the net work. Thus:

$$\frac{wk(\text{net})}{(J)} = \frac{wk_{12}}{(J)} + \frac{wk_{23}}{(J)} + \frac{wk_{34}}{(J)} + \frac{wk_{41}}{(J)}$$
$$= \frac{R(T_2 - T_1)}{(J)(1-k)} + \frac{RT_2}{(J)} \ln\left(\frac{v_3}{v_2}\right) + \frac{R(T_4 - T_3)}{(J)(1-k)}$$
$$+ \frac{RT_4}{(J)} \ln\left(\frac{v_1}{v_4}\right) \text{ Btu/lb (J/kg)} \tag{7-6}$$

For the isentropic processes 1-2 and 3-4, the following relationships are true:

$$\frac{v_3}{v_4} = \left(\frac{T_4}{T_3}\right)^{\frac{1}{k-1}} \quad \text{and} \quad \frac{v_2}{v_1} = \left(\frac{T_1}{T_2}\right)^{\frac{1}{k-1}}$$

But
$$T_4 = T_1 \quad \text{and} \quad T_3 = T_2$$

Substitution of these equalities in the previous equation gives

$$\frac{v_3}{v_4} = \frac{v_2}{v_1}$$

and rearranging gives

$$\frac{v_3}{v_2} = \frac{v_4}{v_1}$$

Substitution of this result in Equation (7-6) and collecting terms gives

THE SECOND LAW OF THERMODYNAMICS

$$\frac{wk(\text{net})}{(J)} = \frac{R}{(J)}\left[\left(\ln\frac{v_3}{v_2}\right)(T_2 - T_4)\right.$$
$$\left. + \frac{(T_2 - T_1 + T_4 - T_3)}{(1-k)}\right] \text{Btu/lb (J/kg)}$$

Since $T_2 = T_3$ and $T_4 = T_1$, the last term drops out, indicating that the work of isentropic compression is exactly equal and opposite in sign to the isentropic work of expansion. Applying again the equality $T_1 = T_4$, the above equation becomes

$$\frac{wk(\text{net})}{(J)} = \frac{R}{(J)}\left(\ln\frac{v_3}{v_2}\right)(T_2 - T_1) \text{ Btu/lb (J/kg)}$$

Since the work equations must be equal, this last equation must be equivalent to Equation (7-4). $(T_2 - T_1)$ is the equivalent of $(T_A - T_B)$. The change of entropy for the isothermal process 2-3 is the heat transferred divided by the absolute temperature. The heat transferred is equal to the work, and the absolute temperature is $T_A = T_2$. Then,

$$s_3 - s_2 = \frac{q_{23}}{T_2} = \frac{wk_{23}}{(J)T_2} = \frac{RT_2}{(J)T_2}\ln\frac{v_3}{v_2} = \frac{R}{(J)}\ln\frac{v_3}{v_2}$$

Finally,

$$\frac{wk(\text{net})}{(J)} = (T_A - T_B)(s_3 - s_2) \text{ Btu/lb (J/kg)} \qquad (7\text{-}7)$$

which is identical to Equation (7-4) developed from the *T-s* coordinates.

Applying the proper algebraic sign for the work represented by each area, there results:

$wk(\text{net}) = +$ area $23ca +$ area $34dc -$ area $41bd -$ area $12ab$
$\qquad\quad = +$ area 1234

Example 7-2: One kilogram of a perfect gas (air) is used as a working substance in a Carnot power cycle. At the beginning of isentropic compression, the temperature is 326 K and the absolute pressure is 359 kPa. The absolute pressure at the end of the isentropic compression is 1373 kPa. For this cycle, the isothermal expansion ratio, (v_3/v_2), is 2.0. For the cycle (a) sketch the *p-v* and *T-s* diagrams; then calculate:
 (b) the pressures, temperatures, and specific volumes at each process termination point
 (c) the heat supplied, kJ/kg
 (d) the heat rejected, kJ/kg
 (e) the net work by the *T-s* and *p-v* area methods
 (f) the thermal efficiency, percent

THE SECOND LAW OF THERMODYNAMICS

Solution:

(a)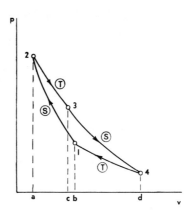

$p_1 = 359 \text{ kPa abs} \qquad p_2 = 1373 \text{ kPa abs}$
$T_1 = 326 \text{ K} \qquad v_3/v_2 = 2.0$
$R = 287 \text{ J/kg} \cdot \text{K}$

(b) $v_1 = \dfrac{RT_1}{p_1} = \dfrac{287 \times 326}{359 \times 10^3} = \underline{0.2606 \text{ m}^3/\text{kg}}$

$T_2 = T_1 \left(\dfrac{p_2}{p_1}\right)^{\frac{k-1}{k}} = 326\left(\dfrac{1373}{359}\right)^{\frac{1.4-1}{1.4}} = \underline{478.3 \text{ K}}$

$v_2 = \dfrac{RT_2}{p_2} = \dfrac{287 \times 478.3}{1373 \times 10^3} = \underline{0.100 \text{ m}^3/\text{kg}}$

$p_3 = p_2 \left(\dfrac{v_2}{v_3}\right) = 1373\left(\dfrac{1}{2}\right) = \underline{686.5 \text{ kPa abs}}$

$T_3 = T_2 = \underline{478.3 \text{ K}}$

$v_3 = 2v_2 = 2 \times 0.100 = \underline{0.200 \text{ m}^3/\text{kg}}$

$T_4 = T_1 = \underline{326 \text{ K}}$

$p_4 = p_3 \left(\dfrac{T_4}{T_3}\right)^{\frac{k}{k-1}} = 686.5 \left(\dfrac{326}{478.3}\right)^{\frac{1.4}{1.4-1}} = \underline{179.5 \text{ kPa abs}}$

$v_4 = \dfrac{RT_4}{p_4} = \dfrac{287 \times 326}{179.5 \times 10^3} = \underline{0.5212 \text{ m}^3/\text{kg}}$

(c) $q_S = p_2 v_2 \ln\left(\dfrac{v_3}{v_2}\right) = (1373)(0.100)\ln 2 = \underline{95.17 \text{ kJ/kg}}$

(d) $q_R = p_4 v_4 \ln\left(\dfrac{v_4}{v_1}\right) = (179.5)(0.5212)\ln\left(\dfrac{0.5212}{0.2606}\right) = \underline{64.85 \text{ kJ/kg}}$

(e) $wk(\text{net}) = q_S - q_R = 95.17 - 64.85 = \underline{30.32 \text{ kJ/kg}}$

$wk(\text{net}) = wk_{23} + wk_{34} + wk_{41} + wk_{12}$

$= p_2 v_2 \ln\left(\dfrac{v_3}{v_2}\right) + \dfrac{p_4 v_4 - p_3 v_3}{(1-k)} + p_4 v_4 \ln\left(\dfrac{v_1}{v_4}\right) + \dfrac{p_2 v_2 - p_1 v_1}{(1-k)}$

THE SECOND LAW OF THERMODYNAMICS

$$= (1373)(0.10)\ln 2 + \frac{(179.5 \times 0.5212) - (686.5 \times 0.200)}{(1 - 1.4)}$$

$$+ (179.5)(0.5212)\ln\left(\frac{0.2606}{0.5212}\right)$$

$$+ \frac{(1373 \times 0.100) - (359 \times 0.2606)}{(1 - 1.4)}$$

$$= 95.17 + 109.36 - 64.85 - 109.36 = \underline{30.32 \text{ kJ/kg}}$$

7-5. Pump Work and Compression Work. From the preceding analysis of the Carnot cycle, it becomes clear that the engine not only derives positive mechanical work from the expansion of the working substance, but also supplies a smaller amount of mechanical work to the fluid to return it from the lowest pressure in the cycle to the highest. The net work of such a cycle is the difference between the positive work of expansion and the negative work of the heat rejection and compression processes. Any cycle employing a separate pump for raising the pressure of the fluid from the lowest to the highest value of the cycle such as the one shown in Fig. 7-1, has its net work expressed as

$$wk(\text{net}) = wk(\text{engine}) - wk(\text{pump}) \tag{7-8}$$

Evaluated according to the sign convention adopted for processes, pump work is always negative. However, **when combined with engine work to give net cycle work, pump work is always used as a positive numerical quantity,** acquiring its negative sign from the equation above. Power cycle analyses of the next two chapters will illustrate this point, as well as the similar point made earlier concerning the cyclic heat rejected term, q_R.

7-6. The Reversed Carnot Cycle. The Carnot cycle as originally conceived was a thermodynamic *power* cycle consisting entirely of reversible processes, taking energy from a source, converting part of the energy into mechanical work, and rejecting the remainder to a receiver. Now consider that same series of processes traversed in the opposite direction from that of the power cycle. The reversed cycle on p-v and T-s coordinates will then appear as shown in Fig. 7-5.

The operation of the reversed Carnot cycle is performed as follows: from state 1 to state 2 of the figure, the cold body is in place on the cylinder head, acting now as a heat source in giving up heat to the working substance rather than as a receiver in the power cycle; at state 2 the nonconducting cover is brought into contact with the cylinder head of the engine, and compression takes place isentropically until state 3 is reached; next, the hot body is brought in contact with the cylinder head and heat is transferred isothermally from the working substance to the hot body,

THE SECOND LAW OF THERMODYNAMICS

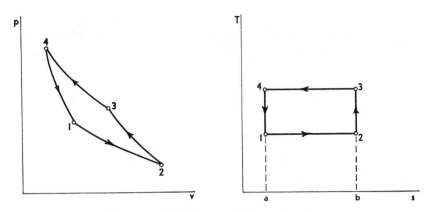

FIGURE 7-5. Reversed Carnot cycle.

which is now acting as a receiver, until the working substance reaches state 4; finally, the nonconducting cover is once more placed on the cylinder head, and isentropic expansion takes place from state 4 to state 1. The net effect of this reversed Carnot cycle is the removal of a quantity of energy from the cold body at a temperature corresponding to that at states 1 and 2, and delivery of a larger quantity of energy to the hot body at a higher temperature corresponding to that at states 3 and 4. As is properly expected in a reversed cycle of this type, mechanical work is supplied to the engine in an amount sufficient to raise the temperature of the working substance from that of the cold body to that of the hot body. Referring again to Fig. 7-5, the heat energy supplied by the cold body to the working substance, q_S, is represented by area 12ba, and the heat rejected by the working substance to the hot body, q_R, is represented by area 34ab. Since this cycle is reversible and no heat is transferred during the isentropic compression and expansion processes, the difference between the heat energy picked up from the cold body and that rejected to the hot body must be the heat equivalent of the net mechanical work supplied to the cycle, area 1234. This reversed cycle is called a refrigerating cycle, and the engine in this situation becomes a *heat pump*.

7-7. The Carnot Principle. The Second Law of Thermodynamics and the Carnot Principle are the foundation, so to speak, of the science of heat engines and their cycles. Nicolas Leonard Sadi Carnot, a brilliant French physicist and member of the French Army Engineer Corps, was so perturbed by what he considered the appalling ineffectiveness of heat engines of his era that he devoted much time and thought to the problem. Starting with the meager store of facts available at that time (about 1816),

THE SECOND LAW OF THERMODYNAMICS

he strove to establish the fundamental nature of heat. In spite of a fallacy concerning the nature of heat (energy) itself, which wasn't discovered until some time later, the final conclusions he drew with respect to thermodynamic cycles were sound, and from them was later formulated the Second Law of Thermodynamics as we know it today. Carnot's views and thoughts were published in 1824 in a book entitled *Reflexions sur la puissance motrice du fou* (Reflections on the motive power of heat). It is interesting to note that Carnot was twenty-eight years of age when his book was published, and the depth of his logic becomes even more amazing when it is realized that the Second Law of Thermodynamics had not been formulated at that time. In his work, Carnot not only presented an analysis of the cycle which now bears his name, but also expressed the thought which has now become known as the Carnot Principle: **The motive power of heat is independent of the agents employed to realize it; its quantity is fixed solely by the temperatures of the bodies between which is effected, finally, the transfer of heat.**

To Rudolph Clausius, a German physicist born just two years before publication of Carnot's historic essay, is attributed one of the more useful and commonly quoted versions of the Second Law: **It is impossible for a self-acting machine, unaided by an external agency, to transfer heat continuously from one body to another at a higher temperature.** Clausius shares with Rankine and Thompson the distinction of establishing the field of thermodynamics on a scientific basis. Much of the motivation for the work of these three scientists in the field of thermodynamics undoubtedly stemmed from the earlier efforts of Carnot.

The Carnot Principle is the root of three basic thermodynamic propositions:

1. No heat engine cycle for production of mechanical work from energy received at a single high temperature and rejected at a single low temperature can be more efficient than a reversible cycle.
2. The efficiencies of all reversible heat engine cycles receiving and rejecting heat at the same higher and lower temperatures are the same.
3. The efficiency of a reversible heat engine cycle depends only on the temperatures of the energy source and sink.

None of the propositions stated above is subject to direct proof. However, situations may be set up that assume they are not true and the fallacy of the assumed situations shown. Throughout this and later discussions, the term heat engine connotes a mechanical device operating in a complete cycle for the sole purpose of continuously converting energy in the form of heat into mechanical work. Thus, *thermal efficiency of a heat*

THE SECOND LAW OF THERMODYNAMICS

engine corresponds to the thermal efficiency of the cycle in which the engine is used. Conversely, a heat pump is a reversible heat engine operating in a reversed power cycle for the sole purpose of the continuous transfer of energy as heat from a low temperature region to a high temperature region by the expenditure of mechanical work. Article 7-9 amplifies these points. In Fig. 7-6 are shown two heat engines, *A* and *B*, both receiving and rejecting energy as heat from constant temperature sources and sinks.

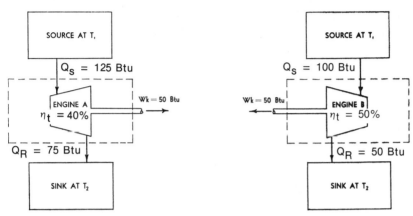

FIGURE 7-6. Heat engines with equivalent source temperatures and equivalent sink temperatures, but different thermal efficiencies.

It is assumed that the temperatures of the two sources are identical and also that the two sink temperatures are the same. Let us arbitrarily assign reversibility and a thermal efficiency of 40 percent to engine *A*, and irreversibility with a thermal efficiency of 50 percent to engine *B*. Referring to Fig. 7-6:

Engine A	Engine B
Assume engine *A* is to develop mechanical work = 50 Btu	Let Q_S from source be 100 Btu
$\eta_{t_A} = 40\%$ (assumed)	$\eta_{t_B} = 50\% = \dfrac{Q_S - Q_R}{Q_S}$
$\dfrac{Wk}{J} = Q_S - Q_R = 50$ Btu	$Q_R = Q_S - \eta_t(Q_S)$
$\eta_{t_A} = \dfrac{Q_S - Q_R}{Q_S} = \dfrac{50}{Q_S}$	$Q_R = 100 - (0.5)(100) = 50$ Btu
$0.4 = \dfrac{50}{Q_S}; \quad Q_S = \dfrac{50}{0.4} = 125$ Btu	$\dfrac{Wk}{J} = Q_S - Q_R = 50$ Btu
$Q_R = Q_S - \dfrac{Wk}{J} = 125 - 50 = 75$ Btu	

With both engines A and B operating as heat engines, their heat added and heat rejected quantities have been established according to their efficiencies. Now let engine A be reversed to operate as a heat pump. At the same time, let engine B supply its 50 Btu mechanical shaft work to drive engine A in its reversed cycle. Since engine A is reversible and operating in a completely reversible cycle, its heat quantities must remain the same so long as its efficiency and shaft work remain the same.

By combining engine B with heat pump A, as in Fig. 7-7, the dashed lines represent a discrete system which is continuously removing net energy of 25 Btu as heat from the sink at temperature T_2, and delivering it to the source at a higher temperature T_1. Since the assumption that

FIGURE 7-7. Carnot engine used to drive heat pump.

engine B was more efficient than reversible engine A led to this violation of the Second Law of Thermodynamics, the assumption is false, and the first of the three propositions is judged true.

Proof of the second proposition may be obtained by a reasoning process similar to the first. Assume this time that both engines are reversible, and engine A is operating as a heat pump as before. Assign a thermal efficiency higher than the 40 percent of A to engine B. If engine A maintains its 40 percent efficiency and is driven by B, the same contradiction as before will arise.

Justification of Carnot's third proposition may be achieved by replacing engine B of Fig. 7-7 with two Carnot engines in series, B_1 and B_2 as shown

THE SECOND LAW OF THERMODYNAMICS

in Fig. 7-8. B_1 receives energy directly from the source at T_1, but rejects its residual to B_2, thereby acting as a source for B_2. Residual energy from B_2 goes to the sink at temperature T_2. Let engines B_1 and B_2 combine to drive engine A reversed as before. Assume that engines B_1 and B_2 have thermal efficiencies of 25 percent and 20 percent respectively. Engine A retains its 40 percent thermal efficiency as before.

FIGURE 7-8. Two Carnot engines in series driving heat pump.

Then:

Engine B_1

$Q_S = 100$ Btu to engine B_1

$\eta_{t_{B_1}} = 25\% = \dfrac{Q_S - Q_R}{Q_S}$

$Q_S - Q_R = \eta_{t_{B_1}}(Q_S)$

$Q_S - Q_R = \dfrac{Wk_{B_1}}{J} = 0.25(100)$

$\qquad = 25$ Btu

$Q_R = Q_S - \dfrac{Wk_{B_1}}{J} = 100 - 25$

$\qquad = 75$ Btu

Engine B_2

$Q_S = Q_{R_{B_1}}$

$Q_{R_B} = Q_{S_B} - \dfrac{Wk_{B_1}}{J} = 100 - 25$

$\qquad = 75$ Btu

$Q_{S_{B_2}} = 75$ Btu

$\eta_{t_{B_2}} = 20\%$

$(Q_S - Q_R)_{B_2} = \dfrac{Wk_{B_2}}{J} = \eta_{t_{B_2}} Q_{S_B}$

$\qquad = 0.20(75) = 15$ Btu

$Q_{R_{B_2}} = \left(Q_S - \dfrac{Wk}{J}\right)_{B_2} = 75 - 15$

$\qquad = 60$ Btu

203

THE SECOND LAW OF THERMODYNAMICS

Engine A

As a Carnot engine with $Q_S = 100$ Btu:

$\eta_{t_A} = 40\%$

$\dfrac{Wk}{J} = Q_S - Q_R = Q_S(\eta_{t_A})$

$= 100(0.40) = 40$ Btu

$Q_R = Q_S - \dfrac{Wk}{J} = 100 - 40$

$= 60$ Btu

As a reversed Carnot engine (heat pump):

$\dfrac{Wk}{J} = 40$ Btu

$Q_S = 60$ Btu (same as Q_R as an engine)

$Q_R = Q_S + \dfrac{Wk}{J} = 60 + 40$

$= 100$ Btu

(same as Q_S as an engine)

The preceding analysis shows that the two reversible engines replacing single engine B and sharing the temperature drop between T_1 and T_2 must each have a thermal efficiency less than that of A, but equivalent to it in the aggregate, because they develop the same work between them when operating with the same source and sink temperatures.

$$Wk_{B_1} + Wk_{B_2} = Wk_{B_1} + Wk_{B_2} = Wk_A$$
$$(Q_S - Q_R)_{B_1} + (Q_S - Q_R)_{B_2} = \eta_t Q_{S_{B_1}} + \eta_t Q_{S_{B_2}} = \eta_t Q_{S_A} \text{(as an engine)}$$
$$25 \text{ Btu} + 15 \text{ Btu} = 0.25(100) + 0.20(75) = 0.40(100)$$
$$40 = 40 = 40$$

That this thermodynamic system did not violate the Second Law of Thermodynamics substantiates Carnot's third proposition that the efficiency of a cycle using a reversible engine is determined solely by the source and sink temperatures between which it operates.

7-8. Available Energy. Earlier discussion has shown that, for given source and sink temperatures, the Carnot cycle has the maximum possible thermal efficiency. This results from heat reception at a high constant temperature equal to that of the source and heat rejection at a low constant temperature equal to that of the sink. None of the actual operating cycles attains both the Carnot heat reception and rejection processes because of practical limitations which will be explained presently. They do, however, try to keep the average temperature of the working substance as high as possible during heat reception and as low as possible during heat rejection. In other words, the advantages of the Carnot cycle are well recognized and provide a level of performance to be approached as closely as possible in the formulation of practical cycles. In the discussion which

THE SECOND LAW OF THERMODYNAMICS

follows, all cycles will be treated as ideal, as is the Carnot, in that they are made up of a sequence of reversible processes.

Most of the idealized cycles follow the Carnot in having both their compression and expansion processes isentropic. The reasons for this choice are outlined in Art. 7-4. Reference to the *T-s* diagram for this situation shows that the change in entropy during both the cyclic heat reception and heat rejection processes is the same numerically. Typically, the heat rejected goes to the great natural sink provided by the earth's atmosphere, the oceans, or other large bodies of water. We have at our disposal, then, the substantially infinite constant temperature sink demanded by the Carnot concept. The temperature of this sink, which we will refer to as T_0, has an average value in the order of 60°F (16°C). There is no physical deterrent to making full use of this sink and rejecting heat at a constant temperature very nearly equal to T_0. As a matter of fact, most steam cycles do just this. That other cycles do not stems from mechanical limitations rather than from any limitations imposed by nature. The heat reception process is an altogether different proposition. There is no universal high temperature source on the earth. The energy required for many of our usual cycles derives from the burning of hydrocarbon fuels. Maximum cyclic temperatures in the order of 3000°F are commonly attained, but there is no known way to take advantage of even this temperature level on a continuing basis. We are seriously limited in the high temperature portion of the cycle by the properties of both engineering materials and working substances, as will be discussed more fully later.

While the Carnot efficiency provides a general yardstick for evaluating the *potential performance* of a cycle once its limiting temperatures are established, it imposes an assumed constant temperature heat reception process on many of the actual cycles that do not and can not utilize such a process. It is desirable to have some other criterion of comparison between cycles that will indicate the theoretical capability of a cycle to convert the heat supplied it into mechanical work. That criterion is available energy. *The available energy of a thermodynamic cycle is defined as that portion of the thermal energy supplied to such a cycle which is ideally capable of ultimate conversion into mechanical work.* Conversely, the unavailable energy of a thermodynamic cycle is that energy which must be rejected to the universal sink at the average minimum temperature, T_0, prevailing. The unavailable energy can be evaluated in terms of T_0 and the entropy change for the cyclic heat supply process. This leads to the following general expression

$$A.E. = q_S - T_0(\Delta s) \text{ Btu/lb (J/kg)} \tag{7-9}$$

THE SECOND LAW OF THERMODYNAMICS

where

A.E. is the available energy of the cycle, Btu/lb (J/kg)
q_S is the heat reversibly supplied, Btu/lb (J/kg)
T_0 is the lowest available sink temperature, °R (K)
Δs is the entropy change occurring during the heat supply process, Btu/lb · °R (J/kg · K)
$T_0(\Delta s)$ is the unavailable energy (*U.A.E.*), Btu/lb (J/kg)

Although cyclic available energy is a defined quantity, it has its logic in the Carnot cycle, but with the heat supply process left unspecified. If the concept is applied to a Carnot cycle, then the work will equal the available energy, and the heat rejected will equal the unavailable energy. These quantities are shown as areas on *T-s* coordinates in Fig. 7-9. The Carnot

FIGURE 7-9. Available and unavailable energies of Carnot cycle.

cycle is, of course, a special case, and in the general case q_R and $T_0(\Delta s)$ as well as *A.E.* and $wk/(J)$ need not be identical. An example problem will serve to amplify certain points concerning available and unavailable energy.

Example 7-3: Calculate the available energy in Btu/lb for a Carnot cycle with source temperature of 3460°R, a sink temperature of 520°R and an energy supply as heat of 100 Btu to one pound of a working substance. Then calculate the reduction in available energy Btu/lb, for a similar Carnot cycle in which all conditions remain the same as before except that the working substance is limited to a maximum temperature of 1960°R.

THE SECOND LAW OF THERMODYNAMICS

Solution:

$T_1 = 3460°R$
$T_1' = 1960°R$
$T_0 = 520°R$
$q_S = 100$ Btu/lb ($aBCb$ or $aB'C'c$)

For the heat addition process $B - C$:

$$ds = \frac{\delta q}{T} \quad \text{and} \quad s_C - s_B = \frac{q_S}{T_1} = \frac{100}{3460} = 0.0289 \text{ Btu/lb} \cdot °R$$

unavailable energy $= T_0(\Delta s)_{BC} = 520(0.0289) = 15.0$ Btu/lb

$A.E. = q_S - T_0(\Delta s)_{BC} = 100.0 - 15.0 = \underline{85.0 \text{ Btu/lb}}$

From Equation 7-5,

$$\frac{q_S - q_R}{q_S} = \frac{T_1 - T_0}{T_1} \quad \text{and} \quad 1 - \frac{q_R}{q_S} = 1 - \frac{T_0}{T_1}$$

$$q_R = q_S \left(\frac{T_0}{T_1}\right) = 100 \left(\frac{520}{3460}\right) = 15.0 \text{ Btu/lb}$$

$$\frac{wk}{J} = q_S - q_R = 100 - 15 = 85.0 \text{ Btu/lb}$$

For 100 Btu supplied to one pound of substance at 1960°R rather than 3460°R:

$$s_{C'} - s_{B'} = \frac{q_S}{T_{1'}} = \frac{100}{1960} = 0.0510 \text{ Btu/lb} \cdot °R$$

unavailable energy $= T_0(\Delta s)_{B'C'} = 520(0.0510) = 26.5$ Btu/lb

$A.E._{B'C'} = q_S - T_0(\Delta s)_{B'C'} = 100 - 26.5 = 73.5$ Btu/lb

Reduction in $A.E. = 85.0 - 73.5 = \underline{11.5 \text{ Btu/lb}}$

For maximum work, the average temperature of the working substance must be kept as high as possible during heat reception and as low as possible during heat rejection. In other words, the external thermal irreversibility of both these heat transfer processes should be minimized. As previously pointed out, it is feasible to reject heat with the working substance at some substantially constant temperature only slightly above T_0. Unfortunately the heat reception process is subject to serious physical limitations. One of these results from the inability of known metals and their alloys to stand high temperatures continuously. Although combustion temperatures of 3000°F are common, the best special alloys currently available cannot sustain **continuous** operating temperatures in excess of about 1500°F without serious deterioration in their properties and even-

tual failure in use. Despite the fact that 3000°F (3460°R) is available, the maximum cyclic temperature of the working substance is limited to about 1500°F (1960°R), and an immediate reduction in available energy results. This is shown in Fig. 7-10, and the preceding example places some numerical values on the resultant loss of available energy. Note that the

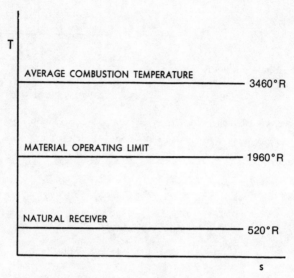

FIGURE 7-10. Cyclic temperature limitations.

decrease in available energy in the example is really a minimum, since constant temperature heat addition is assumed to occur at the top material temperature of 1960°R.

The physical properties of the working fluid sometimes restrict the energy that can be added as heat because of pressure, volume or other considerations. In the example above, if it were impractical to carry the heating process at 1960°R to C', stopping at D instead, then the heat added would be limited to 56.6 Btu per pound. Of this amount, 41.6 Btu per pound would be available energy, and 15.0 would be unavailable. In this case the thermal efficiency would be $41.6/56.6 = 73.5$ percent, compared with 85.0 percent for the original cycle $ABCE$.

In many of our operating cycles, all or part of the heat addition process occurs at other than constant temperature. A reversible cycle formed around the particular heat addition process selected will necessarily have a lower thermal efficiency and less available energy than a Carnot cycle using the same maximum temperature and the same sink temperature. Consider a reversible constant pressure heat addition process for air, as

THE SECOND LAW OF THERMODYNAMICS

shown between 1 and 2 in Fig. 7-11. With sink temperature T_0 and reversible adiabatic (isentropic) expansion and compression processes 2-b and 4-1, a completely reversible cycle can be formed around process 1-2.

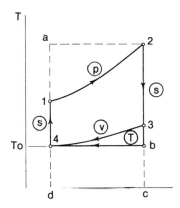

FIGURE 7-11. Heat addition process not at constant temperature.

Comparing this cycle with a Carnot having the same maximum temperature shows these differences:

	Cycle 12b4 (areas)	Carnot Cycle a2b4 (areas)
q_s	12cd	a2cd
A.E.	12b4	a2b4
wk	12b4	a2b4
U.A.E.	b4dc	b4dc

This comparison indicates that available energy and work are sacrificed by any cycle in which the heat addition process is not constantly maintained at the highest temperature of the cycle.

Cycles using gases as working substances frequently not only are incapable of constant temperature heat addition but also cannot reject heat at constant temperature. Such a cycle is shown in Fig. 7-11 as 1234. The available energy and unavailable energy for cycle 1234 will be the same as for cycle 12b4, but q_R for 1234 will be area 34dc and the work will be area 1234. The area 3b4 represents available energy that cycle 1234 could not use because of its constant volume heat rejection process.

Example 7-4: One pound of a perfect gas (air) is assumed as the working substance of reversible cycles a2b4, 12b4, and 1234 of Fig. 7-11. Let $T_1 = 1580°R$, $T_2 = 3047°R$, $T_3 = 1345°R$, and $T_0 = T_4 = 535°R$. Compare the three cycles as to heat added, heat rejected, work done, available energy, unavailable energy and cycle thermal efficiency.

THE SECOND LAW OF THERMODYNAMICS

Solution:

$T_1 = 1580°R$
$T_2 = 3047°R$
$T_0 = 535°R$
$T_3 = 1345°R$

Using the non-flow process relationships of Chapter 3, Equations (7-3) and (7-9), the following results may be determined.

$$s_2 - s_1 = \int_1^2 \frac{\delta q}{T} = c_p \ln \frac{T_2}{T_1} = 0.24 \ln \frac{3047}{1580}$$
$$= 0.1576 \text{ Btu/lb} \cdot °R$$

	Cycle a2b4		Cycle 12b4		Cycle 1234	
	Btu/lb	areas	Btu/lb	areas	Btu/lb	areas
q_S	480.2	(a2cd)	352.1	(12cd)	352.1	(12cd)
q_R	84.3	(b4dc)	84.3	(b4dc)	138.5	(34dc)
wk/J	395.9	(a2b4)	267.8	(12b4)	213.6	(1234)
A.E.	395.9	(a2b4)	267.8	(12b4)	267.8	(12b4)
U.A.E.	84.3	(b4dc)	84.3	(b4dc)	84.3	(b4dc)
η_t, %	82.4	—	76.1	—	60.7	—

It was stated earlier that Carnot cycle efficiency is independent of the nature of the working substance. The physical characteristics of actual working substances place further limitations on the processes to which they are amenable, resulting in further reductions in cyclic available energy. Suppose water in its several phases is to be used as a working substance at 2000 psia maximum pressure between the allowable temperature limits of 1960°R and 520°R. The shape and location of the 2000 psia line on Fig. 7-12, derived from Fig. 1 of the Steam Tables, show immediately that constant temperature heat addition at the upper *temperature* limit established by the materials is not feasible. The physical properties of the fluid, and certain elements of process control which will be discussed later, preclude the addition of heat by any process other than constant pressure. As a result, the average temperature of heat receipt is far below the maximum allowable material temperature. The reduction in available energy resulting from these physical limitations in the case of water will be shown in the following example.

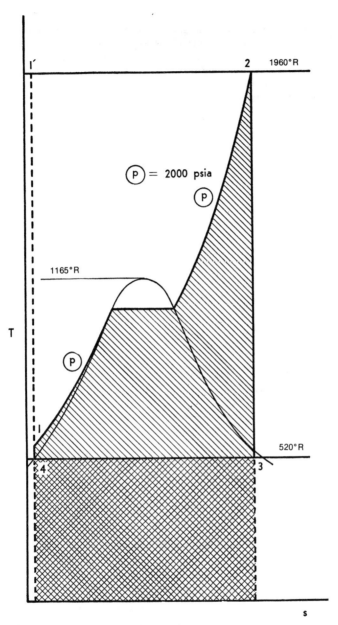

FIGURE 7-12. Physical limitations of fluid on heat addition process.

THE SECOND LAW OF THERMODYNAMICS

Example 7-5: In an ideal steam cycle, heat is added at a constant pressure of 2000 psia. Water entering the boiler has an entropy of 0.0555 Btu/lb · °R and steam leaving the superheater has an entropy of 1.7395 Btu/lb · °R. Superheated steam temperature is 1500°F, and heat added is 1744.5 Btu/lb (equals $h_2 - h_1$, Fig. 7-12). Heat rejection is carried out in a condenser at a constant temperature of 60°F ($h_3 - h_4$, Fig. 7-12). Average temperature of the combustion gases in the boiler furnace is 3000°F. Calculate:
 (a) the available energy of the combustion gases with respect to the sink temperature of 520°R, Btu/lb
 (b) the available energy of an ideal cycle receiving its energy at a constant source temperature equal to the superheated steam temperature, Btu/lb
 (c) the available energy of the ideal steam cycle, Btu/lb
 (d) the mean effective temperature of the working substance during heat receipt in the original steam cycle, degrees F.

Solution:

$$s_1 = s_4 = 0.0555 \text{ Btu/lb} \cdot °R$$
$$s_2 = s_3 = 1.7395 \text{ Btu/lb} \cdot °R$$
$$p_1 = p_2 = 2000 \text{ psia} \qquad t_2 = 1500°F \qquad t_3 = t_4 = 60°F$$
$$t_{2'}(\text{combustion gas temperature}) = 3000°F$$
$$q_s = 1744.5 \text{ Btu/lb}$$

(a)
$$A.E._{2'} = q_{s_{2'}} - T_0(s_3 - s_4) = T_{2'}(s_2 - s_1) - T_0(s_3 - s_4)$$
$$= (T_{2'} - T_0)(s_3 - s_4)$$
$$A.E._{2'} = (3460 - 520)(1.7395 - 0.0555) = \underline{4951 \text{ Btu/lb}}$$

(b)
$$A.E._2 = q_{s_2} - T_0(s_3 - s_4) = T_2(s_3 - s_4) - T_0(s_3 - s_4)$$
$$= (T_2 - T_0)(s_3 - s_4)$$
$$A.E._2 = (1960 - 520)(1.7395 - 0.0555) = \underline{2425 \text{ Btu/lb}}$$

(c)
$$A.E. = q_s - T_0(s_3 - s_4) = 1744.5 - 520(1.7395 - 0.0555)$$
$$A.E. = 1744.5 - 875.7 = \underline{868.8 \text{ Btu/lb}}$$

(d)
$$\text{mean effective temperature} = \frac{q_s}{s_2 - s_1} = \frac{1744.5}{(1.7395 - 0.0555)}$$

$$T_{\text{mean}} = \frac{1744.5}{1.684} = 1035.9°R$$

$$t_{\text{mean}} = T_{\text{mean}} - 460.0 = 1035.9 - 460.0 = \underline{575.9°F}$$

or

$$t_{\text{mean}} = \frac{A.E.}{s_2 - s_1} + t_3 = \frac{868.8}{1.684} + 60 = \underline{575.9°F}$$

7-9. Actual Thermal and Engine Efficiencies.

Only a portion of the heat supplied to any cycle may be converted to work. This fact is derived from the Second Law of Thermodynamics. That portion of the heat supplied converted into work by an ideal engine operating in a reversible cycle and rejecting heat to the lowest mean temperature of the surroundings is called available energy. Because of inherent irreversibilities, the net work of any actual engine operating in a cycle must be less than the available energy for that cycle. In addition, the heat rejection process may not be at constant temperature, which would further reduce the net cyclic work with

THE SECOND LAW OF THERMODYNAMICS

respect to the available energy. It should be apparent, then, that the *actual thermal efficiency* for a cycle, defined by

$$\eta_t = \frac{wk(\text{net})\text{actual}}{(J)(q_s)} \qquad (7\text{-}3)$$

might very reasonably have a rather low value. The mechanical equipment of actual operating cycles is so varied that numerous possibilities for measurement of net work output occur. For the same reason there are several points in an actual cycle at which the heat supply may be determined. This multiplicity of reference points for evaluation of heat supply and work output creates a similar multiplicity of actual thermal efficiencies, each with its particular numerical value, but all based on the fundamental equation stated above. Plant thermal efficiency, which is really an overall thermal efficiency, is a ratio of the ultimate work output per unit time for a plant to the heat supplied during the same unit of time, based on the heating value of the fuel supplied. Overall thermal efficiencies in even the most modern plants rarely exceed 35 percent.

Turbine and engine manufacturers are reluctant to rate their machines on the basis of cyclic thermal efficiency, since many do not directly receive and reject energy as heat. Consequently, to separate heat transfer and other cyclic inefficiencies from the engine and to obtain a more realistic performance rating, the term engine efficiency is defined as

$$\eta_{\text{eng}} = \frac{wk(\text{actual engine})}{wk(\text{ideal engine})} \qquad (7\text{-}10)$$

Equation (7-10) compares the work of an actual turbine or engine with that of an ideal engine, taking a working substance at the same initial conditions and exhausting to the same pressure.

7-10. Entropy Change as an Index of Unavailability and Irreversibility. Every quantity of heat supplied to a cycle has a portion available to do useful work and an unavailable portion which must be rejected ultimately to a receiver. The role of entropy change in indicating the unavailable portion of a quantity of heat supplied to a cycle is shown in Fig. 7-13. Using the same source and receiver temperatures as in Example 7-3, (3460°R and 520°R), assume first that 100 Btu of energy as heat is supplied reversibly to a cycle such as $a34b$ at 3460°R. The entropy change for the heat addition is

$$S_4 - S_3 = \frac{Q_{34}}{T_3} = \frac{100}{3460} = 0.0289 \text{ Btu/°R}$$

FIGURE 7-13. Effect of entropy change on unavailability of energy.

From the diagram it is clear that $(S_8 - S_1) = (S_4 - S_3)$. Therefore, the unavailable portion of Q_S is

$$U.A.E. = T_1(S_8 - S_1) = 520(0.0289) = 15.0 \text{ Btu}$$

Now assume the same quantity of heat is added reversibly, but at a constant temperature of 1960°R. The resulting cycle area is represented by $a26c$ on the figure, and the change of entropy is

$$S_6 - S_2 = \frac{Q_S}{T_2} = \frac{100}{1960} = 0.0510 \text{ Btu/°R}$$

The unavailable portion, as determined by $T_0(\Delta S)$ is

$$U.A.E. = T_1(S_7 - S_1) = T_1(S_6 - S_2) = 520(0.0510) = 26.5 \text{ Btu}$$

The unavailability of the energy supplied increased as the entropy change for the supply process increased. Thus, entropy change for a heat addition process is an index or indicator of the unavailable portion of the heat supplied.

Entropy change plays an important role for those processes in which no heat is transferred, as well as in those just described. Article 7-3 showed that an adiabatic process which was also reversible produced the most work from a given quantity of stored energy. Article 5-8 discussed the

throttling process in which a pressure drop occurred, specific volume was increased and enthalpy remained the same, but no useful work was done. These two processes represent the extremes so far as reversibility is concerned. The throttling process is totally irreversible, with no useful work and a relatively large increase in entropy. The isentropic process has no increase of entropy and a maximum of work for a given set of conditions. Between these two natural extremes of reversibility lie an infinite number of possibilities for irreversible processes. Starting with the reversible adiabatic process 1-2 of Fig. 7-14 and proceeding toward process 1-2′, the throttling process, each of the process lines encountered indicates an expansion process between the pressure limits shown, with decreasing quantities of work developed and increasing values for the change of entropy. This is equivalent to decreasing values of engine efficiency as expressed by Equation (7-10).

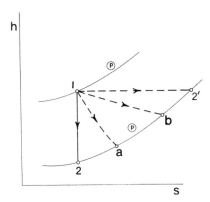

FIGURE 7-14. Processes with varying degrees of irreversibility.

Practical applications of the throttling process and the irreversible adiabatic expansion are shown as processes 2-3 and 3-4, respectively, of Fig. 7-15. Pressure drop from 2 to 3 might be that due to turbine throttling or governor valve action, while the pressure drop from 3 to 4 is the normal drop from throttle or steam chest to turbine exhaust. The degree of irreversibility for each of these processes is shown by entropy increase *a-b* and *b*-4, respectively. Additional heat rejected by the cycle due to throttling is represented by area *abde*, while that due to irreversible turbine expansion is *b4cd*. Both of these processes are adiabatic, so no external transfer of heat occurs; but because of internal friction and turbulence, energy that might have done useful work was dissipated or degraded. The net result was an increase of entropy for the process, an increased quantity

THE SECOND LAW OF THERMODYNAMICS

FIGURE 7-15. Throttling and irreversible adiabatic expansion processes in steam power cycle.

of heat to be rejected by the cycle and a reduced quantity of net cyclic work.

Systems in which one part of the fluid increases in entropy while the other decreases, such as in a mixing process, will always exhibit a net increase of entropy if the system is not allowed to transfer heat with its surroundings. This is demonstrated by an example:

Example 7-6: In a perfectly insulated cylinder and piston arrangement as shown, one pound of saturated steam at 212°F is to be mixed with 9 pounds of water at 60°F. The entire heat exchange process is to take place at standard atmospheric pressure.

Calculate:
 (a) the final mixture temperature, °F
 (b) the change of entropy of the steam, Btu/°R
 (c) the change of entropy of the water, Btu/°R
 (d) the net change of entropy of the system, Btu/°R

Solution:

In this non-flow system the non-flow energy equation will apply to each phase of the fluid involved. By writing the equation separately for each fluid and equating through the heat transferred terms, a heat balance is formed.

THE SECOND LAW OF THERMODYNAMICS

For water $M_w(u_1) + M_w q_{12} = M_w(u_2) + M_w \int_1^2 \dfrac{p\, dv}{J}$

For steam $M_s(u_{1'}) + M_s q_{1'2} = M_s(u_2) + M_s \int_{1'}^2 \dfrac{p\, dv}{J}$

Since the system is thoroughly insulated from its surroundings, all heat transfer must occur between the steam and the water. Therefore,

$$(-)M_s q_{1'2} \text{ (steam)} = M_w q_{12} \text{ (water)}$$

and

$$(-)\left[M_s(u_2 - u_1')_{\text{steam}} + M_s \int_{1'}^2 \dfrac{p\, dv}{J}\right] = M_w(u_2 - u_1)_{\text{water}} + M_w \int_1^2 \dfrac{p\, dv}{J}$$

Finally,

$$(-)M_s\left[\left(u_2 + \dfrac{p_2 v_2}{J}\right) - \left(u_1' + \dfrac{p_1' v_1'}{J}\right)\right] = M_w\left[\left(u_2 + \dfrac{p_2 v_2}{J}\right) - \left(u_1 + \dfrac{p_1 v_1}{J}\right)\right]$$

$$(-)M_s(h_2 - h_1') = M_w(h_2 - h_1)$$

(a) $\quad M_s(h_1' - h_2) = M_w(h_2 - h_{f_1})$
$\quad 1(1150.5 - h_2) = 9(h_2 - 28.08)$
$\quad 1150.5 - h_2 = 9h_2 - 252.7$
$\quad 10h_2 = 1403.2$
$\quad h_2 = 140.32 \text{ Btu/lb}$
$\quad t_2 = \underline{172.4\,°F}$

(b) $\quad 1(s_2 - s_1') = 0.25113 - 1.7567 = -1.5056 \text{ Btu/°R}$

(c) $\quad 9(s_2 - s_1) = 9(0.25113 - 0.05555) = +1.7602 \text{ Btu/°R}$

(d) \quad Net entropy change for system $= +1.7602 - 1.5056$
$\quad\quad\quad = \underline{+0.2546 \text{ Btu/°R}}$

217

THE SECOND LAW OF THERMODYNAMICS

In the example above, the mixing substances are initially at different temperatures which typifies the irreversible transfer of heat. The net increase of entropy for the system demonstrates the irreversibility of the mixing process. Had the initial temperatures been more widely separated, the system increase in entropy would have been greater, and the difference between the two changes of entropy would indicate the greater degree of irreversibility of the second process.

7-11. Closure. To summarize the important concepts and equations developed in this chapter, they will be listed here very briefly.

A cycle is a recurring series of processes for the continuous production of work from energy transferred as heat. The thermal efficiency of any thermodynamic power cycle is

$$\eta_t = \frac{q_S - q_R}{q_S} = \frac{wk(\text{net})}{(J)(q_S)} \tag{7-3}$$

The Carnot cycle has the highest possible thermal efficiency of any cycle operating between constant source and sink temperatures.

Any cycle that does not transfer all its heat at constant temperature, even though completely reversible, will have a lower thermal efficiency than one that does, if source and sink temperatures are comparable.

The **Second Law of Thermodynamics** states: **No engine, actual or ideal, when operating in a cycle can convert all the heat supplied it into work.**

Carnot's principle is the root of three basic thermodynamic propositions:

1. No heat engine cycle for production of mechanical work from energy received at a single high temperature and rejected at a single low temperature can be more efficient than a reversible cycle.
2. The efficiencies of all reversible heat engine cycles receiving and rejecting heat at the same higher and lower temperatures are the same.
3. The efficiency of a reversible heat engine cycle depends only on the temperatures of the energy source and sink.

Pump work, although a recognized negative quantity when calculated as a single process, for purposes of cycle analysis will be used as a positive numerical quantity in the equation

$$wk(\text{net}) = wk(\text{engine}) - wk(\text{pump}) \tag{7-8}$$

For a similar reason, when calculating net cycle work from cyclic heat transfer, q_R is used as a positive numerical quantity in the equation

THE SECOND LAW OF THERMODYNAMICS

$$q_S - q_R = \frac{wk(\text{net})}{(J)} \text{ Btu/lb (J/kg)} \tag{7-2}$$

Available energy is defined as the portion of thermal energy supplied to a thermodynamic cycle or heat engine ideally capable of ultimate conversion into mechanical work. Unavailable energy is that portion of the thermal energy reversibly supplied which ultimately must be discarded at the sink temperature (lowest available temperature of the surroundings). Expressed in equation form,

$$A.E. = q_S - T_0(\Delta s) \text{ Btu/lb (J/kg)} \tag{7-9}$$

and

$$U.A.E. = T_0(\Delta s) \text{ Btu/lb (J/kg)}$$

The change of entropy for a reversible heat transfer process, when combined with the ambient sink temperature, determines unavailability of energy. In this capacity, entropy change is an index of unavailability. For processes with no external heat transfer, the change of entropy indicates the degree of irreversibility of the process.

Overall plant thermal efficiency is expressed by the same equation as actual cycle thermal efficiency, the only difference being the point of measurement for work output and heat input. The equation is

$$\eta_t = \frac{wk(\text{net})(\text{actual})}{(J)(q_S)} \tag{7-3}$$

Engine efficiency is the term used to compare the output of an actual turbine or engine with the output of an ideal machine operating with the same cyclic conditions imposed. It is expressed by

$$\eta_{\text{eng}} = \frac{wk(\text{actual engine})}{wk(\text{ideal engine})} \tag{7-10}$$

Thermal efficiencies of actual cycles are lower than ideal cycles because of methods of heat transfer, imperfections of actual machines, fluid property limitations on process feasibility and, finally, temperature limitations set by the physical properties of the materials involved.

Problems

7-1. Distinguish between the following items:
(a) thermodynamic process and thermodynamic cycle
(b) open cycle and closed cycle
(c) thermal and mechanical irreversibility
(d) internal and external irreversibility

7-2. Which of the four types of irreversibility occasions the major loss in any cycle, and what remedial measures best serve to reduce its magnitude?

THE SECOND LAW OF THERMODYNAMICS

7-3. A Carnot power cycle employing 1 lb of air as a working substance is presumed to operate between temperature limits of 600°F and 70°F. The pressures at the beginning and end of the isothermal expansion process are 510 psia and 170 psia respectively. Determine (a) the change of volume between the end of isentropic expansion and the end of isentropic compression, ft^3/lb; (b) the heat supplied to the cycle, Btu/lb; (c) the heat rejected by the cycle, Btu/lb; (d) the net work of the cycle, Btu/lb; (e) the cycle thermal efficiency, percent.

Ans. (a) 12.3 ft^3/lb; (b) 79.8 Btu/lb; (c) 39.9 Btu/lb; (d) 39.9 Btu/lb; (e) 50.0%

7-4. One kilogram of a perfect gas (air) is used as a working substance in a Carnot power cycle. At the beginning of isentropic compression the temperature is 325.6 K and the pressure is 359.2 kPa abs. The pressure at the end of isentropic compression is 1379.0 kPa abs. For this cycle, the isothermal expansion ratio, (v_3/v_2), is 2. For the cycle (a) sketch the p-v and T-s diagrams; then calculate (b) the pressures, temperatures and specific volumes at each process termination point (SI units, numbered as in Fig. 7-4), (c) the heat supplied, kJ/kg, (d) the heat rejected, kJ/kg, and (e) the thermal efficiency of the cycle, %.

Ans. (b) v_1 = 0.260 m^3/kg p_1 = 359.2 kPa abs T_1 = 325.6 K
v_2 = 0.0995 m^3/kg p_2 = 1379.0 kPa abs T_2 = 478.2 K
v_3 = 0.1990 m^3/kg p_3 = 689.5 kPa abs T_3 = 478.2 K
v_4 = 0.520 m^3/kg p_4 = 179.6 kPa abs T_4 = 325.6 K
(c) 95.1 kJ/kg; (d) 64.8 kJ/kg; (e) 31.9%

7-5. An engine working in a closed thermodynamic cycle produces 400 Btu of work/lb of working substance passing through the machine. The pump in the cycle requires 6 Btu/lb of working substance in performance of its duty. Calculate the net work of the cycle, Btu/lb.

Ans. 394 Btu/lb

7-6. A reversed Carnot cycle using 1 lb of air is assumed to constitute a refrigerating cycle between the temperatures of 140°F and 40°F. The isothermal expansion ratio for the cycle is 3. The pressure at the beginning of isothermal expansion is 50 psia. Calculate (a) the heat added to the working substance, Btu/lb; (b) the heat rejected by the working substance, Btu/lb; (c) the net mechanical work required by the cycle, Btu/lb.

Ans. (a) 37.63 Btu/lb; (b) 45.16 Btu/lb; (c) 7.53 Btu/lb

7-7. A certain working substance receives 100 Btu reversibly as heat at a temperature of 1000°F from an energy source at 3600°R. Referred to a receiver temperature of 80°F, calculate (a) the available energy of the working substance in Btu; (b) the available portion of the 100 Btu added at the source temperature, in Btu; (c) the reduction in available energy between the source temperature and the 1000°F temperature, in Btu.

Ans. (a) 63 Btu; (b) 85 Btu; (c) 22 Btu

7-8. A working substance receives 100 Btu of energy reversibly at a constant temperature of 1000°F in a situation which employs a receiver having a constant temperature of 120°F. Calculate (a) the available energy of the working substance, Btu, and (b) the gain in available energy that could be realized with a receiver maintained at an 80°F temperature.

Ans. (a) 60.3 Btu; (b) 2.7 Btu

7-9. Combustion gases at 3000°F supply 100.0 Btu of energy reversibly as heat to water and its vapor at 190 psia. Water enters the heater at 190 psia and 60°F and leaves as steam at 190 psia and 500°F. A receiver maintained at 60°F is available to condense the steam to a saturated liquid during removal of the unavailable energy. Calculate (a) the available energy of the heat added at combustion gas temperature with respect to the 60°F receiver, Btu; (b) the available portion of the 100.0 Btu if it were all supplied at the 500°F temperature of

THE SECOND LAW OF THERMODYNAMICS

the steam leaving the heater, Btu; (c) the available portion of the 100.0 Btu when added at the constant pressure of 190 psia, Btu.

Ans. (a) 85.0 Btu; (b) 45.8 Btu; (c) 34.0 Btu

7-10. One pound of a certain working substance receives 100 Btu of heat at a constant temperature of 2000°R. Calculate (a) the change of entropy for the substance, Btu/lb · °R, and (b) the available energy in Btu/lb with respect to a 100°F receiver. Then calculate (c) the change of entropy for 100 Btu added to 1 lb of the same substance at 1000°R, and (d) the available energy in Btu/lb at this latter temperature, the receiver temperature remaining 100°F.

Ans. (a) +0.05 Btu/lb · °R; (b) 72 Btu/lb; (c) +0.10 Btu/lb · °R; (d) 44 Btu/lb

7-11. One pound of steam per second at 1000 psia and 600°F expands adiabatically in a turbine to a pressure of 1 inch of mercury absolute where its enthalpy is 800.0 Btu/lb. Calculate (a) the change in entropy of the steam during the process, Btu/lb · °R; (b) the engine efficiency of the turbine, percent; (c) the additional heat rejected because of the irreversibility of the process, Btu/lb, and (d) internal horsepower developed.

Ans. (a) +0.044 Btu/lb · °R; (b) 95.0%; (c) 23.7 Btu/lb; (d) 635 hp.

7-12. One pound of air at 100°F is mixed with another pound of air at 40°F. The mixing process is constant pressure with constant specific heat. Calculate (a) the final mixture temperature, °F; (b) the change of entropy for the lb of gas originally at 100°F, Btu/lb · °R; (c) the change of entropy for the lb of gas originally at 40°F, Btu/lb · °R; (d) the net change of entropy for the final mixture, Btu/lb · °R.

Ans. (a) 70°F; (b) −0.01321 Btu/lb · °R; (c) +0.01398 Btu/lb · °R; (d) +0.00077 Btu/lb · °R

7-13. One kilogram of air at 38°C is mixed with another kilogram of air at 4.0°C. The mixing process is carried out at constant pressure with constant specific heat. Calculate (a) the final mixture temperature, °C; (b) the change of entropy for the air originally at 38°C, kJ/kg · K; (c) the change of entropy for the air originally at 4.0°C, kJ/kg · K; (d) the net change of entropy for the final mixture, kJ/kg · K.

Ans. (a) 21.0°C; (b) −0.0565 kJ/kg · K; (c) +0.0598 kJ/kg · K; (d) +0.0033 kJ/kg · K

7-14. A steam turbine receives steam at 600 psia and 600°F and exhausts it at 1 inch of mercury absolute to a condenser. The turbine develops 400 Btu of work from each pound of steam passing through. Calculate (a) the ideal work of the turbine, Btu/lb; (b) the engine efficiency of the turbine, percent.

Ans. (a) 466.2 Btu/lb; (b) 85.8%

7-15. A heat source is supplying 1000 Btu to each pound of a working substance in a closed thermodynamic power cycle. The sink is receiving 650 Btu/lb of working substance. The system pump requires 6.0 Btu/lb of substance pumped. (a) What is the net work of the cycle, and (b) what is the thermal efficiency of the cycle, percent?

Ans. (a) 350 Btu/lb; (b) 35.0%

7-16. A heat source supplies 2325 kJ/kg to a working substance in a closed thermodynamic power cycle. The sink receives 1510 kJ/kg of working substance. The system pump requires 14 kJ/kg of substance pumped. (a) What is the net work of the cycle, and (b) what is the thermal efficiency of the cycle, percent?

Ans. (a) 815 kJ/kg; (b) 35.0%

THE SECOND LAW OF THERMODYNAMICS

7-17. A closed thermodynamic cycle has a thermal efficiency of 30%. Heat supplied from an energy source is 800 Btu/lb of working substance. Calculate (a) the net work of the cycle and (b) the heat rejected, both in Btu/lb.

Ans. (a) 240 Btu/lb; (b) 560 Btu/lb

7-18. A Carnot power cycle has a maximum cycle temperature of 800°F. The change of entropy for the heat addition process is 0.085 Btu/lb · °R. Heat is rejected at a rate of 45.0 Btu/lb. Determine (a) the heat supplied, Btu/lb; (b) the net cycle work, Btu/lb; (c) the minimum cycle temperature, °F.

Ans. (a) 107.1 Btu/lb; (b) 62.1 Btu/lb; (c) 69.4°F

7-19. A Carnot power cycle operating between the temperature limits of 65°F and 550°F uses air as the working substance. The pressure at the end of the constant temperature heat rejection process is 15.0 psia. The pressure at the beginning of the isentropic expansion process is 34.3 psia. Show the cycle on p-v and T-s coordinates and calculate (a) cycle thermal efficiency; (b) pressure at the end of the isentropic compression process, psia; (c) change of entropy for the heat addition process, Btu/lb · °R; (d) the heat supplied, Btu/lb; (e) the net cycle work, Btu/lb.

Ans. (a) 48.0%; (b) 148.1 psia; (c) 0.1002 Btu/lb · °R; (d) 101.2 Btu/lb; (e) 48.6 Btu/lb

7-20. In a Carnot power cycle using air as the working substance, the temperature and pressure at the end of the isentropic compression process are 1500°F and 800 psia, respectively. The pressure at the beginning of the isentropic expansion is 120 psia. Sink temperature is 70°F. Draw the cycle on p-v and T-s coordinates and determine (a) the change of entropy for the heat addition process, Btu/lb · °R; (b) the heat added, Btu/lb; (c) the cycle thermal efficiency, percent; (d) the available energy for the cycle; Btu/lb.

Ans. (a) 0.1300 Btu/lb · °R; (b) 254.8 Btu/lb; (c) 73.0%; (d) 185.9 Btu/lb

7-21. A reversed Carnot cycle rejects 833 Btu/lb as heat at a temperature of 1140°F. The cycle receives heat at a temperature of 140°F. Sketch the cycle on p-v and T-s coordinates and calculate (a) the change of entropy for the heat rejection process, Btu/lb · °R, and (b) the net cycle work, Btu/lb.

Ans. (a) (−)0.5206 Btu/lb · °R; (b) (−)520.6 Btu/lb (on)

CHAPTER 8

POWER CYCLES (GAS)

8-1. Early Development. It was shown in Chapter 7 that the Carnot cycle, developed in 1824, is the most efficient power cycle possible between any fixed temperature limits. Because of the severe limitations imposed by its processes, however, no engine was ever built to operate on this cycle. A few years later the Stirling cycle was introduced along with the Stirling hot-air engine. Together with the Ericsson cycle proposed about 1853, these were the only important operating cycles until Nicholas A. Otto built his four-stroke-cycle, spark ignition, engine in 1862. Because of its greater simplicity and higher speed, the Otto engine soon outclassed other combustion engines on the market and reigned supreme over the field until Dr. Rudolph Diesel perfected the compression-ignition engine some thirty years later. From then until the present, the two types of engines have shared popularity, each having advantages in particular applications. Recent years have seen tremendous strides in the development of the gas turbine which had been held back by poor compressor efficiency and lack of metals capable of withstanding high, *continuous* temperatures.

The choice of a prime mover for a particular power requirement is often determined more by practical considerations than by the theoretical ones to which we shall limit our discussion. However, the practical considerations are so many and varied that their inclusion in this chapter is not feasible.

The operating cycle for each of the foregoing engines derives from a theoretical cycle, and it is these cycles which it is our purpose to investigate.

8-2. Methods of Analysis. Except for the open cycle gas turbine, our analyses will be based on the "closed" cycle wherein the working substance never leaves the engine. This will require the working substance to receive heat from an external source rather than from the combustion of fuel and air, and to reject heat by some suitable means to another external agency rather than by exhausting the working substance to the atmosphere.

The analysis of a gas cycle may be based on any one of three well known theoretical standards, depending upon the degree of absolute accuracy required of the results. The *Cold Air Standard* specifies air as the working substance and disregards the change in specific heat values with temperature. The specific heats are assumed to remain constant at normal

room temperature values regardless of the temperature range encountered in the cycle. The *Hot Air Standard* is similar to the Cold Air Standard but introduces corrections for the variation of the specific heat with temperature. The *Real Mixture Standard* most nearly approaches actual engine conditions, but is laborious and requires an accurate knowledge of the composition of the working substance. Calculations are carried forward for each constituent of the working substance both before and after combustion, and allowances are made for the variation in specific heat of the several constituents with temperature throughout the cycle. The Cold Air Standard yields less realistic results in determining state points and thermal efficiencies than the other two standards. However, it has the advantage of simplicity and gives correct qualitative results, so that any conclusions drawn from trends or comparisons are entirely valid. Consequently, *all subsequent analyses of gas cycles will be based on the Cold Air Standard* with the exception of the material presented in Art. 8-10.

8-3. The Stirling and Ericsson Cycles. The Stirling cycle is pictured on p-v and T-s coordinates in Fig. 8-1. The cycle consists of two constant volume processes separated by two isothermal processes. The mechanical

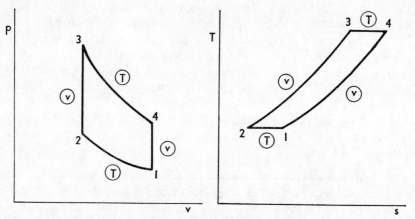

FIGURE 8-1. Ideal Stirling cycle.

equipment of the cycle consists of two separate cyclinders with a relatively small communicating passageway between, filled with a wire coil or mesh, to act as a regenerator or energy-retaining medium when the working substance is passed through from one cylinder to the other by the pistons. A source of heat energy and a heat reservoir complete the equipment of

POWER CYCLES (GAS)

the cycle. The actual engine is designed to use air as a working substance.

In operation, heat from the regenerator is added to the air at constant volume from 2 to 3 while the air moves from one cylinder to the other. Then heat is added from the external source while the air expands isothermally from 3 to 4 in the working cylinder. At 4 the air starts its return to the other cylinder, giving up a portion of its energy reversibly to the regenerator, before coming into contact with an external cold body at 1. Then heat is rejected from the air during an isothermal compression process until 2 is again reached.

Engines operating on the Stirling (constant volume regenerative) cycle were built and used during the latter part of the nineteenth century until displaced by the Otto cycle engine. However, new designs with much better overall efficiencies have appeared on the market in the past few years in what has been termed a "rebirth" of the hot-air engine.

The most interesting and important feature of the Stirling cycle is the provision for transferring heat from the working substance to the regenerator and back again, reversibly, but at a constantly changing temperature as shown by the process lines 2 to 3 and 4 to 1. The result of this storage of energy in the regenerator and subsequent acquisition by the air is that the amount of heat required from the source for a given amount of useful work produced will be less than without the regenerator.

It should be carefully noted with respect to the Stirling cycle that the heat transferred by the air to and from the regenerator during the constant volume processes 2 to 3 and 4 to 1 does not leave the system and does not, therefore, enter into the calculations of the heat added from the source or of that rejected to the receiver. Furthermore, the heat transferred to the regenerator is exactly equal to that regained from the regenerator, and the area on the T-s diagram under line 4-1 is equal and opposite in sign to that under 2-3.

The heat added from the source, then, becomes

$$q_s = q_{34} = T_3(s_4 - s_3)$$

$$q_s = T_3 \left[\frac{R}{(J)} \ln \frac{v_4}{v_3} \right] \text{Btu/lb (J/kg)}$$

The heat rejected by the cycle is

$$q_R = q_{21} = T_2(s_1 - s_2)$$

$$q_R = T_2 \left[\frac{R}{(J)} \ln \frac{v_1}{v_2} \right] \text{Btu/lb (J/kg)}$$

The net work of the cycle is

$$\frac{wk(\text{net})}{(J)} = q_S - q_R$$

$$\frac{wk(\text{net})}{(J)} = T_3 \left[\frac{R}{(J)} \ln \frac{v_4}{v_3} \right] - T_2 \left[\frac{R}{(J)} \ln \frac{v_1}{v_2} \right]$$

But $\qquad v_4 = v_1$ and $v_2 = v_3$

Then $\qquad \dfrac{wk(\text{net})}{(J)} = \dfrac{R}{(J)} \left[\ln \dfrac{v_4}{v_3} \right] (T_3 - T_2)$ Btu/lb (J/kg)

The thermal efficiency of the cycle is the net work divided by the heat supplied, giving

$$\eta_t = \frac{q_S - q_R}{q_S} = \frac{\dfrac{R}{(J)} \left[\ln \dfrac{v_4}{v_3} \right] (T_3 - T_2)}{\dfrac{R}{(J)} \left[\ln \dfrac{v_4}{v_3} \right] (T_3)}$$

$$\eta_t = \frac{T_3 - T_2}{T_3}$$

The important feature of the last equation is that it has the same form as the thermal efficiency of the Carnot cycle. This is to be expected, since all heat received from an external source and rejected to an external sink by the cycle was transferred isothermally.

The Ericsson engine and regenerative cycle are similar to the Stirling engine and cycle. The same mechanical equipment as listed for the Stirling cycle will suffice for the Ericsson. The Ericsson cycle is composed of two isothermal processes separated by two constant pressure processes. It employs regenerative heating in a manner similar to that of the Stirling cycle except the regenerative processes are carried out at constant pressure rather than at constant volume.

8-4. The Otto Cycle. Present-day automotive engines, reciprocating aircraft engines and a host of other small internal combustion spark ignition engines operate on the Otto cycle with gasoline as their fuel. For comparative purposes, an indicator card from an actual Otto cycle engine is shown in Fig. 8-2, along with the p-v and T-s diagrams for the ideal cycle. Note that the suction and exhaust strokes of the actual cycle are omitted on the ideal diagram, since the ideal cycle is closed with the working substance remaining in the cyclinder, while the actual engine works on an open cycle where the charge must be admitted and discharged. Furthermore, the rejection of heat which accompanies the

POWER CYCLES (GAS)

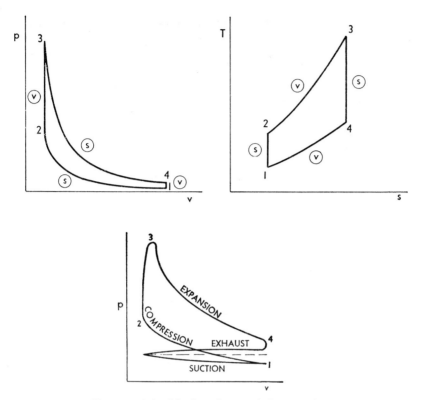

FIGURE 8-2. Ideal and actual Otto cycles.

exhaust of the working substance from the actual engine is accomplished in the ideal cycle by a constant volume process which returns the working substance to the initial state. The indicator diagram above is that of a four-stroke-cycle engine, but the difference between the four-stroke and the two-stroke-cycle diagrams is negligible. The suction and exhaust of the four-stroke-cycle are usually accomplished by an outside agency in the two-stroke-cycle engine, leaving the remainder of the indicator diagram much the same as for the four-stroke-cycle.

For the ideal Otto cycle using air as the working substance, heat is added during a constant volume, non-flow process from 2 to 3 on the T-s diagram of Fig. 8-2. In this case the work term goes to zero ($\nearrow °$) giving:

$$u_2 + q_{23} = u_3 + \frac{wk_{23}^{\;\nearrow°}}{(J)}$$

$$q_s = q_{23} = (u_3 - u_2) = c_v(T_3 - T_2) \text{ Btu/lb (J/kg)} \qquad (8\text{-}1)$$

POWER CYCLES (GAS)

Heat is rejected at constant volume, non-flow, from 4 to 1, giving

$$u_4 + q_{41} = u_1 + \cancel{\frac{wk_{41}}{(J)}}^{\!\!\!\circ}$$

$$q_R = q_{14} = (u_4 - u_1) = c_v(T_4 - T_1) \text{ Btu/lb (J/kg)} \quad (8\text{-}2)$$

Then,

$$\frac{wk(\text{net})}{(J)} = q_S - q_R \quad (7\text{-}2)$$

$$= c_v(T_3 - T_2) - c_v(T_4 - T_1) \text{ Btu/lb (J/kg)} \quad (8\text{-}3)$$

Finally, the thermal efficiency of the cycle is

$$\eta_t = \frac{q_S - q_R}{q_S} = \frac{c_v(T_3 - T_2) - c_v(T_4 - T_1)}{c_v(T_3 - T_2)}$$

$$\eta_t = 1 - \frac{T_4 - T_1}{T_3 - T_2}$$

By virtue of the isentropic processes between states 3 and 4, and 2 and 1, we may write for constant specific heat ratio, k:

$$\frac{T_3}{T_4} = \left(\frac{v_4}{v_3}\right)^{k-1} \quad \text{and} \quad \frac{T_2}{T_1} = \left(\frac{v_1}{v_2}\right)^{k-1}$$

and since $v_4 = v_1$ and $v_3 = v_2$,

$$\frac{T_3}{T_4} = \frac{T_2}{T_1}$$

Now, by rearranging terms and subtracting unity from each side of the resulting equation

$$\frac{T_4}{T_1} - 1 = \frac{T_3}{T_2} - 1$$

which becomes

$$\frac{T_4 - T_1}{T_1} = \frac{T_3 - T_2}{T_2}$$

and, finally,

$$\frac{T_4 - T_1}{T_3 - T_2} = \frac{T_1}{T_2}$$

The thermal efficiency of the ideal Otto cycle then becomes

$$\eta_t = 1 - \frac{T_1}{T_2} \quad (8\text{-}4)$$

POWER CYCLES (GAS)

The isentropic relationships for a perfect gas between states 1 and 2 of the Otto cycle give us the related expressions for thermal efficiency in terms of pressures and specific volumes as

$$\eta_t = 1 - \left(\frac{p_1}{p_2}\right)^{\frac{k-1}{k}} = 1 - \left(\frac{v_2}{v_1}\right)^{k-1}$$

$$\eta_t = 1 - \frac{1}{(r_v)^{k-1}} \tag{8-5}$$

where $r_v = v_1/v_2$, and that ratio is defined as the compression ratio for the cycle. **In the case of gas power cycles generally, the volume at the beginning of compression divided by the volume at the end of compression is called the compression ratio for the cycle.** Either total volumes or specific volumes may be used, since their ratios will be the same. The expansion ratio, which is occasionally useful, is defined for a gas power cycle as the volume at the end of the expansion stroke divided by the volume at the end of addition of heat to the cycle. In the ideal Otto cycle, the compression and expansion ratios are identical as can be seen from the p-v diagram. Equation (8-5) expresses the thermal efficiency of the ideal Otto cycle in terms of compression ratio alone and indicates that an increase in compression ratio will result in an increase in efficiency.

The net work of the Otto cycle may also be obtained by subtracting the work of compression (area under line 1-2 on the p-v diagram of Fig. 8-2) used as a positive numerical value from the work of expansion (area under line 3-4), to obtain area 1234. This area is equivalent to area 1234 of the T-s diagram, one being in Btu per pound and the other being in foot pounds force per pound. In SI units both areas are expressed in joules per kilogram.

Example 8-1: At the beginning of the compression stroke an ideal Otto cycle has an air pressure of 15 psia, a temperature of 75°F, and a specific volume of 13.2 cubic feet per pound. At the end of compression the specific volume is 1.76 cubic feet per pound. The heat supplied to the cycle is 352 Btu/lb. For the cycle (a) sketch the p-v and T-s diagrams. Then calculate:
 (b) the compression ratio
 (c) the highest temperature and pressure of the cycle
 (d) the temperature and pressure at the end of expansion of the air
 (e) the heat rejected, Btu/lb
 (f) the net work of the cycle, Btu/lb
 (g) the thermal efficiency of the cycle, percent
 (h) the horsepower developed by an ideal engine operating on this cycle using 0.5 pound of air per second.

POWER CYCLES (GAS)

Solution:

$$p_1 = 15 \text{ psia} \quad v_1 = 13.2 \text{ ft}^3/\text{lb}$$
$$t_1 = 75°F \quad v_2 = 1.76 \text{ ft}^3/\text{lb}$$
$$q_S = 352 \text{ Btu/lb}$$

(a) See Fig. 8-2.

(b) $$\text{compression ratio} = \frac{v_1}{v_2} = \frac{13.2}{1.76} = \underline{7.5}$$

(c) $$p_2 = p_1 \left(\frac{v_1}{v_2}\right)^k = 15 \left(\frac{13.2}{1.76}\right)^{1.4} = \underline{252 \text{ psia}}$$

$$T_2 = \frac{p_2 v_2}{R} = \frac{252(144)1.76}{53.3} = \underline{1198°R}$$

$$T_3 = \frac{q_s}{c_v} + T_2 = \frac{352}{0.171} + 1198 = \underline{3256°R}$$

$$p_3 = \frac{RT_3}{v_3} = \frac{53.3(3256)}{1.76(144)} = \underline{685 \text{ psia}}$$

(d) $$T_4 = T_3 \left(\frac{v_3}{v_4}\right)^{k-1} = 3256 \left(\frac{1}{7.5}\right)^{0.4} = \underline{1454°R}$$

$$p_4 = p_1 \left(\frac{T_4}{T_1}\right) = 15 \left(\frac{1454}{535}\right) = \underline{40.8 \text{ psia}}$$

(e) $$q_R = c_v(T_4 - T_1) = (0.171)(1454 - 535)$$
$$q_R = \underline{157 \text{ Btu/lb}}$$

(f) $$\frac{wk(\text{net})}{J} = q_S - q_R = 352 - 157 = \underline{195 \text{ Btu/lb}}$$

(g) $$\eta_t = \frac{q_S - q_R}{q_S} = \frac{195}{352} = \underline{55.4\%}$$

or $$\eta_t = 1 - \frac{1}{(r_v)^{k-1}} = 1 - \frac{1}{(7.5)^{0.4}} = \underline{55.3\%}$$

(h) $$P = \dot{W}k = \frac{\dot{M}(wk)60}{J(42.42)} = \frac{0.5(195)60}{42.42} = \underline{138 \text{ hp}}$$

$$1 \text{ hp} = 42.42 \text{ Btu/min}$$

8-5. The Diesel Cycle. In the low horsepower field, many truck, bus, tractor and boat engines operate on a compression ignition cycle. In the medium horsepower field, arbitrarily taken as 500 to 5,000 horsepower, small municipal power plants, locomotives, ships and craft, earth-moving equipment and many other industrial applications are served by Diesel engines. At the upper end of the power spectrum are the low-speed Diesel engines specifically designed for marine propulsion, with full power ratings ranging from 7500 to 20,000 shaft horsepower per unit.

POWER CYCLES (GAS)

The ideal Diesel cycle is shown on p-v and T-s coordinates in Fig. 8-3. The differentiating characteristic between the ideal Otto cycle and the ideal Diesel cycle is the process during which heat is added. In the Otto cycle heat is added at constant volume, while in the Diesel cycle heat is added at constant pressure, process 2 to 3 of Fig. 8-3. Otherwise the two cycles are similar.

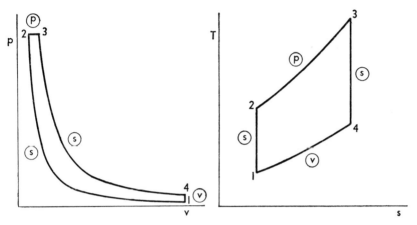

FIGURE 8-3. Ideal Diesel cycle.

The actual engines have other characteristic differences. In the Diesel engine air alone is compressed in the cylinder, while the actual Otto engine compresses a mixture of air and gasoline. Compression of air alone in the Diesel engine allows considerably higher compression ratios without the danger of preignition than are presently allowable for the Otto engine. Also, the heavier fuel used by the Diesel engine ignites more slowly than gasoline, which makes the true Diesel inherently a *relatively* slow-speed machine. Because of higher average pressures, Diesel engines have generally been heavier per horsepower developed than Otto engines.

In the ideal Air Standard Diesel cycle shown, process 1 to 2 represents isentropic compression of air, process 2 to 3 constant pressure addition of heat, 3 to 4 isentropic expansion of the heated air and finally 4 to 1 the constant volume cooling of the air back to the initial state.

Analysis of the cycle, Fig. 8-3, with the non-flow energy equation gives

$$u_2 + q_{23} = u_3 + \frac{wk_{23}}{(J)}$$

$$q_{23} = u_3 - u_2 + \frac{p_3 v_3}{(J)} - \frac{p_2 v_2}{(J)}$$

POWER CYCLES (GAS)

$$q_S = q_{23} = (h_3 - h_2)$$
$$q_S = c_p(T_3 - T_2) \text{ Btu/lb (J/kg)} \tag{8-6}$$

and

$$u_4 + q_{41} = u_1 + \frac{wk_{41}}{(J)}$$
$$q_{41} = u_1 - u_4$$
$$q_R = q_{14} = (u_4 - u_1)$$
$$q_R = c_v(T_4 - T_1) \text{ Btu/lb (J/kg)} \tag{8-7}$$

Then

$$\frac{wk(\text{net})}{(J)} = q_S - q_R \tag{7-2}$$

$$= c_p(T_3 - T_2) - c_v(T_4 - T_1) \text{ Btu/lb (J/kg)} \tag{8-8}$$

The thermal efficiency of the cycle is

$$\eta_t = \frac{q_S - q_R}{q_S} \tag{7-3}$$

$$\eta_t = \frac{c_p(T_3 - T_2) - c_v(T_4 - T_1)}{c_p(T_3 - T_2)}$$

$$\eta_t = 1 - \frac{c_v(T_4 - T_1)}{c_p(T_3 - T_2)}$$

$$\eta_t = 1 - \frac{(T_4 - T_1)}{k(T_3 - T_2)} \tag{8-9}$$

In discussing the factors affecting the thermal efficiency of the Air Standard Diesel cycle, it is convenient to convert this last equation into another form:

$$\eta_t = 1 - \frac{1}{r_v^{k-1}}\left[\frac{r_L^k - 1}{k(r_L - 1)}\right] \tag{8-10}$$

A complete derivation is given in Appendix C. The fuel cut-off ratio, r_L, is defined as the ratio of the specific volume after constant pressure addition of heat to the specific volume at the beginning of the process. For the Diesel cycle, then, $r_L = v_3/v_2 = T_3/T_2$. With fixed inlet conditions and a definite compression ratio, the fuel cut-off ratio will vary directly with the amount of heat added at constant pressure, as indicated by the temperature quotient.

Inspection of Equation (8-10) shows the thermal efficiency of the Diesel cycle increases with an increase in compression ratio, as for the Otto cycle.

POWER CYCLES (GAS)

The immediate effect of the bracketed term, however, is not so readily apparent. If the equation is differentiated with respect to r_L for a constant compression ratio and set equal to zero, it will be found that the maximum thermal efficiency of the Diesel cycle occurs when the fuel cut-off ratio, r_L, has a value of unity. Furthermore, as r_L is allowed to approach unity as a limit, the thermal efficiency approaches the Otto cycle efficiency for that compression ratio. Actually, the bracketed factor always has a value greater than unity, with the result that as more heat is added at constant pressure, the thermal efficiency falls off. However, as more heat is added per cycle, the useful work per cycle is increased, contributing toward a smaller engine for a given horsepower output. In practice, a compromise must be reached between efficiency and engine size. The increased thermal efficiency, which accrues from high compression ratios as compared to the Otto cycle, is largely overcome in actual engines by the deleterious effect of the usual fuel cut-off ratios. The following example will demonstrate some of the concepts presented in this discussion.

Example 8-2: At the beginning of compression an ideal Diesel cycle using air has a pressure of 15 psia, a temperature of 75°F and a specific volume of 13.2 cubic feet per pound. For a compression ratio of 15 and a heat addition of 352 Btu per pound (a) sketch the p-v and T-s diagrams for the cycle, and then calculate:
 (b) the temperatures and pressures at the end of compression, at the end of addition of heat and at the end of the expansion process
 (c) the heat rejected, Btu/lb
 (d) the net work, Btu/lb
 (e) the thermal efficiency, percent
 (f) the horsepower developed by an ideal engine operating on the cycle and using 0.5 pound per second of air
 (g) the thermal efficiency of a cycle having the same initial conditions and compression ratio, but with a constant pressure heat addition of 500 Btu/lb

Solution:

$$p_1 = 15 \text{ psia} \qquad v_4 = v_1 = 13.2 \text{ ft}^3/\text{lb}$$
$$t_1 = 75°F \qquad v_1/v_2 = 15.0$$
$$q_s = 352 \text{ Btu/lb}$$

(a) See Fig. 8-3.

(b) $$v_2 = \frac{v_1}{r_v} = \frac{13.2}{15.0} = 0.88 \text{ ft}^3/\text{lb}$$

$$p_2 = p_1 \left(\frac{v_1}{v_2}\right)^{1.4} = 15(15)^{1.4} = \underline{664.7 \text{ psia}}$$

$$T_2 = \frac{p_2 v_2}{R} = \frac{664.7(144)0.88}{53.3} = \underline{1580°R}$$

$$p_3 = p_2 = \underline{664.7 \text{ psia}}$$

$$T_3 = T_2 + \frac{q_s}{c_p} = 1580 + \frac{352}{0.24} = 1580 + 1467 = \underline{3047°R}$$

233

POWER CYCLES (GAS)

$$v_3 = \frac{RT_3}{p_3} = \frac{53.3(3047)}{664.7(144)} = 1.697 \text{ ft}^3/\text{lb}$$

$$T_4 = T_3 \left(\frac{v_3}{v_4}\right)^{k-1} = 3047 \left(\frac{1.697}{13.2}\right)^{0.4}$$

$$= 3047(0.44) = \underline{1341°R}$$

$$p_4 = p_1 \left(\frac{T_4}{T_1}\right) = 15 \left(\frac{1341}{535}\right) = \underline{37.6 \text{ psia}}$$

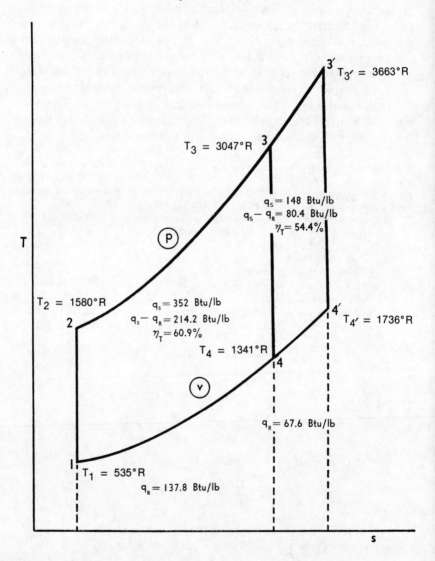

POWER CYCLES (GAS)

(c) $\quad q_R = c_v(T_4 - T_1) = 0.171(1341 - 535) = \underline{137.8 \text{ Btu/lb}}$

(d) $\quad \dfrac{wk(\text{net})}{J} = q_S - q_R = 352 - 137.8 = \underline{214.2 \text{ Btu/lb}}$

(e) $\quad \eta_t = \dfrac{q_S - q_R}{q_S} = \dfrac{214.2}{352} = \underline{60.9\%}$

(f) $\quad P = \dot{W}k = \dfrac{(wk)\dot{M}}{J(42.42)} = \dfrac{214.2(0.5)60}{42.42} = \underline{151.5 \text{ hp}}$

(g) $\quad T_3' = T_2 + \dfrac{q_S'}{c_p} = 1580 + 2083 = 3663°R$

$\quad v_3' = \dfrac{RT_3'}{p_3'} = \dfrac{53.3(3663)}{664.7(144)} = 2.04 \text{ ft}^3/\text{lb}$

$\quad T_4' = T_3'\left(\dfrac{v_3'}{v_4'}\right)^{k-1} = 3663\left(\dfrac{2.04}{13.2}\right)^{0.4} = 1736°R$

$\quad q_R' = c_v(T_4' - T_1) = 0.171(1736 - 535) = 205.4 \text{ Btu/lb}$

$\quad \eta_t = \dfrac{q_S' - q_R'}{q_S'} = \dfrac{500 - 205.4}{500} = \dfrac{294.6}{500} = \underline{58.9\%}$

8-6. The Dual Combustion Cycle. The dual combustion cycle, also known as the Sabathé cycle and the limited pressure cycle, is really a combination of the Otto and the Diesel cycles. It consists of the usual isentropic compression and expansion processes, but part of the heat is added at constant volume, as in the Otto cycle, and the remainder is added at constant pressure, as in the Diesel cycle. Most of the high-speed Diesel engines of today operate on this dual combustion cycle and are more properly called compression ignition engines, as opposed to the spark ignition engines. Piston speed and fuel characteristics are largely responsible for the difference between the actual Diesel and dual combustion engines.

The dual combustion cycle is pictured on p-v and T-s coordinates in Fig. 8-4. Compression of air alone, in the ideal cycle, takes place isentropically from 1 to 2, heat is added at constant volume from 2 to 3, more heat is added but at constant pressure from 3 to 4 and expansion of the air takes place from 4 to 5. Finally, heat is rejected at constant volume from 5 to 1.

From Fig. 8-4 and the non-flow energy equation, the heat added is

$$q_S = q_{23} + q_{34} = (u_3 - u_2) + (h_4 - h_3)$$
$$q_S = c_v(T_3 - T_2) + c_p(T_4 - T_3) \text{ Btu/lb (J/kg)} \quad (8\text{-}11)$$

FIGURE 8-4. Ideal dual combustion cycle.

and the heat rejected is

$$q_R = q_{15} = (u_5 - u_1)$$
$$q_R = c_v(T_5 - T_1) \text{ Btu/lb (J/kg)} \quad (8\text{-}12)$$

The net work of the cycle is

$$\frac{wk(\text{net})}{(J)} = q_S - q_R \quad (7\text{-}2)$$

$$= c_v(T_3 - T_2) + c_p(T_4 - T_3)$$
$$- c_v(T_5 - T_1) \text{ Btu/lb (J/kg)} \quad (8\text{-}13)$$

The thermal efficiency of the cycle is

$$\eta_t = \frac{q_S - q_R}{q_S} \quad (7\text{-}3)$$

$$\eta_t = \frac{c_v(T_3 - T_2) + c_p(T_4 - T_3) - c_v(T_5 - T_1)}{c_v(T_3 - T_2) + c_p(T_4 - T_3)}$$

$$\eta_t = 1 - \frac{(T_5 - T_1)}{(T_3 - T_2) + k(T_4 - T_3)} \quad (8\text{-}14)$$

The thermal efficiency of the dual combustion cycle may be expressed in terms of the compression ratio and the fuel cut-off ratio plus a third ratio, which is the pressure ratio, r_p, across the constant volume heating process. In this form the thermal efficiency is

POWER CYCLES (GAS)

$$\eta_t = 1 - \frac{1}{r_v^{k-1}} \left[\frac{r_L^k r_p - 1}{(r_p - 1) + k r_p(r_L - 1)} \right] \qquad (8\text{-}15)$$

where compression ratio $r_v = v_1/v_2$, fuel cutoff ratio $r_L = v_4/v_3$ and pressure ratio $r_p = p_3/p_2$. The complete development of this equation appears in Appendix C. For a fixed compression ratio, the equation permits a comparison with the Otto and Diesel cycles. Letting r_L approach unity as a limit puts the above equation in a form representing the thermal efficiency of the Otto cycle, while letting r_p approach unity as a limit gives an expression for Diesel cycle thermal efficiency. Furthermore, for a fixed quantity of heat added to the cycle, the equation shows that any increase in r_L, which effects a decrease in r_p, will cause the bracketed factor to increase, with a corresponding decrease in cycle thermal efficiency.

Example 8-3: At the beginning of compression an ideal dual combustion cycle using air has a pressure of 15 psia, a temperature of 75°F and a specific volume of 13.2 cubic feet per pound. For a compression ratio of 12 and a heat addition of 176 Btu/lb at constant volume and 176 Btu/lb at constant pressure, (a) sketch the cycle on p-v and T-s coordinates. Then, calculate for the cycle:
 (b) the pressure and temperature at the end of isentropic compression
 (c) the pressures and temperatures at the end of each heat addition process
 (d) the temperature at the beginning of heat rejection, °R
 (e) the heat rejected, Btu/lb
 (f) the net work, Btu/lb
 (g) the thermal efficiency, percent
 (h) the horsepower developed by an ideal engine operating on the cycle using 0.5 pound of air per second.

Solution:

$p_1 = 15$ psia $v_5 = v_1 = 13.2$ ft³/lb
$t_1 = 75°F$ $v_1/v_2 = 12.0$

heat added
 176 Btu/lb at constant volume
 176 Btu/lb at constant pressure

(a) See Fig. 8-4.

(b) $p_2 = p_1 \left(\dfrac{v_1}{v_2}\right)^k = 15(12)^{1.4} = 15(32.4) = \underline{486 \text{ psia}}$

$T_2 = T_1 \left(\dfrac{v_1}{v_2}\right)^{k-1} = 535(12)^{0.4} = \underline{1446°R}$

(c) $T_3 = T_2 + \dfrac{q_{23}}{c_v} = 1446 + \dfrac{176}{0.171} = \underline{2475°R}$

$p_3 = p_2 \left(\dfrac{T_3}{T_2}\right) = 486 \left(\dfrac{2475}{1446}\right) = \underline{832 \text{ psia}}$

POWER CYCLES (GAS)

$$p_4 = p_3 = \underline{832 \text{ psia}}$$

$$T_4 = T_3 + \frac{q_{34}}{c_p} = 2475 + \frac{176}{0.24} = \underline{3208°R}$$

(d) $$v_4 = v_3 \left(\frac{T_4}{T_3}\right) = \frac{13.2}{12}\left(\frac{3208}{2475}\right) = 1.426 \text{ ft}^3/\text{lb}$$

$$T_5 = T_4 \left(\frac{v_4}{v_5}\right)^{k-1} = 3208 \left(\frac{1.426}{13.2}\right)^{0.4} = \underline{1317°R}$$

(e) $$q_R = c_v(T_5 - T_1) = 0.171(1317 - 535)$$
$$q_R = \underline{134 \text{ Btu/lb}}$$

(f) $$\frac{wk(\text{net})}{J} = q_S - q_R = 352 - 134 = \underline{218 \text{ Btu/lb}}$$

(g) $$\eta_t = \frac{q_S - q_R}{q_S} = \frac{218}{352} = \underline{61.9\%}$$

(h) $$P = \dot{W}k = \frac{218(0.5)60}{42.42} = \underline{154.2 \text{ hp}}$$

At first glance this thermal efficiency may cause concern since it is higher than either the Otto cycle or the Diesel cycle so far considered, with the same heat added and the same initial conditions. However, with a higher compression ratio than the Otto cycle and a smaller quantity of heat added at constant pressure than the Diesel cycle, this result is to be expected.

8-7. The Joule or Brayton Cycle. The Joule cycle originally was proposed in the late nineteenth century for a reciprocating piston-cylinder configuration, while the Brayton cycle is currently associated with steady flow machines such as the gas turbine. Although the mechanical components for the two gas power cycles are considerably different, the ideal processes are identical. The ideal cycle will be treated first as a closed cycle, with the working substance remaining in the engine, and second as an open cycle.

The Joule or Brayton cycle representation on p-v and T-s coordinates is shown in Fig. 8-5 in the same manner as for the other gas cycles in this chapter. Air at state 1 is compressed isentropically to state 2. From state 2 to state 3, heat is added to the air at constant pressure. From state 3 to state 4 the air expands isentropically, after which rejection of heat proceeds at constant pressure from state 4 to state 1. From the steady flow

POWER CYCLES (GAS)

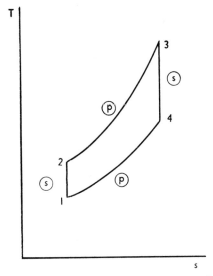

FIGURE 8-5. Ideal basic Brayton cycle.

energy equation applied to heat addition process 23,

$$q_S = q_{23} = (h_3 - h_2) = c_p(T_3 - T_2) \text{ Btu/lb (J/kg)} \quad (8\text{-}16)$$

Similarly, the heat rejected from 4 to 1 is

$$q_R = q_{14} = (h_4 - h_1) = c_p(T_4 - T_1) \text{ Btu/lb (J/kg)} \quad (8\text{-}17)$$

The net work of the cycle is

$$\frac{wk(\text{net})}{(J)} = q_S - q_R \quad \underline{(7\text{-}2)}$$

$$= c_p(T_3 - T_2) - c_p(T_4 - T_1) \text{ Btu/lb (J/kg)} \quad (8\text{-}18)$$

POWER CYCLES (GAS)

The thermal efficiency is

$$\eta_t = \frac{q_S - q_R}{q_S} = \frac{c_p(T_3 - T_2) - c_p(T_4 - T_1)}{c_p(T_3 - T_2)}$$

$$\eta_t = 1 - \frac{T_4 - T_1}{T_3 - T_2}$$

Both the compression and expansion processes are isentropic, allowing us to write for constant specific heat ratio, k:

$$\frac{T_4}{T_3} = \left(\frac{p_4}{p_3}\right)^{\frac{k-1}{k}} \quad \text{and} \quad \frac{T_1}{T_2} = \left(\frac{p_1}{p_2}\right)^{\frac{k-1}{k}}$$

Then, with $p_4 = p_1$ and $p_3 = p_2$,

$$\frac{T_4}{T_3} = \frac{T_1}{T_2}$$

Rearranging the latter expression and subtracting unity from each side of the resulting equation, we have

$$\frac{T_4}{T_1} - 1 = \frac{T_3}{T_2} - 1$$

Now

$$\frac{T_4 - T_1}{T_1} = \frac{T_3 - T_2}{T_2}$$

and

$$\frac{T_4 - T_1}{T_3 - T_2} = \frac{T_1}{T_2}$$

Finally, substituting the above in the last expression for thermal efficiency, we have

$$\eta_t = 1 - \frac{T_1}{T_2} \tag{8-19}$$

which corresponds to Equation (8-4) for the thermal efficiency of the Otto cycle. In the case of the Otto cycle and the associated piston-cylinder arrangement, the compression ratio (r_v) has clear-cut physical significance, is readily measurable and serves as a performance indicator as shown in Equation (8-5). In the Brayton cycle, with its separate steady flow components, the pressure ratio ($r_p = p_2/p_1$) is a more convenient criterion, easily determined by measuring the pressures on either side of the compressor. Introducing r_p into Equation (8-19) yields another useful expres-

sion for Brayton cycle thermal efficiency:

$$\eta_t = 1 - \frac{T_1}{T_2} = 1 - \left(\frac{p_1}{p_2}\right)^{\frac{k-1}{k}} = 1 - \left(\frac{1}{r_p}\right)^{\frac{k-1}{k}} \qquad (8\text{-}19\text{A})$$

As can be seen from this equation, the right-hand term decreases in value as the pressure ratio increases, and the thermal efficiency consequently also increases.

Although the equations for the thermal efficiencies of the Otto and Brayton cycles (Equations 8-4 and 8-19) appear identical, the performances of the actual devices (engines) based on these ideal cycles are at considerable variance because of certain physical differences and constraints. As pointed out in Chapter 7, even the best alloys cannot sustain *continuous* operating temperatures in excess of about 1500°F without serious problems. In the Diesel, Otto and Combined cycles, the compression, heat addition and expansion all take place within the confines of a piston-cylinder arrangement. Although the instantaneous temperature of the combustion products may momentarily exceed 3000°F, the temperature immediately begins to fall as the expansion process proceeds. In addition, the cylinders are cooled externally to reduce the skin temperature of the surrounding cylinder walls and heads. In the larger Diesel engines, the pistons are also water-cooled. Thus, in these cycles under actual conditions a considerable portion of the heat supplied is lost to the cooling water (or air in some cases) with a corresponding reduction in thermal efficiency—a fact not reflected in the analysis of the cycles on an ideal basis. In the case of the gas turbine operating on the Brayton cycle, the penalty associated with maximum allowable temperature at the turbine inlet is an inherent part of the ideal analysis as well as that of the actual machine. The development of designs that incorporate air cooling of the turbine blades and nozzles at the hot end has alleviated this handicap considerably. Inlet temperatures in the range of 1800 to 2200°F are attainable with excursions up to 2300°F permissible. As further advances in gas turbine technology are made, a steady increase in the allowable maximum inlet temperatures can be anticipated.

As might be expected, the upper and lower temperature limits imposed on the Brayton cycle by nature and state-of-the-art technology have significant effects on the theoretical performance of the cycle. An important performance factor, in addition to thermal efficiency, is specific work, a measure of useful output in relation to the mass flow rate of working substance involved. The higher the specific work, the smaller the unit and its several components will be for a given useful power output. As will be shown, maximum theoretical specific work is directly related to the limiting temperatures of the Brayton cycle.

POWER CYCLES (GAS)

Referring to Fig. 8-5, T_1 (at inlet to the compressor) is the lowest temperature in the cycle and is substantially fixed by the ambient conditions. Similarly, T_3 (at inlet to the turbine) is the highest temperature in the cycle and it is substantially fixed by the material constraints. From Equation (8-18), the net work of the cycle is:

$$\frac{wk(\text{net})}{(J)} = c_p(T_3 - T_2) - c_p(T_4 - T_1)$$

But as shown earlier under the constant specific heat assumption:

$$\frac{T_4}{T_3} = \frac{T_1}{T_2} \quad \text{or} \quad T_4 = \frac{T_1 T_3}{T_2}$$

Substituting for T_4 in the preceding work equation:

$$\frac{wk(\text{net})}{(J)} = c_p\left(T_3 - T_2 - \frac{T_1 T_3}{T_2} + T_1\right)$$

The next step is to determine the value of T_2 for maximum net work by the familiar differentiation technique, recognizing that c_p, T_1 and T_3 are "constants" for the purposes of this development. Then:

$$\frac{d}{dT_2}\left[\frac{wk(\text{net})}{(J)}\right] = 0 = \left[0 - 1 + \left(\frac{T_1 T_3}{T_2^2}\right) + 0\right]$$

and

$$T_2 = \sqrt{T_1 T_3}$$

Finally,

$$\frac{wk(\text{max})}{(J)} = c_p(T_3 + T_1 - 2\sqrt{T_1 T_3})$$

$$= c_p(\sqrt{T_3} - \sqrt{T_1})^2 \text{ Btu/lb (J/kg)}$$

From the foregoing it can be seen that the maximum net work per pound of working substance is realized when T_3 is as high as possible, T_1 is as low as possible, and $T_2 = \sqrt{T_1 T_3}$. Since T_2, the temperature at the end of isentropic compression, depends on the pressure ratio (r_p), there must be a particular value of r_p which will yield maximum net work. It can be determined as follows:

$$\frac{T_2}{T_1} = \left(\frac{p_2}{p_1}\right)^{\frac{k-1}{k}} = (r_p)^{\frac{k-1}{k}} = \left(\frac{p_3}{p_4}\right)^{\frac{k-1}{k}} = \frac{T_3}{T_4}$$

For maximum net work, $T_2 = \sqrt{T_1 T_3} = T_4$. Then:

$$(r_p)^{\frac{k-1}{k}} = \frac{T_2}{T_1} = \frac{\sqrt{T_1 T_3}}{T_1} = \left(\frac{T_3}{T_1}\right)^{1/2}$$

POWER CYCLES (GAS)

Finally:

$$r_p \text{ (for max work)} = \left(\frac{T_3}{T_1}\right)^{\frac{k}{2(k-1)}} \qquad (8\text{-}20)$$

Substituting for r_p in Equation (8-19A) yields the expression for thermal efficiency under the maximum net work condition:

$$\eta_t(\text{max net work}) = 1 - \sqrt{T_1/T_3}$$

Note that if the pressure ratio is increased beyond the value dictated by Equation (8-20) for maximum work, wk(net) will decrease but thermal efficiency will increase as deduced from Equation (8-19A). This inherent trade-off is typical of many engineering situations.

Example 8-4: In an ideal Brayton cycle, air enters the compressor at 15 psia and 75°F. The temperature of the air at turbine inlet is 1600°F. For maximum theoretical net work, find: (a) the temperature of the air leaving the compressor, °F; (b) the pressure ratio; (c) the net work, Btu/lb; (d) the thermal efficiency.

If the end conditions remain unchanged but the pressure ratio is increased to 15, find: (e) the net work, Btu/lb, and (f) the thermal efficiency.

Solution: Refer to Fig. 8-5.

$$p_1 = 15 \text{ psia} \qquad T_3 = 1600 + 460 = 2060°R$$
$$T_1 = 75 + 460 = 535°R$$

(a) $T_2 = \sqrt{T_1 T_3} = \sqrt{535 \times 2060} = 1050°R = \underline{590°F}$

(b) $r_p = \dfrac{p_2}{p_1} = \left[\dfrac{T_2}{T_1}\right]^{\frac{k}{k-1}} = \left[\dfrac{1050}{535}\right]^{\frac{1.4}{0.4}} = \underline{10.6}$

(c) $\dfrac{wk(\text{max})}{J} = c_p(T_3 + T_1 - 2\sqrt{T_1 T_3})$

$= 0.24(2060 + 535 - 2\sqrt{535 \times 2060}) = \underline{118.9 \text{ Btu/lb}}$

(d) $\eta_t(\text{max work}) = 1 - \sqrt{T_1/T_3} = 1 - \sqrt{535/2060} = 0.49 = \underline{49\%}$

(e) $r_p = 15$

$\dfrac{T_2}{T_1} = (r_p)^{\frac{k-1}{k}}; \quad T_2 = T_1(r_p)^{\frac{k-1}{k}} = 535(15)^{\frac{0.4}{1.4}} = 1160°R$

$T_4 = \dfrac{T_1 T_3}{T_2} = \dfrac{535 \times 2060}{1160} = 950°R$

$\dfrac{wk}{J} = c_p(T_3 - T_2) - c_p(T_4 - T_1) = q_S - q_R$

$= 0.24(2060 - 1160) - 0.24(950 - 535)$
$= 216.0 - 99.6 = \underline{116.4 \text{ Btu/lb}}$

(f) $\eta_t = \dfrac{q_S - q_R}{q_S} = \dfrac{116.4}{216.0} = 0.539 = \underline{53.9\%}$

or

$\eta_t = 1 - \dfrac{1}{(r_p)^{\frac{k-1}{k}}} = 1 - \dfrac{1}{(15)^{\frac{0.4}{1.4}}} = \underline{53.9\%}$

POWER CYCLES (GAS)

The four processes making up the basic Brayton cycle take place in separate but interconnected devices—compressor, heater, turbine and cooler. Hence it is physically possible to modify the basic cycle by interjecting other pieces of equipment which will affect the performance parameters. One such modification involves the use of a two-stage compressor with an intercooler between stages as shown schematically in Fig. 8-6a.

FIGURE 8-6a. Brayton cycle with intercooling.

Staging and intercooling result in reduced work of compression, as pointed out in Chapter 6. Under ideal conditions, the pressure ratios across each stage are equal (as is the work done in each stage) and the air leaving the intercooler is cooled back down to the temperature of the air entering the first stage. For the modified Brayton cycle shown in Fig. 8-6a, the ideal is met when:

$$\frac{p_2}{p_1} = \frac{p_{2'}}{p_{1'}}; \quad wk_{LP} = wk_{HP}; \quad T_{1'} = T_1; \quad T_{2'} = T_2$$

But ideally

$$p_2 = p_{1'} \text{ and } p_{2'} = p_3$$

244

Then
$$p_2 = p_{1'} = \sqrt{p_1 p_{2'}}$$

The basic Brayton cycle T-s diagram shown in Fig. 8-5 when modified by intercooling under optimum compressor criteria presented above would appear as shown in Fig. 8-6a. For the same initial conditions, overall pressure ratio and maximum temperature, the intercooled cycle will produce more specific work than the basic Brayton cycle because the work of compression is reduced while the total turbine work remains the same. Remember that

$$wk(\text{net}) = wk(\text{eng}) - wk(\text{pump}) = wk(\text{turb}) - wk(\text{comp})$$

Although the net work output is increased by intercooling as seen in the T-s diagram, the required heat addition will be greater than for the basic cycle operating under the same limiting conditions. The thermal efficiency of the ideal intercooled cycle is:

$$\eta_t = 1 - \frac{q_R}{q_S} = 1 - \left[\frac{(T_2 - T_{1'}) + (T_4 - T_1)}{(T_3 - T_{2'})} \right]$$

The thermal efficiency of the intercooled cycle compares unfavorably with that of the basic Brayton cycle, particularly at the higher pressure ratios. Furthermore, the additional equipment required will obviously add to the first cost and physical complexity of the plant. For these reasons, despite the increase in net work and the corresponding reduction in air flow for a given power output, the introduction of intercooling alone is generally unattractive.

A second modification to the basic Brayton cycle separates the turbine into two distinct units and utilizes a reheater between them as shown in Fig. 8-6b. The thermodynamic logic for maximum turbine work is the mirror image of that for minimum compressor work in the intercooled cycle. Thus for the reheat cycle, ideal conditions are realized when:

$$wk_{\text{LP}} = wk_{\text{HP}}; \quad \frac{p_4}{p_3} = \frac{p_{4'}}{p_{3'}}; \quad T_{3'} = T_3; \quad T_{4'} = T_4$$

Also
$$p_2 = p_1 r_p; \quad p_3 = p_{4'} r_p; \quad p_4 = p_{3'} = \sqrt{p_3 p_{4'}}$$

The basic Brayton cycle T-s diagram shown in Fig. 8-5, when modified for ideal reheating under the maximum turbine work conditions developed above, would appear as shown in Fig. 8-6b. Again assuming the same initial conditions, overall pressure ratio and maximum allowable temperature, the heat added in the basic heater and the compressor work will be the same as in the basic cycle. The heat added in the reheater is

POWER CYCLES (GAS)

FIGURE 8-6b. Brayton cycle with reheat.

above and beyond that of the basic cycle, resulting in an increase in the total heat supplied to the cycle and in the net specific work delivered. The heat rejected will also increase, since the temperature and entropy of the gas leaving the low-pressure turbine will be higher than that of the gas leaving the single turbine of the basic cycle. The thermal efficiency of the reheat cycle is given by:

$$\eta_t = 1 - \frac{q_R}{q_S} = 1 - \left[\frac{(T_{4'} - T_1)}{(T_3 - T_2) + (T_{3'} - T_4)} \right]$$

The relationship between q_S and q_R is such that the end result is similar to that for the intercooled cycle. Net work is increased, but thermal efficiency suffers. As in the case of intercooling, reheat by itself is not an attractive alternative.

A third modification to the basic Brayton cycle introduces the concept of regeneration. The regenerator, commonly a two-fluid heat exchanger, is used to preheat the air between the compressor discharge and the heater inlet as shown in Fig. 8-6c. The heating is accomplished by the hot exhaust from the turbine, thus utilizing part of the "waste" heat, which would be

POWER CYCLES (GAS)

FIGURE 8-6c. Brayton cycle with regeneration.

rejected via the cooler in the basic cycle. The heat exchanged within the regenerator is internal to the cycle, so the only external heat transfer to and from the regenerative cycle occurs in the heater and cooler, just as in the basic cycle.

In the ideal regeneration process, the two-fluid heat balance requires that:

$$(-)\dot{M}_4 q_{44'} = \dot{M}_2 q_{22'} \quad \text{or} \quad (-)q_{44'} = q_{22'}$$

since the mass flow rate is the same throughout the system and $\dot{M}_2 = \dot{M}_4 \cdots = \dot{M}$.

For the hot exhaust gases to preheat the air, it is necessary that the turbine exhaust temperature T_4 be greater than the compressor discharge temperature T_2. Ideal regenerator heating of the air would require $T_{2'} = T_4$ and $T_{4'} = T_2$. Actual regenerator performance is expressed in terms of a regenerator effectiveness, ϵ, as

$$\epsilon = \frac{\text{actual heat transfer to the air}}{\text{ideal (maximum) heat transfer to the air}}$$

$$= \frac{c_p(T_{2'} - T_2)}{c_p(T_4 - T_2)} = \frac{c_p(T_4 - T_{4'})}{c_p(T_4 - T_2)}$$

POWER CYCLES (GAS)

With the same initial conditions, maximum temperature and pressure ratio, the ideal regenerative and basic Brayton cycles will produce exactly the same amount of specific net work:

$$\frac{wk(\text{basic})}{(J)} = q_S - q_R = c_p(T_3 - T_2) - c_p(T_4 - T_1)$$

$$= c_p(T_3 + T_1) - c_p(T_2 + T_4)$$

$$\frac{wk(\text{regen})}{(J)} = q_S - q_R = c_p(T_3 - T_{2'}) - c_p(T_{4'} - T_1)$$

$$= c_p(T_3 + T_1) - c_p(T_{2'} + T_{4'})$$

For the comparative conditions chosen, T_1, T_2, T_3 and T_4 are coincident in both cycles. Ideally the "regenerator effectiveness" is 100%, simply meaning that $T_{2'} = T_4$ and $T_{4'} = T_2$. Thus the two expressions for net work are, in fact, equal. By inference, the conditions established for maximum net work for the basic cycle incident to the development of Equation (8-20) will also apply to the regenerative cycle. Since $T_4 = T_2$ in the maximum work case, there will be no heat exchanged in the regenerator, and the two cycles are coincident in the thermodynamic sense. The regenerator "floats" on the line and serves no useful purpose.

The thermal efficiency of the regenerative cycle, as developed in Appendix C, is given by the following equation:

$$\eta_t = 1 - \left(\frac{T_1}{T_3}\right)\left(\frac{P_2}{P_1}\right)^{\frac{k-1}{k}} = 1 - \left(\frac{T_1}{T_3}\right)(r_p)^{\frac{k-1}{k}} \qquad (8\text{-}21)$$

The first point to note is that the thermal efficiency improves as the maximum temperature, T_3, goes up and the minimum temperature, T_1, goes down. These are the same conditions that increase maximum theoretical net work, for this and the basic Brayton cycle. But the effect of pressure ratio on the thermal efficiency of the two cycles is diametrically opposite. As can be seen by comparing Equations (8-19A) and (8-21), any increase in r_p will increase the thermal efficiency of the basic cycle but decrease that of the regenerative cycle. Since the two cycles are coincident under maximum net work conditions, their thermal efficiencies will naturally become equal at that point. This can be verified by substituting the expression for r_p (maximum net work) from Equation (8-20) in Equations (8-19A) and (8-21). The existence of this crossover point suggests that regenerative plants should be designed to operate at relatively low pressure ratios, whereas the basic cycle plants should be designed to operate at the upper end of the pressure-ratio spectrum. The regenerative cycle has considerable merit, displaying good thermal effi-

ciency characteristics at low pressure ratios with reasonable specific work delivery. Increasing interest in such cycles can be anticipated.

The regenerative cycle can be further modified to include reheat or intercooling or both. However, the added complexity and initial investment inherent in such cycles outweigh the theoretical thermodynamic advantages that may accrue. Note that both reheat and intercooling require additional rotating components, whereas regeneration alone does not.

8-8. Open Cycle and Split Shaft Gas Turbine Plants. In the preceding article, the basic Brayton cycle and several variations were presented on the ideal basis. Compression and expansion were assumed to occur isentropically in the compressor(s) and turbine(s) with no pressure drop in the other components or in the interconnecting duct work and with the individual components operating at 100% efficiency. Further, the cycles were considered to be "closed," as opposed to "open," with all rotating elements tied together mechanically in what is known as the "single shaft" configuration. Many gas turbine plants operate on the open cycle, and many are of the "split shaft" type. This, however, does not affect the basic method of analysis. Similarly, the ideal analyses provide a bench mark for the evaluation of actual performance under less-than-perfect conditions.

The mechanical elements of an open cycle gas turbine are pictured diagrammatically in Fig. 8-7. The Brayton cycle, which is the thermodynamic cycle for most gas turbines, is shown in Fig. 8-7 with a broken line for the heat rejection process to indicate that the cycle is actually exhausting the working substance to the atmosphere at state 4 and taking in atmospheric air at state 1. Whether the turbine completes the cycle from 4 to 1 by a rejection of heat as in the ideal cycle, or by rejecting the working substance as in most actual cycles, the effect insofar as net work and thermal efficiency of the cycle are concerned is exactly the same. Since all of the elements of the gas turbine cycle are steady flow devices, the steady flow energy equation applies. The following analysis is based on application of the energy equation to each device in the cycle and negligible changes in the potential and kinetic energy quantities.

For the compressor,

$$h_1 + q_{12} = h_2 + \frac{wk_{12}}{(J)}$$

If the compression be assumed isentropic,

$$\frac{wk_{12}}{(J)} = h_1 - h_2 \text{ Btu/lb (J/kg)}$$

POWER CYCLES (GAS)

FIGURE 8-7. Ideal open Brayton cycle.

Since there will always be an increase of enthalpy through the compressor, $wk_{12}/(J)$ will always have a negative sign. However, the compressor work is the pump work for this cycle, and by the convention established in Chapter 7 whereby pump work is considered positive we may write

$$\frac{wk(\text{comp})}{(J)} = h_2 - h_1 \text{ Btu/lb (J/kg)} \tag{8-22}$$

For the combustion chamber,

$$q_S = q_{23} = h_3 - h_2 \text{ Btu/lb (J/kg)}$$

POWER CYCLES (GAS)

Finally, the work of the turbine with isentropic expansion is

$$\frac{wk\text{(turbine)}}{(J)} = \frac{wk_{34}}{(J)} = h_3 - h_4 \text{ Btu/lb (J/kg)} \qquad (8\text{-}23)$$

The net work of the cycle, then, will be

$$\frac{wk\text{(net)}}{(J)} = \frac{wk\text{(turbine)}}{(J)} - \frac{wk\text{(comp)}}{(J)}$$

$$\frac{wk\text{(net)}}{(J)} = (h_3 - h_4) - (h_2 - h_1) \text{ Btu/lb (J/kg)}$$

The thermal efficiency of the cycle is

$$\eta_t = \frac{\dfrac{wk\text{(net)}}{(J)}}{q_s}$$

$$\eta_t = \frac{(h_3 - h_4) - (h_2 - h_1)}{(h_3 - h_2)} \qquad (8\text{-}24)$$

This last expression for thermal efficiency may be developed into any of the thermal efficiency expressions for the basic closed Brayton cycle. In similar fashion the analysis presented in Art. 8-7 for the more complicated closed cycles can be directly applied to corresponding open Brayton cycles.

Actual compressor and turbine exhaust conditions will differ from values based on the ideal analysis, as shown in Fig. 8-8. The ideal analysis can be modified to account for real compressor behavior by introducing the isentropic efficiency as defined in Art. 6-9:

$$\eta_{c\circledS} = \frac{\text{isentropic compressor work}}{\text{actual compressor work}} < 1.0$$

FIGURE 8-8. Open Brayton cycle with actual compression and expansion processes.

POWER CYCLES (GAS)

For constant specific heat,

$$\eta_{c_\circledS} = \frac{c_p(T_{2_\circledS} - T_1)}{c_p(T_{2_a} - T_1)} \qquad (8\text{-}25)$$

where
- T_a = actual temperature
- T_\circledS = corresponding isentropic temperature

Similarly, turbine performance is evaluated by comparing the actual work produced with the ideal, represented by an isentropic expansion process between the same pressure limits. In gas turbine parlance, this ratio is referred to as the *isentropic turbine efficiency*, η_{t_\circledS}. In other usage, the same quantity is known as engine or internal engine efficiency. See Equation (7-10).

$$\eta_{\text{eng}} = \eta_{t_\circledS} = \frac{\text{actual turbine work}}{\text{isentropic turbine work}} < 1.0$$

Again for constant specific heat,

$$\eta_{t_\circledS} = \frac{c_p(T_3 - T_{4_a})}{c_p(T_3 - T_{4_\circledS})} \qquad (8\text{-}26)$$

Fig. 8-8 indicates the influence of both compressor and turbine efficiencies on the open gas turbine cycle.

Operating flexibility, essential for certain applications, can be realized by utilizing the split shaft configuration as shown in Fig. 8-9. Here the single turbine of the basic cycle is split into two separate and distinct units; a compressor drive turbine and an output power turbine. This arrangement permits the power turbine to respond to varying speed/load demands at the output shaft regardless of the compressor drive turbine speed, since the two units are mechanically independent.

FIGURE 8-9. Split shaft open Brayton cycle.

POWER CYCLES (GAS)

Example 8-5: A split shaft marine gas turbine has a power turbine rated at 15,000 internal horsepower. Typical operating conditions for the unit are: compressor inlet 14.5 psia and 60°F; compressor discharge 174 psia and 716°F; compressor turbine inlet 171 psia and 1630°F; and power turbine exhaust 14.8 psia and 760°F. The compressor turbine at the above rating has an 85% isentropic turbine efficiency. For these rated conditions, (a) sketch the T-s diagram for the cycle; then calculate:
 (b) the compressor isentropic efficiency, %
 (c) the compressor turbine discharge pressure and temperature, psia, °R
 (d) the power turbine isentropic efficiency, %
 (e) the compressor turbine power output, hp
 (f) the cycle thermal efficiency

Solution:

$$p_1 = 14.5 \text{ psia} \qquad T_1 = 60 + 460 = 520°R$$
$$p_2 = 174 \text{ psia} \qquad T_{2_a} = 716 + 460 = 1176°R$$
$$p_3 = 171 \text{ psia} \qquad T_3 = 1630 + 460 = 2090°R$$
$$p_4 = 14.8 \text{ psia} \qquad T_{4_a} = 760 + 460 = 1220°R$$

(a) [T-s diagram sketch showing states 1, 2(s), 2a, 3, 3'(s), 3'a, 4(s), 4a]

(b)
$$T_{2_{(s)}} = T_1 \left(\frac{p_2}{p_1}\right)^{\frac{k-1}{k}} = 520 \left(\frac{174}{14.5}\right)^{\frac{0.4}{1.4}} = 1057.6°R$$

$$\eta_{c_{(s)}} = \frac{\text{isentropic compressor work}}{\text{actual compressor work}} = \frac{T_{2_{(s)}} - T_1}{T_{2_a} - T_1}$$

$$= \frac{1057.6 - 520}{1176 - 520} = 0.8195 = \underline{82.0\%}$$

(c)
$$\dot{W}k_{12} = \dot{P}_{12} = -\dot{P}_{33'_a} = -\dot{W}k_{33'_a}$$
$$\dot{M}c_p(T_1 - T_{2_a}) = -\dot{M}c_p(T_3 - T_{3'_a})$$
$$T_{3'_a} = T_1 - T_{2_a} + T_3$$
$$= 520 - 1176 + 2090 = \underline{1434°R}$$

$$\eta_{t_{(s)}} = \frac{\text{actual turbine work}}{\text{isentropic turbine work}} = \frac{T_3 - T_{3'_a}}{T_3 - T_{3'_{(s)}}} = 0.85$$

$$T_{3'_{(s)}} = T_3 - \left(\frac{T_3 - T_{3'_a}}{0.85}\right)$$

253

POWER CYCLES (GAS)

$$T_{3'\circledS} = 2090 - \left(\frac{2090 - 1434}{0.85}\right) = 1318°R$$

$$p_{3'} = p_3 \left(\frac{T_{3'\circledS}}{T_3}\right)^{\frac{k}{k-1}} = 171 \left(\frac{1318}{2090}\right)^{\frac{1.4}{0.4}} = 34.06 \text{ psia}$$

(d) For the power turbine:

$$T_{4\circledS} = T_{3'_a}\left(\frac{P_4}{P_{3'}}\right)^{\frac{k-1}{k}} = 1434 \left(\frac{14.8}{34.06}\right)^{\frac{0.4}{1.4}}$$

$$= 1130°R$$

$$\eta_{t\circledS} = \frac{\text{actual expansion work}}{\text{isentropic expansion work}} = \frac{T_{3'_a} - T_{4_a}}{T_{3'_a} - T_{4\circledS}}$$

$$\eta_{t\circledS} = \frac{1434 - 1220}{1434 - 1130} = 0.704 = \underline{70.4\%}$$

(e) For the power turbine:

$$P = \dot{W}k = \dot{M}c_p(T_{3'_a} - T_{4_a})$$

or

$$\dot{M} = \frac{(15{,}000)(42.42)}{(0.240)(1434 - 1220)} = 12{,}389 \text{ lb/min}$$

For the compressor turbine:

$$P = \dot{W}k = \dot{M}c_p(T_3 - T_{3'_a})$$

$$= \frac{(12{,}389)(0.240)(2090 - 1434)}{42.42} = \underline{45{,}980 \text{ hp (by)}}$$

or

$$P = \dot{W}k = \dot{M}c_p(T_{2_a} - T_1)$$

$$= \frac{(12{,}389)(0.240)(1176 - 520)}{42.42} = \underline{45{,}980 \text{ hp (by)}}$$

(f) Thermal efficiency:

$$\eta_t = \frac{wk_{33'_a} + wk_{3'_a 4_a} - wk_{12_a}}{q_{23}} = \frac{wk_{3'_a 4_a}}{q_{23}}$$

$$= \frac{c_p(T_{3'_a} - T_{4_a})}{c_p(T_3 - T_{2_a})} = \frac{(1434 - 1220)}{(2090 - 1176)}$$

$$= 0.234 = \underline{23.4\%}$$

8-9. Brayton Cycle Aircraft Gas Turbine Analysis. Aircraft gas turbines, or turbojets, differ from stationary gas turbines in two distinct features. First is the diffuser, which efficiently transforms the velocity of the entering air stream to stored energy in form of pressure, and second is the nozzle, which converts the energy stored in the gases leaving the system to kinetic energy. Unlike the gas turbine whose end product is shaft power, the turbojet engine produces only enough shaft power (ideally) to drive its compressor. Thrust is its end product, and both the diffuser and the nozzle play an important part in the development of this thrust.

POWER CYCLES (GAS)

The mechanical elements of a simple Brayton cycle aircraft gas turbine engine are shown in Fig. 8-10. V_a is the aircraft velocity of flight and V_j is the velocity of the propulsive jet. Neglecting change of momentum due to the fuel introduced for combustion and assuming complete expansion of the combustion products to ambient pressure by the exhaust nozzle, the net thrust obtained from an aircraft gas turbine is:

$$F_T = \rho Q \Delta V = \frac{\dot{M}}{g_c}(V_j - V_a) \text{ lbf (N)} \tag{8-27}$$

where F_T = net thrust, lbf (N)
ρ = fluid density, lb/ft³ (kg/m³)
Q = volume rate of flow, ft³/sec (m³/sec)
\dot{M} = mass rate of flow, lb/sec (kg/sec)
V_j = exit jet velocity, ft/sec (m/sec)
V_a = entering air velocity, ft/sec (m/sec)

FIGURE 8-10. Mechanical elements of the Brayton cycle aircraft gas turbine engine.

The propulsive efficiency for an ideal aircraft gas turbine is expressed as

$$\eta_p = \frac{\text{output}}{\text{input}} = \frac{\text{output}}{\text{output} + \text{losses}}$$

$$\eta_p = \frac{\text{thrust power}}{\text{thrust power} + \text{wake loss power}} \tag{8-28}$$

Thrust power is the product of net thrust and flight velocity, while wake loss power is the power associated with the absolute kinetic energy of the propulsive jet:

$$\text{thrust power} = F_T V_A = \frac{\dot{M}}{g_c}(V_j - V_a) V_a \text{ ft} \cdot \text{lbf/sec (J/sec)} \tag{8-29}$$

$$\text{wake loss power} = \frac{1}{2}\frac{\dot{M}}{g_c}(V_j - V_a)^2 \text{ ft} \cdot \text{lbf/sec (J/sec)} \tag{8-30}$$

POWER CYCLES (GAS)

Substitution of these two expressions in the equations for propulsive efficiency and simplifying yield:

$$\eta_p = \frac{2}{1 + \dfrac{V_j}{V_a}} \qquad (8\text{-}31)$$

Inspection of this equation shows that when propulsive jet velocity equals flight speed, propulsive efficiency is one hundred percent. However, under these circumstances, the net thrust would be zero. Therefore, any net thrust realized will be accompanied by a certain wake loss and the greater the thrust produced, the larger will be the wake loss. As shown by Equation (8-30), this wake loss is mechanical energy which is degraded to unavailable thermal energy in the turbulent wake of the engine.

The preceding equations indicate that an analysis of the aircraft gas turbine engine requires a knowledge of exhaust conditions and velocity of propulsive jet in order to evaluate thrust. Consequently, to determine the thrust for any given combination of flight speed and altitude, it will be necessary to analyze the processes undergone by the air as it flows through the engine from the atmosphere to the compressor, the combustors, the turbine and the exhaust nozzle.

Figure 8-11 shows the processes of the Brayton aircraft gas turbine

FIGURE 8-11. Brayton aircraft gas turbine cycle.

POWER CYCLES (GAS)

cycle on temperature-entropy coordinates. The indicated state points are numbered and lettered in accordance with the mechanical equipment diagram, Fig. 8-10.

The reversible processes employed in the Brayton cycle yield maximum performance characteristics for the altitudes and flight speeds involved. Because of thermal and mechanical irreversibilities, the actual engine approaches, but never attains, these conditions.

Atmospheric air enters the diffuser at a velocity V_a (aircraft flight velocity), temperature T_a and pressure p_a (aircraft environmental pressure). Ram compression occurs in the diffuser, process a-1, isentropically converting the kinetic energy of the air to stored thermal energy at increased temperature and pressure. Neglecting potential energy changes for the steady flow, isentropic diffuser process, the equation is

$$h_a + \frac{V_a^2}{2g_c(J)} = h_1 \text{ Btu/lb (J/kg)}$$

For the mechanical compressor,

$$h_1 = h_2 + \frac{wk(\text{comp})}{(J)} \text{ Btu/lb (J/kg)}$$

Given V_a and T_a and using the perfect gas relationship $c_p(T_1 - T_a) = h_1 - h_a$, air temperature leaving the diffuser and entering the mechanical compressor may be determined. Applying the pressure-temperature relationship for isentropic perfect gas processes will yield the pressure at diffuser exit:

$$p_1 = p_a \left(\frac{T_1}{T_a}\right)^{\frac{k}{k-1}}$$

With a knowledge of the mechanical compressor pressure ratio, a similar procedure will yield the air temperature leaving the compressor.

In the combustors, heat is added (ideally) at constant pressure. For the combustion process, $p_3 = p_2$ and

$$q_s = q_{23} = h_3 - h_2 = c_p(T_3 - T_2) \text{ Btu/lb (J/kg)}$$

As pointed out previously, the turbine of the turbojet engine produces exactly the shaft power required by its compressor. Since they are on the same shaft and subject to the same mass rate of air flow, their work expressions may be equated, giving

$$\frac{wk(\text{comp})}{(J)} = \frac{wk(\text{turb})}{(J)} = h_3 - h_4 \text{ Btu/lb (J/kg)}$$

POWER CYCLES (GAS)

Turbine outlet pressure may be calculated from

$$p_4 = p_3 \left(\frac{T_4}{T_3}\right)^{\frac{k}{k-1}}$$

Assuming the air has negligible velocity leaving the turbine and entering the exhaust nozzle, the energy equation for the isentropic process across the nozzle gives

$$h_4 = \frac{V_j^2}{2g_c(J)} + h_j$$

$$\frac{V_j^2}{2g_c(J)} = h_4 - h_j = c_p(T_4 - T_j)$$

For complete expansion of the air in the exhaust nozzle, $p_j = p_a$. T_j may then be calculated from the isentropic relationship

$$T_j = T_4 \left(\frac{p_j}{p_4}\right)^{\frac{k-1}{k}}$$

Using this result in solving for jet velocity yields

$$V_j = \sqrt{2g_c(J)c_p(T_4 - T_j)} \text{ ft/sec (m/sec)} \qquad (8\text{-}32)$$

The net thrust may now be calculated from the previously developed relationship

$$F_T = \frac{\dot{M}}{g_c}(V_j - V_a) \text{ lbf (N)} \qquad (8\text{-}27)$$

As in any thermodynamic power cycle, the thermal efficiency of the Brayton cycle may be expressed as useful output divided by heat supplied. In terms of mechanical energy, the useful output of the engine is the energy available for propulsion of the aircraft, or in terms of energy per unit mass of working substance,

$$wk(\text{net}) = \frac{V_j^2 - V_a^2}{2g_c} \text{ ft} \cdot \text{lbf/lb (J/kg)}$$

the difference in the kinetic energies of the working substance leaving and entering the engine. In terms of thermal energy, net cycle work is equal to $q_S - q_R$, and cycle thermal efficiency is equal to $(q_S - q_R)/q_S$.

$$\frac{wk(\text{net})}{(J)} = q_S - q_R$$

$$\eta_t = \frac{q_S - q_R}{q_S}$$

POWER CYCLES (GAS)

For the Brayton cycle shown in Fig. 8-11,

$$q_S = h_3 - h_2 = c_p(T_3 - T_2) \text{ Btu/lb (J/kg)}$$

and

$$q_R = c_p(T_j - T_a) \text{ Btu/lb (J/kg)}$$

By a development exactly analogous to that shown by Equations (8-18) and (8-19) for the closed cycle, the thermal efficiency of this ideal Brayton cycle may be expressed as

$$\eta_t = 1 - \frac{T_a}{T_2} = 1 - \frac{T_j}{T_3}$$

Since the compression processes a-1 and 1-2 are both isentropic, the temperature ratio T_a/T_2 may be replaced by its equivalent $(p_a/p_2)^{\frac{k-1}{k}}$ giving,

$$\eta_t = 1 - \frac{1}{\left(\frac{p_2}{p_a}\right)^{\frac{k-1}{k}}} = 1 - \frac{1}{(r_p)^{\frac{k-1}{k}}} \quad (8\text{-}33)$$

where r_p is the cycle *overall* pressure ratio.

The cycle overall pressure ratio may be increased either by increasing the ram compression effect (through increased flight velocity) or by increasing the mechanical compressor pressure ratio. Figure 8-12 shows three Brayton cycles with varying overall pressure ratios, but all having the same maximum cycle temperature and the same initial conditions of pressure and temperature. It is apparent from the diagram and the

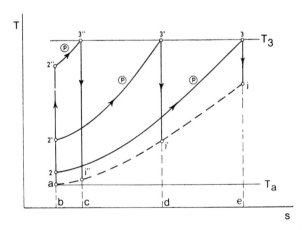

FIGURE 8-12. Brayton cycle with varying pressure ratios.

equations already developed that cycles with the lower overall pressure ratios have the lower thermal efficiencies. Also, as the overall pressure ratio is increased, thermal efficiency increases and propulsive efficiency decreases, but it may be shown that the net cycle work reaches a maximum value when

$$\frac{p_2}{p_a} = \left(\frac{T_3}{T_a}\right)^{\frac{k}{2(k-1)}}$$

A development of this expression is given in Appendix C and is analogous to that of Equation (8-20) for the basic closed Brayton cycle.

In Fig. 8-12, cycle $a23j$ may be considered a cycle with a low overall pressure ratio, low thermal efficiency, and less than maximum cycle work. Cycle $a2'3'j'$ might be considered a cycle with moderate overall pressure ratio, moderate thermal efficiency, and maximum net work. Cycle $a2''3''j''$ would then be characterized by high overall pressure ratio, high thermal efficiency, and less than maximum net cycle work.

Power considerations for combat aircraft propelled by thrust-producing gas turbines frequently necessitate the provision of thrust augmentation for critical periods of operation such as take-off, rapid climb and combat. One such method of thrust augmentation widely used to obtain high thrust for short duration is afterburning. Because of the low fuel-air ratios used in present gas turbine engines, the exhaust products leaving the turbine contain sufficient oxygen to support further combustion. By adding a set of fuel nozzles and a flame holder after the turbine wheel of the basic equipment shown in Fig. 8-10, a Brayton cycle aircraft gas turbine engine may be equipped for afterburning. Thus fuel may be added to the turbine exhaust gases and burned at a flame holder to provide thrust augmentation. Because the nozzle contains no high speed rotating parts, temperatures considerably in excess of the maximum in the cycle without afterburning may be maintained at the afterburner. Figure 8-13 shows a T-s diagram for the ideal Brayton gas turbine cycle with afterburning. Process 4-5 indicates the constant pressure heat addition process of the afterburner following the turbine expansion process 3-4. Process 5-j indicates the isentropic expansion in the nozzle to ambient atmospheric pressure.

Analysis of the cycle with afterburning proceeds as in the basic cycle with the further assumption that the kinetic energy of the gases entering the afterburner is negligible. The jet velocity is found to be

$$V_j = \sqrt{2g_c(J)c_p(T_5 - T_j)}$$

Afterburning is a simple but expensive means of thrust augmentation.

POWER CYCLES (GAS)

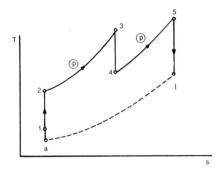

FIGURE 8-13. Ideal Brayton gas turbine cycle with afterburning.

Example 8-6: A turbojet engine is equipped with a 12-stage, axial flow compressor having a mechanical compression pressure ratio of 12 to 1. The engine is designed to handle 90 lb of air per second under sea level static conditions and 116 lb of air per second at a flight velocity of 850 ft/sec at sea level. Material design considerations limit the temperature of the working substance entering the turbine to 2000°R at 7950 rpm. The ambient temperature and pressure are 15 psia and 530°R. (a) Estimate the thrust, propulsive efficiency and thermal efficiency of the engine, assuming all processes to be ideal and the engine frictionless, for a flight velocity of 850 ft/sec at sea level. (b) When the above engine is equipped for afterburning, the nozzle inlet temperature is 2500°R. Assuming all of the stated performance parameters remain constant, estimate the thrust with afterburning for a flight velocity of 850 ft/sec at sea level.

Solution:

$$T_{amb} = T_a = 530°R$$
$$T_3 = 2000°R$$
$$p_a = 15 \text{ psia}$$
$$M = 116 \text{ lb/sec}$$
$$V_a = 850 \text{ ft/sec}$$

Ram compression—

$$h_a + \frac{V_a^2}{2g_c J} = h_1 + \frac{V_1^{2^0}}{2g_c J}$$

$$\frac{V_a^2}{2g_c J} = h_1 - h_a$$

$$\frac{V_a^2}{2g_c J} = c_p(T_1 - T_a)$$

$$\frac{(850)^2}{2(32.2)778} = 0.24(T_1 - 530)$$

$$T_1 = 590°R$$

$$p_1 = p_a \left(\frac{T_1}{T_a}\right)^{\frac{k}{k-1}} = 15\left(\frac{590}{530}\right)^{\frac{1.4}{0.4}} = 21.8 \text{ psia}$$

POWER CYCLES (GAS)

Compressor—

$$p_2 = 12\, p_1 = 261.6 \text{ psia}$$
$$T_2 = T_1 \left(\frac{p_2}{p_1}\right)^{\frac{k-1}{k}} = 590(12)^{\frac{0.4}{1.4}} = 1200°\text{R}$$

Combustor—

$$h_2 + q_{23} = h_3$$
$$q_{23} = h_3 - h_2 = c_p(T_3 - T_2)$$
$$q_{23} = 0.24(2000 - 1200) = 192.0 \text{ Btu/lb}$$
$$p_3 = p_2 = 261.6 \text{ psia}$$

Turbine—

$$h_3 = h_4 + \frac{wk_{34}}{J}$$

$$\frac{wk_{34}}{J} = h_3 - h_4$$

$$\frac{wk(\text{turb})}{J} = \frac{wk(\text{comp})}{J}$$

$$\therefore h_2 - h_1 = h_3 - h_4$$
$$c_p(T_2 - T_1) = c_p(T_3 - T_4)$$
$$T_4 = T_3 - (T_2 - T_1)$$
$$T_4 = 2000 - (1200 - 590)$$
$$T_4 = 1390°\text{R}$$

$$p_4 = p_3 \left(\frac{T_4}{T_3}\right)^{\frac{k}{k-1}}$$

$$p_4 = 261.6 \left(\frac{1390}{2000}\right)^{\frac{1.4}{0.4}}$$

$$p_4 = 73.2 \text{ psia}$$

Nozzle—

$$h_4 + \frac{V_4^2}{2g_cJ}^0 = h_j + \frac{V_j^2}{2g_cJ} \qquad T_j = T_4 \left(\frac{p_j}{p_4}\right)^{\frac{k-1}{k}} = 1390 \left(\frac{15}{73.2}\right)^{\frac{0.4}{1.4}}$$

$$T_j = 883.7°\text{R}$$

$$\frac{V_j^2}{2g_cJ} = h_4 - h_j = c_p(T_4 - T_j)$$

$$V_j = \sqrt{2g_c J c_p(T_4 - T_j)}$$
$$V_j = 223.8 \sqrt{0.24(1390 - 883.7)} = 2467 \text{ ft/sec}$$

Thrust—

$$F_T = \frac{\dot{M}}{g_c}(V_j - V_a) = \frac{116}{32.2}(2467 - 850) = \underline{5825 \text{ lbf}}$$

Propulsive efficiency—

$$\eta_p = \frac{2}{1 + \frac{V_j}{V_a}} = \frac{2}{1 + \frac{2467}{850}} = \underline{51.3\%}$$

Thermal efficiency—

$$q_R = c_p(T_j - T_a) = 0.24(883.7 - 530)$$
$$= 84.9 \text{ Btu/lb}$$

POWER CYCLES (GAS)

$$\eta_t = \frac{q_s - q_R}{q_s} = 1 - \frac{q_R}{q_s} = 1 - \frac{84.9}{192.0} = 55.8\%$$

Part (b)

$$T_{j'} = T_{4'}\left(\frac{p_{j'}}{p_{4'}}\right)^{\frac{k-1}{k}} = 2500\left(\frac{15}{73.2}\right)^{0.4/1.4} = 1589°R$$

$$V_{j'} = \sqrt{2g_c J c_p (T_{4'} - T_{j'})} = 223.8\sqrt{0.24(2500 - 1589)} = 3309 \text{ ft/sec}$$

$$F_T = \frac{\dot{M}}{g_c}(V_{j'} - V_a) = \frac{116}{32.2}(3309 - 850) = 8860 \text{ lbf}$$

8-10. Air Table Analysis of Gas Power Cycles. The thermodynamic performance of various gas power cycles was analyzed in previous sections assuming that air has constant specific heat. It was shown in Art. 3-11 that the specific heat of air is actually a function of temperature. This effect can be important over the wide temperature range encountered in most gas power cycles. Analyses using thermodynamic properties from the Air Tables, as discussed in Chapter 3, include the effects of variable specific heat.

The compression ratio, r_v, for an isentropic process can be expressed using the relative volume expression, Equation (3-20), as

$$r_v = \left[\frac{v_2}{v_1}\right]_{(S)} = \frac{v_{r_2}}{v_{r_1}} \qquad (8\text{-}34)$$

The pressure ratio, r_p, for an isentropic process can be written in similar fashion using the relative pressure relation, Equation (3-18), as

$$r_p = \left[\frac{p_2}{p_1}\right]_{(S)} = \frac{p_{r_2}}{p_{r_1}} \qquad (8\text{-}35)$$

Changes in internal energy and enthalpy can be determined as

$$h_2 - h_1 = h\langle T_2\rangle - h\langle T_1\rangle \qquad (8\text{-}36)$$

and

$$u_2 - u_1 = u\langle T_2\rangle - u\langle T_1\rangle \qquad (8\text{-}37)$$

The following examples, reworked versions of Examples 8-3 and 8-5, will illustrate the use of Air Tables in gas power cycle analysis and permit comparison with the constant specific heat solutions.

Example 8-7: At the beginning of compression an ideal dual combustion cycle using air has a pressure of 15 psia, a temperature of 75°F and a specific volume of 13.2 ft³/lb. For a compression ratio of 12 and a heat addition of 176 Btu/lb at constant volume and 176 Btu/lb at constant pressure, (a) sketch the cycle on p-v and T-s coordinates. Then, using the Air Tables, calculate for the cycle:
 (b) the pressure and temperature at the end of isentropic compression

POWER CYCLES (GAS)

(c) the pressures and temperatures at the end of each heat addition process
(d) the temperature at the beginning of heat rejection, °R
(e) the heat rejected, Btu/lb
(f) the net work. Btu/lb
(g) the thermal efficiency, percent
(h) the horsepower developed by an ideal engine operating on the cycle using 0.5 pound of air per second

Solution: Note that this solution is based on Table 1 of *Gas Tables* (Keenan and Kaye) with interpolation to the nearest degree.

$$P_1 = 15 \text{ psia} \qquad v_1 = 13.2 \text{ ft}^3/\text{lb}$$
$$t_1 = 75°\text{F} \qquad v_1/v_2 = 12.0$$

heat added

$$176 \text{ Btu/lb at constant volume}$$
$$176 \text{ Btu/lb at constant pressure}$$

(a) See Fig. 8-4.

(b) From the Gas Tables:

	T	h	p_r	u	v_r
state 1	535	127.86	1.3416	91.19	147.72

$$v_{r_2} = v_{r_1} \left(\frac{v_2}{v_1}\right)_{\text{\textcircled{S}}} = 147.72 \left(\frac{1}{12}\right) = 12.310$$

	T	h	p_r	u	v_r
state 2 \textcircled{S}	1391	340.55	41.84	245.20	12.315

$$T_{2_{\text{\textcircled{S}}}} = \underline{1391°\text{R}}$$

$$p_{2_{\text{\textcircled{S}}}} = p_1 \left(\frac{p_{r_2}}{p_{r_1}}\right) = 15 \left(\frac{41.84}{1.3416}\right) = \underline{467.8 \text{ psia}}$$

(c) $$u\langle T_3\rangle = u\langle T_2\rangle + q_{23} = 245.20 + 176 = 421.2$$

	T	h	p_r	u	v_r
state 3	2253	575.53	283.0	421.09	2.949

$$T_3 = \underline{2253°\text{R}}$$

$$p_3 = p_2 \left(\frac{T_3}{T_2}\right) = 467.80 \left(\frac{2253}{1391}\right) = \underline{757.7 \text{ psia}}$$

$$p_4 = p_3 = \underline{757.7 \text{ psia}}$$

$$h\langle T_4\rangle = h\langle T_3\rangle + q_{34} = 575.53 + 176 = 751.53$$

	T	h	p_r	u	v_r
state 4	2866	751.54	774.9	555.08	1.3700

$$T_4 = \underline{2866°\text{R}}$$

POWER CYCLES (GAS)

$$v_4 = v_3 \left(\frac{T_4}{T_3}\right) = \frac{13.2}{12}\left(\frac{2866}{2253}\right) = 1.40 \text{ ft}^3/\text{lb}$$

(d)
$$v_{r_5} = v_{r_4}\left(\frac{v_5}{v_4}\right)_{(S)} = 1.3700\left(\frac{13.2}{1.4}\right) = 12.92$$

	T	h	p_r	u	v_r
state 5	1367	334.31	39.16	240.60	12.931

$$T_5 = \underline{1367°R}$$

$$p_5 = p_4\left(\frac{p_{r_5}}{p_{r_4}}\right) = 757.7\left(\frac{39.16}{774.9}\right) = \underline{38.3 \text{ psia}}$$

(e) $\quad q_R = u\langle T_5\rangle - u\langle T_1\rangle = 240.60 - 91.19 = \underline{149.41 \text{ Btu/lb}}$

(f) $\quad \dfrac{wk \text{ (net)}}{J} = q_S - q_R = 352 - 149.41 = \underline{202.59 \text{ Btu/lb}}$

(g) $\quad \eta_t = \dfrac{q_S - q_R}{q_S} = \dfrac{202.59}{352} = \underline{57.6\%}$

(h) $\quad P = \dot{W}k = \dfrac{(202.59)(0.5)(60)}{42.42} = \underline{143 \text{ hp}}$

Example 8-8: A split-shaft marine gas turbine has a power turbine rated at 15,000 internal horsepower. Typical operating conditions for the unit are: compressor inlet, 14.5 psia and 60°F; compressor discharge, 174 psia and 716°F; compressor turbine inlet, 171 psia and 1630°F; and power turbine exhaust, 14.8 psia and 760°F. The compressor turbine has an 85% isentropic turbine efficiency at the above power rating. (a) Sketch the *T-s* diagram for the cycle; then, using the Air Tables, calculate:
 (b) the compressor isentropic efficiency, %
 (c) the compressor turbine discharge pressure and temperature, psia, °R
 (d) the power turbine isentropic efficiency, %
 (e) the compressor turbine power output, hp
 (f) the cycle thermal efficiency, %
Solution:

$$p_1 = 14.5 \text{ psia} \qquad T_1 = 60 + 460 = 520°R$$
$$p_2 = 174 \text{ psia} \qquad T_{2_a} = 716 + 460 = 1176°R$$
$$p_3 = 171 \text{ psia} \qquad T_3 = 1630 + 460 = 2090°R$$
$$p_4 = 14.8 \text{ psia} \qquad T_{4_a} = 760 + 460 = 1220°R$$

(a) See Ex. 8-5.
(b) From the Gas Tables:

	T	h	p_r	u	v_r
state 1	520	124.27	1.2147	88.62	158.58

$$p_{r_2} = p_{r_1}\left(\frac{p_2}{p_1}\right)_{(S)} = 1.2147\left(\frac{174}{14.5}\right) = 14.576$$

POWER CYCLES (GAS)

	T	h	p_r	u	v_r
state 2_{\circledS}	1048	252.95	14.584	181.11	26.62

$$\eta_{c_{\circledS}} = \frac{\text{isentropic compression work}}{\text{actual compression work}} = \frac{h\langle T_{2_{\circledS}}\rangle - h\langle T_1\rangle}{h\langle T_{2_a}\rangle - h\langle T_1\rangle}$$

	T	h	p_r	u	v_r
state 2_a	1176	285.20	22.28	204.59	19.556

$$\eta_{c_{\circledS}} = \frac{252.95 - 124.27}{285.20 - 124.27} = 0.7996 = \underline{80.0\%}$$

(c)

$$\dot{W}k_{12} = P_{12} = -P_{33'_a} = -\dot{W}k_{33'_a}$$

	T	h	p_r	u	v_r
state 3	2090	529.75	208.06	386.48	3.721

$$\dot{M}[h\langle T_1\rangle - h\langle T_{2_a}\rangle] = -\dot{M}[h\langle T_3\rangle - h\langle T_{3'_a}\rangle]$$
$$h\langle T_{3'_a}\rangle = h\langle T_1\rangle - h\langle T_{2_a}\rangle + h\langle T_3\rangle$$
$$= 124.27 - 285.20 + 529.75$$
$$= 368.82$$

	T	h	p_r	u	v_r
state $3'_a$	1499	368.91	55.71	266.14	9.967

$$T_{3'_a} = \underline{1499°\text{R}}$$

$$\eta_{t_{\circledS}} = \frac{\text{actual expansion work}}{\text{isentropic expansion work}} = \frac{h\langle T_3\rangle - h\langle T_{3'_a}\rangle}{h\langle T_3\rangle - h\langle T_{3'_{\circledS}}\rangle} = 0.85$$

$$h\langle T_{3'_{\circledS}}\rangle = h\langle T_3\rangle - \frac{h\langle T_3\rangle - h\langle T_{3'_a}\rangle}{0.85}$$
$$= 529.75 - \frac{529.75 - 368.91}{0.85} = 340.53$$

	T	h	p_r	u	v_r
state $3'_{\circledS}$	1391	340.55	41.84	245.20	12.315

$$p_{3'} = p_3\left(\frac{p_{r_{3'_{\circledS}}}}{p_{r_3}}\right) = 171\left(\frac{41.84}{208.06}\right) = \underline{34.39 \text{ psia}}$$

(d)

$$p_{r_4} = p_{r_{3'_a}}\left(\frac{p_4}{p_{3'}}\right)_{\circledS} = 55.71\left(\frac{14.8}{34.39}\right) = 23.98$$

	T	h	p_r	u	v_r
state 4_a	1220	296.41	25.53	212.78	17.700
state 4_{\circledS}	1200	291.30	24.01	209.05	18.514

POWER CYCLES (GAS)

$$\eta_{t\circledS} = \frac{\text{actual expansion work}}{\text{isentropic expansion work}} = \frac{h\langle T_{3'_a}\rangle - h\langle T_{4_a}\rangle}{h\langle T_{3'_a}\rangle - h\langle T_{4\circledS}\rangle}$$

$$\eta_{t\circledS} = \frac{368.91 - 296.41}{368.91 - 291.30} = 0.934 = \underline{93.4\%}$$

(e) For the power turbine:

$$P = \dot{W}k = \dot{M}(h\langle T_{3'_a}\rangle - h\langle T_{4_a}\rangle)$$

or:

$$\dot{M} = \frac{(15{,}000)(42.42)}{(368.91 - 296.41)} = 8{,}780 \text{ lb/min}$$

For the compressor turbine:

$$P = \dot{W}k = \dot{M}(h\langle T_3\rangle - h\langle T_{3'_a}\rangle]$$

$$= \frac{(8{,}780)(529.75 - 368.91)}{42.42} = \underline{33{,}300 \text{ hp (by)}}$$

or:

$$P = \dot{W}k = M(h\langle T_{2_a}\rangle - h\langle T_1\rangle)$$

$$= \frac{(8780)(285.20 - 124.27)}{42.42}$$

$$= \underline{33{,}300 \text{ hp (by)}}$$

(f) Thermal efficiency:

$$\eta_t = \frac{wk_{33'_a} + wk_{3'_a 4_a} - wk_{12_a}}{q_{23}} = \frac{wk_{3'_a 4_a}}{q_{23}}$$

$$= \frac{h\langle T_{3'_a}\rangle - h\langle T_{4_a}\rangle}{h\langle T_3\rangle - h\langle T_{2_a}\rangle}$$

$$= \frac{368.91 - 296.41}{529.75 - 285.2} = \underline{29.6\%}$$

Problems

In all problems sketch cycle on appropriate coordinates and use constant specific heat unless Gas (Air) Tables are specified.

8-1. A Stirling cycle operating with air as a working substance has a pressure of 80 psia, a temperature of 250°F, and a specific volume of 3.28 ft^3/lb at the beginning of the isothermal heat rejection process. The ratio of volumes at the beginning and end of isothermal heat rejection is 2. The highest temperature involved in the cycle is 1200°F. Calculate (a) the heat supplied, Btu/lb; (b) the heat rejected, Btu/lb; (c) the thermal efficiency of the cycle, percent.

Ans. (a) 78.8 Btu/lb; (b) 33.7 Btu/lb; (c) 57.2%

8-2. In an ideal Otto cycle the pressure at the beginning of the compression process is 15 psia, the temperature is 70°F, and the volume is 13.08 ft^3. The maximum temperature of the cycle is 2804°F, and the heat supplied is 372 Btu. Using constant specific heat, calculate (a)

POWER CYCLES (GAS)

the temperature at the beginning of combustion, °R; (b) the compression ratio; (c) the temperature at the end of the expansion process, °R; (d) the net work of the cycle, Btu; (e) the thermal efficiency of the cycle, percent.

Ans. (a) 1089°R; (b) 6.05; (c) 1589°R; (d) 191 Btu; (e) 51.3%

8-3. At the beginning of the compression process in an ideal Diesel cycle the pressure is 15 psia, the temperature is 70°F, and the volume is 6.54 ft^3. The compression ratio for the cycle is 15, and the heat supplied is 188 Btu. At the end of the compression process the pressure is 665 psia, and the temperature is 1108°F. Using constant specific heat, calculate (a) the maximum temperature in the cycle, °F; (b) the temperature at the end of the expansion process, °F; (c) the heat rejected, Btu; (d) the net work of the cycle, Btu; (e) the thermal efficiency of the cycle, percent; (f) the horsepower of an engine operating on such a cycle using 0.5 lb of air/second.

Ans. (a) 2675°F; (b) 939°F; (c) 74.3 Btu; (d) 113.7 Btu; (e) 60.5%; (f) 160.8 hp

8-4. At the beginning of the isentropic compression process in an ideal dual combustion cycle the pressure is 15 psia, the temperature is 70°F and the volume is 0.5 ft^3. The compression ratio is 12 to 1, and the temperature after constant volume receipt of heat is 1800°F. During constant pressure heat receipt, 5 Btu are supplied to the cycle. Using constant specific heat, calculate (a) the total heat added to the cycle, Btu; (b) the heat rejected by the cycle, Btu; (c) the net work of the cycle, Btu; (d) the thermal efficiency of the cycle, percent.

Ans. (a) 10.41 Btu; (b) 3.93 Btu; (c) 6.48 Btu; (d) 62.2%

8-5. An ideal Joule cycle using air operates with a maximum pressure of 600 kPa abs. and a maximum temperature of 725°C. At the beginning of the isentropic compression process, the pressure is 102 kPa abs. and the temperature is 20°C. Calculate (a) the heat supplied, kJ/kg; (b) the heat rejected, kJ/kg; (c) the thermal efficiency of the cycle, percent.

Ans. (a) 514.3 kJ/kg; (b) 310.0 kJ/kg; (c) 39.7%

8-6. A Brayton cycle with a pressure ratio of 6 has compressor inlet conditions of 70°F and 14.7 psia. The turbine inlet temperature is equal to 1500°F. Calculate (a) the isentropic compressor discharge temperature, °R; (b) the isentropic turbine discharge temperature, °R; (c) the thermal efficiency of the basic cycle, percent; (d) the thermal efficiency of the cycle used as an ideal regenerative cycle.

Ans. (a) 884.3°R; (b) 1174.7°R; (c) 40.1%; (d) 54.9%

8-7. A Brayton cycle with a pressure ratio of 6 has compressor inlet conditions of 70°F and 14.7 psia. The turbine inlet temperature is equal to 1500°F. The basic cycle is to be modified by adding ideal reheat. Calculate (a) the HP turbine discharge pressure, psia; (b) the HP turbine discharge temperature, °R; (c) the thermal efficiency of the ideal reheat cycle.

Ans. (a) 36.0 psia; (b) 1517.4°R; (c) 35.0%

8-8. A gas turbine operating in an ideal open cycle is served by a compressor which receives air at 15 psia, 80°F and 13.32 ft^3/lb and delivers it to the combustion chamber at 60 psia. In the combustion chamber 300 Btu are added at constant pressure to each pound of air. Steady flow conditions exist throughout the cycle. Using constant specific heat, calculate (a) the compressor work, Btu/lb; (b) the turbine work, Btu/lb; (c) the net cycle work, Btu/lb; (d) the cycle thermal efficiency, percent; (e) the net horsepower developed by an ideal turbine operating in this cycle and using 36.1 lb of air per second.

Ans. (a) 63.0 Btu/lb; (b) 161.1 Btu/lb; (c) 98.1 Btu/lb; (d) 32.7%; (c) 5000 hp

POWER CYCLES (GAS)

8-9. An ideal Otto cycle operates with a temperature of 370°C at the end of the compression process, a maximum temperature of 1510°C and a minimum temperature of 20°C. What is the temperature at the beginning of the heat rejection process, °C?

Ans. 539.5°C

8-10. An open Brayton cycle using air operates with a maximum cycle temperature of 1300°F. The mechanical compressor pressure ratio is 6.0. Heat supplied in the combustion chamber is 200 Btu/lb. The ambient temperature before the compressor is 95°F and the atmospheric pressure is 14.17 psia. Using constant specific heat, calculate (a) the maximum pressure of the cycle, psia; (b) the temperature of the air leaving the turbine, °F; (c) the cycle thermal efficiency, percent.

Ans. (a) 85.0 psia; (b) 595°F; (c) 40.0%

8-11. In an ideal Otto cycle the heat supplied is 920 kJ/kg and the highest temperature in the cycle is 1600°C. Calculate (a) the temperature at the end of the isentropic compression process, °C, and (b) change of entropy of the heat rejection process, kJ/kg·K.

Ans. (a) 315°C; (b) 0.829 kJ/kg·K

8-12. In an ideal Diesel cycle the change of entropy for the heat rejection process is 0.1662 Btu/lb·°R. The temperature and pressure at the beginning of the isentropic compression are 75°F and 15 psia, respectively. Pressure at the end of the isentropic compression process is 555 psia. Determine (a) the temperature at the beginning of the heat addition process, °F, and (b) the highest temperature in the cycle, °F.

Ans. (a) 1040°F; (b) 2540°F

8-13. In an ideal Otto cycle, the temperature at the end of the heat addition process is 2600°F, and the temperature at the end of the expansion process is 883°F. What is the compression ratio?

Ans. 7.84

8-14. A Brayton cycle aircraft gas turbine engine has an axial flow compressor which provides a pressure ratio of 10 to 1. Material design conditions limit the temperature of the working substance entering the turbine to 2200°R. The engine is designed to handle 90 lb of air/sec at static conditions and 120 lb/sec at a flight speed of 400 knots when sea level ambient pressure and temperature are 15.0 psia and 530°R, respectively. (a) Estimate the sea level static thrust, lbf, and (b) the thrust, lbf, and propulsive efficiency, percent, for a flight speed of 400 knots at sea level. (1 knot = 1.69 ft/sec)

Ans. (a) 7300 lbf; (b) 7550 lbf, 40%

8-15. A Brayton cycle aircraft gas turbine has an axial flow compressor providing a pressure ratio of 12 to 1. Material design conditions limit the temperature of the working substance entering the turbine to 2000°R. The engine is designed to handle 90 lb of air/sec at static conditions and 120 lb/sec at a flight speed of 500 knots when sea level ambient pressure and temperature are 15.0 psia and 530°R, respectively. Estimate (a) the sea level static thrust, lbf, and (b) the thrust, lbf, propulsive efficiency, percent, and cycle thermal efficiency, percent, for a flight speed of 500 knots at sea level. (1 knot = 1.69 ft/sec)

When equipped for afterburning, nozzle inlet temperature is limited to 2500°R. Assuming all stated performance parameters remain constant, estimate (c) the thrust, lbf, and thermal efficiency, percent, with afterburning for a flight speed of 500 knots at sea level.

Ans. (a) 6635 lbf; (b) 6045 lbf, 51%, 55.8%; (c) 9180 lbf, 44.5%

POWER CYCLES (GAS)

8-16. A split-shaft gas turbine has its power unit receiving 140 lb/sec of combustion products at 45 psia and 1600°R. Exhaust temperature and pressure for the power turbine are 1245°R and 0.2″ Hg gauge. Barometric pressure is 29.9″ Hg abs. Ambient temperature is 59°F. Using the Air Tables determine (a) the isentropic turbine outlet temperature, °R; (b) the isentropic turbine work, Btu/lb; (c) the actual turbine work, Btu/lb; (d) the isentropic turbine efficiency; (e) the turbine internal power, hp.

Ans. (a) 1194°R; (b) 106.0 Btu/lb; (c) 92.9 Btu/lb; (d) 87.7%; (e) 18,400 hp

8-17. A gas turbine operating in an ideal open cycle is served by a compressor which receives air at 15 psia, 80°F, and delivers it to the combustion chamber at 60 psia. Three hundred Btu are added at constant pressure to each pound of air in the combustion chamber. Using the Air Tables calculate (a) the compressor work, Btu/lb; (b) the turbine work, Btu/lb; (c) the net work, Btu/lb; (d) thermal efficiency, percent; (e) the net horsepower, hp (air supplied at 36.1 lb/sec).

Ans. (a) 63.0 Btu/lb; (b) 156.7 Btu/lb; (c) 93.7 Btu/lb; (d) 31.23%; (e) 4784 hp

8-18. A marine split-shaft gas turbine has its power turbine supplied with 150 lb/sec of gas at 50 psia and 1100°F, and exhausts the gas from the power turbine at 16 psia and 800°F, respectively. The power turbine exhaust is led through a counter-flow regenerator where the gas temperature is lowered another 40°F. The compressor pressure ratio is 12.0 and the compressor inlet conditions are 14.6 psia and 60°F. Compressor discharge temperature is 740°F. Using the Air Tables calculate (a) the power turbine engine efficiency, percent; (b) the compressed air temperature leaving the regenerator, °F; and (c) regenerator effectiveness, %.

Ans. (a) 74.5%; (b) 780°F; (c) 66.7%

CHAPTER 9

POWER CYCLES (STEAM)

9-1. Introduction. During the discussion of thermodynamic cycles in Chapter 7, the Carnot cycle was shown to be the most efficient cycle possible between any fixed source and receiver temperatures. The natural and physical limitations under which power cycles in general must operate were presented there in the discussion of available energy. One of these limitations, the physical properties of the working substance, is responsible for the absence of the Carnot cycle from the field of steam power cycles. The inability to compress steam isentropically from a low-pressure, wet vapor to a saturated liquid economically, and the difficulty of arranging a constant temperature superheating process obviate the use of a Carnot steam power cycle. These two impractical processes are shown in Fig. 9-1

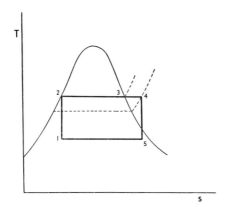

FIGURE 9-1. A Carnot cycle for steam.

as process lines 1–2 and 3–4 respectively. This is just one of many Carnot cycles which might have been proposed for use with steam, but the impracticalities are typical of those generally encountered. However, the high thermal efficiency of that cycle is most desirable, and all the ideal power cycles presented in the following articles will reflect an attempt by various means and devices to attain the *highest possible average temperature of heat receipt,* while utilizing the practical constant pressure processes for vaporization of a liquid and condensation of a vapor. Thermodynamic cycle efficiencies are, of course, based on the temperature difference

between reception and rejection of heat; but since heat rejection temperatures are governed by the naturally available receivers, atmospheric air and large bodies of water, nearly all attention is focused on raising the average temperature of heat receipt for each power cycle.

9-2. The Rankine Cycle. The equipment necessary for the Rankine steam power cycle is shown in Fig. 9-2. The feed pump takes water from

FIGURE 9-2. Apparatus for Rankine cycle.

the condenser, raises its pressure isentropically to boiler pressure and delivers it to the boiler. Next, heat is added in the boiler at constant pressure until the steam becomes superheated. The steam then expands isentropically through the turbine until it reaches condenser pressure. Finally, heat is removed and the steam becomes a saturated liquid at the condenser discharge, where it is finally picked up once more by the feed pump. The T-s and h-s diagrams for the cycle followed by the steam in moving through the equipment of Fig. 9-2 are shown in Fig. 9-3. Since few, if any, modern steam power plants operate without benefit of superheat for the steam, the basic Rankine cycle including superheat is shown.

The most convenient method of analysis for the cycle is based on the areas of the T-s diagram representing heat supplied and heat rejected. The heat supplied to the cycle from an outside source is that supplied by the boiler, calculated by applying the steady flow energy equation across the boiler. This yields

$$q_S = h_2 - h_1 \text{ Btu/lb (J/kg)} \tag{9-1}$$

POWER CYCLES (STEAM)

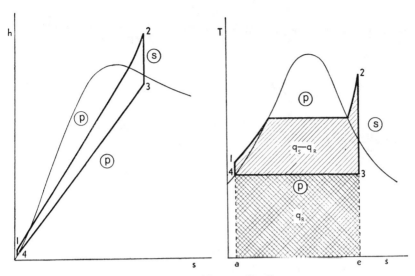

FIGURE 9-3. Rankine cycle diagram.

The heat rejected from the cycle is the heat removed by the condenser,

$$q_R = h_3 - h_4 \text{ Btu/lb (J/kg)} \quad (9\text{-}2)$$

The net work of any thermodynamic power cycle is

$$\frac{wk(\text{net})}{(J)} = q_S - q_R \text{ Btu/lb (J/kg)} \quad (9\text{-}3)$$

In the notation of Fig. 9-3,

$$\frac{wk(\text{net})}{(J)} = (h_2 - h_1) - (h_3 - h_4) \text{ Btu/lb (J/kg)}$$

$$= (h_2 - h_3) - (h_1 - h_4) \text{ Btu/lb (J/kg)} \quad (9\text{-}4)$$

The net work for this cycle is also expressed by the difference between the turbine work and the pump work. Thus,

$$\frac{wk(\text{net})}{(J)} = \frac{wk(\text{turbine})}{(J)} - \frac{wk(\text{pump})}{(J)}$$

and in the notation of Fig. 9-3

$$\frac{wk(\text{turbine})}{(J)} = h_2 - h_3 \text{ Btu/lb (J/kg)}$$

POWER CYCLES (STEAM)

and
$$\frac{wk(\text{pump})}{(J)} = h_1 - h_4 \text{ Btu/lb (J/kg)}$$

Then
$$\frac{wk(\text{net})}{(J)} = (h_2 - h_3) - (h_1 - h_4) \text{ Btu/lb (J/kg)}$$

The thermal efficiency of any thermodynamic power cycle is the net work divided by the heat supplied, giving

$$\eta_t = \frac{wk(\text{net})}{(J)q_S} = \frac{q_S - q_R}{q_S} \tag{7-3}$$

In the notation of Fig. 9-3,

$$\eta_t = \frac{(h_2 - h_1) - (h_3 - h_4)}{(h_2 - h_1)} \tag{9-5}$$

The Rankine cycle is an *ideal* cycle, and since there is no heat transferred to or from the working substance in the turbine or in the pump, the associated processes are the isentropic expansion 2 to 3 and the isentropic compression 4 to 1. The available energy of this Rankine cycle is that represented by $q_S - q_R$ of Fig. 9-3, since $q_R = T_3(s_3 - s_4)$, which is equal to $T_0(s_2 - s_1)$.

From an inspection of Fig. 9-3 it becomes clear that the only means available for raising the average temperature of heat receipt, and thereby raising the thermal efficiency of the cycle, without addition of extra equipment is raising the boiler pressure (and thereby the saturation temperature), raising the superheat temperature, T_2, or both. Condenser pressures in modern power plants are maintained between one and two inches of mercury absolute. Further reductions are usually not justified by the diminishing gains in cycle thermal efficiency.

Example 9-1: A Rankine steam power cycle operates with steam at 600 psia and 850°F from the boiler and a condenser pressure of one inch of mercury absolute. Sketch the cycle on h-s and T-s coordinates and determine for the cycle:
 (a) enthalpies for the steam leaving the boiler, leaving the turbine, leaving the condenser and leaving the pump, Btu/lb
 (b) the pump work, Btu/lb
 (c) the heat supplied, Btu/lb
 (d) the heat rejected, Btu/lb
 (e) the net work, Btu/lb
 (f) the turbine work, Btu/lb
 (g) the thermal efficiency, percent
 (h) the average temperature of heat receipt as determined by the heat added divided by the change of entropy during addition of heat, °F.

POWER CYCLES (STEAM)

Solution:

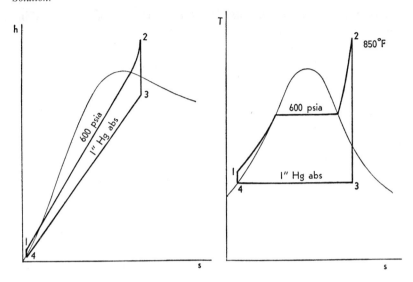

(a) From Steam Tables and Mollier chart,

$$h_1 = 49.4 \text{ (from Table 4)}$$
$$h_2 = 1435.4 \quad s_2 = 1.6559$$
$$h_3 = 890.0$$
$$h_4 = 47.1 \quad s_4 = 0.09146 = s_1$$

(b) $$\frac{wk(\text{pump})}{J} = h_1 - h_4 = 49.4 - 47.1$$
$$= 2.3 \text{ Btu/lb}$$

(c) $\quad q_S = h_2 - h_1 = 1435.4 - 49.4 = \underline{1386.0 \text{ Btu/lb}}$
(d) $\quad q_R = h_3 - h_4 = 890.0 - 47.1 = \underline{842.9 \text{ Btu/lb}}$
(e) \quad net work $= q_S - q_R = 1386.0 - 842.9 = \underline{543.1 \text{ Btu/lb}}$
(f) \quad turbine work $= h_2 - h_3 = 1435.4 - 890.0 = \underline{545.4 \text{ Btu/lb}}$

(g) $$\eta_t = \frac{q_S - q_R}{q_S} = \frac{1386.0 - 842.9}{1386.0} = \underline{39.2\%}$$

$$\eta_t = \frac{\text{output}}{\text{input}} = \frac{\text{turbine work} - \text{pump work}}{q_S} = \frac{545.4 - 2.3}{1386.0} = \underline{39.2\%}$$

(h) $$t_{AV} = \frac{q_S}{\Delta_S} - 460 = \frac{1386.0}{1.5644} - 460 = 886 - 460 = \underline{426°F}$$

9-3. The Rankine Reheat Cycle. In order to gain the practical advantage of drier steam at the turbine exhaust, as well as realize the advancement in average temperature of heat receipt with its attendant increase in thermal efficiency, the Rankine reheat cycle is used in many power plants. It

consists of a Rankine cycle in which the steam is removed from the turbine at a desirable condition, usually near the saturation point, reheated in the boiler and returned to the turbine to complete its expansion to exhaust pressure.

The mechanical equipment required and the flow path of the steam are shown diagrammatically in Fig. 9-4. Steam is generated in the boiler,

FIGURE 9-4. Rankine reheat cycle apparatus.

passes to the high pressure turbine where it expands until it reaches the extraction point, and returns to the boiler to be resuperheated. After resuperheating, the steam flows to the low pressure turbine, completing its expansion. Finally it is condensed and returned to the boiler by the feed pump. In the interest of simplicity, feed water heaters and other auxiliary equipment have been omitted from the diagram.

On h-s and T-s coordinates the Rankine reheat cycle appears as shown in Fig. 9-5. Process line 1–2 shows the heating from compressed liquid to superheated steam by the boiler, 2–3 the first expansion in the turbine, 3–4 the reheating process in the boiler, 4–5 the second expansion in the turbine, and 5–6 the change of phase in the condenser. Process line 6–1 shows the isentropic process in the feed pump.

For this ideal Rankine reheat cycle, the heat added is

$$q_S = (h_2 - h_1) + (h_4 - h_3) \text{ Btu/lb (J/kg)}$$

The heat rejected is

$$q_R = h_5 - h_6 \text{ Btu/lb (J/kg)}.$$

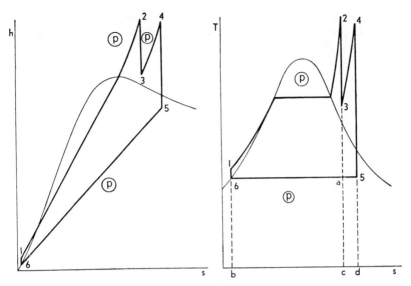

FIGURE 9-5. Rankine reheat cycle diagrams.

The net cycle work is $q_S - q_R$ and is given by

$$\frac{wk(\text{net})}{(J)} = q_S - q_R \qquad (7\text{-}2)$$

$$= (h_2 - h_1) + (h_4 - h_3) - (h_5 - h_6) \text{ Btu/lb (J/kg)} \qquad (9\text{-}6)$$

The net cycle work may also be calculated from the difference between the turbine work and the pump work. Thus:

$$\frac{wk(\text{turbine})}{(J)} = (h_2 - h_3) + (h_4 - h_5) \text{ Btu/lb (J/kg)}$$

$$\frac{wk(\text{pump})}{(J)} = (h_1 - h_6) \text{ Btu/lb (J/kg)}$$

and

$$\frac{wk(\text{net})}{(J)} = (h_2 - h_3) + (h_4 - h_5) - (h_1 - h_6) \text{ Btu/lb (J/kg)} \qquad (9\text{-}7)$$

The thermal efficiency of the cycle is given by

$$\eta_t = \frac{q_S - q_R}{q_S} = \frac{(h_2 - h_1) + (h_4 - h_3) - (h_5 - h_6)}{(h_2 - h_1) + (h_4 - h_3)} \qquad (9\text{-}8)$$

The moisture in the steam at point 5 of the $T\text{-}s$ diagram of Fig. 9-5 is considerably less than it would have been had the original turbine ex-

POWER CYCLES (STEAM)

pansion been allowed to terminate at *a*. Furthermore, the added equivalent work area 345*a* is a larger percentage of the equivalent work area 123*a*6 than the additional heat-rejected area *a*5*dc* is of the original heat-rejected area 6*acb*. This results in a net gain in thermal efficiency for the cycle due to an increase in the average temperature of heat receipt over that of Rankine cycle 123*a*6.

Example 9-2: An ideal Rankine reheat cycle operates with steam at 1200 psia and 1060°F from the boiler. After expansion in the turbine to 90 psia, the steam is returned to the boiler and reheated to a temperature of 950°F. Condenser pressure is maintained at one inch of mercury absolute. Sketch the cycle on *h-s* and *T-s* coordinates and calculate for the cycle:
 (a) pump work, Btu/lb
 (b) heat added, Btu/lb
 (c) heat rejected, Btu/lb
 (d) turbine work, Btu/lb
 (e) net cycle work, Btu/lb
 (f) cycle thermal efficiency, percent
 (g) the average temperature of heat receipt for the cycle, °F
Solution:
For sketches of cycle refer to Fig. 9-5.
From the Mollier Chart and Steam Tables the following enthalpies and entropies are determined. They are numbered in conformity with Figs. 9-4 and 9-5.

$h_2 = 1534.7$ Btu/lb $h_4 = 1506.5$ Btu/lb
$h_3 = 1219.3$ Btu/lb $h_5 = 1029.0$ Btu/lb
$h_6 = 47.1$ Btu/lb $h_1 = 51.2$ Btu/lb (Table 4)
$s_2 = 1.6532 = s_3$
$s_4 = 1.9141 = s_5$
$s_6 = 0.09146 = s_1$

(a) $$\frac{wk(\text{pump})}{J} = h_1 - h_6 = 51.2 - 47.1 = \underline{4.1 \text{ Btu/lb}}$$

(b) $$q_S = (h_2 - h_1) + (h_4 - h_3) = (1534.7 - 51.2) + (1506.5 - 1219.3) = \underline{1770.7 \text{ Btu/lb}}$$

(c) $$q_R = (h_5 - h_6) = (1029.0 - 47.1) = \underline{981.9 \text{ Btu/lb}}$$

(d) $$\frac{wk}{J}(\text{turbine}) = (h_2 - h_3) + (h_4 - h_5) = (1534.7 - 1219.3) + (1506.5 - 1029) = \underline{792.9 \text{ Btu/lb}}$$

(e) net cycle work $= q_S - q_R = 1770.7 - 981.9 = \underline{788.8 \text{ Btu/lb}}$

(f) $$\eta_t = \frac{q_S - q_R}{q_S} = \frac{788.8}{1770.7} = \underline{44.5\%}$$

(g) $$t_{AV} = \frac{q_S}{\Delta s} - 460 = \frac{1770.7}{1.9141 - 0.0915} - 460 = \underline{512°F}$$

9-4. The Rankine Regenerative Cycle.
The regenerative process was first mentioned in Chapter 8 in connection with the Stirling and Ericsson gas

POWER CYCLES (STEAM)

power cycles. In steam power cycles, regeneration refers to the feedwater heating done by steam extracted expressly for that purpose from various stages of the turbine. Advantages are a higher cycle thermal efficiency through lower heat rejection to the condenser, and a reduced steam volume flowing through the large low pressure section of the turbine. Disadvantages are a higher system cost and complexity coupled with a certain reduction in operating flexibility.

The equipment necessary for a regenerative steam cycle with two stages of regenerative feedwater heating is shown in Fig. 9-6. The boiler fur-

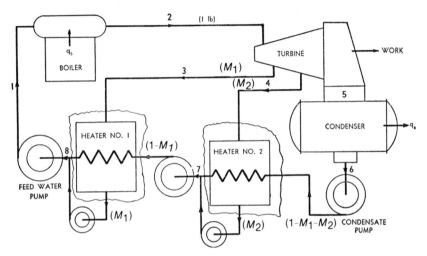

FIGURE 9-6. Apparatus for Rankine regenerative cycle.

nishes steam for the turbine which expands all of it isentropically down to the first extraction point, where M_1 is shown leaving the turbine. At this point, a quantity of steam, M_1, determined by the feed heating requirements of heater No. 1, is extracted from the total turbine flow to be used as a source of heat in heater No. 1. In condensing to a saturated liquid at its extraction pressure, this M_1 pounds of vapor will give up any superheat it may have plus its heat of evaporation, thereby raising the temperature of the $(1 - M_1)$ pounds of feedwater being pumped through heater No. 1 to the saturation temperature for the extraction pressure of M_1.

The steam remaining in the turbine after M_1 has been extracted continues to expand isentropically until the second extraction point is reached, where another quantity of steam, M_2, is extracted. This second quantity, like the first, has its magnitude determined by the feed heating requirements of heater No. 2. As M_2 gives up its remaining heat of

evaporation in heater No. 2, it serves to raise the temperature of $(1 - M_1 - M_2)$ pounds of feedwater being pumped through heater No. 2 to the saturation temperature corresponding to the extraction pressure of M_2.

After the second turbine-extraction point $(1 - M_1 - M_2)$ pounds per pound of throttle steam remains to expand isentropically to condenser pressure. After liquefying at condenser pressure, the main condensate pump starts the liquid on its way to heater No. 2, where it is heated until it becomes a saturated liquid at the saturation temperature corresponding to the extraction pressure of M_2. The heat is supplied, as pointed out previously, by the condensation of the steam, M_2, taken from the turbine at extraction point 2. From heater No. 2 the condensate is pumped to heater No. 1 after having the condensed steam, M_2, added to it by an auxiliary pump, making the quantity of condensate entering heater No. 1 $(1 - M_1)$ pounds. This $(1 - M_1)$ pounds of condensate is heated by M_1 pounds of extracted steam, as it condenses, to the saturation temperature corresponding to the extraction pressure of M_1. Finally the condensed steam from heater No. 1 is pumped by an auxiliary pump into the main body of condensate, and the entire quantity is returned to the boiler by the feed pump.

FIGURE 9-7. Double-extraction Rankine regenerative cycle.

POWER CYCLES (STEAM)

The *T-s* diagram for the above cycle is shown in Fig. 9-7. The points of major importance are numbered to correspond to the apparatus shown in Fig. 9-6. All the pump work is assumed to be done by the feed pump, as shown by process line 8-1 of Fig. 9-7. Actually, the work of each individual pump may be calculated, but the pressure ratio across the lower pressure pumps is low, the quantity of water handled is not the total cycle flow, and the energy quantities involved are small. Therefore, only the work of the last (feed) pump, which handles the total cycle flow, and which accounts for approximately 85 percent of the total pump work, will be calculated.

Test data, borne out by rather involved calculations, show that the best efficiency for a given number of heaters is obtained when the temperature range per heater is approximately equal. Accordingly, for the cycle of Fig. 9-7, dividing the *temperature* interval between the *saturated* liquid at boiler pressure and the *saturated* liquid at condenser pressure into three equal parts determines the temperatures for the two extractions. The saturation pressures corresponding to these temperatures, then, are the optimum extraction pressures. For a cycle containing a single extraction point and a single heater, the temperature which divides that same range into two equal parts would determine the extraction pressure.

The mass of steam to be removed from the turbine at each extraction point may be calculated by setting up a heat balance across each heater, starting with the heater nearest the boiler, No. 1, as shown in Fig. 9-6. Equating the total heat given up by the extracted steam to the total heat gained by the feedwater gives

$$|Q_{\text{steam}}| = |Q_{\text{feedwater}}|$$

and for unit mass of steam at the turbine throttle,

$$M_1(h_3 - h_8) = (1 - M_1)(h_8 - h_7)$$

In a similar manner, a heat balance may be drawn across the second heater to give

$$M_2(h_4 - h_7) = (1 - M_1 - M_2)(h_7 - h_6)$$

By determining enthalpies from the Steam Tables and solving for M_1 first, the result may be used to solve for M_2. For three heaters, the same plan would be followed, solving for each mass in order.

For the ideal cycle the heat supplied is

$$q_S = h_2 - h_1 \text{ Btu/lb (J/kg)}$$

The heat rejected is

$$q_R = (1 - M_1 - M_2)(h_5 - h_6) \text{ Btu/lb throttle steam (J/kg)}$$

POWER CYCLES (STEAM)

The net work of the cycle is

$$\frac{wk(\text{net})}{(J)} = q_S - q_R \tag{7-2}$$

$$= (h_2 - h_1) - (1 - M_1 - M_2)(h_5 - h_6) \text{ Btu/lb (J/kg)} \tag{9-9}$$

Also,

$$\frac{wk(\text{net})}{(J)} = \frac{wk(\text{turbine})}{(J)} - \frac{wk(\text{pump})}{(J)} \text{ Btu/lb (J/kg)}$$

$$= (h_2 - h_3) + (1 - M_1)(h_3 - h_4)$$
$$+ (1 - M_1 - M_2)(h_4 - h_5) - (h_1 - h_8) \text{ Btu/lb (J/kg)}$$

The thermal efficiency of the cycle is

$$\eta_t = \frac{q_S - q_R}{q_S} = \frac{(h_2 - h_1) - (1 - M_1 - M_2)(h_5 - h_6)}{(h_2 - h_1)} \tag{9-10}$$

Example 9-3: A Rankine regenerative steam cycle employs two stages of steam extraction for feedwater heating. Boiler pressure and temperature are 1200 psia and 1050°F respectively. Saturated steam at condenser pressure has a temperature of 79°F. Sketch the cycle on *T-s* coordinates and calculate:

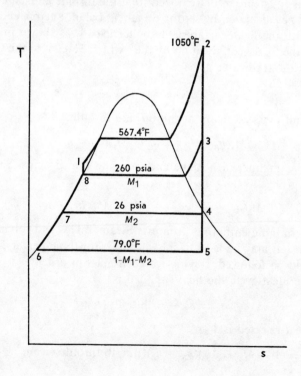

POWER CYCLES (STEAM)

(a) the optimum extraction pressures to the nearest pound per square inch
(b) the mass of steam removed from the turbine at each extraction point, pounds per pound of throttle steam
(c) pump work, Btu/lb
(d) the heat supplied to the cycle, Btu/lb
(e) the heat rejected by the cycle, Btu/lb throttle steam
(f) the turbine work, Btu/lb throttle steam
(g) the net cycle work, Btu/lb throttle steam
(h) the thermal efficiency of the cycle, percent

Solution:

$$p_1 = 1200 \text{ psia} \qquad t_5 = t_6 = 79.0°F$$
$$t_2 = 1050°F$$

(a) $t_{\text{sat.}}$ at 1200 psia = 567.4°F $\quad h_1 = 383.4$ Btu/lb $\quad s_1 = s_8 = 0.5723$
$t_{\text{sat.}}$ at condenser = 79.0°F $\quad h_2 = 1528.9$ Btu/lb $\quad s_2 = 1.6494$
$\quad\quad\quad\quad\quad \Delta t = 488.4°F \quad h_3 = 1322.7$ Btu/lb
$\quad\quad\quad\quad\quad\quad\quad\quad\quad\quad h_4 = 1118.3$ Btu/lb
$\dfrac{\Delta t}{3} = \dfrac{488.4}{3} = 162.8°F \quad h_5 = 886.4$ Btu/lb
$\quad\quad\quad\quad\quad\quad\quad\quad\quad\quad h_6 = 47.1$ Btu/lb
$\quad 567.4 - 162.8 = 404.6°F \quad h_7 = 210.7$ Btu/lb
$\quad\quad\quad\quad\quad\quad\quad\quad\quad\quad h_8 = 380.0$ Btu/lb

Nearest whole psia = $\underline{260 \text{ psia}} = p_3 = p_8$
$404.6 - 162.8 = 241.8°F$
Nearest whole psia = $\underline{26 \text{ psia}} = p_4 = p_7$

(b)
$$M_1(h_3 - h_8) = (1 - M_1)(h_8 - h_7)$$
$$M_1(1322.7 - 380.0) = (1 - M_1)(380.0 - 210.7)$$
$$942.7 M_1 = 169.3 - 169.3 M_1$$
$$M_1 = \underline{0.1522 \text{ lb per lb throttle steam}}$$
$$M_2(h_4 - h_7) = (1 - M_1 - M_2)(h_7 - h_6)$$
$$M_2(1118.3 - 210.7) = (1 - 0.1522 - M_2)(210.7 - 47.1)$$
$$907.6 M_2 = 0.8478(163.6) - 163.6 M_2$$
$$M_2 = \underline{0.1295 \text{ lb per lb throttle steam}}$$

(c) $\dfrac{wk(\text{pump})}{J} = h_1 - h_8 = 383.4 - 380.0 = \underline{3.4 \text{ Btu/lb}}$

(d) $q_S = (h_2 - h_1) = (1528.9 - 383.4) = \underline{1145.5 \text{ Btu/lb}}$

$q_R = (1 - M_1 - M_2)(h_5 - h_6) = 0.7183(886.4 - 47.1)$
$\quad = \underline{602.9 \text{ Btu/lb}}$

(f) $\dfrac{wk}{J}(\text{turbine}) = (h_2 - h_3) + (1 - M_1)(h_3 - h_4) + (1 - M_1 - M_2)(h_4 - h_5)$

$\quad\quad\quad = 206.2 + 173.3 + 166.6 = \underline{546.1 \text{ Btu/lb}}$

(g) $\dfrac{wk(\text{net})}{J} = q_S - q_R = 1145.5 - 602.9 = \underline{542.6 \text{ Btu/lb}}$

(h) $\eta_t = \dfrac{q_S - q_R}{q_S} = \dfrac{542.6}{1145.5} = \underline{47.4\%}$

POWER CYCLES (STEAM)

Practical considerations, such as simplicity of operation, ease of maintenance, space requirements and initial dollar investment, must be balanced off against the fuel economy realized from regenerative feed heating. As the number of heaters and matching turbine extraction points is increased, the thermal efficiency of the plant increases, as do the mechanical complexity and first cost. The gain is most pronounced in plants operating at the upper practical limits of pressure; but, in any case, the *rate* of increase in thermal efficiency declines with each added heater. In large central station generating plants, where lifetime fuel costs are of great importance, as many as seven heaters may be employed. In the usual naval ship propulsion plant, utilizing only moderately high pressures and subject to widely varying loads, the advantages of regeneration are currently insufficient to overcome the added space requirements and operating complexity. However, merchant vessels, operating at substantially constant sea speeds over long distances, are frequently designed with up to four stages of regenerative feed heating in the interests of fuel economy. With steadily rising fuel costs, increasing attention to both reheat and regenerative cycles, singly or in combination, can be anticipated in marine practice.

9-5. The Rankine Regenerative Reheat Cycle. For certain types of steam plant operating conditions, advantage may be taken of the features included in both the reheat and regenerative cycles. Such a cycle, which includes both feed heating extraction and reheating, is called a regenerative reheat cycle. Since it is merely a combination of the features of the regenerative and the reheat cycles, it will be best explained by an example.

Example 9-4: A Rankine regenerative reheat cycle employs steam at 2000 psia and 1050°F leaving the boiler. The cycle includes one steam extraction stage for feedwater heating and one steam reheat to 950°F. The saturated steam temperature in the condenser is 101.7°F. Sketch the cycle on T-s coordinates and calculate:
 (a) the optimum extraction pressure, psia, for feedwater heating
 (b) the weight of steam extracted for feedwater heating, lb per lb of throttle steam
 (c) the pump work, Btu/lb
 (d) the heat supplied to the cycle, Btu/lb of throttle steam
 (e) the heat rejected by the cycle, Btu/pound of throttle steam
 (f) the turbine work, Btu/lb throttle steam
 (g) the net cycle work, Btu/lb throttle steam
 (h) the cycle thermal efficiency

POWER CYCLES (STEAM)

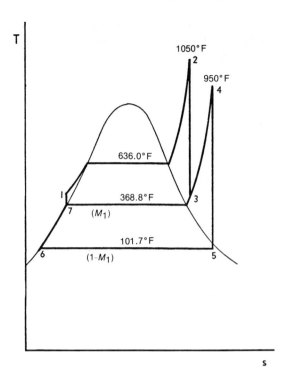

Solution:

$$p_2 = p_1 = 2000 \text{ psia} \qquad t_2 = 1050°\text{F}$$
$$p_6 = p_5 = 1.0 \text{ psia} \qquad t_4 = 950°\text{F}$$
$$\qquad\qquad\qquad\qquad\qquad t_5 = 101.7°\text{F}$$

(a) sat. temp. at 2000 psia = 636.0°F $h_1 = 347.9$ Btu/lb
sat. temp. at condenser = 101.7°F $h_2 = 1506.0$ Btu/lb $s_2 = 1.5813$
$\Delta t = 534.3°$F $h_3 = 1214.8$ Btu/lb
 $h_4 = 1504.0$ Btu/lb $s_4 = 1.8427$
$$\frac{\Delta t}{2} = \frac{534.3}{2} = 267.2$$ $h_5 = 1029.9$ Btu/lb
 $h_6 = 69.7$ Btu/lb
$636.0 - 267.2 = 368.8°$F $h_7 = 341.3$ Btu/lb $s_7 = 0.5270$

Nearest whole psia = 170 psia = $p_3 = p_4 = p_7$
170 psia will also be reheat pressure.

(b) Heat balance across feedheater:
$$M_1(h_3 - h_7) = (1 - M_1)(h_7 - h_6)$$
$$M_1(1214.8 - 341.3) = (1 - M_1)(341.3 - 69.7)$$
$$1145.1 M_1 = 271.6$$
$$M_1 = \frac{271.6}{1145.1} = \underline{0.2372 \text{ lb/lb throttle steam}}$$

285

(c) $\quad wk_p/J = h_1 - h_7 = 347.9 - 341.3 = \underline{6.6 \text{ Btu/lb}}$

(d) $\quad q_S = (h_2 - h_1) + (1 - M_1)(h_4 - h_3)$
$\qquad = (1506.0 - 347.9) + (1 - 0.2372)(1504.0 - 1214.8)$
$\quad q_S = \underline{1378.7 \text{ Btu/lb throttle steam}}$

(e) $\quad q_R = (1 - M_1)(h_5 - h_6) = 0.7628(1029.9 - 69.7)$
$\quad q_R = \underline{732.4 \text{ Btu/lb throttle steam}}$

(f) $\quad \dfrac{wk(\text{turbine})}{J} = (h_2 - h_3) + (1 - M_1)(h_4 - h_5)$
$\qquad = (1506.0 - 1214.8) + (0.7628)(1504.0 - 1029.9)$
$\qquad = 291.2 + 361.6 = \underline{652.8 \text{ Btu/lb throttle steam}}$

(g) $\quad \dfrac{wk(\text{net cycle})}{J} = q_S - q_R = 1378.7 - 732.4$
$\qquad = \underline{646.3 \text{ Btu/lb throttle steam}}$

(h) $\quad \eta_t = \dfrac{q_S - q_R}{q_S} = \dfrac{646.3}{1378.7} = \underline{46.9\%}$

The steam temperature of this example is near the maximum value in use in large steam power stations at the present time. Higher pressures are employed, but usually with lower steam temperatures.

9-6. Nuclear Reactor Steam Plants and Cycles. The various Rankine steam cycles in Arts. 9-2 through 9-5 employed a boiler of some kind as an energy source for the system. In recent years, the nuclear reactor has emerged as an alternate heat source for these cycles. Unfortunately, time and space are not available for an investigation of all the many possible types of reactors, but a summary of some of the more promising will be given, along with a few of the equipment and property variations imposed by each one on the basic steam cycle.

For conventional power generation systems, overall cycle or plant efficiency is the final determinant with respect to operating cost. Each of the Rankine cycles analyzed so far has been investigated for cycle thermal efficiency, but little or nothing has been said about operating cost, which would include fuel cost and heating value as well as thermal efficiency. For a nuclear plant, operating cost is strongly affected by fuel core life, which is unrelated to efficiency. In fact, if lower reactor temperature (which could mean lower efficiency) would contribute to longer core life, it might be economically sound to sacrifice efficiency. The cost of the uranium burned and the recovery of unburned uranium per net kW-hr is small relative to the cost of fabricating that material into fuel elements or cores. In order to bring the design of a nuclear reactor steam plant into the

POWER CYCLES (STEAM)

closest possible economic competition with the large, fossil-fuel-fired central power stations, it is necessary to base the design on material economy as well as thermal efficiency. Material economy is the reactor output per unit of the fuel loading of the reactor. In a nuclear reactor, high material economy and high thermal efficiency lead to low overall power cost.

Historically, one of the first, and perhaps the broadest, classifications of nuclear reactors is according to the energy level of their neutrons. On this basis there are the thermal-, intermediate- and fast-neutron reactors. Thermal-neutron and some intermediate-neutron reactors require a neutron moderator, a substance that slows the neutrons to a velocity at which the probability for the occurrence of fission is greater.

Fast-neutron reactors generally suffer low material economies in comparison with thermal-neutron reactors, but have the potential advantage of "breeding" or producing more fissionable material than is actually used. Thermal reactors also give some production of plutonium, although not generally a 1:1 ratio. The better figures that are quoted for breeding ratio in fast reactors are still hard to realize in actual practice because of practical difficulties in reprocessing and in design for high neutron economy.

Form of fuel used is another method of classification. The fissionable fuel may be isolated from the moderator in slugs, plates, pebbles, alloys or ceramics, or dispersed throughout in form of solutions, slurries and fused salts. The first type takes the name heterogeneous reactor and the latter homogeneous reactor.

Advantages of the homogeneous reactor are good heat transfer characteristics, good nuclear control and simple chemical processing. Balanced against these are the disadvantages of internal fuel corrosion, greater fuel loading, and high radioactivity outside the reactor core where it might leak.

The heterogeneous reactor contains radioactivity fairly easily but involves higher costs of chemical and metallurgical reprocessing and fabrication of new cores.

Finally, reactors may be classified as to coolant, that substance which transports the thermal energy from the fission process to the heat exchangers or turbines. Some of the coolants used are water, gases and liquid metals. Organic fluids have been used as coolants to a limited degree, but are currently out of favor. The most widely used coolant is water, in either the pressurized water reactor (PWR) or the boiling water reactor (BWR) configurations. Gases, particularly helium and carbon dioxide, and liquid metals, such as sodium, have attractive theoretical

attributes, but the associated technology is much less developed than that for the water-cooled systems. Most existing systems, ashore and afloat, are of the PWR variety and enjoy an enviable record of safety and reliability, going back to the first nuclear submarine, the *Nautilus*.

The pressurized water systems are temperature-limited, which is a thermodynamic disadvantage. The water in the primary (coolant) loop must be kept in the liquid phase, for vaporization would mean radical changes in density and heat transfer characteristics within the reactor. Thus, the absolute maximum upper limit is 705.4°F and 3203.6 psia, the critical temperature and pressure for steam. Remember that if the temperature exceeds the critical value, the vapor cannot be liquefied regardless of the pressure imposed. Table 2 of the Steam Tables shows that for equal increments of pressure, the corresponding temperature increments diminish as the critical conditions are approached. Current design practice compromises the issue with a primary loop pressure of about 2000 psia and a corresponding saturation (maximum) temperature of 636°F. This in

FIGURE 9-8. Division of heat transfer in nuclear steam generating system.

POWER CYCLES (STEAM)

turn limits the steam temperature (and associated saturation pressure) in the secondary loop to still lower values because an appreciable Δt between the primary and secondary loops is necessary for effective heat transfer. This is shown graphically on the temperature versus percent of heat transfer diagram in Fig. 9-8. The practical end result is wet steam at the turbine throttle unless a separately fired superheater is employed. In any case the thermal efficiency of the cycle will not compare favorably with those employing fossil fuel steam generators with higher steam temperatures and pressures.

Gas-cooled reactors are not limited by pressure-temperature characteristics of the gas and so may be operated at higher temperatures. However, gases do not have the specific heat capacity of water, which means that much larger volumes of gas must be circulated to transport an equivalent quantity of heat.

Liquid metal reactor cooling combines some of the advantages of both gases and water. High boiling points of liquid metals allow high temperatures in the primary (reactor) loop without unduly high pressures. In addition, the high heat capacity of molten metals requires a relatively small quantity of coolant compared with gases. Sodium is one of the better metals for this service, but it, too, has disadvantages. With a melting point of 208°F, it must be preheated on starting up a unit. Furthermore, it becomes highly radioactive and reacts violently with water and oxygen. Advantages for sodium are its high boiling point of 1620°F, allowing high primary loop temperatures at low pressure, and its general lack of a tendency to corrode in metal systems. Because of radioactivity and reactivity with water and oxygen, primary sodium systems are usually separated from the steam generators and turbines. This is accomplished by a secondary sodium system, taking heat from the primary system but not acquiring its radioactivity. The secondary sodium system then exchanges heat with the feedwater and steam which operate the turbines. Current fast breeder reactors are sodium-cooled.

Figure 9-9* shows the general arrangement of one of the early successful central station nuclear power plants. It is of the pressurized water reactor (PWR) type and was built by the Consolidated Edison Company at Indian Point, New York, on the Hudson River. Note the separate oil-fired superheater and the four stages of regenerative feed heating.

*Indian Point Pressurized-Water Reactor, G. R. Milne and F. R. Ward, *Mechanical Engineering*, March, 1957.

FIGURE 9-9. Indian Point pressurized water reactor plant.

POWER CYCLES (STEAM)

Example 9-5. In a PWR nuclear plant, the primary loop coolant water, pressurized to 2000 psia, enters the reactor at 500°F and leaves at 600°F. It then enters the steam generator (boiler) section at the same temperature where energy in the form of heat is transferred to the secondary steam loop. The secondary loop operates on a simple Rankine cycle without regenerative feed heating or any external heat supply. Steam leaves the boiler at 420 psia with a moisture content of 1% and enters the turbine at the same conditions. The condenser pressure is 2 psia. The pressure of the water leaving the feed pump and entering the steam generator is 500 psia. The flow rate in the steam plant (secondary) side is 500,000 lb/hr. Assuming no losses other than the pressure drop in the steam generator given above, find: (a) the required heat exchange rate in the steam generator, Btu/hr; (b) the primary coolant flow rate necessary to satisfy the heat transfer rate of part (a); (c) the heat supplied, Btu/lb; (d) the heat rejected, Btu/lb; (e) the thermal efficiency, %.

Solution: Refer to Fig. 9-8.

$$p_1 = 500 \text{ psia} \qquad p_3 = p_4 = 2 \text{ psia}$$
$$p_2 = 420 \text{ psia} \qquad m_2 = 0.01$$
$$\dot{M}_2 = 500{,}000 \text{ lb/hr}$$
$$t_{HR} = 600°F \qquad t_{CR} = 500°F$$

(a) $\dot{Q}_2 = \dot{M}_2(h_2 - h_1) \qquad h_4 = 94.0 \ ; \ s_4 = 0.17499 = s_1$
$$h_1 = 96.0$$
$$h_2 = 1197.7; \ s_2 = 1.4725 = s_3$$
$$h_3 = 854.1$$

$$\dot{Q}_2 = 500{,}000 \ (1197.7 - 96) = 550.85 \times 10^6 \text{ Btu/hr}$$

(b) $\dot{Q}_2 = \dot{Q}_R = \dot{M}_R(t_{HR} - t_{CR})c_p$

From Fig. 3 of the Steam Tables for an average temperature in the reactor of 550°F and 2000 psia, $c_p = 1.25$ Btu/lb · °F.

$$\dot{M}_R = \frac{\dot{Q}_R}{c_p(t_{HR} - t_{CR})} = \frac{550.85 \times 10^6}{1.25(600 - 500)} = \underline{4.407 \times 10^6 \text{ lb/hr}}$$

(c) $q_S = h_2 - h_1 = 1197.7 - 96.0 = \underline{1101.7 \text{ Btu/lb}}$

(d) $q_R = h_3 - h_4 = 854.1 - 94.0 = \underline{760.1 \text{ Btu/lb}}$

(e) $\eta_t = (q_S - q_R)/q_S = (1101.7 - 760.1)/1101.7 = \underline{31.0\%}$

9-7. The Incomplete-Expansion and Non-Expansion Cycles. The Rankine cycle of Art. 9-2 is sometimes called a complete-expansion cycle because the steam expands to the lowest pressure in the cycle while doing useful work. The steam turbine is capable of such complete expansion, while the reciprocating steam engine is not. Actually, a steam engine could be built with sufficient piston area and travel to accommodate the large volume occupied by steam after expansion to condenser pressure. However, the size of such an engine would result in more friction loss in its working parts than could be overcome by the useful work produced in the latter stages of the expansion. The end result for an actual reciprocating engine could very well be a net loss of useful work compared with an engine whose expansion process is incomplete, but whose size and friction losses are smaller. For these reasons, the ideal incomplete-expansion cycle establishes the optimum conditions for reciprocating steam engines.

As in certain of the ideal gas cycles, a non-flow, constant volume cooling process is substituted in the ideal incomplete-expansion steam cycle for the exhaust process of the actual engine. Therefore the ideal incomplete-expansion cycle is a closed cycle always containing the working substance.

In Fig. 9-10, the incomplete-expansion steam cycle is shown on *T-s* coordinates. Heat is added by the boiler from 1 to 2, isentropic expansion

FIGURE 9-10. Incomplete-expansion steam cycle.

POWER CYCLES (STEAM)

in the engine takes place from 2 to 3, constant volume cooling of the steam to condenser pressure and temperature occurs from 3 to 4, and constant pressure cooling in the condenser to a saturated liquid takes place from 4 to 5. The process 5 to 1 is the isentropic pump process between condenser pressure and boiler pressure. If the pressure differential between condenser and boiler is small, the pump work will also be small and may be neglected.

The heat added to the cycle, in the notation of Fig. 9-10 is

$$q_S = h_2 - h_1 \text{ Btu/lb (J/kg)} \tag{9-11}$$

The heat rejected is

$$q_R = (u_3 - u_4) + (h_4 - h_5) \text{ Btu/lb (J/kg)} \tag{9-12}$$

The net work of the cycle is

$$\frac{wk(\text{net})}{(J)} = q_S - q_R = (h_2 - h_1) - (u_3 - u_4) - (h_4 - h_5) \text{ Btu/lb (J/kg)}$$

By adding a positive and a negative $p_3 v_3/(J)$ term to the right side of the above equation, substituting $-u_4 - p_4 v_4/(J)$ for $-h_4$ and rearranging terms, there results

$$\frac{wk(\text{net})}{(J)} = h_2 - h_1 - u_3 - \frac{p_3 v_3}{(J)} + \frac{p_3 v_3}{(J)} + u_4 - u_4 - \frac{p_4 v_4}{(J)} + h_5$$

$$= h_2 - h_1 - h_3 + \frac{p_3 v_3}{(J)} - \frac{p_4 v_4}{(J)} + h_5$$

Finally, using $v_3 = v_4$,

$$\frac{wk(\text{net})}{(J)} = (h_2 - h_3) - (h_1 - h_5) + \frac{v_3(p_3 - p_4)}{(J)} \text{ Btu/lb (J/kg)} \tag{9-13}$$

The thermal efficiency of the cycle is

$$\eta_t = \frac{q_S - q_R}{q_S} = \frac{(h_2 - h_3) - (h_1 - h_5) + v_3(p_3 - p_4)/(J)}{(h_2 - h_1)} \tag{9-14}$$

The non-expansion steam cycle is similar to the incomplete-expansion cycle except for the isentropic expansion process, which does not occur in the non-expansion engine. Steam engines direct-connected to reciprocating pumps are required to develop a maximum force throughout their stroke. This is accomplished by allowing steam at boiler pressure to enter the actual engine cylinder during the entire working stroke, and then opening the exhaust valve immediately with no time for expansion of the steam.

POWER CYCLES (STEAM)

The *T-s* diagram of Fig. 9-11 shows the ideal non-expansion steam cycle with a non-flow constant volume cooling process, 2-3, replacing the exhaust process of the actual engine. Process 1-2 shows the heat added by

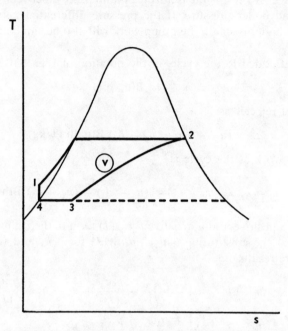

FIGURE 9-11. Non-expansion steam cycle.

the boiler; 2-3, the constant volume cooling in the ideal engine cylinder; 3-4, the constant pressure cooling in the condenser; and 4-1, the isentropic pump process from condenser pressure to boiler pressure. Since there is no opportunity for the steam to expand isentropically, there is no advantage in superheating the steam.

The heat added to the cycle is

$$q_S = h_2 - h_1 \text{ Btu/lb (J/kg)} \tag{9-15}$$

The heat rejected is

$$q_R = (u_2 - u_3) + (h_3 - h_4) \text{ Btu/lb (J/kg)} \tag{9-16}$$

The net cycle work is

$$q_S - q_R = (h_2 - h_1) - (u_2 - u_3) - (h_3 - h_4) \text{ Btu/lb (J/kg)}$$

Adding a negative and a positive $p_2 v_2/(J)$ term to the right side of the above equation, substituting $-u_3 - p_3 v_3/(J)$ for $-h_3$ and rearranging

POWER CYCLES (STEAM)

gives

$$\frac{wk(\text{net})}{(J)} = h_2 - h_1 - u_2 - \frac{p_2 v_2}{(J)} + \frac{p_2 v_2}{(J)} + u_3 - u_3 - \frac{p_3 v_3}{(J)} + h_4$$

$$\frac{wk(\text{net})}{(J)} = h_2 - h_1 - h_2 + \frac{p_2 v_2}{(J)} - \frac{p_3 v_3}{(J)} + h_4$$

Finally, using $v_2 = v_3$ we have

$$\frac{wk(\text{net})}{(J)} = (h_4 - h_1) + \frac{v_2(p_2 - p_3)}{(J)} \text{ Btu/lb (J/kg)} \qquad (9\text{-}17)$$

The thermal efficiency of the cycle is

$$\eta_t = \frac{q_S - q_R}{q_S} = \frac{(h_4 - h_1) + v_2(p_2 - p_3)/(J)}{h_2 - h_1} \qquad (9\text{-}18)$$

This cycle has a basically low thermal efficiency, but engines that operate in the cycle are simple, small in size for the horsepower developed and capable of producing maximum force throughout their length of working stroke.

The reciprocating steam engines associated with the incomplete and non-expansion cycles are discussed in some detail in Chapter 11.

Problems

9-1. In a Rankine cycle, steam enters the turbine at 750 psia and 800°F. The condenser pressure is 1 psia. Show the cycle on T-s and h-s coordinates and find (a) the moisture content of the steam entering the condenser; (b) the turbine work, Btu/lb; (c) the pump work, Btu/lb; (d) the heat supplied, Btu/lb; (e) the heat rejected, Btu/lb; (f) the net work of the cycle, Btu/lb; (g) the thermal efficiency of the cycle.

Ans. (a) 20.2%; (b) 503.9 Btu/lb; (c) 2.3 Btu/lb; (d) 1328.6 Btu/lb; (e) 827.0 Btu/lb; (f) 501.6 Btu/lb; (g) 37.8%

9-2. A Rankine cycle has a boiler pressure of 1200 psia and a condenser pressure of 1.0 inches Hg abs, and the temperature of the steam entering the turbine is 1000°F. Sketch the cycle on T-s and h-s coordinates and calculate (a) moisture content of the steam entering the condenser, percent; (b) the turbine work, Btu/lb; (c) the pump work, Btu/lb; (d) the cycle net work, Btu/lb; (e) the heat rejected to the condenser, Btu/lb; (f) the thermal efficiency of the cycle, percent.

Ans. (a) 21.0%; (b) 623.9 Btu/lb; (c) 4.1 Btu/lb; (d) 619.8 Btu/lb; (e) 828.7 Btu/lb; (f) 42.8%

9-3. A Rankine steam power cycle operates with a top pressure of 8,274 kPa abs and a top temperature of 593°C. Condenser pressure is 6.895 kPa absolute. Calculate for the cycle, in kJ/kg, (a) the heat added; (b) the heat rejected; (c) the net work; (d) the pump work. Then calculate (e) the cycle thermal efficiency, percent.

Ans. (a) 3453 kJ/kg; (b) 2006 kJ/kg; (c) 1447 kJ/kg; (d) 8.6 kJ/kg; (e) 41.9%

POWER CYCLES (STEAM)

9-4. Heat is added to the working substance in an ideal Rankine cycle between the states of a compressed liquid at 1000 psia and 60.54°F and superheated steam at 1000 psia and 1000°F. For the cycle determine (a) the heat added, Btu/lb; (b) the heat rejected from the cycle for a condenser temperature of 60°F, Btu/lb; (c) the available energy for the cycle, Btu/lb.

Ans. (a) 1474.5 Btu/lb; (b) 830.2 Btu/lb; (c) 644.3 Btu/lb

9-5. Sketch on T-s and h-s coordinates a Rankine steam power cycle and label the process lines and intersection points correspondingly. Then, using an exhaust pressure of 2.0 psia, a heat addition pressure of 1200 psia and a maximum cycle entropy of 1.7040 Btu/lb·°R, determine the enthalpies at the four process intersection points, Btu/lb.

Ans. Using process intersection numbers as in Fig. 9-3: $h_1 = 98.1$ Btu/lb, $h_2 = 1615.5$ Btu/lb, $h_3 = 989.7$ Btu/lb, $h_4 = 94.0$ Btu/lb

9-6. A Rankine steam power cycle has a heat addition process pressure of 1000 psia and maximum steam temperature of 800°F. Condenser pressure is 1.0 psia. Sketch and label the cycle on T-s coordinates and calculate for the cycle (a) pump work, Btu/lb; (b) heat added, Btu/lb; (c) heat rejected, Btu/lb; (d) net cycle work; (e) cycle thermal efficiency, percent; (f) the power output of such a cycle when circulating 20 lb/sec of steam, hp.

Ans. (a) 3.1 Btu/lb; (b) 1315.7 Btu/lb; (c) 805.1 Btu/lb; (d) 510.6 Btu/lb; (e) 38.8%; (f) 14,444 hp

9-7. In an ideal reheat cycle, steam enters the high pressure turbine at 750 psia and 800°F and leaves at 100 psia. It is then reheated to 800°F, passes through the low pressure turbine and exhausts to a condenser at 1 psia. Show the cycle on T-s and h-s coordinates and find (a) the moisture content of the steam entering the condenser; (b) the pump work, Btu/lb; (c) the work of the high pressure turbine, Btu/lb; (d) the total turbine work, Btu/lb; (e) the heat supplied, Btu/lb; (f) the heat rejected, Btu/lb; (g) the net work of the cycle, Btu/lb; (h) the thermal efficiency of the cycle.

Ans. (a) 7.2%; (b) 2.3 Btu/lb; (c) 211.1 Btu/lb; (d) 609.6 Btu/lb; (e) 1568.7 Btu/lb; (f) 961.4 Btu/lb; (g) 607.3 Btu/lb; (h) 38.7%

9-8. In an ideal reheat cycle the pressure and temperature entering the high pressure turbine are 1200 psia and 1000°F, the reheat pressure is 72 psia, the reheat temperature is 800°F and the condenser pressure is 1.0 inches of mercury absolute. Sketch the cycle on h-s and T-s coordinates and determine (a) the high pressure turbine work, Btu/lb; (b) the moisture content of the steam entering the condenser, percent; (c) the low pressure turbine work, Btu/lb; (d) the pump work, Btu/lb; (e) the net work of the cycle, Btu/lb; (f) the heat added to the cycle, Btu/lb; (g) the heat rejected to the condenser, Btu/lb; (h) the thermal efficiency of the cycle, percent.

Ans. (a) 318.2 Btu/lb; (b) 8.0%; (c) 418.6 Btu/lb; (d) 4.1 Btu/lb; (e) 732.7 Btu/lb; (f) 1697.7 Btu/lb; (g) 965.0 Btu/lb; (h) 43.2%

9-9. In an ideal regenerative cycle, steam is generated at 1200 psia and 1000°F. The condenser pressure is 1 inch of mercury absolute. Consider only the pump work involved in pumping the total cycle flow from the single extraction heater to the boiler. Sketch the cycle on T-s coordinates and calculate (a) the optimum extraction pressure to the nearest pound, psia; (b) the mass of steam extracted, lb/lb of throttle steam; (c) the heat added to the cycle, Btu/lb; (d) the heat rejected by the cycle, Btu/lb of throttle steam; (e) the net cycle work, Btu/lb of throttle steam; (f) the cycle thermal efficiency, percent.

POWER CYCLES (STEAM)

Ans. (a) 94 psia; (b) 0.2135 lb/lb throttle steam; (c) 1201.7 Btu/lb; (d) 651.8 Btu/lb throttle steam; (e) 549.9 Btu/lb throttle steam; (f) 45.8%

9-10. Steam is generated in an ideal reheat-regenerative cycle at 1500 psia and delivered at 1050°F. After expansion in the turbine to the optimum pressure for a single extraction, the steam is removed from the turbine, one portion to be used for feedwater heating and the remainder to be reheated to 960°F at the extraction pressure. Condenser pressure is 1 inch of mercury absolute. Sketch the cycle on T-s coordinates and calculate (a) the extraction pressure, to the nearest pound, psia; (b) the mass of steam extracted, lb/lb of throttle steam; (c) the heat added to the cycle, Btu/lb of throttle steam; (d) the heat rejected by the cycle, Btu/lb of throttle steam; (e) the net cycle work, Btu/lb throttle steam; (f) the cycle thermal efficiency, percent.

Ans. (a) 114 psia; (b) 0.2244 lb/lb throttle steam; (c) 1438.3 Btu/lb throttle steam; (d) 752.0 Btu/lb throttle steam; (e) 686.3 Btu/lb throttle steam; (f) 47.7%

9-11. A Rankine steam cycle operates with steam leaving the superheater at 600 psia and 800°F. Condenser pressure for the cycle is 1 inch of mercury absolute. (a) Sketch the cycle on h-s and T-s coordinates; (b) list the enthalpies at each of the four significant state points of the cycle in Btu per pound. Then determine (c) the pump work, Btu/lb; (d) the temperature in the condenser, °F; (e) turbine work, Btu/lb; (f) heat supplied to the cycle, Btu/lb; (g) thermal efficiency of the cycle, percent; (h) the degrees of superheat of steam entering the turbine, °F.

Ans. (c) 2.3 Btu/lb; (d) 79°F; (e) 529.3 Btu/lb; (f) 1358.2 Btu/lb; (g) 38.8%; (h) 313.67°F

9-12. A Rankine reheat cycle operates with a top pressure of 600 psia, an entropy at the first superheater outlet of 1.6517 Btu/lb · °R, a reheat pressure of 100 psia and an entropy at the condenser entrance of 1.8203 Btu/lb · °R.

The condenser pressure is 1 inch of mercury absolute. There is no pressure drop during reheating. Sketch the cycle on h-s and T-s coordinates and calculate (a) pump work, Btu/lb; (b) high pressure turbine work, Btu/lb; (c) low pressure turbine work, Btu/lb; (d) heat added to the steam, Btu/lb; (e) heat rejected by the steam, Btu/lb; (f) the cycle thermal efficiency, percent.

Ans. (a) 2.3 Btu/lb; (b) 202.3 Btu/lb; (c) 420.8 Btu/lb; (d) 1552.2 Btu/lb; (e) 931.4 Btu/lb; (f) 40.0%

9-13. A modified Rankine reheat cycle (irreversible turbine expansion processes) operates with a top pressure of 600 psia, an entropy at the first superheater outlet of 1.6067 Btu/lb · °R, an enthalpy leaving the high pressure turbine of 1210 Btu/lb, a reheat pressure of 100 psia (no pressure drop in reheater), an entropy at the entrance of the low pressure turbine of 1.8203 Btu/lb · °R, a condenser pressure of 1 inch of mercury absolute and an enthalpy at entrance to the condenser of 1050 Btu/lb. Sketch h-s and T-s diagrams and calculate (a) pump work, Btu/lb; (b) total turbine work, Btu/lb; (c) heat added to the steam, Btu/lb; (d) heat rejected by the steam, Btu/lb; (e) cycle thermal efficiency, percent.

Ans. (a) 2.3 Btu/lb; (b) 513.0 Btu/lb; (c) 1513.6 Btu/lb; (d) 1002.9 Btu/lb; (e) 33.7%

9-14. A regenerative steam power cycle with boiler pressure of 922.1 psia (saturation temperature 535°F) and discharge temperature of 1000°F employs two stages of extraction for feedwater heating. Condenser pressure is 0.9503 psia (saturation temperature 100°F). Steam removed at the first and second extraction points is designated M_1 and M_2, pounds per pound of throttle steam for these pressures, respectively. (a) Show the cycle on T-s coordinates, indicating steam flow extractions for each process in terms of M_1 and M_2; (b)

POWER CYCLES (STEAM)

determine the optimum extraction pressures to the nearest psia; (c) determine the mass of steam removed at the first extraction point, lb/lb of throttle steam.

Ans. (b) 220 psia, 27 psia; (c) 0.1369 lb/lb throttle steam.

9-15. Sketch a Rankine steam power cycle on T-s and h-s coordinates; label the process lines and number the process intersection points correspondingly on both diagrams. Then sketch and number an equipment diagram to correspond with the coordinate diagrams. Then calculate the thermal efficiency, percent, for the cycle if it operates with high and low pressures of 1200 psia and 1.0 psia with a superheat temperature of 900°F.

Ans. 40.2%

CHAPTER 10

REFRIGERATION CYCLES

10-1. Background. Refrigeration may be defined as the maintenance of the temperature of a body below that of its surrounding atmosphere. The "body" may be, for example, the air being cooled for an air conditioning system, the contents of a cold storage room or an ice-making tank. In its practical application, *refrigeration includes the production of the lower temperature as well as its maintenance.* A relatively low temperature may be produced by placing a can of warm water on a block of ice. Heat, of course, flows from the water to the ice and the temperature of the water is lowered. We may maintain the temperature of the water at a lower value than that of the surrounding atmosphere as long as our supply of ice holds out. However, even if we had an unlimited supply of ice available, the lowest temperature which the water could be made to reach would be the temperature of the ice. When the water reached that temperature, heat would no longer flow from it to the ice because a temperature gradient is essential to heat flow. Therefore we could not freeze the water in this manner because to do so would require further extraction of heat from the water, i.e., its latent heat of fusion. It is apparent that this system of refrigeration would be useless when the object was either the manufacture of ice or the production of temperatures lower than that of the ice already available. Both of these objectives can be realized by the use of a suitable refrigeration cycle.

10-2. Ideal Refrigeration—The Reversed Carnot Cycle. The Second Law of Thermodynamics asserts, in effect, that heat will not, of itself, flow from one region to another which is at a higher temperature. *The object of a refrigeration cycle is to cause heat to be extracted from a body at a low temperature and discharged to a body at a higher temperature* through the expenditure of energy either as mechanical work or, in some cases, through the application of heat. This can be accomplished by various cycles in which the working substance is alternately at low and high temperatures. When the temperature of the working substance is below that of the body being cooled, heat flows from the body to the working substance; and when the temperature of the working substance is greater than that of the surrounding atmosphere, heat is given up to the atmosphere. In many refrigeration systems this change of temperature of the working substance is effected by the expenditure of mechanical energy,

REFRIGERATION CYCLES

i.e., by doing work upon the working substance. Thus the energy exchanges in any refrigeration cycle occur exactly in reverse order to the corresponding exchanges of a power cycle; and for this reason a refrigerating system is sometimes referred to as a "heat pump." Reversal of the sequence of processes of the Carnot cycle provides an excellent illustration of the fundamental requirements for any mechanical refrigeration system.

Figure 10-1 represents the conventional Carnot cycle on the T-s plane,

FIGURE 10-1. The Carnot (power) cycle.

heat being received by the working substance at temperature T_A along the process line 2-3 and rejected by the working substance at a lower temperature, T_B, along the process line 4-1. As established in Chapter 7, the useful work of such a cycle would be,

$$\frac{wk(\text{net})}{(J)} = q_{23} - q_{14} = q_S - q_R \text{ Btu/lb (J/kg)}$$

Figure 10-2 represents the same cycle with the thermodynamic processes taking place in reverse sequence. In the reversed cycle heat is received by the working substance at the original temperature, T_B, under the process line 1-2, "pumped" to the higher temperature, T_A, by the isentropic process 2-3 and rejected by the working substance at the original high temperature, T_A, under the process line 3-4. The process 4-1

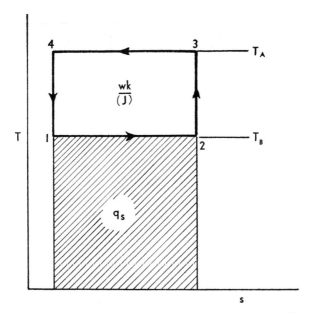

FIGURE 10-2. Reversed Carnot (refrigeration) cycle.

represents the isentropic expansion necessary to complete the cycle. It should be noted that in both cycles the terms q_S and q_R indicate the receipt and rejection of heat *by the working substance* despite the fact that the areas are interchanged by the reversal of the cycle. Useful work is done by the working substance during the isentropic expansion, 4-1, and the isothermal expansion, 1-2, but the work thus produced is less than that required for the corresponding compression processes. Hence work from an external source is required for the operation of the reversed cycle. Figure 10-2a, b, c shows the relative magnitudes of the energy quantities, Fig. 10-2b being the net external work required by the cycle. Thus for the reversed cycle,

$$\frac{wk(\text{net})}{(J)} = q_{43} - q_{12} = q_R - q_S \text{ Btu/lb (J/kg)} \quad (10\text{-}1)$$

10-3. Refrigerating Effect, Capacity, Coefficient of Performance. The useful effect of refrigeration is the removal of heat from a cold body. The amount of heat thus removed is absorbed by *the working substance*, i.e., by *the refrigerant.* Thus in Figs. 10-2 and 10-2a, q_S (area 1-2-6-5) is the useful cooling provided by the cycle. This useful cooling is defined by The American Society of Heating, Refrigerating and Air Conditioning Engi-

REFRIGERATION CYCLES

FIGURE 10-2a, b, c

neers (ASHRAE) on the basis of unit mass of the refrigerant, as **the refrigerating effect**. Thus *"refrigerating effect is the heat gain per pound of refrigerant"* circulated in the cycle *and is designated by the abbreviation R.E.*

The function of power cycles and machinery is the production of useful work, and the capacity of such systems is measured in horsepower, or work output per unit time. Since the function of any refrigerating system is the absorption of heat from a low temperature source, it is logical that the capacity of such a system be measured by the rate at which it will remove heat from a cold body. As defined by ASHRAE, **the ton of refrigeration** is *"a unit of capacity representing the removal of heat at the rate of **200 Btu per minute or 12,000 Btu per hour.**"* This is approximately equivalent to the production of one ton of ice per day, from and at a temperature of 32 degrees Fahrenheit. Verification of this approximation using the heat of fusion of ice (sat. solid) found in Table 6 of the Steam Tables is left to the student as an exercise.

The performance of any cycle or system is judged on the basis of the effectiveness with which it accomplishes its purpose. The performance of a power cycle is conventionally stated in terms of the efficiency of the cycle; that is, the ratio of the useful work produced to the energy required for the operation of the cycle:

$$\text{Thermal Efficiency} = \frac{\text{work done}}{\text{heat supplied}} = \frac{Wk(\text{net})}{(J) \times Q_S}$$

Although expressed as a percentage figure, thermal efficiency is essentially a coefficient which may be applied to the energy supplied to the cycle, the

REFRIGERATION CYCLES

product being the useful output that may be expected from the cycle. The performance of a refrigerating cycle is stated on the basis of a similar coefficient, i.e., on the basis of the ratio of the useful effect produced to the energy required for its accomplishment. *This ratio is defined as the coefficient of performance and designated as* **C.P.** Thus for a refrigerating cycle or system,

$$C.P. = \frac{refrigerating\ effect}{work\ required} = \frac{(J)(R.E.)}{wk(net)} \qquad (10\text{-}2)$$

For the reversed Carnot cycle, the coefficient of performance is a function of temperatures only. This may be established by referring to Fig. 10-2, in which the refrigerating effect is $T_B(s_2 - s_1)$ and the net work required is $(T_A - T_B)(s_2 - s_1)$; then

$$C.P._{\text{Carnot}} = \frac{T_B(s_2 - s_1)}{(T_A - T_B)(s_2 - s_1)} = \frac{T_B}{T_A - T_B} \qquad (10\text{-}3)$$

Other ideal cycles which will be developed in succeeding articles, together with actual operating cycles, employ other means of determining both the refrigerating effect and the work required.

Example 10-1: A refrigerating plant for an air conditioning system removes 10,000 Btu/min from the air. The plant circulates 170 lb of refrigerant/min and the internal power delivered by its compressor is 60 horsepower. The refrigerant evaporation temperature is 40°F and its condensation temperature is 100°F. Calculate (a) the capacity of the plant, tons; (b) the refrigerating effect, Btu/lb; (c) the coefficient of performance of the actual plant; (d) the coefficient of performance of the equivalent Carnot cycle.

Solution:

heat absorbed by refrigerant = 10,000 Btu/min

(a) \quad Capacity $= \dfrac{\text{heat absorbed, Btu/min}}{200\ \text{Btu/min per ton}} = \dfrac{10{,}000}{200} = \underline{50\ \text{tons}}$

(b) $\quad R.E. = \dfrac{\text{heat absorbed, Btu/min}}{\text{mass refrigerant, lb/min}} = \dfrac{10{,}000}{170} = \underline{58.8\ \text{Btu/lb}}$

(c) $\quad \dfrac{wk(net)}{J} = \dfrac{hp \times 2545}{\text{mass refrigerant} \times 60} = \dfrac{60 \times 2545}{170 \times 60} = \underline{14.97\ \text{Btu/lb}}$

$\quad C.P. = \dfrac{J \times R.E.}{wk(net)} = \dfrac{58.8}{14.97} = \underline{3.93}$

(d) $\quad T_A = 100 + 460 = 560°R \qquad T_B = 40 + 460 = 500°R$

$\quad C.P._{\text{Carnot}} = \dfrac{T_B}{T_A - T_B} = \dfrac{500}{560 - 500} = \underline{8.33}$

10-4. Vapor Compression Refrigeration.

The reversed Carnot cycle establishes the conditions necessary for the accomplishment of refrigera-

tion most effectively, that is, with maximum coefficient of performance. The reversed Rankine cycle is the basis from which the practical vapor compression refrigeration cycle is developed. Such a cycle is illustrated in Fig. 10-3, wherein the absorption of heat by the refrigerant is accom-

FIGURE 10-3. Reversed Rankine cycle.

plished *through the evaporation of a suitable vapor* (process 1-2), the compression of the refrigerant remaining isentropic as in the Carnot cycle. However, it is impossible to accomplish the rejection of heat required for the reversed Rankine cycle in entirety because the temperature at point 4 is less than that of the atmosphere surrounding the system.

Ideally the heat rejection process may be stopped when the refrigerant becomes a saturated liquid, and the change from high to low pressures (and corresponding temperatures) may be accomplished by an isentropic expansion, as shown in Fig. 10-4. However, the expansion of a liquid in a cylinder produces only a very small amount of work, and such an expansion is difficult to accomplish. Consequently, the isentropic process, 4-a, is discarded in favor of the throttling process, 4-1. Since for any liquid the enthalpy decreases with temperature, this process will terminate in the wet region, and *the temperature after throttling will be that of saturation at the terminal pressure.* The low temperature is thus fixed in terms of a controllable pressure; and a simple pressure-reducing valve replaces the piston and cylinder required by process 4-a.

REFRIGERATION CYCLES

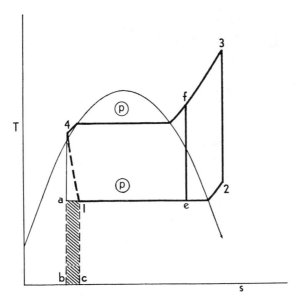

FIGURE 10-4. Modified Rankine refrigeration cycle.

If the conventional power cycle is reversed, compression will follow line e-f of Fig. 10-4. However, *wet compression* (in which the process begins in the wet region, as at "e") is very difficult to control, and the possibility of large amounts of liquid entering the compressor cannot be tolerated. *Dry compression* is therefore in general use. In this case the low pressure refrigerant becomes slightly superheated before entering the compressor, and compression is represented by the process line 2-3.

With the basic working processes established, the performance of the system can be analyzed in terms of the enthalpy changes of the refrigerant. Based on the cyclic concept (Chapter 7) and the steady flow energy equation,

$$R.E. = q_S = h_2 - h_1 \text{ Btu/lb (J/kg)} \qquad (10\text{-}4)$$

$$q_R = h_3 - h_4 \text{ Btu/lb (J/kg)} \qquad (10\text{-}5)$$

$$\frac{wk(\text{net})}{(J)} = q_R - q_S = q_R - R.E.$$

$$= (h_3 - h_4) - (h_2 - h_1)$$

But,
$$h_1 = h_4$$

Then
$$\frac{wk(\text{net})}{(J)} = (h_3 - h_4) - (h_2 - h_4)$$

REFRIGERATION CYCLES

and
$$\frac{wk(\text{net})}{(J)} = h_3 - h_2 \text{ Btu/lb (J/kg)} \tag{10-6}$$

and from these energy relationships,

$$C.P. = \frac{(J)(R.E.)}{wk} = \frac{h_2 - h_1}{h_3 - h_2} \tag{10-7}$$

The equipment required for a basic vapor compression system is shown in Fig. 10-5, and Fig. 10-6 illustrates a convenient working diagram for vapor refrigerants. Centrifugal as well as reciprocating compressors are employed. The vapor compression system operates between two characteristic pressures. A high pressure (with attendant high condensation temperature) is maintained from the compressor discharge through the condenser to the expansion valve inlet. A relatively low pressure (to produce a low evaporation temperature) is imposed from the expansion valve exit through the evaporator coil to the compressor suction. Pressure is therefore selected as the ordinate of the new diagram, with enthalpy as

FIGURE 10-5. Vapor compression refrigeration system.

REFRIGERATION CYCLES

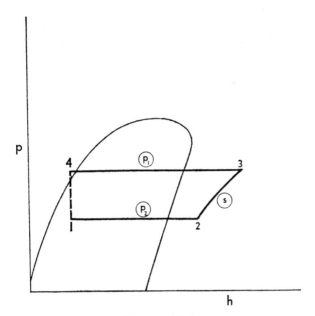

FIGURE 10-6

the abscissa, since this latter property is used directly in any analysis of the cycle. A standard *p-h* diagram for Freon-12, a generally used refrigerant, together with tables containing the thermodynamic properties of this fluid, will be found in Appendix B.

It should be noted that although the refrigerant leaving the condenser is a subcooled liquid, its properties are taken as equal to those of a saturated liquid *at the condenser discharge temperature.*

Example 10-2: A refrigerating plant circulates 23 lb Freon-12 per minute and is assumed to operate on a cycle similar to that of Fig. 10-6. The pressure in the evaporator coil is 50 psia, the temperature of the Freon-12 entering the compressor is 50°F, the pressure in the condenser is 120 psia and the temperature of the liquid refrigerant entering the expansion valve is 86°F. Calculate or determine (a) the evaporation temperature, °F; (b) the condensation temperature, °F; (c) the refrigerating effect, Btu/lb; (d) the capacity of the plant, tons; (e) the power required to compress the Freon-12, hp; (f) the coefficient of performance.

Solution: Refer to Fig. 10-6 and Appendix B.

$$h_4 = h_1 = 27.72 \text{ Btu/lb} \qquad h_2 = 84.24 \text{ Btu/lb}$$
$$s_2 = s_3 = 0.17187 \qquad h_3 = 91.31 \text{ Btu/lb}$$

(a) evaporation temperature = 38.3°F (Appendix B)

(b) condensation temperature = 93.4°F (Appendix B)

(c) R.E. = $h_2 - h_1$ = 84.24 − 27.72 = 56.52 Btu/lb

(d) $\quad\text{capacity} = \dfrac{(\dot{M}\text{ Freon-12})(R.E.)}{200} = \dfrac{(23)(56.52)}{200} = \underline{6.5\text{ tons}}$

(e) $\quad P = \dot{W}k = \dfrac{(\dot{M}\text{ Freon-12})(h_3 - h_2)}{42.42} = \dfrac{(23)(91.31 - 84.24)}{42.42} = \underline{3.83\text{ hp}}$

(f) $\quad C.P. = \dfrac{R.E.}{wk/J} = \dfrac{h_2 - h_1}{h_3 - h_2} = \dfrac{84.24 - 27.72}{91.31 - 84.24} = \underline{7.99}$

10-5. The Refrigerant Rating Cycle. Theoretically, any vapor may be used in the modified Rankine cycle; but practical considerations establish very specific requirements for a suitable refrigerant. These requirements, together with the characteristics of various substances which fulfill them, will be discussed in a later article. In addition, a comparison of the relative performance of the various refrigerants is desirable.

A refrigerant rating cycle may be established to provide a basis for comparison of refrigerants from a performance standpoint. Such a cycle, illustrated in Fig. 10-7, is similar to the modified Rankine cycle of Fig. 10-6, the operating conditions being established on the basis of selected temperatures. This refrigerant rating cycle specifies an evaporation tem-

FIGURE 10-7. Refrigerant rating cycle with Freon-12.

perature of 5°F and a condensing temperature of 86°F, with 9°F superheat at the compressor suction and an equal amount of subcooling leaving the condenser. *It is emphasized that these four temperatures establish a cycle for any refrigerant,* the high and low pressures being the saturation values corresponding to the respective condensation and evaporation temperatures; and a careful study of Fig. 10-7 by the student will verify these observations.

REFRIGERATION CYCLES

Example 10-3: A cooling plant using Freon-12 as the refrigerant is to have a capacity of 50 tons when operating on the refrigerant rating cycle. For this ideal plant cycle determine (a) the refrigerating effect, Btu/lb; (b) the rate of Freon-12 circulation, lb/min; (c) the net work required per pound of Freon-12 circulated, Btu/lb; (d) the coefficient of performance; (e) the power required per ton of refrigeration, hp/ton; (f) the heat rejected by the condenser, Btu/min; (g) the compressor piston displacement, ft³/min·ton of refrigeration.

Solution: (from Freon-12 tables in Appendix B)

$$p_1 = p_2 = 26.51 \text{ psia}, \quad p_3 = p_4 = 107.9 \text{ psia}$$
$$t_2 = 14°F \quad\quad t_4 = 77°F$$
$$s_2 = s_3 = 0.17317 \text{ Btu/lb·°R}$$
$$h_1 = h_4 = 25.56 \text{ Btu/lb}, \; h_2 = 80.04 \text{ and } h_3 = 91.13 \text{ Btu/lb}$$

(a) $\quad R.E. = h_2 - h_1 = 80.04 - 25.56 = \underline{54.48 \text{ Btu/lb}}$

(b) $\quad \dot{M} \text{ Freon-12} = \dfrac{\text{heat absorbed}}{R.E.} = \dfrac{200 \times \text{capacity}}{R.E.}$

$$= \dfrac{50 \times 200}{54.48} = \underline{183.6 \text{ lb/min}}$$

(c) $\quad \dfrac{wk(\text{net})}{J} = h_3 - h_2 = 91.13 - 80.04 = \underline{11.09 \text{ Btu/lb}}$

(d) $\quad C.P. = \dfrac{h_2 - h_1}{h_3 - h_2} = \dfrac{54.48}{11.09} = \underline{4.91}$

(e) $\quad P = \dot{W}k_{\text{ton}} = \dfrac{\dot{M} \text{ Freon-12} \times \dfrac{wk(\text{net})}{J} \times 60}{2545 \times \text{capacity}}$

$$= \dfrac{183.6 \times 11.09 \times 60}{2545 \times 50} = \underline{0.96 \text{ hp/ton}}$$

(f) $\quad \dot{Q}_R = (\dot{M} \text{ Freon-12})(h_3 - h_4)$
$$= 183.6(91.13 - 25.56) = \underline{12{,}040 \text{ Btu/min}}$$

(g) $\quad \text{piston displacement} = \dfrac{\dot{M} \text{ Freon-12} \times v_2}{\text{capacity}}$

$$v_2 = 1.516 \text{ ft}^3/\text{lb}$$
$$= \dfrac{183.6 \times 1.516}{50} = \underline{5.57 \text{ ft}^3/\text{min·ton}}$$

10-6. Practical Aspects of Vapor Compression Systems.

From an operating standpoint the vapor compression system of refrigeration is ideal because of its inherent simplicity. In addition the system is readily adaptable to variations in operating conditions, and the machinery required together with associated control devices may be of rugged, durable construction which ensures a high degree of reliability under all conditions of service.

The processes of this refrigeration system are equally divided between two essential pressures. The high pressure side of the system begins at the

REFRIGERATION CYCLES

compressor discharge and continues through the condenser to the expansion valve. The low pressure side of the system begins at the discharge side of the expansion valve and continues through the evaporator to the compressor suction. These high and low side pressures may be varied to suit the temperatures desired and encountered in a given circumstance. For example, the 5°F evaporating temperature of the rating cycle is unsuitable for air conditioning applications because the water vapor entering such systems in the air will freeze on the coil surfaces at this temperature. However, if the low side pressure is increased to 60 psia, for Freon-12 the evaporator temperature is 48.7°F and the freezing difficulty is eliminated. Again, if the system is to operate in a tropical region, the condensing temperature must be raised to keep it above the temperature of the available cooling medium. Increasing the high side pressure to 130 psia raises the condensing temperature for Freon-12 to 99.1°F.

To study the effect of such changes upon the performance of the system, each may be separately compared with the rating cycle. Such a comparison, illustrated in Fig. 10-8a, b, shows that raising the low side pressure to 60 psi absolute increases the *C.P.* to 12.78 and decreases the power

FIGURE 10-8. Effect of operating conditions upon vapor refrigeration cycle.

required to 0.37 horsepower per ton. Corresponding values for the rating cycle are 4.91 and 0.96 respectively. As shown in Fig. 10-8b, increasing the high side pressure to 130 psi absolute decreases the *C.P.* to 4.05 and increases the power requirement to 1.16 horsepower per ton. The combined effect of raising both high and low side pressures, illustrated in Fig. 10-8c, results in a *C.P.* of 9.09 with a power requirement of 0.52 horsepower per ton.

The conclusions that may be drawn from the above comparisons are of practical value and should be thoroughly understood by the student. Of major importance is the fact that the system may be adapted to a variety of applications by raising or lowering the low side pressure, thus varying the temperature at which refrigeration is accomplished. Secondly, the minimum high side pressure is fixed by the temperature of the available coolant. Finally, for best performance the difference between low and high side pressures must be the minimum permitted by the required temperature of refrigeration and the encountered temperature of the cooling fluid.

Industrial and commercial refrigeration plants frequently require several low temperatures for different processes. For example, frozen foods are quick-frozen at approximately $-50°F$ and are stored after freezing at approximately $0°F$. A single compressor may accommodate both freezing and storage loads by using the dual expansion system illustrated in Fig. 10-9. So long as the total load remains substantially constant, this system is practical and economical. Where there are large variations in the amount of refrigeration required, two or more compressors operating in parallel should be used. Figure 10-9 shows one possible method of accomplishing dual expansion with two compressors in parallel to accommodate large variations in total load.

10-7. Vapor Refrigerants. From the thermodynamic standpoint a refrigerant should have a high latent heat of vaporization at low temperatures and be readily condensable with atmospheric fluids (such as air and water) as coolants. In addition, the freezing point must be well under the low temperature working range, the critical point should be well above the condensing temperatures and there should be minimum pressure difference between evaporating and condensing temperatures. From practical and safety considerations a refrigerant should have a low specific volume at the compressor suction, be chemically stable under all service and handling conditions, be nontoxic and nonirritating, and should not form combustible or explosive mixtures either with lubricants or with air or have corrosive properties for ordinary engineering materials.

There are a variety of refrigerants in use today, but none completely fills all of the above requirements. Ammonia, an extensively used com-

REFRIGERATION CYCLES

FIGURE 10-9. Vapor compression system for dual expansion and variable load.

mercial refrigerant because of its high refrigerating effect, is flammable and will form low-order explosive mixtures under certain conditions. Furthermore very small amounts of ammonia in air are extremely irritating, and a mixture containing as little as 0.6 percent by volume is dangerously toxic. Following ammonia (because of smaller refrigerating effect, hence greater mass requirements per ton of refrigeration) are chlorinated and fluorinated hydrocarbon refrigerants such as the Freon group. Of these, Freon-12 is of particular importance to this discussion because of its extensive use and because it is classified as nontoxic, noncombustible nonflammable, and nonirritating. In high concentrations it may act as an asphyxiant.

In comparing the performance of vapor refrigerants the standard cycle used in rating refrigerating plants may be employed, and Table 10-1 is based upon this cycle.

This comparison shows that there is no practical difference between the two refrigerants from a thermodynamic standpoint, since there is no

REFRIGERATION CYCLES

TABLE 10-1

Refrigerant	R.E. Btu/lb	C.P.	Mass lb/min-ton	Power hp/ton	v_2 ft³/lb	V_2 ft³/min-ton
Ammonia	490	4.76	0.408	0.991	8.36	3.41
Freon-12	54.5	4.91	3.67	0.96	1.516	5.57

practical difference in the power requirement per ton of refrigeration produced. However the required piston displacement for ammonia (based on the rating cycle and neglecting volumetric efficiency) is 3.41 cubic feet per minute per ton. On the other hand Freon-12 requires a piston displacement of 5.57 cubic feet per minute for the same capacity. For many uses the larger Freon-12 compressor is readily acceptable in view of the fire hazard and toxic properties of ammonia.

10-8. Air Compression System—Reversed Joule and Brayton Cycles. In any study of refrigeration the use of air as a refrigerant is suggested from an observation of the low temperatures obtained as air expands and does work in a cylinder. Refrigerating systems using reversed Air Standard Joule and Brayton cycles have been built and operated. The equipment essential to one such system is illustrated in Fig. 10-10.

The sequence of processes accomplished by the air refrigeration system shown, beginning at the inlet to the compression cylinder, are as follows. Low-temperature air leaving the brine tank at a relatively low pressure (2) is compressed to a relatively high pressure and accompanying high temperature (3) in the compression cylinder. The air is next cooled by passing it through a suitable heat exchanger (air cooler) and enters the expansion cylinder (4) under high pressure but only moderate temperature. In this cylinder the air expands to the original low pressure and to the lowest temperature encountered in the cycle. Work is accomplished during this process at the expense of the internal energy of the expanding air, and this work assists in driving the compression cylinder. The low temperature, low pressure air (1) then flows through a second heat exchanger (brine tank) where heat is absorbed and the temperature rises to the temperature of admission to the compression cylinder. These processes are illustrated in Fig. 10-11 on p-v and on T-s coordinates, and the cycle may be analyzed as follows:

$$R.E. = q_S = \text{area } 1\text{-}2\text{-}b\text{-}a = c_p(T_2 - T_1) \text{ Btu/lb (J/kg)} \quad (10\text{-}8)$$
$$q_R = \text{area } 3\text{-}4\text{-}a\text{-}b = c_p(T_3 - T_4) \text{ Btu/lb (J/kg)} \quad (10\text{-}9)$$

Net work required by the cycle is the difference between the heat rejected

REFRIGERATION CYCLES

FIGURE 10-10. Air compression refrigeration system.

FIGURE 10-11. Reversed Joule (Brayton) cycle.

REFRIGERATION CYCLES

by the cycle and the refrigerating effect (heat absorbed by the air during the cycle); thus,

$$\frac{wk(\text{net})}{(J)} = c_p(T_3 - T_4) - c_p(T_2 - T_1) \text{ Btu/lb (J/kg)} \quad (10\text{-}10)$$

and,

$$C.P. = \frac{c_p(T_2 - T_1)}{c_p(T_3 - T_4) - c_p(T_2 - T_1)} \quad (10\text{-}11)$$

Performance of the air compression system of refrigeration may be compared with that of the vapor compression systems on the basis of the refrigerant rating conditions, provided the pressure and the allowable increase in air temperature through the low temperature heat exchanger are established. On the basis of such a comparison (see Example 10-4 and Art. 10-7) the reversed Joule cycle requires a piston displacement approximately four times greater (for both compression and expansion cylinders) than a Freon-12 system of the same capacity, and the coefficient of performance is materially smaller.

Example 10-4: An air compression refrigeration system is to have an air pressure of 100 psia in the brine tank and an allowable air temperature increase of 60°F. For standard vapor compression cycle temperatures of 77°F entering the expansion cylinder and 14°F entering the compression cylinder, calculate: (a) the coefficient of performance; (b) the mass of air circulated per ton of refrigeration; (c) the required piston displacement of the compressor cylinder, neglecting volumetric efficiency.

Solution:

$$T_4 = 77 + 460 = 537°R \quad T_2 = 14 + 460 = 474°R$$
$$T_2 - T_1 = 60°R \quad \text{and} \quad T_1 = 474 - 60 = 414°R$$

Since processes 3-4 and 1-2 are at constant pressure,

$$\frac{p_3}{p_2} = \frac{p_4}{p_1} \quad \text{and} \quad \left(\frac{p_4}{p_1}\right)^{\frac{k-1}{k}} = \frac{T_4}{T_1} = \frac{T_3}{T_2}$$

$$T_3 = T_2\left(\frac{T_4}{T_1}\right) = 474\left(\frac{537}{414}\right) = 615°R$$

(a) $$C.P. = \frac{c_p(T_2 - T_1)}{c_p(T_3 - T_4) - c_p(T_2 - T_1)} = \frac{60}{(615 - 537) - 60} = \underline{3.33}$$

(b) $$R.E. = c_p(T_2 - T_1) = 0.24(60) = 14.4 \text{ Btu/lb}$$

$$\dot{M} \text{ air} = \frac{200 \text{ Btu/min} \cdot \text{ton}}{14.4 \text{ Btu/lb}} = \underline{13.9 \text{ lb/min} \cdot \text{ton}}$$

(c) $$\text{piston displacement} = \dot{V}_2 = \frac{\dot{M}RT_2}{p_2}$$

$$= \frac{13.9 \times 53.3 \times 474}{144 \times 100}$$

$$= \underline{24.4 \text{ ft}^3/\text{min} \cdot \text{ton}}$$

REFRIGERATION CYCLES

Refrigerating systems using the reversed Joule cycle with air as the working substance are seldom used but of continuing thermodynamic interest. However, the reversed open Brayton cycle, similar in thermodynamic principle to the reversed Joule cycle, is widely used for cooling purposes in aircraft. A portion of the total air flow may be bled from any of several convenient stages of a multi-stage, axial-flow compressor, expanded and cooled to a predetermined temperature in an air turbine or simply cooled directly in a heat exchanger in an aircraft at altitude. By either means a desired supply of cool air for direct application or for mixing with recycled air may be made available for environmental control systems. Various modifications of such axial-flow compressor cooling systems, from the very simple to the very complex, are presently in use for aircraft cockpit and cabin cooling applications.

Example 10-5: A simplified line diagram and *T-s* plot for one section of a cooling system for a large aircraft are shown below. When it is used for cooling on the ground the following Fahrenheit temperatures are experienced at the numbered points on the diagrams:

point	1	2	3	4	5
$t(°F)$	342	142	252	145	35

For a situation where the air flow rate through the system is 65 lb/min, specific heat of the air is assumed constant and the compressor and expander processes are assumed isentropic, calculate the:
 (a) heat transferred to the atmospheric air supply, Btu/min
 (b) power developed by the expander, hp
 (c) heat transferred from auxiliary power unit compressor bleed, expressed in tons of refrigeration

Solution:

(a)
$$\dot{Q}_{1-4} = \dot{M}c_p[(t_1 - t_2) + (t_3 - t_4)]$$
$$= 65(0.24)[(342 - 142) + (252 - 145)]$$
$$= \underline{4789.2 \text{ Btu/min}}$$

316

REFRIGERATION CYCLES

(b) $$P = \dot{W}k = \frac{\dot{M}(h_4 - h_5)}{42.42} = \frac{65(0.24)(145 - 35)}{42.42}$$
$$= 40.45 \text{ hp}$$

(c) $$\text{REFRIG} = \frac{\dot{Q}_{1-4}}{200} = \frac{4789.2}{200} = 23.95 \text{ tons}$$

10-9. Vapor Absorption Systems. The essential difference between vapor compression and vapor absorption refrigeration is the form in which the motivating energy is supplied. Whereas vapor compression utilizes mechanical work to produce refrigeration, *heat is the primary source of energy for the absorption system.* In this system the low temperature refrigerant leaving the refrigerating coil is dissolved in *a secondary working substance known as the* **absorbent**, and the *liquid solution* raised to a suitable condenser pressure through a liquid pump. Although mechanical energy is required to drive the pump, the amount is very small in comparison with the power required by a vapor compressor.

Water may be used as the absorbent with ammonia as the refrigerant, the solution being known as **aqua ammonia**. The basic aqua ammonia system is illustrated in Fig. 10-12, and the operation is as follows. Ammonia vapor leaving the brine tank coils enters an **absorbing chamber.** *Weak*

FIGURE 10-12. Basic vapor absorption refrigeration system.

aqua solution is sprayed into the chamber, absorbs ammonia vapor and leaves the chamber as a *strong solution*. A centrifugal pump is used to remove the strong aqua solution from the low pressure absorber and pump it into the high pressure **generator**. In the generator a steam coil or other source is used to heat the solution. The boiling temperature of the solution is materially less than that of water, although greater than that of ammonia under the same pressure. Therefore, ammonia vapor is driven out of the solution. Weak aqua solution from the generator returns to the absorber through a pressure-reducing valve; and the ammonia vapor liberated in the generator is condensed and enters the refrigerating coil through an expansion valve to complete the cycle.

In practice various additional pieces of equipment are required, depending upon the individual properties of the absorbent and the refrigerant. Other combinations of fluids such as dimethylether or tetraethylenglycol with methylene chloride as the refrigerant and some solids such as lithium nitrate with ammonia may be used in vapor absorption systems. Usually the combination employed depends upon the temperature at which refrigeration is desired, and for air conditioning refrigeration salt solutions of lithium chloride and of lithium bromide with water as the refrigerant are practical.

10-10. Heating with a Refrigeration System. The use of the refrigerating machine as a heating device or heat pump in the literal sense was suggested by Lord Kelvin in 1852. In modern usage the refrigerating plant required for air conditioning during the summer may be operated as a heating system during the winter season. For this purpose the coil used for summer cooling functions as a condenser during the heating season, the heat rejected by the refrigerant being absorbed by air which is circulated through the spaces to be warmed. On the low pressure side of the system the source of heat is usually the atmosphere, although evaporator coils may be placed in wells or buried in the ground.

Because of its inherent flexibility the vapor compression system is easily adapted to such a heating system. Since the work of compression increases the heating capacity of the plant, a low coefficient of performance is acceptable. Both high and low pressures may be varied to suit individual needs. However, the average operating temperature of the heating system will in all cases be relatively low, usually not exceeding 120°F.

Example 10-6: An ideal Freon-12 refrigerating system has a capacity of 50 tons. The condenser pressure is 180 psia, and the Freon-12 temperature leaving the condenser is 120°F. The pressure leaving the expansion valve is 44 psia, and the temperature of the Freon-12 leaving the succeeding coil is 40°F. Circulating water enters the condenser at a temperature of 100°F and leaves it at 115°F. Determine:

REFRIGERATION CYCLES

(a) the mass of Freon-12 circulated, lb/hr
(b) the compressor power for isentropic compression, Btu/hr
(c) the heat capacity of the system, Btu/hr
(d) the mass of water circulated through the condenser and heating system, lb/hr
(e) the useful heat furnished per Btu of compressor work (Heating Performance Ratio)

Solution: (See Freon-12 tables in Appendix B and Fig. 10-6.)

$$h_4 = h_1 = 36.16 \text{ Btu/lb}, \quad h_2 = 83.03 \text{ Btu/lb}$$
$$s_2 = s_3 = 0.17142 \text{ Btu/lb} \cdot {}^\circ\text{R}, \quad h_3 = 94.31 \text{ Btu/lb}$$

(a) $$\dot{M} \text{ Freon-12} = \frac{12{,}000 \times \text{capacity}}{h_2 - h_1} = \frac{12{,}000 \times 50}{83.03 - 36.16} = \underline{12{,}800 \text{ lb/hr}}$$

(b) Compressor Power $= \dfrac{\dot{W}k_{\text{comp}}}{J} = (\dot{M} \text{ Freon-12})(h_3 - h_2)$

$$P = \frac{\dot{W}k_{\text{comp}}}{J} = (12{,}800)(94.31 - 83.03) = \underline{144{,}400 \text{ Btu/hr}}$$

(c) Heating Effect $= q_R = h_3 - h_4 = 94.31 - 36.16 = 58.15$ Btu/lb

Heat Capacity $= H.C. = (\dot{M} \text{ Freon-12})(\text{heating effect})$
$= 12{,}800(58.15) = \underline{744{,}300 \text{ Btu/hr}}$

(d) $$\text{circulating water} = \frac{H.C.}{c(t_{\text{out}} - t_{\text{in}})}$$

$$= \frac{744{,}300}{1(115 - 100)} = \underline{49{,}620 \text{ lb/hr}}$$

(e) Heating Performance Ratio $= \dfrac{H.C.}{\text{compressor power}} = \dfrac{744{,}300}{144{,}400} = \underline{5.15}$

Problems

10-1. Calculate the horsepower required per ton of refrigeration produced by the reversal of a Carnot cycle having a thermal efficiency of (a) 50 percent; (b) 25 percent; (c) 12.5 percent.

Ans. (a) 4.71 hp; (b) 1.57 hp; (c) 0.673 hp

10-2. A refrigerating plant for an air conditioning system is to have a capacity of 10 tons and a coefficient of performance of 2.50 when operating with a refrigerating effect of 61.4 Btu per lb of refrigerant. Calculate (a) the refrigerant flow rate, lb/min; (b) the work done on the refrigerant by the compressor, Btu/lb; (c) the compressor internal horsepower; (d) the rate of heat rejection from the system, Btu/min.

Ans. (a) 32.6 lb/min; (b) 24.6 Btu/lb; (c) 18.9 hp; (d) 2800 Btu/min

10-3. A refrigeration system has a capacity of 25 tons and rejects heat at the rate of 6560 Btu/min. Calculate (a) the rate of heat absorption by the refrigerant, Btu/min; (b) the power required as input to the system, Btu/min; (c) the coefficient of performance for the system.

Ans. (a) 5,000 Btu/min; (b) 1560 Btu/min; (c) 3.2

10-4. A modified Rankine refrigerating cycle operates with an evaporator pressure of 21.4 psia and a condenser pressure of 141 psia. Refrigerant is Freon-12 circulating through

REFRIGERATION CYCLES

the system at 30 lb/min. Liquid refrigerant at 141 psia and 100°F enters the expansion valve, and superheated vapor at 21.4 psia and 5°F enters the compressor. Calculate (a) the refrigerating effect, Btu/lb; (b) the plant capacity in tons of refrigeration; (c) power required to compress the refrigerant, hp; and (d) the plant coefficient of performance.

Ans. (a) 47.86 Btu/lb; (b) 7.18 tons; (c) 10.66 hp; (d) 3.17

10-5. A Freon-12 refrigerating unit for an air conditioning system has an evaporation pressure of 60 psia and a condensing pressure of 110 psia. The temperature of the Freon-12 entering the expansion valve is 80°F, the temperature of the vapor entering the compressor (at 60 psia) is 60°F and the enthalpy after compression is 90.18 Btu/lb. Calculate (a) the refrigerating effect, Btu/lb; (b) the coefficient of performance; (c) the mass of refrigerant circulated if the capacity of the unit is 10 tons, lb/min; (d) the internal power required, hp.

Ans. (a) 59.05 Btu/lb; (b) 12.18; (c) 33.9 lb/min; (d) 3.87 hp

10-6. A refrigerating system for an ice plant circulates 213.5 lb Freon-12 per minute. The evaporation pressure is 30 psia, the condensing pressure is 110 psia, the Freon-12 temperature leaving the coil and entering the compressor is 20°F and the temperature entering the expansion valve is 80°F. The plant produces ice at 30°F from fresh water at 55°F. Assume isentropic compression and calculate or determine (a) the enthalpies of the Freon-12 entering and leaving the freeze box and the enthalpy leaving the compressor, Btu/lb; (b) the refrigerating effect, Btu/lb; (c) the internal compressor work, Btu/lb; (d) the coefficient of performance; (e) the internal power required, hp; (f) the refrigerating capacity of the plant, tons; (g) the ice output of the plant for continuous operations, tons/day.

Ans. (a) 26.28 Btu/lb, 80.73 Btu/lb, 91.02 Btu/lb; (b) 54.45 Btu/lb; (c) 10.29 Btu/lb; (d) 5.29; (e) 51.8 hp; (f) 58.1 tons; (g) 50 tons ice/day

10-7. A Freon-12 refrigerating unit was tested by circulating the low pressure refrigerant through a commercial water cooler, and the following data were obtained: Freon-12 pressure and temperature at compressor suction, 46 psia and 78.4°F; at compressor discharge, 131 psia and 160.4°F; at condenser inlet, 131 psia and 159.6°F; at condenser outlet, 131 psia and 97.6°F; at expansion valve, 129 psia and 96.4°F; entering water chiller, 50 psia (and 38.3°F); leaving water chiller, 50 psia and 79.8°F: temperature of fresh water entering chiller, 91.4°F; leaving chiller, 69.2°F: Freon-12 flow rate, 21 lb/min; water flow rate, 56 lb/min: power to compressor motor, 5.8 kW: motor and belt efficiency, 82 percent. Calculate (a) the capacity of the unit as established by the Freon-12 measurements, tons; (b) the capacity of the unit as determined from the fresh water measurements, tons; (c) the power input to the unit at the compressor shaft, Btu/min; (d) the shaft input per ton of refrigeration based on Freon-12, hp; (e) the coefficient of performance based on Freon-12 capacity and compressor shaft input.

Ans. (a) 6.13 tons; (b) 6.21 tons; (c) 270.5 Btu/min; (d) 1.04 hp/ton; (e) 4.5

10-8. In an ideal (reversed Joule cycle) air refrigerating system the temperature of the air entering the compression cylinder is 50°F, the temperature entering the after-cooler is 160°F and the temperature entering the brine tank is 0°F. Calculate (a) the temperature of the air leaving the after-cooler, °F; (b) the coefficient of performance; (c) the mass of air which must be circulated per ton of refrigeration, lb/min.

Ans. (a) 99°F; (b) 4.54; (c) 16.7 lb/min

CHAPTER 11

RECIPROCATING STEAM ENGINES

11-1. Foreword. The expansive power of steam had been a matter of record for many hundreds of years before Savery, at the end of the seventeenth century, applied it practically to the duty of mine drainage. Savery's apparatus would not fit our present-day definition of a steam engine, but it undoubtedly was a direct and most important antecedent. In 1705, James Newcomen patented what is now recognized as the first steam engine. Engines of this type, together with the numerous and far-reaching refinements added by James Watt, were, to a certain degree, responsible for the Industrial Revolution. By the start of the nineteenth century, application of the steam engine had reached the locomotive and ship industries, with successful commercial results finally credited to Stephenson and Fulton, respectively.

The modern steam engine is characterized by rugged, simple construction and low first cost. For many years the reciprocating steam engine has been unable to compete with the steam turbine for many applications having large power requirements, because the large capacity turbine is more efficient, smaller, and lighter, and requires less upkeep than its steam engine counterpart. Except for difficulties arising from contamination of boiler feedwater by cylinder and stuffing-box lubricant accompanying the use of superheated steam, the reciprocating prime mover is ideally suited to those applications requiring wide flexibility of speed control and full-power reversed operation. Other highly suitable applications include direct drive for reciprocating pumps, compressors and numerous auxiliaries.

During World War II over one hundred naval vessels including tugs, landing ships (dock) and escort aircraft carriers used the Uniflow steam engine for main propulsion power. On these ships the engine capacities ranged from 600 to 4500 internal horsepower. Triple-expansion, multicylinder engines in sizes up to 1600 internal horsepower can still be found in older ships plying the Great Lakes.

11-2. Apparatus for Steam Engine Cycles. Apparatus for the cycle using the reciprocating steam engine as prime mover is similar in all respects to that shown for the Rankine cycle in Fig. 9-2, except for the substitution of the reciprocating engine in place of the turbine as shown in Fig. 11-1.

Treating the engine as a steady-flow machine with no heat transfer and

FIGURE 11-1. Apparatus for Rankine cycle with reciprocating steam engine.

negligible changes in potential and kinetic energies, the steady-flow energy equation gives

$$\frac{wk_{23}}{(J)} = h_2 - h_3 \text{ Btu/lb (J/kg)}$$

For an ideal engine, that is one having no losses, mechanical or thermal, $h_2 - h_3$ represents the isentropic drop in enthalpy for steam between admission and exhaust conditions, and consequently the greatest possible work that could be realized from such an engine. However, an actual engine designed for complete expansion of steam to the usual exhaust pressures would involve such a large piston and long stroke that much of the work realized from the complete expansion would be lost because of increased mechanical friction.

The reciprocating steam engine derives work from the force of the steam pressure acting on the piston face throughout the piston travel against a resisting external load applied to the piston rod. The steam admission valve, A, and the exhaust valve, B, control the quantity and timing of steam admitted to and exhausted from the cylinder, and thereby the degree of expansion and compression of the steam remaining in the cylinder during the periods when both valves are closed. For fixed steam conditions at the engine entrance and exhaust points, variations in the

timing of valves *A* and *B* produce a wide range of energy supply to, and work output from, the engine. In view of this multiplicity of operating conditions for a single engine with fixed conditions of steam supply and exhaust, only two of the conditions most frequently met in practice, and for which the complete cycle analysis appears in Chapter 9, will be analyzed for the *ideal* engine.

11-3. Work of the Ideal Incomplete-Expansion Steam Engine. Basically, the mechanical construction of the steam engine is similar to that of an air compressor and, aside from differences in working substance involved, the same general approach may be used in obtaining equations for work derived *from* an ideal steam engine as were used to determine work required *for* an ideal air compressor.

Figure 11-2 shows an indicator diagram for an ideal incomplete-expansion steam engine with no clearance. In the diagram, process line 1-2 represents constant pressure admission of steam to the cylinder until "cutoff" occurs, at which time the inlet valve closes. Process line 2-3 represents *non-flow* isentropic expansion of steam in the engine cylinder with both inlet and exhaust valves closed. During the constant volume process 3-4 the cylinder pressure is reduced to exhaust pressure. Process line 4-5 indicates constant pressure exhaust of steam from the cylinder. Since the ideal engine needs no compressed steam to cushion the effect of

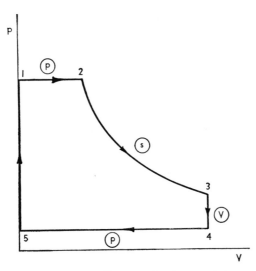

FIGURE 11-2. Indicator diagram for incomplete-expansion steam engine without clearance.

reversing the direction of motion of its piston, piston rod and crosshead, it has no clearance. That is, there is no space remaining between the piston face and cylinder head when the piston completes its exhaust stroke. Furthermore, the exhaust valve remains open until the piston reaches the end of its exhaust stroke, point 5 of the diagram, and then instantaneously the exhaust valve closes and the admission valve opens, producing the constant volume pressure rise shown as 5-1. It must constantly be borne in mind that this ideal indicator diagram is a *pressure-total volume* diagram showing the state of a varying mass of steam *in the engine cylinder* from which the ideal engine work per cycle may be computed.

Processes 1-2 and 4-5 of the ideal indicator diagram represent the work done *by* the steam on the piston during admission and *on* the steam by the piston during exhaust, respectively. The isentropic expansion process 2-3 and the constant volume process 3-4 both occur with the inlet and exhaust valves of the engine closed and are therefore non-flow processes. By evaluating the work of each process and taking into account the proper algebraic sign, the engine work may be determined. Thus,

$$\frac{Wk_{eng}}{(J)} = \frac{Wk_{12}}{(J)} + \frac{Wk_{23}}{(J)} + \frac{Wk_{34}}{(J)} + \frac{Wk_{45}}{(J)} + \frac{Wk_{51}}{(J)} \text{ Btu (J)}$$

and

$$\frac{Wk_{eng}}{(J)} = \frac{p_2 V_2}{(J)} + (U_2 - U_3) + 0 - \frac{p_4 V_4}{(J)} + 0 \text{ Btu (J)}$$

The total work from 1 to 2 is $p_2 V_2/(J)$ and from 4 to 5 is $-p_4 V_4/(J)$ since the total volume of steam in the engine at points 1 and 5 is zero. The work of the non-flow isentropic expansion is expressed by the change of internal energy, $U_2 - U_3$, and the work of the constant volume processes 3-4 and 5-1 is zero. Adding a negative and a positive $p_3 V_3/(J)$ to the right side of the last equation gives

$$\frac{Wk_{eng}}{(J)} = \left(\frac{p_2 V_2}{(J)} + U_2\right) - U_3 - \frac{p_3 V_3}{(J)} + \frac{p_3 V_3}{(J)} - \frac{p_4 V_4}{(J)}$$

which rearranges to

$$\frac{Wk_{eng}}{(J)} = H_2 - H_3 + \frac{V_3(p_3 - p_4)}{(J)} \text{ Btu (J)}$$

or

$$\frac{wk_{eng}}{(J)} = h_2 - h_3 + \frac{v_3(p_3 - p_4)}{(J)} \text{ Btu/lb (J/kg)} \qquad \underline{(11\text{-}1)}$$

Comparison of this equation with the net work of the incomplete-

expansion *cycle* from Equation (9-13) shows that they differ only by the pump work term for the cycle, $h_1 - h_5$.

11-4. Work of the Ideal Non-Expansion Steam Engine. Figure 11-3 shows an indicator diagram for an ideal non-expansion steam engine without clearance, and represents the ultimate diagram area obtainable with a given cylinder. Steam is admitted at constant pressure throughout

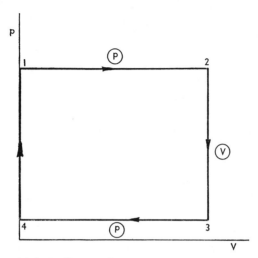

FIGURE 11-3. Indicator diagram for ideal non-expansion steam engine without clearance.

the piston stroke, shown by process line 1-2. At point 2 the admission valve closes and the exhaust valve opens simultaneously, producing the constant volume process 2-3, which reduces the pressure in the cylinder from admission to exhaust pressure. During the constant pressure process 3-4, the steam is exhausted from the cylinder. With no clearance, the constant volume process 4-1 is produced by simultaneous closing of the exhaust valve and opening of the admission valve with no movement of the piston.

Analyzing this ideal non-expansion indicator diagram for the work of the ideal engine gives

$$\frac{Wk_{eng}}{(J)} = \frac{Wk_{12}}{(J)} + \frac{Wk_{23}}{(J)} + \frac{Wk_{34}}{(J)} + \frac{Wk_{41}}{(J)} \text{ Btu (J)}$$

The work of each of the two constant volume processes, 2-3 and 4-1, is, of

RECIPROCATING STEAM ENGINES

course, zero, and the two constant pressure processes evaluated give

$$\frac{Wk_{eng}}{(J)} = \frac{p_2 V_2}{(J)} + 0 - \frac{p_3 V_3}{(J)} + 0 \text{ Btu (J)}$$

With V_2 equal to V_3, there results

$$\frac{Wk_{eng}}{(J)} = \frac{V_2(p_2 - p_3)}{(J)} \text{ Btu (J)}$$

and

$$\frac{wk_{eng}}{(J)} = \frac{v_2(p_2 - p_3)}{(J)} \text{ Btu/lb (J/kg)} \qquad (11\text{-}2)$$

Again, in comparison with the net work of the ideal non-expansion *cycle* as expressed by Equation (9-17), the only difference shown is the pump work of the ideal cycle.

Example 11-1: An ideal single cylinder, single-acting steam engine has a stroke of 2 ft and, a piston cross-sectional area of 1 sq ft. For an initially saturated steam supply at 230 psia and a condenser pressure of 2 psia, calculate (a) the steam supplied when operating as an ideal incomplete-expansion engine with expansion terminal pressure of 20 psia, lb/cycle; (b) the ideal incomplete-expansion engine work, Btu/cycle; (c) the steam supplied when operating as an ideal non-expansion engine, lb/cycle; (d) the ideal non-expansion engine work, Btu/cycle.

Solution:

$p_1 = p_2 = 230$ psia
$p_3 = 20$ psia = expansion terminal pressure
$p_4 = p_5 = 2$ psia
piston cross-sectional area = 1 sq ft
piston stroke = 2 ft
$V_3 = V_4 = 2 \text{ ft}^3$
$s_3 = s_2 = 1.5346$

(a) $$m_3 = \frac{s_{g_3} - s_3}{s_{fg_3}} = \frac{1.7320 - 1.5346}{1.3962} = 0.1414$$

$$v_3 = v_g - mv_{fg} = 20.09 - (0.1414)(20.07) = 17.25 \text{ ft}^3/\text{lb}$$

$$M = \frac{V_3}{v_3} = \frac{2.00}{17.25} = \underline{0.116 \text{ lb/cycle}}$$

(b) $$\frac{wk_{eng}}{J} = (h_2 - h_3) + \frac{v_3(p_3 - p_4)}{J}$$

$$\frac{wk_{eng}}{J} = 1201.1 - 1020.6 + \frac{17.25(20.0 - 2.0)144}{778}$$

$$\frac{wk_{eng}}{J} = 1201.1 - 1020.6 + 57.5 = 238.0 \text{ Btu/lb}$$

$$\frac{Wk_{eng}}{J} = 238.0(0.116) = \underline{27.6 \text{ Btu/cycle}}$$

RECIPROCATING STEAM ENGINES

(c) $$M = \frac{V_2'}{v_2'} = \frac{2.0}{2.000} = 1.0 \text{ lb/cycle}$$

(d) $$\frac{wk_{eng}}{J} = \frac{v_2'(p_2 - p_4)}{J} = \frac{2.000(230 - 2.0)144}{778}$$

$$\frac{wk_{eng}}{J} = 84.40 \text{ Btu/lb}$$

$$\frac{Wk_{eng}}{J} = 84.40(1.0) = 84.40 \text{ Btu/cycle}$$

In the above example, answers (a) and (c) indicate the ideal steam consumption per engine cycle for an incomplete-expansion engine and a non-expansion engine with the same steam conditions and cylinder size, while answers (b) and (d) reflect the quantity of work per engine cycle that an engine of given size will produce ideally.

11-5. The Actual Indicator Diagram. Figure 11-4 shows an actual indicator diagram superimposed on an ideal diagram for the same cylinder.

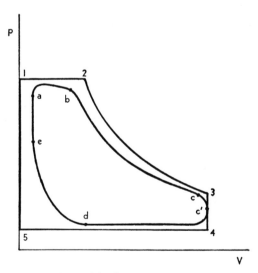

FIGURE 11-4. Actual indicator diagram superimposed on ideal diagram.

The events occurring in the actual engine cylinder and the points on the diagram at which they begin are: admission, e; cutoff, b; release, c; and compression, d. Admission refers to the entrance of steam to the cylinder

when the admission valve opens. Cutoff is the point in the piston stroke at which steam stops flowing into the cylinder and is controlled by closing of the admission valve. Release is the point at which the exhaust valve opens, completing the reduction of cylinder pressure to exhaust pressure. Compression is initiated by closing of the exhaust valve. The steam trapped in the cylinder at d and compressed to e serves as a cushion for the reversal in direction of the relatively heavy piston and piston rod.

The *clearance volume* of an actual steam engine is the volume between the piston and valve faces when the piston is at the head end of its stroke. This volume is represented on the diagram by the horizontal distance between lines 1-5 and a-e.

The *expansion ratio* for an actual steam engine is a ratio of the volume change effected by the piston during its stroke to the volume change during expansion while admission and exhaust valves are closed. In terms of the diagram, the expansion ratio is ec'/bc.

A comparison of the actual and ideal diagrams indicates the nature and magnitudes of some of the losses occurring in an actual steam engine. The pressure drop during admission of steam is caused by throttling of the steam in the admission valves and ports. Inability to obtain instantaneous admission-valve closing at b causes the rounded portion of the diagram in that region. The expansion line, b-c, has a flatter slope than the ideal curve due to initial condensation and later re-evaporation of the steam in the cylinder. Time required for opening of the exhaust valve and throttling in the exhaust valve and port are responsible for the rounding of the curve beyond point c, and for the exhaust pressure being higher than the ideal.

The actual indicator diagram is obtained by means of a steam engine indicator (see Fig. 6-1a). The indicator includes a marking device, usually a pen or stylus actuated by cylinder pressure, and a reducing mechanism which rotates a cylindrical recording drum. The vertical scale of the diagram is determined by the strength of the indicator spring, which is accurately calibrated in psi per inch deflection. The horizontal scale is determined by the reducing mechanism operating the recording drum. Since each position of the piston determines a definite cylinder volume between piston face and cylinder head, the indicator produces to scale a pressure-volume diagram of events in the cylinder.

Indicator diagram areas represent to scale the work done in an engine cylinder during one engine cycle. Measurement of this area, usually in square inches, is easily accomplished with a planimeter. Dividing the diagram area by the diagram length produces the mean ordinate. If the ordinate is determined in inches, multiplication by the spring scale (psi per inch deflection) gives the **mean effective pressure** in psi for the engine cylinder during the period of the diagram.

RECIPROCATING STEAM ENGINES

11-6. Indicated Horsepower. The work shown by the area of the indicator diagram is called *indicated work*. It is a measure of the net energy transferred between the steam in the cylinder and one face of the piston for each complete revolution of the engine (engine cycle). Multiplying this by the number of revolutions per unit time and converting to horsepower results in the *indicated horsepower* for that end of the cylinder. Expressed in equation form, in Engineers' units

$$IHP = \frac{p_m LAN}{33,000}$$

where p_m = mean effective pressure, psi
L = length of piston stroke, ft
A = area of piston face, sq in
N = engine cycles (revolutions) per minute.

For double-acting engines, that is, engines which have steam admission to each end of the cylinder, indicator diagrams must be taken for each end of all cylinders. For single-acting, multiple-expansion engines, in which two or more cylinders are employed to complete an expansion, diagrams must be taken on each cylinder. In any case, the engine indicated horsepower is the sum of the values calculated for each cylinder.

Example 11-2: The piston diameter of a single-cylinder, double-acting steam engine is 4.25 in., and the length of piston stroke is 9 in. Piston rod diameter is 1.13 in. extending on only one side of the piston. Indicator diagrams were taken on each end of the cylinder at an engine speed of 95 rpm. Diagram areas were 4.5 sq in for the head end and 4.6 sq in for the crank end. Reduction ratio for piston stroke to diagram length was 3 to 1, and the indicator spring scale was 90 psi/in. Calculate the indicated horsepower of the engine under these operating conditions.
Solution:

piston dia = 4.25 in. indicator spring scale = 90 psi/in.
piston stroke = 9 in. diagram areas
piston rod dia = 1.13 in. head end = 4.5 in.²
engine speed = 95 rpm crank end = 4.6 in.²
piston stroke to diagram length reduction ratio = 3 to 1

Head end:

$$\text{length of diagram} = \frac{9}{3} = 3 \text{ in.}$$

$$\text{mep} = \frac{4.5}{3}(90) = 135 \text{ psi}$$

$$\text{piston area} = \frac{\pi d^2}{4} = \frac{3.14(4.25)^2}{4} = 14.19 \text{ in.}^2$$

$$IHP = \frac{135(9)14.19(95)}{(12)(33,000)} = 4.14 \text{ IHP}$$

329

Crank end:

$$\text{mep} = \frac{4.6}{3}(90) = 138 \text{ psi}$$

$$\text{effective piston area} = \text{piston area} - \text{rod area}$$

$$= \frac{\pi}{4}[(4.25)^2 - (1.13)^2] = 13.18 \text{ in.}^2$$

$$IHP = \frac{138(9)13.18(95)}{(12)(33,000)} = 3.93 \text{ IHP}$$

Total $IHP = 4.14 + 3.93 = \underline{8.07 \text{ IHP}}$

11-7. Engine Efficiency and Steam Rate. Preceding articles have established methods for calculation of the work of both an ideal and an actual steam engine. The ratio of the work of an actual engine to the work of an ideal engine under the same imposed conditions is termed the *engine efficiency* for the actual engine. Since an actual engine may have its work measured at the output shaft coupling by a brake or torsion-meter, as well as internally by an indicator, there arise two engine efficiencies, one in terms of the indicated work and the other in terms of brake work. Therefore,

$$\eta_{ei} = \frac{wk(\text{indicated})}{wk(\text{ideal})} \quad \underline{(11\text{-}3)}$$

and

$$\eta_{eb} = \frac{wk(\text{brake})}{wk(\text{ideal})} \quad \underline{(11\text{-}4)}$$

It is a simple matter with most actual steam engines to provide means for measuring the steam flow rate. This information may be combined with the power output to give the steam rate, SR, as

$$SR = \frac{\dot{M}}{P} \text{ lb/hp} \cdot \text{hr (kg/kW} \cdot \text{hr)} \quad \underline{(11\text{-}5)}$$

where \dot{M} = steam flow, lb/hr (kg/hr)
P = power output, hp (kW)

Steam rate values will be influenced by the point of power measurement in the same manner as the engine work, and to be fully descriptive must be labeled brake or indicated steam rate. Engine steam rates are also called steam engine economies.

11-8. Steam Engine Losses. Steam engine losses may be broadly classified as follows:

RECIPROCATING STEAM ENGINES

A. Thermal losses
 1. Radiation and conduction
 2. Cylinder condensation and re-evaporation
B. Friction losses
 1. Mechanical
 2. Fluid friction and throttling
C. Loss of working substance
 1. Leakage past valves, pistons and glands

Radiation and conduction losses are primarily a matter of cylinder temperatures and may be controlled by proper jacketing of the cylinders.

Cylinder condensation and re-evaporation represents the major loss in counter-flow steam engines, that is, those having inlet and exhaust ports at the same end of the cylinder. Factors affecting this loss are the temperature difference between cylinder walls and steam in the cylinder, the ratio of cylinder surface to volume of steam in the cylinder and the length of time the steam is in contact with the cylinder walls, which is a function of engine speed. As the high temperature admission steam contacts the cylinder head and walls, which have just been cooled by the exhaust steam from the previous cycle, a portion of the steam condenses immediately, and more continues to condense until shortly after the cutoff point, when the temperature of the expanding steam drops below that of the surrounding cylinder walls. Heat then flows from the cylinder walls to the steam, re-evaporating some liquid that has condensed on the walls. Such action continues until compression of the clearance steam raises the steam temperature once more above that of the cylinder walls. With saturated steam at admission, as much as 40 percent of the entering steam may condense up to the time cutoff occurs. Cylinder insulation, steam-jacketed cylinder heads, use of superheated steam for admission, and compounding of cylinders all tend to reduce this loss. The most practical answer to this particular loss is the Uniflow steam engine, which will be discussed separately.

Mechanical friction occurs in the main and cross-head bearings, cross-head guides, stuffing boxes, valves and valve mechanisms, and from piston and piston ring contact with the cylinder wall. Bearing friction is the major of these mechanical friction losses and may be minimized by good design with proper lubrication and maintenance.

Fluid friction and throttling losses which occur in admission and exhaust valves and passages have their effect shown by Fig. 11-4, where a large part of the difference between the ideal and the actual indicator diagrams is due to these losses. Although the energy content of the working substance is not decreased by throttling, the available energy is lowered and the entropy is increased, resulting in a net loss of work from

the engine. Throttling losses are inherent in any valve action, but certain valve types cause far less throttling than others. Typical of these low-loss valves are the Corliss rotary valve and the poppet valve. Engine governing is sometimes accomplished by throttling the admission steam to accommodate the engine load, but this is particularly costly at low-load operation. Under cutoff governing, where the point of cutoff is variable, the pressure drop from admission pressure to cutoff pressure is completely due to throttling of the steam in the admission valves and ports.

Loss of working substance in the steam engine may occur through valves, past piston rings and piston and through stuffing boxes and glands. High pressure steam which leaks either to the atmosphere or ultimately to a condenser represents a loss of energy which must be charged against the engine. Fortunately, these losses may be controlled and minimized by proper engine maintenance.

11-9. The Uniflow Steam Engine. The Uniflow engine was developed in order to overcome some of the inherent losses in the single-cylinder, counter-flow steam engine. Specifically, the Uniflow steam engine, as its name implies, passes the steam in one direction only through the cylinder, thereby reducing materially the initial condensation and re-evaporation losses in the cylinder. To accomplish this, steam is admitted alternately at each end of the cylinder by quick-acting valves of the poppet type, and is exhausted through ports at the center of the cylinder when they are uncovered by the piston. This action is obtained by making the length of the Uniflow piston just slightly less than its length of stroke. Without having cool exhaust steam in contact with the cylinder walls on the return piston stroke, it is possible to have the wall temperature approach the steam temperature throughout the length of the cylinder. Furthermore, the cylinder head and inlet passages, which cause a large portion of the initial condensation, may be steam-jacketed with high temperature steam to further reduce condensation losses.

11-10. Multi-Cylinder Engines. Compound engines consisting of two cylinders and triple-expansion engines of either three or four cylinders constitute the majority of engines of this class. Condensation losses in compound and triple-expansion engines are correspondingly less than in single-cylinder engines, but losses in valves and steam passages may be increased. Expansion ratios for minimum cylinder condensation loss should not exceed 5, and with high initial steam pressure and condensing operation, expansion of the steam in two or more successive cylinders is mandatory. Multiple-cylinder engines reduce the steam temperature drop per cylinder, thereby reducing the condensation loss.

RECIPROCATING STEAM ENGINES

11-11. Effects of Superheat and Back Pressure. Although superheated steam increases steam engine cyclic efficiencies, its biggest justification lies in the reduction of cylinder condensation. Because of a lower coefficient of heat transfer, less heat is lost by superheated than by wet steam in contact with cylinder walls. Sufficient superheat should be employed to keep the steam dry down to the point of cutoff for minimum cylinder condensation. Lubrication difficulties accompany any appreciable amount of superheat, because of the lack of lubricating effect realized from condensed steam film on cylinder walls. Supplying lubricating oil to cylinder walls greatly increases the difficulties caused by lubricant fouling of condenser and boiler heat transfer surfaces. Highly superheated steam also has a tendency to warp valves and valve stems, increasing the difficulty of proper engine maintenance.

Decreasing the exhaust, or back pressure, below that of the surrounding atmosphere is accomplished by exhausting to a condenser. The low back pressure induced by a condenser increases the temperature drop through an engine cylinder (for single-cylinder engines, or the low pressure cylinder on multiple-cylinder engines) and increases the cylinder condensation loss; but the increased available energy from the reduction in back pressure more than offsets the added losses, resulting in a net gain in operating economy for the engine. The reduction in steam consumption for condensing over noncondensing operation of the same engine extends as high as 30 percent, the individual value depending on the type of engine and its operating conditions.

11-12. Mechanical Efficiency. The mechanical losses described earlier are responsible for a reduction in horsepower available at the engine output shaft coupling (brake horsepower) from the indicated horsepower developed in the engine and determined from the actual indicator diagram. These mechanical losses are substantially constant at all loads for a constant speed, but increase in direct proportion to an increase in speed. The ratio of brake horsepower to indicated horsepower is termed the **mechanical efficiency** of the engine. In equation form,

$$\eta_m = \frac{BHP}{IHP} = \frac{IHP - FHP}{IHP} \qquad (11\text{-}6)$$

where η_m = mechanical efficiency
BHP = brake horsepower
IHP = indicated horsepower
FHP = friction (mechanical loss) horsepower.

RECIPROCATING STEAM ENGINES

Mechanical efficiencies of 90 to 96 percent at full capacity are common for well-designed engines when properly adjusted and lubricated.

Problems

11-1. An ideal single-cylinder, single-acting, incomplete-expansion steam engine having a 23-in. bore and a 24-in. stroke operates with saturated steam at 240 psia at the throttle and an absolute exhaust pressure of 4 in. Hg. For an expansion terminal pressure of 27.0 psia and an engine speed of 210 rpm, calculate (a) the steam supplied, lb/cycle; (b) the ideal engine work, Btu/cycle; (c) the total steam flow to the engine, lb/hr; (d) the horsepower developed by the ideal engine.

Ans. (a) 0.438 lb/cycle; (b) 98.6 Btu/cycle; (c) 5520 lb/hr; (d) 488 hp

11-2. An ideal single-cylinder, double-acting, non-expansion steam engine is supplied with saturated steam at 120 psia and exhausts at 15 psia. The cylinder volume is 5.77 ft^3. The engine completes 80 strokes (40 cycles) per minute. Calculate (a) the steam supplied, lb/cycle; (b) the ideal engine work, Btu/cycle; (c) the total steam flow to the engine, lb/hr; (d) the horsepower developed by the ideal engine.

Ans. (a) 3.1 lb/cycle; (b) 224.3 Btu/cycle; (c) 7440 lb/hr; (d) 211.5 hp

11-3. A Uniflow steam engine (double-acting) having a piston diameter of 23 in., a stroke of 24 in. and a piston rod diameter of 4.75 in. (crank end only) was tested at a speed of 210 rpm. Indicator diagram areas for the test averaged 3.5 sq in for the head end and 3.6 sq in for the crank end with an indicator spring scale of 80 psi per inch. The length of each indicator diagram was 4.0 in. Calculate (a) the mep for both the head end and crank end of the cylinder, psi; (b) the indicated horsepower of the head end; (c) the indicated horsepower of the crank end; (d) the total indicated horsepower of the engine.

Ans. (a) 70 psi, 72 psi; (b) 370 ihp; (c) 364 ihp; (d) 734 ihp

11-4. A single cylinder, double-acting steam engine has a 23-in. bore, 24-in. stroke and a piston rod diameter of 4.75 in. (crank end only). At full load and a speed of 210 rpm the mep of both ends of the cylinder is 70 psi. Saturated steam at 240 psia is supplied to the engine. Terminal expansion pressure is 27 psia, and the exhaust pressure is 4 in. Hg abs. Calculate (a) the ideal horsepower for the engine; (b) the indicated horsepower; (c) the indicated engine efficiency, percent; (d) the ideal steam flow, lb/hr; (e) the ideal steam rate, lb/hp·hr.

Ans. (a) 955.1 hp; (b) 724.5 ihp; (c) 75.9%; (d) 10,800 lb/hr; (e) 11.3 lb/hp·hr

11-5. A five-cylinder, double-acting steam engine develops 3700 indicated horsepower when running at 210 rpm. At this speed the mechanical losses consume 300 hp. Calculate (a) the brake horsepower of the engine, and (b) the mechanical efficiency of the engine, percent.

Ans. (a) 3400 bhp; (b) 91.9%

11-6. A three-cylinder, 24.5-in. bore by 24.0-in. stroke, double-acting steam engine on test developed 1200 indicated horsepower when supplied with 15,120 lb steam/hr at 196 psia and 480°F. The condenser vacuum was 26 in. Hg and the barometer read 30 in. Hg. Mechanical losses amounted to 110 hp. Calculate (a) the ideal horsepower developed if the work of an ideal engine for these operating conditions is 243.8 Btu/lb; (b) the brake horsepower developed by the actual engine; (c) the indicated engine efficiency, percent; (d) the brake engine efficiency, percent; (e) the indicated steam rate, lb/ihp·hr; (f) the brake steam rate, lb/bhp·hr; (g) the mechanical efficiency of the engine, percent.

Ans. (a) 1448 hp; (b) 1090 bhp; (c) 82.9%; (d) 75.3%; (e) 12.60 lb/ihp·hr; (f) 13.87 lb/bhp·hr; (g) 90.8%

CHAPTER 12

NOZZLES

12-1. Foreword. The analysis of problems dealing with the flow of fluids through nozzles and diffusers involves specific applications of the theory of steady flow developed in Chapter 5. However, in that chapter only ideal conditions were considered, whereas it is the purpose of this chapter to demonstrate the development and practical application of that theory to actual apparatus.

In investigating actual flow the conventional method initially assumes that the processes involved are reversible and the devices ideal or perfect. This has the advantage of establishing the limits of performance within which the actual nozzle or diffuser can operate. The ideal flow will neither gain nor lose heat to its surroundings, no work will be done and all of the energy available under a given condition will be converted to the desired form. The actual device is not perfect, nor is the working substance; and fluid friction and turbulence are invariably encountered in actual flow. In all cases the problem is to produce maximum energy conversion with minimum loss.

12-2. Definition, Function and Uses of Nozzles. From a thermodynamic standpoint a nozzle may be defined as *an apparatus for converting energy within a fluid into kinetic energy*. Its primary function is one of transformation of energy and not one of merely transferring the working substance from one apparatus to another. In a similar fashion a diffuser may be defined as *an apparatus for converting kinetic energy within a flow into an increase in pressure head*. As energy converters, nozzles and diffusers have a variety of uses and are essential to steam and gas turbines, injectors, ejectors, eductors and jets.

12-3. The Nozzle Energy Equation. As discussed in Chapter 5, the application of the steady flow energy equation to any particular apparatus is based upon continuity of flow as well as upon conservation of energy. This condition is expressed mathematically as

$$\dot{M} = \dot{M}_1 = \dot{M}_2 = \dot{M}_x \text{ lb/sec (kg/sec)}$$

where mass \dot{M} = flow rate and the numerical subscripts denote successive sections of the channel. Then if

NOZZLES

A = cross-sectional area of channel, sq ft (m²)
V = fluid velocity, ft/sec (m/sec)
v = fluid specific volume, ft³/lb (m³/kg)
ρ = fluid density, lb/ft³ (kg/m³)
$\dot{M} = AV/v = AV\rho$ = constant

or, using the same subscripts as before,

$$\dot{M} = \frac{A_1 V_1}{v_1} = \frac{A_2 V_2}{v_2} = \frac{A_x V_x}{v_x} \text{ lb/sec (kg/sec)}$$

In the ideal case the flow process is isentropic; in the actual device flow is essentially adiabatic but the process is irreversible because of attendant friction and turbulence. In either case (see Fig. 12-1):

$$\frac{Z_1}{(J)} + \frac{V_1^2}{2g_c(J)} + u_1 + \frac{p_1 v_1}{(J)} + q_{12} = \frac{Z_2}{(J)} + \frac{V_2^2}{2g_c(J)} + u_2 + \frac{p_2 v_2}{(J)} + \frac{wk_{12}}{(J)}$$
(1-20)

FIGURE 12-1

For both the ideal and the actual case, no work is done on or by the working substance, no heat is transferred and the inlet and outlet may be considered to be at the same level. Further, since the function of a nozzle is the production of kinetic energy from other available forms, and since the internal and displacement energy terms may be combined as enthalpy, the nozzle energy equation having general utility is

$$\frac{V_1^2}{2g_c(J)} + h_1 = \frac{V_2^2}{2g_c(J)} + h_2 = \text{constant}$$

where subscript 1 refers to the entering section and 2 to any following section of the nozzle.

A useful parameter, the stagnation enthalpy, can be obtained from the

nozzle energy equation, as follows:

$$\frac{V^2}{2g_c(J)} + h = h_0 \qquad (12\text{-}1)$$

where
$$h_0 = \text{stagnation state enthalpy.}$$

The stagnation state is the condition that a flow *would* achieve if isentropically brought to rest or which the fluid is actually in if already at rest. Under *ideal* conditions, the stagnation enthalpy and other stagnation properties remain constant for any particular defined flow through a nozzle or diffuser. The actual enthalpy will be approximately the same as the stagnation enthalpy at the entrance to a nozzle or at the exit from a diffuser when the actual local velocities are small enough to be neglected. At any other location in the flow stream, the stagnation enthalpy will be greater than the actual enthalpy by an amount equal to the local kinetic energy.

12-4. Velocity of Gases in Ideal Nozzles. For gases the energy equation becomes

$$\frac{V_2^2}{2g_c(J)} - \frac{V_1^2}{2g_c(J)} = h_1 - h_2 = c_p(T_1 - T_2) \qquad (12\text{-}2)$$

in which T_1 = absolute temperature at entering section, °R (K)
T_2 = absolute temperature at some following section, °R (K)
V_1 and V_2 = fluid velocities at the respective sections, ft/sec (m/sec).

In the ideal case the isentropic relationships apply; thus

$$T_2 = T_1 \left(\frac{p_2}{p_1}\right)^{\frac{k-1}{k}}$$

Example 12-1: Air enters an ideal nozzle at a pressure of 60 psia with a temperature of 1340°F. The pressure at the nozzle exit is 15 psia. Determine the increase of kinetic energy of 1 lb of air produced by an isentropic expansion through the nozzle.
Solution:

$$T_1 = 1340 + 460 = 1800°\text{R}$$

$$T_2 = T_1 \left(\frac{p_2}{p_1}\right)^{\frac{k-1}{k}} = 1800 \left(\frac{15}{60}\right)^{\frac{1.4-1}{1.4}} = 1211°\text{R}$$

$$\frac{V_2^2}{2g_cJ} - \frac{V_1^2}{2g_cJ} = c_p(T_1 - T_2) = 0.24(1800 - 1211)$$

$$= 141.4 \text{ Btu/lb}$$

In the usual case V_1 is very small in comparison with V_2 and the

NOZZLES

corresponding kinetic energy term of Equation (12-1) is neglected. The assumption of low entrance velocity (stagnation conditions at inlet) will be made for all problems in this chapter. For this condition *the velocity of a gas at any section* within a nozzle may be determined from

$$V_2 = \sqrt{2g_c(J)c_p(T_1 - T_2)} \text{ ft/sec (m/sec)} \quad (12\text{-}3)$$

and the ideal or maximum velocity is obtained when T_2 is the temperature of isentropic expansion as above defined.

Equation (12-2) may be further developed in terms of pressures; thus

$$\frac{V_2^2 - V_1^2}{2g_c} = (J)c_p(T_1 - T_2)$$

and for a gas,

$$(J)c_p = R\left(\frac{k}{k-1}\right) \quad \text{and} \quad T_2 = T_1\left(\frac{p_2}{p_1}\right)^{\frac{k-1}{k}}$$

$$\frac{V_2^2 - V_1^2}{2g_c} = \left[R\left(\frac{k}{k-1}\right)\right]\left[T_1 - T_1\left(\frac{p_2}{p_1}\right)^{\frac{k-1}{k}}\right]$$

$$= \frac{kRT_1}{k-1}\left[1 - \left(\frac{p_2}{p_1}\right)^{\frac{k-1}{k}}\right] \quad (12\text{-}4)$$

Equation (12-4) indicates that for a given initial condition the kinetic energy conversion in a nozzle becomes a function of the ratio of the entrance and exit pressures. In addition, since pressures are readily adaptable to mechanical regulation, Equation (12-4) points out a general method of nozzle control. For the solution of problems the enthalpy approach has the advantage of general applicability irrespective of the characteristics of the working substance and will be used throughout this chapter. But it is significant that for *any* specified entrance condition to *any* nozzle, *the final enthalpy is directly controlled by the exit pressure*.

Example 12-2: Air enters an ideal nozzle at a pressure of 60 psia with a temperature of 1340°F and is isentropically expanded to an exit pressure of 15 psia. What is the velocity of the air leaving the nozzle, ft/sec?
Solution: From Ex. 12-1, $c_p(T_1 - T_2) = 141.4$ Btu/lb.

$$V_2 = \sqrt{2g_cJ}\sqrt{c_p(T_1 - T_2)} = 223.8\sqrt{c_p(T_1 - T_2)}$$
$$= 223.8\sqrt{141.4} = \underline{2662 \text{ ft/sec}}$$

An important quantity in nozzle flow applications is the velocity of sound, or sonic velocity. The velocity of sound is defined as the speed at which a small pressure disturbance propagates through a medium. The nozzle flow conditions of sonic velocity of a gas can be determined using the continuity of flow equation, the nozzle energy equation, the isentropic

NOZZLES

relationships and the ideal gas equation of state. Let c represent gas flow traveling at the speed of sound. The continuity of flow equation becomes:

$$\rho c A = \text{constant}$$

and differentiating:

$$\rho c dA + \rho A dc + c A d\rho = 0$$

When sonic velocity occurs in a nozzle, it will always occur at the point of minimum cross-sectional area, the so-called throat, as will be verified later. In the converging portion of the nozzle, A is decreasing and dA is negative ($dA < 0$). In the diverging portion, if any, A is increasing and dA is positive ($dA > 0$). But at the throat, the area A is instantaneously invariant and $dA = 0$. Under this condition the preceding equation becomes:

$$\rho dc + c d\rho = 0$$

and

$$dc = -\frac{c}{\rho} d\rho \tag{12-5}$$

The corresponding nozzle flow energy equation using the concept of stagnation enthalpy is:

$$\frac{c^2}{2g_c(J)} + h = h_0 \tag{12-1}$$

Differentiating for a fixed set of inlet conditions (h_0 = a constant) yields:

$$\frac{cdc}{g_c(J)} + dh = 0$$

For a gas with constant specific heat, $dh = c_p dT$ and

$$\frac{cdc}{g_c} = -(J)c_p dT = \frac{-Rk}{k-1} dT \tag{12-6}$$

Combining Equations (12-5) and (12-6) yields a relationship for c:

$$\left(\frac{c}{g_c}\right)\left(\frac{c}{\rho} d\rho\right) = \frac{Rk}{k-1} dT$$

or

$$c^2 = \left[\frac{Rkg_c}{k-1}\right] \rho \frac{dT}{d\rho} \tag{12-7}$$

The isentropic relationship can be written as:

$$pv^k = \frac{p}{\rho^k} = \text{constant}$$

NOZZLES

while the ideal gas relationship can be expressed as:

$$pv = \frac{p}{\rho} = RT$$

Combining these relationships yields the expression

$$RT\rho^{1-k} = \text{constant}$$

Differentiating then yields:

$$R\rho^{1-k}\,dT + RT(1-k)\rho^{-k}\,d\rho = 0$$

giving

$$\frac{dT}{d\rho} = -(1-k)\frac{T}{\rho}$$

Substitution of the above expression for the term $dT/d\rho$ in the expression for sonic velocity, Equation (12-7), gives:

$$c^2 = \left[\frac{Rkg_c}{k-1}\right]\left[\frac{-T}{\rho}(1-k)\right]\rho$$
$$= kRTg_c$$

or

$$c = \sqrt{kRTg_c} \text{ ft/sec (m/sec)} \tag{12-8}$$

Gas velocity (V) relative to the sonic velocity (c) is given by the Mach number:

$$N_m = \frac{V}{c} \tag{12-9}$$

where

$N_m < 1.0$ subsonic flow
$N_m = 1.0$ sonic flow
$N_m > 1.0$ supersonic flow.

Example 12-3: Helium, $k = 1.66$, at a temperature of 130°C is flowing at a local speed of 1500 m/sec. Determine the following:
(a) the local sonic velocity, m/sec
(b) the local Mach number, N_m
(c) if the flow is subsonic or supersonic
Solution: See Table 3-I.

(a) $$c = \sqrt{kRTg_c} = \sqrt{(1.66)\left(2080\,\frac{\text{N}\cdot\text{m}}{\text{kg}\cdot\text{K}}\right)(403\text{K})\left(1\,\frac{\text{kg}\cdot\text{m}}{\text{N}\cdot\text{sec}^2}\right)}$$
$$= \underline{1179.6 \text{ m/sec}}$$

(b) $$N_m = \frac{V}{c} = \frac{1500.0}{1179.6} = \underline{1.272}$$

(c) $N_m > 1.0$; thus the local flow is supersonic.

340

12-5. Fluid Flow and Nozzle Shape. In the preceding articles it has been assumed that the process accompanying the passage of the fluid through a nozzle or diffuser might be controlled. This is only partly true and then only when the nozzle or diffuser shape suits the fluid and its flow characteristics. Where the fluid is a liquid, i.e., for incompressible flow in which the change of specific volume with pressure is negligible, the proper shape may be deduced from the basic equations and certain physical observations. For this purpose Equation (1-20) may be written as

$$\frac{V_2^2 - V_1^2}{2g_c} = p_1 v_1 - p_2 v_2 = v(p_1 - p_2)$$

under the conditions of Art. 12-3 and since, for frictionless, incompressible flow,

$$u_2 = u_1 \quad \text{and} \quad v_2 = v_1 = v$$

Thus for incompressible flow the increase in kinetic energy across a nozzle is produced solely by the imposed pressure difference. Liquid nozzles usually discharge against atmospheric pressure; hence the pressure difference is increased by raising the pressure at the nozzle entrance and results in increased exit velocity. The continuity equation may take the form

$$\dot{M}v = AV$$

in which \dot{M} and v are constant. It is apparent that for a liquid nozzle the area A must decrease in order to increase the fluid velocity, V. A liquid nozzle must therefore be convergent, as shown in Fig. 12-2. Also, for any

FIGURE 12-2

particular liquid nozzle the exit area is fixed; thus the fluid velocity and the mass flow rate will increase with increased pressure drop across the nozzle. For a liquid diffuser the area must increase in order to decrease the fluid velocity, V, with a resulting increase in discharge pressure.

In the case of *compressible flow*, where the fluid is a gas or a vapor, the density will vary widely with temperature and pressure, so the analysis is inherently more complex than in the case of incompressible flow. The shape of a gas or vapor nozzle may be deduced by considering continuity, the nozzle energy equation and the isentropic relationship for an ideal gas. Differentiating the continuity equation yields:

$$\rho A dV + AV d\rho + \rho V dA = 0$$

or

$$\frac{dA}{A} = -\frac{dV}{V} - \frac{d\rho}{\rho} \qquad (12\text{-}10)$$

The derivative term for density can be written using Equation (12-7) as:

$$\frac{d\rho}{\rho} = \left(\frac{Rkg_c}{k-1}\right)\frac{dT}{c^2} \qquad (12\text{-}7)$$

Using the nozzle flow energy equation (12-2) and the definition of Mach number, Equation (12-9), the above density expression becomes:

$$\frac{d\rho}{\rho} = -\frac{N_m{}^2}{V} dV$$

Substituting this result back into Equation (12-10) yields:

$$\frac{dA}{A} = -[1 - N_m{}^2]\frac{dV}{V} \qquad (12\text{-}11)$$

A complete development of the above can be found in Appendix D.

Equation (12-11) expresses the relationship between nozzle or diffuser shape and corresponding flow velocity. For an increasing flow velocity $dV > 0$ and for a decreasing velocity $dV < 0$, while for a decreasing area $dA < 0$ and for an increasing area $dA > 0$. When the local Mach number is less than unity (subsonic flow), the velocity and area will vary in opposite sense. For this case, an increase in downstream velocity requires a decrease in area and vice versa. When the local Mach number is greater than unity (supersonic flow), the velocity and area vary in the same sense. Hence an increase in velocity requires an increase in area—and the converse. Table 12-I summarizes the conditions for various flow and area combinations.

For sonic conditions the Mach number N_m is unity, and Equation (12-11) becomes

$$\frac{dA}{A} = -[1 - 1]\frac{dV}{V} = 0$$

NOZZLES

which verifies the statement made without proof in Art. 12-4 to the effect that, when sonic velocity occurs, it characteristically occurs at the point of minimum cross-sectional area where $dA = 0$.

For a subsonic flow to be accelerated by a nozzle to a supersonic flow, Table 12-I demonstrates that the area must first decrease and then increase. In addition the flow at the minimum, or throat, area will have to be sonic, i.e., $N_m = 1.0$, for supersonic flow to occur downstream.

TABLE 12-I

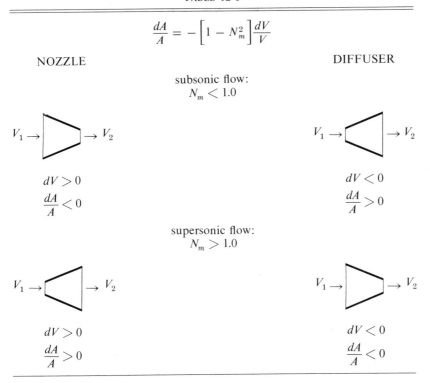

12-6. Nozzle Flow and Stagnation Conditions. It is convenient to express nozzle or diffuser flow for gases in terms of the stagnation state. The nozzle flow energy equation for ideal gas flow is written as:

$$c_p(T_2 - T_1) = \frac{1}{2g_c(J)}[V_1^2 - V_2^2] \qquad (12\text{-}2)$$

or

$$T_2 - T_1 = \frac{V_1^2}{2g_c c_p(J)} - \frac{V_2^2}{2g_c c_p(J)}$$

NOZZLES

Since $(J)c_p = \dfrac{Rk}{k-1}$, $c^2 = Rkg_cT$ and $N_m = \dfrac{V}{c}$,

$$T_2 - T_1 = \dfrac{V_1^2(k-1)}{2g_cRk} - \dfrac{V_2^2(k-1)}{2g_cRk}$$

$$= \dfrac{V_1^2(k-1)T_1}{2g_cRkT_1} - \dfrac{V_2^2(k-1)T_2}{2g_cRkT_2}$$

$$= N_{m_1}^2\left(\dfrac{k-1}{2}\right)T_1 - N_{m_2}^2\left(\dfrac{k-1}{2}\right)T_2$$

Dividing by T_1:

$$\dfrac{T_2}{T_1} - 1 = \left(\dfrac{k-1}{2}\right)N_{m_1}^2 - \dfrac{T_2}{T_1}\left(\dfrac{k-1}{2}\right)N_{m_2}^2$$

Letting the stagnation state be represented by the condition at 1 ($T_1 = T_0$, $V_1 = 0$, and $N_m = 0$) and any condition in the flow by state 2 gives:

$$\dfrac{T}{T_0} - 1 = -\dfrac{T}{T_0}\left(\dfrac{k-1}{2}\right)N_m^2$$

and

$$\dfrac{T_0}{T} = \left[1 + \left(\dfrac{k-1}{2}\right)N_m^2\right] \tag{12-12}$$

Using the isentropic relations, similar expressions for pressure and density can be written as:

$$\dfrac{p_0}{p} = \left(\dfrac{T_0}{T}\right)^{\frac{k}{k-1}} = \left[1 + \left(\dfrac{k-1}{2}\right)N_m^2\right]^{\frac{k}{k-1}} \tag{12-13}$$

$$\dfrac{\rho_0}{\rho} = \left(\dfrac{T_0}{T}\right)^{\frac{1}{k-1}} = \left[1 + \left(\dfrac{k-1}{2}\right)N_m^2\right]^{\frac{1}{k-1}} \tag{12-14}$$

The values of p, T, and ρ in a flow can be determined with a knowledge of k, the local Mach number, N_m, and the flow stagnation conditions.

Example 12-4: Helium has a value of $k = 1.66$. For a Mach number, N_m equal to 1.0, determine:
(a) T/T_0
(b) p/p_0
(c) ρ/ρ_0
Solution:

(a)
$$\dfrac{T_0}{T} = \left[1 + \left(\dfrac{k-1}{2}\right)N_m^2\right] \tag{12-12}$$

$$\dfrac{T_0}{T} = \left[1 + \dfrac{1.66 - 1}{2}\right] = 1.330$$

NOZZLES

(b)

$$\frac{T}{T_0} = 0.7519$$

or

$$\frac{p_0}{p} = \left[1 + \left(\frac{k-1}{2}\right)N_m^2\right]^{\frac{k}{k-1}} \qquad (12\text{-}13)$$

$$\frac{p_0}{p} = \left[1 + \left(\frac{1.66-1}{2}\right)1\right]^{\frac{1.66}{0.66}} = 2.0488$$

or

$$\frac{p}{p_0} = 0.4881$$

(c)

$$\frac{\rho_0}{\rho} = \left[1 + \left(\frac{k-1}{2}\right)N_m^2\right]^{\frac{1}{k-1}} \qquad (12\text{-}14)$$

$$\frac{\rho_0}{\rho} = \left[1 + \left(\frac{1.66-1}{2}\right)\right]^{\frac{1}{0.66}} = 1.540$$

or

$$\frac{\rho}{\rho_0} = 0.6494$$

12-7. Critical Flow with Compressible Fluids. To accelerate a subsonic compressible gas flow with a nozzle, it is necessary to reduce the area in accordance with the discussion of Art. 12-5. If sonic velocity is attained, it will occur at the throat and only at the throat. The pressure that accompanies sonic velocity is designated the **critical pressure** and given the symbol p_c. The corresponding ratio p_c/p_0 *is the critical pressure ratio.* Although the value of p_c is variable, the critical pressure ratio is observed

TABLE 12-II

Name of working substance	$k = \dfrac{c_p}{c_v}$	$\dfrac{p_c}{p_0} = \left(\dfrac{2}{k+1}\right)^{k/(k-1)}$
Gases:		
Helium	1.66	.49
Air	1.40	.53
Oxygen	1.40	.53
Hydrogen	1.40	.53
Vapors:		
Ammonia (NH_3)	1.31	.54
Carbon dioxide (CO_2)	1.28	.55
Sulphur dioxide (SO_2)	1.25	.55
Methyl chloride (CH_3Cl)	1.20	.56
Steam, initial state, wet	1.13	.58
Steam, initial state, superheated	1.30	.55

to be a characteristic property of any compressible fluid; and this ratio may be expressed mathematically using Equation (12-13) with $N_m = 1.0$ as

$$\frac{p_c}{p_0} = \left(\frac{2}{k+1}\right)^{\frac{k}{k-1}} \qquad (12\text{-}15)$$

(Numerical values of the critical pressure ratio for several common fluids are listed in Table 12-II.) The lowest pressure to which a compressible fluid can be expanded in a convergent nozzle is, therefore,

$$p_c = p_0 \left(\frac{2}{k+1}\right)^{\frac{k}{k-1}}$$

and the corresponding sonic velocity of a gas in such a nozzle is

$$c = \sqrt{kg_c RT_c} \text{ ft/sec (m/sec)} \qquad (12\text{-}8)$$

in which

$$T_c = T_0 \left(\frac{2}{k+1}\right) \qquad (12\text{-}12)$$

A normalized plot of pressure through a converging-diverging nozzle is shown in Fig. 12-3. For the case of no flow there will be no change in

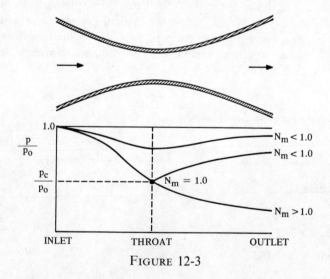

FIGURE 12-3

pressure through the nozzle. If the pressure at the throat area is less than the stagnation pressure but greater than the critical pressure, the flow will be subsonic in the throat, and flow through the downstream diverging

NOZZLES

section will be decelerated with a corresponding increase in pressure. If the throat pressure is equal to the critical pressure, the flow is termed critical.

For either convergent or convergent-divergent nozzles, **critical flow** is *established when the nozzle exit pressure is equal to or less than the critical pressure.* If the nozzle is convergent, the critical pressure will occur in the exit section, i.e., the section of minimum area. If the nozzle is convergent-divergent, the critical pressure *is found to occur in the* **throat,** again at the section of minimum area. In either case critical flow establishes the maximum mass rate of flow through the nozzle. And in either case the critical condition is a function of the stagnation state of the fluid at the nozzle entrance. As stated above, the convergent nozzle cannot induce velocities exceeding an imposed sonic value; and at one time it was thought that this sonic velocity was the ultimate for any compressible fluid. However, when expansion is continued in a divergent section, the velocity can continue to increase to supersonic values as the pressure decreases from the critical value to the exit pressure. It is recognized that critical flow may occur in a convergent-divergent nozzle when the exit pressure is greater than the critical pressure with a return to the subsonic regime as indicated in Fig. 12-3. However, this condition is not encountered in any applications of this text and will not be considered.

Example 12-5: Air at a pressure of 60 psia with a temperature of 1340°F is to be expanded isentropically to a pressure of 15 psia in an ideal nozzle. (A) Calculate or determine (a) the type nozzle required; (b) the critical velocity, ft/sec; (c) the mass rate of flow for a minimum nozzle diameter of 0.50 in., lb/min. (B) If the nozzle requires a divergent section, determine (d) the maximum (exit) velocity developed in the nozzle, ft/sec; (e) the required exit diameter in inches.

Solution:

$$p_1 = p_0$$

$$\frac{p_c}{p_0} = 0.53 \text{ (Table 12-II)}$$

$$p_t = p_c = 0.53 p_1 = 0.53 \times 60 = 31.8 \text{ psia}$$

(a) The nozzle must be convergent-divergent because p_2 is less than p_c.

(b)
$$T_c = T_1 \left(\frac{p_c}{p_1}\right)^{\frac{k-1}{k}} = 1800(0.53)^{\frac{1.4-1}{1.4}} = 1501°R$$

$$V_c = \sqrt{2g_c J} \sqrt{c_p(T_1 - T_c)} = 223.8 \sqrt{0.24(1800 - 1501)}$$
$$= 1896 \text{ ft/sec}$$

(c)
$$v_t = \frac{RT_c}{p_c} = \frac{53.3 \times 1501}{144 \times 31.8} = 17.47 \text{ ft}^3/\text{lb}$$

$$A_t = \frac{\pi d_t^2}{4 \times 144} = \frac{\pi (0.50)^2}{4 \times 144} = 0.001364 \text{ ft}^2$$

$$\dot{M} = \frac{A_t V_t}{v_t} = 0.001364 \times \frac{1896}{17.47} \times 60 = \underline{8.9 \text{ lb/min}}$$

NOZZLES

(d)
$$T_2 = T_1\left(\frac{p_2}{p_1}\right)^{\frac{k-1}{k}} = 1800\left(\frac{15}{60}\right)^{\frac{1.4-1}{1.4}} = 1211°R$$

$$V_{\text{exit}} = \sqrt{2g_cJ}\sqrt{c_p(T_1 - T_2)} = 223.8\sqrt{0.24(1800 - 1211)}$$
$$= \underline{2661 \text{ ft/sec}}$$

(e)
$$v_{\text{exit}} = \frac{RT_2}{p_2} = \frac{53.3 \times 1211}{144 \times 15} = 29.88 \text{ ft}^3/\text{lb}$$

$$d_{\text{exit}} = \sqrt{\frac{4\dot{M}v}{\pi V}} \times 144 = \sqrt{\frac{4 \times 8.9 \times 29.88 \times 144}{\pi \times 2661 \times 60}} = \underline{0.553 \text{ in.}}$$

12-8. Performance of the Actual Nozzle. Velocity Coefficient and Nozzle Efficiency. The actual nozzle must be as short as possible for two principal reasons: (1) because the total effect of friction is directly influenced by the length of the nozzle, and (2) because it is desirable to keep nozzle spaces in machinery to a minimum in the interests of overall space and weight requirements. In addition, the exact curves of an ideal nozzle are very difficult to produce with available cutting machinery. For these reasons the profile of the actual nozzle usually consists of a radial section from **entrance** to **throat** (or section of minimum area) and a straight line taper from throat to exit. To prevent excessive turbulence in the divergent section the angle of divergence should not be greater than 12 degrees. The cross-sectional shape of the actual nozzle illustrated in Fig. 12-4 may be circular, square or rectangular.

No matter how perfectly designed or how carefully finished a nozzle may be, friction is always present. The convergent portion of the nozzle is relatively short (Fig. 12-4), and the frictional effect in this section is usually small. In the divergent section the effect of friction is more pronounced because of the progressively increasing velocities and the possibility of

FIGURE 12-4

NOZZLES

separation of the fluid from the channel with subsequent turbulence or burble effect. The resultant effect upon exit velocity and kinetic energy is estimated on the basis of empirical coefficients.

One method of estimating the probable performance of a nozzle employs a **velocity coefficient** which *is applied to the ideal velocity* for the given flow conditions. Thus,

$$C_n = \frac{V_{actual}}{V_{ideal}} \tag{12-16}$$

or, for a gas with negligible entrance velocity,

$$V_{actual} = C_n V_{ideal} = C_n \sqrt{2g_c(J)c_p(T_1 - T_2)} \tag{12-3a}$$

A second generally employed method utilizes a nozzle efficiency that is based upon the actual kinetic energy at the nozzle exit. The energy available to the nozzle is the isentropic change of enthalpy of the fluid between nozzle entrance conditions and the nozzle exit pressure. That is, the maximum performance of any nozzle is the kinetic energy which would be produced by an ideal nozzle under identical flow conditions. Thus,

$$A.E. = K.E._{ideal} = h_1 - h_2 \text{ Btu/lb (J/kg)} \tag{12-17}$$

in which h_1 = enthalpy at the nozzle entrance, Btu/lb (J/kg)
h_2 = enthalpy after isentropic expansion to the nozzle exit pressure, Btu/lb (J/kg)

A measure of the performance of an actual nozzle is the degree to which it is able to approach ideal performance. On this basis, **nozzle efficiency** is defined as

$$\eta_n = \frac{K.E._{actual}}{K.E._{ideal}} = \frac{K.E._{actual}}{A.E.} \tag{12-18}$$

As stated in Art. 12-3, the energy equation for adiabatic flow applies to both the ideal and the actual nozzle. Although in the latter case the flow is adiabatic, it is inherently irreversible. The result is an increase of entropy of the fluid as it passes through the nozzle with an accompanying increase in the exit enthalpy. Thus the change of enthalpy through an actual nozzle will in all cases be less than the energy available in the given situation. Both the ideal and the actual nozzle processes are illustrated on h-s coordinates in Fig. 12-5. (As might be expected, there is a high degree of similarity between illustrations of gas processes on T-s and h-s coordinates because enthalpy is a direct function of temperature.) As shown, the available energy is $h_1 - h_2$, and the actual change of enthalpy is $h_1 - h_2'$.

NOZZLES

FIGURE 12-5

Since from Chapter 5

$$K.E._{\text{actual}} = h_1 - h_2' \tag{12-19}$$

the nozzle efficiency may be conveniently stated in terms of fluid properties as

$$\eta_n = \frac{K.E._{\text{actual}}}{A.E.} = \frac{h_1 - h_2'}{h_1 - h_2} \tag{12-20}$$

The difference between the actual change of enthalpy and the change of enthalpy for the ideal nozzle is known as the nozzle loss, or reheat. Reheat is frequently designated by the symbol y; thus,

$$\begin{aligned} y_n &= (h_1 - h_2) - (h_1 - h_{2'}) \\ &= h_{2'} - h_2 \text{ Btu/lb (J/kg)} \end{aligned} \tag{12-21}$$

A third method which is sometimes employed is primarily applicable to gases and is based upon the irreversible adiabatic process. In this case the polytropic relationship, $pv^n = $ constant, is employed, and n replaces k in the flow equations.

For the purpose of this text either a velocity coefficient or a nozzle efficiency will be used, since there is a direct relationship between these performance coefficients. For negligible entrance velocity,

$$V_{2(\text{actual})} = C_n \sqrt{2g_c(J)(h_1 - h_2)}$$

and,

$$h_1 - h_{2'} = \eta_n(h_1 - h_2) = K.E._{\text{actual}} = \frac{V_{2(\text{actual})}^2}{2g_c(J)}$$

NOZZLES

Combining the above relationships,

$$V_{2(\text{actual})} = \sqrt{2g_c(J)\eta_n(h_1 - h_2)} = C_n\sqrt{2g_c(J)(h_1 - h_2)}$$

and,

$$\eta_n = C_n^2$$

Actual values of both velocity coefficient and nozzle efficiency are based upon experience and result from innumerable tests over a wide range of operating conditions. In general for a well-made nozzle operating under the conditions for which it was designed, C_n will vary between 0.96 and 0.98, and 0.97 is a representative average value. Corresponding values of η_n range from 92 to 96 percent with 94 percent as a good average for general use.

Example 12-6: The nozzles of a gas turbine receive air at a pressure of 54 psia with a temperature of 1480°F and discharge to the impulse blades at a pressure of 18 psia. For a nozzle velocity coefficient of 0.97 calculate (a) the air velocity leaving the nozzles, ft/sec; (b) the corresponding nozzle efficiency, percent; (c) the kinetic energy delivered to the turbine blades, Btu/lb.

Solution:

$$V_1 \to 0$$

$$T_1 = 1480 + 460 = 1940°R$$

$$T_{2(\text{ideal})} = T_1\left(\frac{p_2}{p_1}\right)^{\frac{k-1}{k}} = 1940\left(\frac{18}{54}\right)^{\frac{1.4-1}{1.4}} = 1417°R$$

$$A.E. = h_1 - h_2 = c_p(T_1 - T_2) = 0.24(1940 - 1417)$$
$$= 125.5 \text{ Btu/lb}$$

(a)
$$V_{2(\text{actual})} = C_n\sqrt{2g_c J c_p(T_1 - T_2)} = 223.8 C_n\sqrt{A.E.}$$
$$= 223.8 \times 0.97\sqrt{125.5} = \underline{2432 \text{ ft/sec}}$$

(b)
$$\eta_n = C_n^2 = (0.97)^2 = \underline{94.1\%}$$

(c)
$$K.E._{\text{actual}} = \eta_n(A.E.) = 0.941 \times 125.5$$
$$= \underline{118 \text{ Btu/lb}}$$

12-9. Flow of Vapors in Nozzles. The flow of superheated vapors in nozzles is similar in all respects to the flow of gases. However, when the steam is initially saturated, or when the initial degree of superheat is relatively small such that the expansion process crosses the saturation line and continues into the wet region, *the performance of the nozzle is materially affected by the phenomenon of* **supersaturation.**

Supersaturation occurs when a vapor expands into the wet region with such rapidity that condensation, i.e., the actual formation of liquid particles, lags behind the pressure change, and the vapor exists temporarily in an unstable condition. In this supersaturated condition the vapor will have a greater velocity and a lower temperature than for a corresponding stable

expansion. For this reason the mass rate of flow may be greater than that provided by stable isentropic expansion in those cases of supersaturation in which the critical pressure is in the wet region. It is generally believed that supersaturation persists to the throat of a nozzle with the expansion in the divergent section progressively approaching equilibrium conditions. In any event, *supersaturation increases the degree of irreversibility,* i.e., contributes to the increase of entropy during the actual expansion. For this reason the nozzle efficiency with supersaturated flow is somewhat less than that for gases or superheated vapors.

12-10. Steam Flow in Nozzles. The general equations developed in earlier sections of this chapter are directly applicable to steam flow. However, the fluid properties require special attention and the working equations are repeated in this article in correlation with useful vapor diagrams (see Fig. 12-6).

Figure 12-6

For ideal expansion in a nozzle,

$$\frac{V_2^2 - V_1^2}{2g_c(J)} = h_1 - h_2 \text{ Btu/lb (J/kg)}$$

NOZZLES

Similarly for any actual nozzle expansion the increase in kinetic energy is

$$\frac{V_{2'}^2 - V_1^2}{2g_c(J)} = h_1 - h_{2'} = \eta_n(h_1 - h_2) \text{ Btu/lb (J/kg)} \quad (12\text{-}19a)$$

and, for negligible entrance velocity, the exit velocity is

$$V_{\text{actual}} = V_{2'} = \sqrt{2g_c(J)\eta_n(h_1 - h_2)} \text{ ft/sec (m/sec)} \quad (12\text{-}3b)$$

In all cases the equation of continuity applies. That is,

$$\dot{M} = \frac{A_1 V_1}{v_1} = \frac{A_t V_t}{v_t} = \frac{A_2 V_2}{v_2} \cdots \text{ lb/sec (kg/sec)}$$

Steam, like air, is a compressible fluid and exhibits similar compressible flow characteristics. The steam nozzle, too, may be either convergent or convergent-divergent. For either nozzle type maximum mass flow accompanies the establishment of critical pressure in the throat section. Although the equation $p_c/p_0 = (2/k + 1)^{(k/k-1)}$ was initially developed for gases, it is applied to steam and other vapors on the basis of average values of k. For steam that is highly superheated entering the nozzle the critical pressure ratio is approximately 0.55, based upon an average value of k of 1.30. *When the steam entering the nozzle contains even a small amount of moisture the critical pressure ratio becomes substantially 0.58* for a corresponding value of k of 1.13. However, the intermediate situation in which the entering steam is saturated or only moderately superheated requires particular attention. When the expansion is such that the inception of a phase change would normally precede the establishment of the critical pressure, supersaturation is established in the nozzle throat. Experiments by Stodola indicate that under these circumstances there is little if any moisture formed ahead of the throat, but that sudden condensation immediately follows the establishment of critical flow. Since the physical characteristics of supersaturated and superheated steam are closely similar (because of the absence of moisture in both conditions), $k = 1.30$ is presumed by this text to apply to the supersaturated condition. Thus *for both superheated and supersaturated steam the critical pressure ratio which will be used herein is 0.55.*

Having established the critical flow criteria for any steam nozzle it remains for us to calculate the mass flow rate and the discharge velocity and kinetic energy. Although polytropic equations can be developed for these calculations, they have only limited applicability. For this reason the solution of all problems in this text will employ the general equations in conjunction with steam table data.

NOZZLES

Example 12-7: A converging-diverging nozzle receives steam at a pressure of 380 psia with a temperature of 480°F and expands it to a pressure of 50 psia. Assuming a velocity coefficient of 0.98 for the supersaturated throat condition and an overall nozzle efficiency of 92 percent, calculate (a) the actual throat velocity, ft/sec; (b) the mass rate of flow for a throat diameter of 0.50 in., lb/min; (c) the actual kinetic energy available at the nozzle exit, Btu/lb; (d) the actual nozzle exit velocity, ft/sec; (e) the required nozzle exit diameter, inches; (f) the length of the divergent section of the nozzle, inches, for an included angle of 12 degrees between the nozzle sides.

Solution: From steam tables for a pressure of 380 psia and a temperature of 480°F:

$$h_1 = 1234.4 \text{ Btu/lb}, \quad s_1 = 1.5220 \text{ Btu/lb} \cdot °R$$

$$p_t = p_c = 0.55 p_1 = 0.55 \times 380 = 209 \text{ psia}$$

$$s_{t(\text{ideal})} = s_1, \quad m = \frac{s_g - s_t}{s_{fg}} = \frac{1.5427 - 1.5220}{0.9941} = 2.08\%$$

$$h_{t(\text{ideal})} = h_g - m h_{fg} = 1199.9 - 0.0208 \times 840.3 = 1182.4 \text{ Btu/lb}$$

(a)
$$V_t = C_n V_{\text{ideal}} = 223.8 C_n \sqrt{h_1 - h_{t(\text{ideal})}}$$
$$= 223.8 \times 0.98 \sqrt{1234.4 - 1182.4} = \underline{1582 \text{ ft/sec}}$$

(b)
$$\eta_n' = C_n^2 = (0.98)^2 = 0.9604$$

$$h_1 - h_{t(\text{actual})} = \eta_n'(h_1 - h_{t\text{ideal}}) = 0.9604(1234.4 - 1182.4)$$
$$= 49.9 \text{ Btu/lb}$$

$$h_{t(\text{actual})} = h_1 - 49.9 = 1234.4 - 49.9 = 1184.5 \text{ Btu/lb}$$

$$m_{t(\text{actual})} = \frac{h_g - h_{t(\text{actual})}}{h_{fg}} = \frac{1199.9 - 1184.5}{840.3} = 1.83\%$$

$$v_{t(\text{actual})} = v_g - m_t v_{fg} = 2.174 - 0.0183 \times 2.156$$
$$= 2.135 \text{ ft}^3/\text{lb}$$

$$A_t = \frac{0.1964}{144} \text{ sq ft (see Ex. 12-5, Art. 12-7)}$$

$$\dot{M} = \frac{A_t V_t}{v_t} = \frac{0.1964 \times 1582}{144 \times 2.135} \times 60 = \underline{60.64 \text{ lb/min}}$$

(c)
$$s_{2(\text{ideal})} = s_1, \quad m_2 = \frac{s_g - s_{2(\text{ideal})}}{s_{fg}} = \frac{1.6589 - 1.5220}{1.2476} = 10.97\%$$

$$h_{2(\text{ideal})} = h_g - m_2 h_{fg} = 1174.4 - 0.1097 \times 924.2 = 1073.0 \text{ Btu/lb}$$

$$K.E._{\text{actual}} = h_1 - h_{2'} = \eta_n(h_1 - h_2) = 0.92(1234.4 - 1073.0)$$
$$= \underline{148.5 \text{ Btu/lb}}$$

(d)
$$V_{\text{actual}} = V_{2'} = 223.8 \sqrt{h_1 - h_{2'}} = 223.8 \sqrt{148.5}$$
$$= \underline{2727 \text{ ft/sec}}$$

(e)
$$h_{2'} = h_1 - K.E._{\text{actual}} = 1234.4 - 148.5 = 1085.9 \text{ Btu/lb}$$

$$m_{2'} = \frac{h_g - h_{2'}}{h_{fg}} = \frac{1174.4 - 1085.9}{924.2} = 9.58\%$$

$$v_{2'} = v_g - m_2 v_{fg} = 8.518 - 0.0958 \times 8.501 = 7.704 \text{ ft}^3/\text{lb}$$

$$d_{\text{exit}} = \sqrt{\frac{4 \dot{M} v_{2'}}{\pi V_{\text{actual}}}} = \sqrt{\frac{4 \times 60.64 \times 7.704}{60 \times \pi \times 2727} \times 144} = \underline{0.724 \text{ in.}}$$

NOZZLES

(f)
$$\tan\frac{\alpha}{2} = \frac{r_{exit} - r_t}{l}; \quad l = \frac{r_{exit} - r_t}{\tan\frac{\alpha}{2}}$$

$$= \frac{0.362 - 0.25}{0.1051} = 1.07 \text{ in.}$$

12-11. Overexpansion and Underexpansion in Nozzles. It is recognized that the use of a convergent nozzle is limited to those circumstances in which the nozzle discharge pressure is not less than the critical pressure as established by the fluid properties entering the nozzle. Within this limitation the discharge velocity and kinetic energy and the mass rate of flow all vary directly with the change of enthalpy across the nozzle, i.e., with the available energy, the nozzle efficiency remaining substantially constant under all operating conditions. Consequently, a convergent nozzle will give satisfactory performance under a variety of imposed conditions.

In the case of the convergent-divergent nozzle, the length of the divergent section must be carefully considered. For any given operating conditions the flow rate will be established by the throat area (since this type nozzle will not be used unless the discharge pressure is less than critical). However, since the angle of divergence cannot exceed 12 degrees, the proper area for full expansion may or may not be provided, depending upon the length of this section of the nozzle. If the divergent section is too short, the pressure in the exit section will be greater than the pressure against which the nozzle discharges, since the area has not increased sufficiently to allow full expansion within the nozzle. *Such a nozzle is said to be **underexpanded**;* and while it will deliver the required mass of fluid, discharge will be accompanied by free expansion and loss of available energy. If the length of the divergent section is greater than necessary, the fluid first expands below the nozzle discharge pressure and then recompresses ahead of the nozzle exit. *Such a nozzle is **overexpanded**,* and the recompression is accompanied by fluid shock and turbulence. As the lesser of two evils it is preferable to have a slightly underexpanded nozzle and thus avoid the shock losses of recompression. But whether it is underexpanded or overexpanded, the result of improper divergent length is a loss of efficiency and a decrease in the amount of useful kinetic energy produced.

NOZZLES

Problems

12-1. Air enters an ideal converging-diverging nozzle at a pressure of 73.5 psia with a temperature of 1400°F and negligible approach velocity. For isentropic expansion to an exit pressure of 14.7 psia calculate (a) the temperature of the air leaving the nozzle, °F; (b) the kinetic energy of the air leaving the nozzle, Btu/lb; (c) the velocity of the air leaving the nozzle, ft/sec.

Ans. (a) 714°F; (b) 164.6 Btu/lb; (c) 2870 ft/sec

12-2. The pressure of the air entering an ideal convergent nozzle is 73.5 psia, the temperature is 1400°F and the velocity of approach is negligible. The nozzle discharges against a pressure of 14.7 psia. What is the nozzle exit velocity when expansion in the nozzle is isentropic, ft/sec?

Ans. 1930 ft/sec

12-3. Air enters a convergent-divergent nozzle having stagnation conditions of 65°C and 285 kPa. The Mach number of the throat is 0.77. For isentropic expansion through the nozzle calculate (a) the throat pressure, kPa; (b) the throat temperature, K; (c) the throat velocity, m/sec. All pressures are absolute.

Ans. (a) 192.5 kPa; (b) 302.2 K; (c) 268.3 m/sec

12-4. Air enters a convergent-divergent nozzle at a pressure of 73.5 psia with a temperature of 1400°F and negligible approach velocity. The nozzle throat area is 1.373 in.2 and the exit pressure is 14.7 psia. For isentropic expansion throughout the nozzle calculate (a) the throat pressure, psia; (b) the velocity in the throat, ft/sec; (c) the rate of air flow through the nozzle, lb/sec; (d) the exit area necessary for the isentropic expansion, in.2

Ans. (a) 39.0 psia; (b) 1930 ft/sec; (c) 1.25 lb/sec; (d) 1.854 in.2

12-5. A convergent-divergent nozzle receives air at 88.2 psia with a temperature of 1340°F and negligible approach velocity, and expands it to a discharge pressure of 14.7 psia. If the nozzle efficiency is 94 percent, calculate (a) the energy available to the nozzle, Btu/lb; (b) the actual change of enthalpy across the nozzle, Btu/lb; (c) the nozzle exit velocity, ft/sec; (d) the nozzle velocity coefficient.

Ans. (a) 173.1 Btu/lb; (b) 162.7 Btu/lb; (c) 2855 ft/sec; (d) 0.97

12-6. Air enters a diffuser of a jet engine with a velocity of 1800 ft/sec relative to the aircraft. The intake pressure is 1.05 psia and intake temperature is −70°F. Assuming isentropic compression in the diffuser calculate (a) sonic velocity at the inlet conditions, ft/sec; (b) the inlet Mach number; (c) the stagnation temperature, °R; (d) the stagnation pressure, psia.

Ans. (a) 968 ft/sec; (b) 1.86; (c) 659.8°R; (d) 6.61 psia

12-7. Air is accelerated by a convergent nozzle to sonic conditions at the exit. The exit conditions are 14.7 psia and 75°F. For a mass flow rate of 5 lb/sec calculate (a) the exit velocity, ft/sec; (b) the exit density, lb/ft^3; (c) the nozzle exit diameter, in.

Ans. (a) 1134 ft/sec; (b) 0.07423 lb/ft^3; (c) 3.30 in.

12-8. Air enters an ideal diffuser at a speed of 1500 ft/sec at 22 psia and 650°F. For an isentropic compression calculate (a) the inlet Mach number; (b) the stagnation temperature, °R; (c) the stagnation pressure, psia; (d) the ratio of inlet to exhaust area necessary to reduce velocity to 15 ft/sec.

Ans. (a) 0.918; (b) 1297.1°R; (c) 38.0 psia; (d) 0.01476

NOZZLES

12-9. Air is discharged from a convergent-divergent nozzle at a Mach number of 1.732. The exit pressure is 15 psia and the temperature is 150°F. For isentropic expansion and a negligible inlet velocity calculate (a) the stagnation temperature, °R; (b) the stagnation pressure, psia; (c) the throat temperature, °R; (d) the throat pressure, psia.

Ans. (a) 976°R; (b) 77.7 psia; (c) 814°R; (d) 41.2 psia

12-10. A convergent nozzle receives steam at a pressure of 200 psia with a temperature of 480°F and negligible approach velocity and discharges against a pressure of 50 psia. The minimum nozzle area is 1.373 in.2 Assume isentropic expansion in the nozzle and calculate (a) the temperature of the steam leaving the nozzle, °F; (b) the velocity of the steam leaving the nozzle, ft/sec; (c) the rate of steam flow through the nozzle, lb/sec.

Ans. (a) 358°F; (b) 1660 ft/sec; (c) 3.76 lb/sec

12-11. A convergent-divergent nozzle that has a throat area of 1.373 in.2 receives steam at a pressure of 200 psia with a temperature of 480°F and negligible approach velocity, and discharges against a pressure of 50 psia. Assume isentropic expansion through the nozzle and calculate (a) the temperature of the steam leaving the nozzle, °F; (b) the rate of steam flow through the nozzle, lb/sec; (c) the moisture of the steam leaving the nozzle, percent; (d) the velocity of the steam leaving the nozzle, ft/sec.

Ans. (a) 281°F; (b) 3.76 lb/sec; (c) 3.74%; (d) 2430 ft/sec

12-12. A group of turbine nozzles receives steam at 600 psia with a temperature of 800°F and discharges to the blades at a pressure of 300 psia. Assume a nozzle efficiency of 94 percent and calculate (a) the energy available to the nozzles, Btu/lb; (b) the kinetic energy of the steam leaving the nozzle group, Btu/lb; (c) the nozzle reheat, Btu/lb; (d) the velocity of the steam entering the blades, ft/sec; (e) the state of the steam entering the turbine blades, psia and °F.

Ans. (a) 84.9 Btu/lb; (b) 79.8 Btu/lb; (c) 5.1 Btu/lb; (d) 2000 ft/sec; (e) 300 psia, 624.3°F

12-13. A converging nozzle has steam at its inlet with stagnation properties of 580 psia and 530°F. Assuming isentropic expansion and the steam exit conditions to be saturated vapor, calculate (a) the exit velocity, ft/sec; (b) exit pressure, psia; (c) the exit temperature, °F.

Ans. (a) 1376 ft/sec; (b) 381.3 psia; (c) 440°F

12-14. Steam enters a convergent nozzle with negligible velocity at 500 psia and 700°F. The exit area of the nozzle has a 0.5-in. diameter. The conditions of the throat are 275 psia and 560°F. Calculate (a) the available energy, Btu/lb; (b) the nozzle efficiency, percent; (c) the mass flow rate, lb/sec.

Ans. (a) 67.5 Btu/lb; (b) 92.0%; (c) 1.15 lb/sec

12-15. A nozzle has a steam flow rate of 0.8 lb/sec. The exit pressure and temperature are 40 psia and 300°F, respectively. The exit area of the nozzle is 0.45 in.2 and the nozzle efficiency is 92 percent. Determine (a) the actual nozzle exit velocity, ft/sec, (b) the inlet stagnation temperature, °F; (c) the inlet stagnation pressure, psia. (Use Mollier chart.)

Ans. (a) 2826 ft/sec; (b) 650°F; (c) 238 psia

CHAPTER 13
STEAM TURBINES

13-1. Introduction. Steam turbines currently occupy a predominant position in high power applications. They are also used extensively in smaller sizes for auxiliary drive and other special applications. For over a hundred years, the reciprocating steam engine was the sole practical steam-actuated prime mover. But in the 1880s the steam turbine emerged. Development proceeded at an astounding rate, led by Parsons in England and DeLaval in Sweden. At about the same time, Thomas Edison conceived the idea of generating and distributing electric power from large central stations. By the turn of the century, the steam turbine had won general acceptance as the preferred prime mover for the thermal generation of electric power.

Compared to the steam engine, the turbine is not inherently size-limited and can take full advantage of the thermodynamic efficiencies offered by high-temperature, high-pressure, high-vacuum steam cycle conditions, as well as the refinements of reheat and regenerative feed heating. As might be expected, the steam turbine became the prime mover of choice when the nuclear steam generator appeared on the scene. The marriage of the turbine and the constant speed generator was facilitated by the fact that both operate most efficiently at fairly high rotative speeds; thus direct coupling, shaft to shaft, is indicated. The marine turbine, on the other hand, must operate at variable speeds and drives a screw propeller which is limited to a few hundred rpm. The reasons for the marine turbine's present popularity are similar to those mentioned in connection with the land turbine, plus the perfection of reliable high capacity reduction gears to couple the high speed turbine with the low speed propeller.

The reciprocating steam engine persisted into the 1920s and had a resurgence during World War II when turbine and gear manufacturing facilities became saturated. Today the "up-and-down" steam engine has largely disappeared from the oceans of the world. The Diesel engine is enjoying increasing popularity in merchant ships, and the gas turbine is being used in certain special marine applications. However, the steam turbine remains the most commonly encountered prime mover in warships and large merchantmen.

A detailed presentation of the various mechanical features and internal workings of the steam turbine is beyond the scope of this book. In this chapter the discussion will be confined to the basic thermodynamic aspects

leading up to the overall performance of a typical large complete expansion condensing turbine.

13-2. Turbine Types. There are many types of steam turbine in use, each a tribute to the ingenuity of its designer. Many manufacturers loosely classify their products as impulse or reaction; but the dividing line is nebulous, and the nomenclature stems more from construction details than from the presence or absence of reactive or impulse effects. In practice, "reaction" turbines always have some impulse features, and "impulse" turbines often include some utilization of the reaction principle. Pending further discussion, an impulse stage will be defined as one in which the entire pressure drop of the stage takes place in the fixed nozzles. On the other hand, in a reaction stage an appreciable pressure drop occurs in the moving blades as well as the fixed. *In any case the work done stems from the force associated with the acceleration of a mass of steam within a moving vane. Since velocity is a vector quantity, the acceleration may arise from a change of velocity direction or magnitude or both.*

13-3. Impulse Blading. In an impulse turbine, a stage includes the *fixed nozzles* wherein a pressure drop occurs and the succeeding passages wherein the pressure remains constant at the nozzle exit value. An impulse stage begins at the entrance to a set of nozzles and ends at the entrance to the next succeeding set. The simplest type of impulse stage consists of a group of nozzles followed by a single row of moving blades commonly referred to as *buckets*. Since no pressure drop is to occur across the buckets, the net flow area through them ideally remains constant. Such an arrangement is shown schematically in Fig. 13-1.

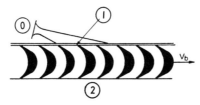

FIGURE 13-1. Simple impulse stage.

The analysis of turbine blading is best carried out by means of a *velocity vector diagram*. By determining the several velocities concerned, we can predict the work done by the buckets. Figure 13-2 shows a velocity vector diagram for the simple impulse stage shown in Fig. 13-1.

In the figure, V_1 is the absolute velocity of the jet leaving the nozzle and

STEAM TURBINES

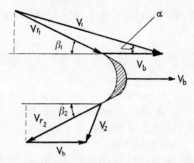

FIGURE 13-2. Velocity vector diagram, impulse blading.

α is the nozzle angle. V_b is variously called the **blade speed, bucket velocity** or **wheel speed**. By graphics V_{r_1} is the entering velocity of the jet relative to the bucket. The required inlet angle of the bucket for smooth entrance conditions will be β_1, and it will be assumed that the actual buckets are so designed. The steam jet is changed in direction by the bucket contour and will leave with a relative velocity of V_{r_2} in a direction established by the bucket exit angle, β_2. The velocity diagram is completed by combining V_{r_2} and V_b vectorially to arrive at the absolute exit velocity, V_2. For the ideal case V_{r_2} will be equal to V_{r_1}, but actually V_{r_2} will be somewhat less because of friction and other losses within the bucket.

In determining the work done in a turbine stage, it must be borne in mind that *all the work is performed in the moving blades and that the only useful force is the component in the direction of motion*. This suggests a determination of work by investigating the velocities relative to the buckets and in their plane of motion. From Newton's Second Law:

$$F = \frac{Ma}{g_c} \tag{1-2}$$

But $\qquad wk = \text{Force } (F) \times \text{Distance } (L)$

And $\qquad a = \dfrac{\Delta V}{\Delta t}; \quad L = V(\Delta t)$

Then

$$wk = FL = \frac{Ma}{g_c}L = \left(\frac{M}{g_c}\right)\left(\frac{\Delta V}{\Delta t}\right)V(\Delta t) = \left(\frac{M}{g_c}\right)(V)(\Delta V)$$

For unit mass of working substance:

$$wk = \frac{(V)(\Delta V)}{g_c} \text{ ft} \cdot \text{lbf/lb (J/kg)} \tag{13-1}$$

For the particular problem under consideration, the velocities are referred

to the blade motion so that ΔV is the total change of tangential velocity relative to the bucket and V is the wheel speed. Then:

$$\Delta V = V_{r_1} \cos \beta_1 + V_{r_2} \cos \beta_2$$

and

$$V = V_b$$

The work equation for the turbine buckets (blade work) becomes:

$$(wk)_b = \frac{V_b(V_{r_1} \cos \beta_1 + V_{r_2} \cos \beta_2)}{g_c} \text{ ft·lbf/lb (J·kg)} \qquad (13\text{-}2)$$

By writing this equation as:

$$(wk)_b = \frac{V_b(V_{r_1} \cos \beta_1) + V_b(V_{r_2} \cos \beta_2)}{g_c}$$

and performing some trigonometric operations, we can derive another valuable work equation. Referring to Fig. 13-3 and using the law of cosines:

$$V_b^2 + V_{r_1}^2 - 2V_b V_{r_1} \cos(180 - \beta_1) = V_1^2$$

But

$$\cos(180 - \beta_1) = -\cos \beta_1$$

So

$$V_b V_{r_1} \cos \beta_1 = \frac{V_1^2 - V_b^2 - V_{r_1}^2}{2}$$

Also

$$V_b^2 + V_{r_2}^2 - 2V_b V_{r_2} \cos \beta_2 = V_2^2$$

And

$$V_b V_{r_2} \cos \beta_2 = \frac{V_b^2 + V_{r_2}^2 - V_2^2}{2}$$

Substituting in the basic equation:

$$(wk)_b = \frac{V_1^2 - V_b^2 - V_{r_1}^2 + V_b^2 + V_{r_2}^2 - V_2^2}{2g_c}$$

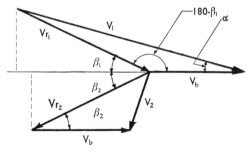

FIGURE 13-3. Velocity vector diagram, impulse blading.

Clearing and rearranging yields the final form:

$$(wk)_b = \frac{(V_1^2 - V_2^2) - (V_{r_1}^2 - V_{r_2}^2)}{2g_c} \text{ ft} \cdot \text{lbf/lb (J/kg)} \quad (13\text{-}3)$$

It should be readily apparent that the right-hand side of Equation (13-3) is made up of four *kinetic energy* terms which, in combination, yield blade work, ft · lbf/lb (J/kg). $V_1^2/2g_c$ is the kinetic energy of the entering jet and represents the available energy to the buckets $(A.E.)_b$. $V_2^2/2g_c$ is the kinetic energy of the jet leaving the buckets and is an unutilized portion of the energy available to the buckets. It represents a loss to the stage but may be partially recouped by succeeding stages if installed. The difference between $V_{r_1}^2/2g_c$ and $V_{r_2}^2/2g_c$ is the bucket loss resulting from friction and shock in the blade passages with some minor effects that are due to leakage over the blade tips. *The bucket loss results in an **increase** in the enthalpy and entropy of the steam as it passes through the buckets.* The work of the buckets can also be determined by deducting the losses from the available energy.

The *diagram efficiency*, η_b, evaluates bucket performance by comparing the blade work, as fixed by a velocity vector diagram analysis, with the energy available to the buckets. Then,

$$\eta_b = \frac{(wk)_b}{(A.E.)_b} = \frac{(V_1^2 - V_2^2) - (V_{r_1}^2 - V_{r_2}^2)}{V_1^2} \quad (13\text{-}4)$$

The friction losses through the buckets may be minimized by proper design and machining. A *velocity coefficient* is conventionally used to indicate performance in this respect. By definition:

$$\textbf{Bucket velocity coefficient} = C_b = \frac{V_{r_2}}{V_{r_1}}$$

No simple relationship exists between C_b and η_b, since the latter depends on the unutilized kinetic energy at exit as well as the bucket frictional losses.

The value of C_b decreases with increasing relative entering velocity and with an increase in the total angle through which the steam jet is turned. As a general rule, well-designed buckets will have velocity coefficients ranging from about 0.85 to 0.90. End spillage results in a slight reduction in C_b for blades below 3 inches in height.

The diagram efficiency will be improved if the velocity of the leaving jet is kept as small as possible. From the geometry of Fig. 13-3, it can be deduced that the magnitude of V_2 will probably depend upon the nozzle angle and on the relationship between the absolute entering jet velocity

and the blade speed. In order to simplify the discussion that follows, let us assume that $\beta_1 = \beta_2$ and $V_{r_1} = V_{r_2}$ as shown in Fig. 13-4. The bucket

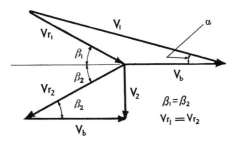

FIGURE 13-4. Simplified velocity vector diagram for impulse blading.

angles are frequently equal in small turbines, although the exit angle is usually made slightly less than the entrance angle in larger machines. For the assumed conditions,

$$V_{r_1} \cos \beta_1 = V_{r_2} \cos \beta_2$$

and Equation (13-2) becomes:

$$(wk)_b = \frac{V_b(2V_{r_1} \cos \beta_1)}{g_c} \text{ ft} \cdot \text{lbf/lb (J/kg)}$$

Also, $$(A.E.)_b = \frac{V_1^2}{2g_c} \text{ ft} \cdot \text{lbf/lb (J/kg)}$$

Then, $$\eta_b = \frac{(wk)_b}{(A.E.)_b} = \frac{2V_b(2V_{r_1} \cos \beta_1)}{V_1^2}$$

Referring to Fig. 13-4, $V_{r_1} \cos \beta_1 = V_1 \cos \alpha - V_b$.

Then, $$\eta_b = \frac{4(V_b V_1 \cos \alpha - V_b^2)}{V_1^2} = 4\left[\frac{V_b}{V_1} \cos \alpha - \left(\frac{V_b}{V_1}\right)^2\right]$$

Introducing the *velocity ratio*, ρ, also known as the *blade speed–steam speed ratio* and defined by the equation:

$$\rho = \frac{V_b}{V_1}$$

we get $$\eta_b = 4(\rho \cos \alpha - \rho^2)$$

For a given velocity ratio, the diagram efficiency will be a maximum when $\cos \alpha = 1$ ($\alpha = 0$). The construction features of a steam turbine preclude the use of zero-angle nozzles, but the angle is kept small, usually

STEAM TURBINES

in the neighborhood of 12 to 15 degrees. Some increase in this angle may be dictated by other considerations in low pressure stages.

The value of the velocity ratio for maximum efficiency can be determined by differentiating the equation above and equating to zero.

Then
$$4 \, d\rho \cos \alpha - 8 \, \rho d\rho = 0$$

And
$$\rho = \frac{\cos \alpha}{2}$$

Thus for a zero-angle nozzle, the optimum velocity ratio is $\frac{1}{2}$, and the optimum ratio for a 15-degree nozzle will be about 0.483. *As a good rule of thumb, the ideal velocity ratio for a single row of impulse blading is about one-half.*

With zero nozzle angle, $\alpha = 0 = \beta_1 = \beta_2$, and the vector diagram, Fig. 13-4, collapses on itself. For ideal conditions, $V_1 = 2V_b$, as just shown, and without bucket loss, $V_{r_1} = V_{r_2}$. From the diagram, $V_1 = V_{r_1} + V_b$.

Then $\quad 2V_b = V_{r_1} + V_b \quad$ and $\quad V_b = V_{r_1} = V_{r_2}$

But also from the diagram,

$$V_2 = V_{r_2} - V_b. \text{ So } \underline{V_2 = 0}.$$

This makes physical sense, since we would expect that, under ideal conditions, all the kinetic energy in the entering jet would be converted to work and none would escape at exit.

It should be noted that blade speed is a function of the rotational speed of the turbine and the mean (pitch) diameter of the blades in any particular stage:

$$V_b = \pi D \times \frac{RPM}{60} \text{ ft/sec (m/sec)}$$

For a single stage and a given set of entering steam conditions, the ideal velocity ratio and the peak diagram efficiency will occur at only one rotational speed.

Example 13-1: In a simple impulse stage, steam leaves the nozzles with a velocity of 1200 ft/sec. The nozzle angle is 15 degrees. Assume the bucket entrance and exit angles are the same and that the bucket velocity coefficient is 0.88. The wheel speed is 580 ft/sec, and steam is supplied to the turbine at the rate of 6000 lb/hr. Find:
 (a) the required bucket entrance angle for the given conditions
 (b) the bucket work, ft·lbf/lb and Btu/lb
 (c) the power developed in the buckets, hp
 (d) the available energy to the buckets, ft·lbf/lb and Btu/lb
 (e) the diagram efficiency

STEAM TURBINES

Solution: Refer to Fig. 13-2:

$$V_1 = 1200 \text{ ft/sec} \qquad V_b = 580 \text{ ft/sec}$$
$$\alpha = 15° \qquad \dot{M} = 6000 \text{ lb/hr}$$
$$C_b = 0.88 \qquad \beta_1 = \beta_2$$

(a) $$\tan \beta_1 = \frac{V_1 \sin \alpha}{V_1 \cos \alpha - V_b} = \frac{1200 \sin 15°}{1200 \cos 15° - 580}$$

$$\beta_1 = \beta_2 = \underline{28.2°}$$

(b) $$(wk)_b = \frac{V_b(V_{r_1} \cos \beta_1 + V_{r_2} \cos \beta_2)}{g_c}$$

$$V_{r_1} \cos \beta_1 = V_1 \cos \alpha - V_b$$

$$V_{r_1} = \frac{V_1 \cos \alpha - V_b}{\cos \beta_1} = \frac{1200 \cos 15° - 580}{\cos 28.2°} = \underline{657.1 \text{ ft/sec}}$$

$$V_{r_2} = C_b V_{r_1} = 0.88 \times 657.1 = \underline{578.2 \text{ ft/sec}}$$

$$(wk)_b = \frac{580(657.1 \cos 28.2° + 578.2 \cos 28.2°)}{32.2} = \underline{19610 \text{ ft} \cdot \text{lbf/lb}}$$

$$\frac{(wk)_b}{(J)} = \frac{19610}{778} = \underline{25.2 \text{ Btu/lb}}$$

(c) $$P = \dot{M}(wk) \qquad \dot{M} = 6000 \text{ lb/hr} = 100 \text{ lb/min}$$

$$P = 100 \times 19610 \text{ ft} \cdot \text{lbf/min} = \frac{100 \times 19610}{33,000} = \underline{59.4 \text{ hp}}$$

or: $$P = 6000 \times 25.2 \text{ Btu/hr} = \frac{6000 \times 25.2}{2545} = \underline{59.4 \text{ hp}}$$

(d) $$(A.E.)_b = \frac{V_1^2}{2g_c} = \frac{(1200)^2}{2 \times 32.2} = \underline{22360 \text{ ft} \cdot \text{lbf/lb}}$$

or: $$(A.E.)_b = \frac{22360}{778} = \underline{28.74 \text{ Btu/lb}}$$

(e) $$\eta_b = \frac{(wk)_b}{(A.E.)_b} = \frac{19610}{22360} = 0.877 = \underline{87.7\%}$$

Example 13-2: Steam enters a simple impulse bucket wheel with an absolute velocity of 450 m/sec and a relative velocity of 270 m/sec. It leaves the blades with a relative velocity of 230 m/sec and an absolute velocity of 105 m/sec. Find:
(a) the bucket velocity coefficient
(b) the available energy, kJ/kg
(c) the bucket loss, kJ/kg
(d) the unused kinetic energy at exit, kJ/kg
(e) the diagram efficiency

Solution:

$$V_1 = 450 \text{ m/sec} \qquad V_{r_2} = 230 \text{ m/sec}$$
$$V_{r_1} = 270 \text{ m/sec} \qquad V_2 = 105 \text{ m/sec}$$

STEAM TURBINES

(a) $$C_b = \frac{V_{r_2}}{V_{r_1}} = \frac{230}{270} = \underline{0.85}$$

(b) $$(A.E.)_b = \frac{V_1^2}{2g_c} = \frac{\overline{450^2} \times 10^{-3}}{2} = \underline{101.25 \text{ kJ/kg}}$$

(c) $$\text{bucket loss} = \frac{V_{r_1}^2 - V_{r_2}^2}{2g_c} = \frac{\overline{270^2} - \overline{230^2}}{2} \times 10^{-3} = \underline{10.00 \text{ kJ/kg}}$$

(d) $$(K.E.)_2 = \frac{V_2^2}{2g_c} = \frac{\overline{105^2} \times 10^{-3}}{2} = \underline{5.51 \text{ kJ/kg}}$$

(e) $$\eta_b = \frac{(V_1^2 - V_2^2) - (V_{r_1}^2 - V_{r_2}^2)}{V_1^2}$$

$$= \frac{(\overline{450^2} - \overline{105^2}) - (\overline{270^2} - \overline{230^2})}{\overline{450^2}} = 0.847 = \underline{84.7\%}$$

or:

$$\eta_b = \frac{(wk)_b}{(A.E.)_b} = \frac{(A.E.)_b - \text{losses}}{(A.E.)_b} = \frac{101.25 - 10.00 - 5.51}{101.25} = \underline{84.7\%}$$

In this article, attention has been focused on the conversion of one form of mechanical energy (*kinetic*) to another (*work*). Thus the basic mechanical energy unit of the Engineers' System, the foot pound force, has made its natural appearance. In the developments which follow, thermal energy considerations come to the fore, and use of the British thermal unit is indicated. Additionally, much of the problem work involves use of the Steam Tables (English Units). Hence, the SI applications will be minimal, and the Btu/lb will be the primary specific energy unit throughout the remainder of this chapter.

13-4. The Simple Impulse Stage. The simple impulse stage (see Fig. 13-1) will be analyzed by combining the nozzle and bucket characteristics already presented. In addition to this, some investigation of the steam conditions through the stage will be made. Figure 13-5 shows a typical impulse stage process on a partial *h-s* diagram. Point 0 represents the state of the steam as it enters the nozzles. The velocity here is assumed to be negligible, i.e., there is no velocity "carryover" from the preceding stage. The steam expands through the nozzles to *stage pressure*. Point 1 represents the state of the steam leaving the nozzles and entering the buckets. During the expansion, a drop in enthalpy occurs with an attendant increase in the kinetic energy of the steam. For the nozzle, as shown in the preceding chapter:

$$\frac{V_1^2}{2g_c(J)} = h_0 - h_1$$

In the ideal case the process would be isentropic and *the available energy*

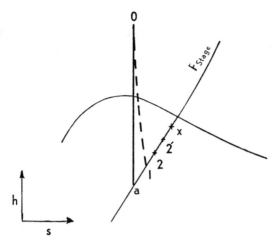

FIGURE 13-5. Partial h-s diagram for impulse stage.

to the nozzles is equal to the isentropic change of enthalpy from the initial state to the stage pressure. The entire pressure drop in the stage occurs in the nozzles and *the available energy to the nozzles is also the available energy to the stage.* Then:

$$(A.E.)_n = (A.E.)_{st} = h_0 - h_a \text{ Btu/lb (J/kg)}$$

Point 2 indicates the state of the steam leaving the buckets. This point must lie on the *stage pressure line* but is displaced from point 1 because of the increase in enthalpy and entropy arising from friction and other losses within the buckets. The velocity of the jet leaving the buckets (V_2) may be appreciable, but in the analysis of a single stage it is convenient to consider that this velocity is destroyed. Then all of the associated kinetic energy will be transformed to enthalpy, and the corresponding final state becomes 2′. It is common practice to refer to $h_{2'}$ as the exit stagnation enthalpy. The energy conversions through the stage can be verified by applying the flow energy equation between applicable points. The actual state of the steam leaving the stage will lie somewhere between 2 and 2′, depending upon the amount of velocity *carryover* to the next succeeding stage and provided there are no other internal losses to be considered. If there are other losses, the final state will be represented by some point x.

An interpretation of Fig. 13-5 can now be made:

$$\text{Nozzle loss or reheat} = (h_1 - h_a)$$
$$\text{Blade loss or reheat} = (h_2 - h_1)$$
$$\text{Exit loss or reheat} = (h_{2'} - h_2)$$

Each of these expressions represents an energy loss to the stage but not a complete loss to the turbine overall, as will be explained later. The increase in enthalpy accompanying a stage loss is frequently referred to as *reheat*. The miscellaneous losses, which give rise to the reheat ($h_x - h_{2'}$), are not considered in the equations developed below but will be discussed briefly at the end of this article.

Since *the only work done in the stage is on the buckets* and the flow energy equation applies:

$$\frac{(wk)_b}{(J)} = \frac{(wk)_{st}}{(J)} = h_0 - h_2 - \frac{V_2^2}{2g_c(J)}$$

$$= h_0 - h_{2'} \text{ Btu/lb (J/kg)} \qquad (13\text{-}5)$$

As previously shown, the diagram efficiency is equal to the bucket work divided by the available energy to the buckets. In terms of enthalpies:

$$\eta_b = \frac{(wk)_b}{(J)(A.E.)_b} = \frac{h_0 - h_{2'}}{h_0 - h_1}$$

But the nozzle efficiency is:

$$\eta_n = \frac{h_0 - h_1}{h_0 - h_a}$$

Multiplying these two efficiency equations:

$$\eta_n \eta_b = \frac{h_0 - h_{2'}}{h_0 - h_a} = \frac{(wk)_{st}}{(J)(A.E.)_{st}} \qquad (13\text{-}6)$$

This is the *nozzle-bucket efficiency* of the stage and includes the effects of the losses associated with nozzle efficiency, bucket velocity coefficient and unutilized kinetic energy at the bucket exit.

The stage efficiency, defined as the ratio of the stage output, as determined by test, to the stage available energy, is the ultimate criterion of actual stage performance. The stage efficiency will always be *less* than the nozzle-bucket efficiency because of additional internal losses included in the former. For an impulse stage these are *disk, diaphragm leakage, idle blade* and *radiation losses*. Disk losses arise from the friction between the rotating wheels carrying the buckets and the surrounding steam. Diaphragm leakage losses result from the leakage of steam through the packing between the shaft and the stationary nozzle diaphragm. This leakage steam undergoes a throttling process and performs no useful work. Idle blade losses occur after the first stage nozzle block and in other stages where the nozzles extend only partially around the periphery of the diaphragm. The buckets not actually in way of an active nozzle "churn" to

STEAM TURBINES

a certain extent, and in addition the dead steam in the buckets must be reaccelerated upon re-entering the active nozzle arc. The heat losses from the steam to the surroundings are primarily due to radiation and are of minor magnitude. In well-designed multi-stage turbines all of these losses in the aggregate are reasonably small. Under optimum conditions, the stage efficiency may be as high as 98 percent of the nozzle-bucket efficiency, but a figure of 90 to 95 percent is more representative of usually encountered conditions. On the h-s diagram, state x is finally established by these additional losses.

The terminology used for the several efficiencies concerned in the analysis of a turbine stage is not entirely standardized in the literature. This presents no particular problem if the conventions are properly explained in each case. The efficiency terms employed in this text are not necessarily the best or only ones in use but have been chosen because they are largely self-explanatory.

Example 13-3: The impulse stage of a turbine receives steam at 220 psia with a temperature of 420°F when the stage pressure is 140 psia. Under these conditions the available energy to the stage is 38.5 Btu/lb, the nozzle-bucket efficiency is 85 percent and the absolute blade entrance and exit velocities are 1350 fps and 350 fps respectively. Assume the stage efficiency is equal to the nozzle-bucket efficiency and calculate:

(a) the enthalpy of the steam after isentropic expansion to the stage pressure, Btu/lb
(b) the enthalpy of the steam entering the blades, Btu/lb
(c) the enthalpy of the steam leaving the blades, Btu/lb
(d) the enthalpy of the steam leaving the stage if there is negligible carryover velocity, Btu/lb
(e) the state of the steam leaving the stage, pressure (psia) and moisture (percent) for the conditions postulated in (d) above.

Solution: Refer to Fig. 13-5.

$p_0 = 220$ psia, $\qquad t_0 = 420°$F
$p_1 = p_2 = p_{2'} = 140$ psia, $\quad (A.E.)_{st} = 38.5$ Btu/lb
nozzle-bucket efficiency = 85 percent
$V_1 = 1350$ fps, $\quad V_2 = 350$ fps

(a) $\quad h_a = h_0 - (A.E.)_{st} = 1219.8 - 38.5 = \underline{1181.3 \text{ Btu/lb}}$

(b) $\quad h_1 = h_0 - \dfrac{V_1^2}{2g_cJ} = 1219.8 - \dfrac{1350^2}{2 \times 32.2 \times 778} = \underline{1183.4 \text{ Btu/lb}}$

(c) $\quad \dfrac{(wk)_{st}}{J} = \eta_n\eta_b(A.E.)_{st} = 0.85 \times 38.5 = 32.7$ Btu/lb

$\quad h_2 = h_0 - \dfrac{(wk)_{st}}{J} - \dfrac{V_2^2}{2g_cJ}$

$\quad\quad = 1219.8 - 32.7 - \dfrac{350^2}{2 \times 32.2 \times 778} = \underline{1184.7 \text{ Btu/lb}}$

(d) $\quad h_{2'} = h_2 + \dfrac{V_2^2}{2g_cJ} = 1184.7 + \dfrac{350^2}{2 \times 32.2 \times 778} = \underline{1187.1 \text{ Btu/lb}}$

(e) $\quad p_{2'} = \underline{140 \text{ psia}}$

$$m_{2'} = \frac{h_g - h_{2'}}{h_{fg}} = \frac{1193.8 - 1187.1}{868.7} = \underline{0.77\%}$$

13-5. The Reaction Stage. A reaction stage consists of one row of fixed blades and the following row of moving blades. Both rows are essentially a series of converging nozzles, and a pressure drop occurs across both with an attendant conversion of enthalpy to kinetic energy. Herein lies the distinction from the impulse stage where the entire pressure drop and energy conversion or "release" occurs across the fixed nozzles. The fixed blades of a reaction stage perform the same function as the nozzles of an impulse stage. The moving blades differ in appearance from impulse buckets in that they show a definite *choke* or reduction in area from entrance to exit. A typical reaction stage is shown in Fig. 13-6. The relative

FIGURE 13-6. Typical reaction stage.

amount of energy released in the moving blades as compared to the fixed blades may be controlled by the designer, and a rough approximation of this apportionment may be obtained by comparing the relative amount of choke in the two sets of blades making up the stage.

The available energy to a reaction stage is established by the isentropic change of enthalpy from the initial state entering the fixed blades to the final pressure at exit from the moving blades. In a 100 percent reaction stage, all the energy release would occur in the moving blades and the fixed blades become merely guides to direct the steam into the moving blades. A pure impulse stage has 0 percent reaction with no energy release in the buckets. *In a 50 percent reaction stage, often called "pure" reaction, half the available energy is assigned for release in the moving blades and half in the fixed blades.* Nearly all reaction turbines are designed for this equal distribution of energy release. On the other hand, it is not uncommon to find as much as 30 percent reaction designed into the low pressure stages of multi-stage "impulse" turbines. Current practice dictates the term

STEAM TURBINES

"impulse" for any stage in which less than half the stage available energy is assigned for release in the moving blades. Conversely, a "reaction" stage is one in which 50 percent or more of the available energy is to be released in the moving blades. The older and possibly more descriptive term "impulse-reaction" is no longer in general use. The discussion in this text is primarily concerned with the 50 percent reaction stage, although the methods used are applicable to a stage with any percent reaction.

The reaction stage may be analyzed by using a velocity vector diagram such as that shown in Fig. 13-7. The nomenclature, procedure and

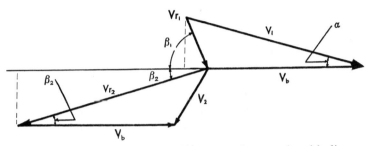

FIGURE 13-7. Velocity vector diagram for reaction blading.

equations are basically the same as those employed for the impulse stage. The major modification arises from the energy release occurring in the moving blades. This is done by introducing a *reactive effectiveness*, η_R, to be applied to the energy available for release in the moving blades. The velocity vector diagram differs in appearance from that for the impulse stage in several respects. The relative velocity of the steam leaving the moving blades, V_{r_2}, is greater than the relative entering velocity and approaches V_1 in magnitude; the blade speed, V_b, becomes proportionately larger in relation to V_1; and the moving blades are opened up at entrance so that β_1 is appreciably greater than β_2.

Referring to Fig. 13-7, the work equation will take the same form as for the impulse stage:

$$\frac{(wk)_b}{(J)} = \frac{(wk)_{st}}{(J)} = \frac{V_b(V_{r_1} \cos \beta_1 + V_{r_2} \cos \beta_2)}{g_c(J)}$$

The work can also be expressed in terms of kinetic energy as before:

$$\frac{(wk)_{st}}{(J)} = \frac{(V_1^2 - V_2^2) + (V_{r_2}^2 - V_{r_1}^2)}{2g_c(J)} \text{ Btu/lb (J/kg)} \qquad (13\text{-}7)$$

This equation is simply a rearrangement of Equation (13-3) but with the mechanical equivalent of heat (J) introduced in order to make the

Engineers' System units Btu/lb rather than ft · lbf/lb. However, a departure from the impulse analysis is necessary in determining the several kinetic energy terms. If $(A.E.)_{st}$ is the stage available energy, $(A.E.)_n$ the energy available for release in the fixed blades and $(A.E.)_R$ the energy available for release in the moving blades, then:

$$(A.E.)_{st} = (A.E.)_n + (A.E.)_R$$

Assuming the velocity entering the fixed blades is negligible:

$$\frac{V_1^2}{2g_c(J)} = \eta_n (A.E.)_n$$

From the geometry of the velocity diagram:

$$V_{r_1}^2 = V_1^2 + V_b^2 - 2V_1 V_b \cos \alpha$$

The kinetic energy associated with V_{r_2} is found from a summation of the entering kinetic energy and the energy available for release and conversion to the kinetic form within the moving blades. Each component is treated separately in regard to losses by use of a moving blade velocity coefficient, C_m, and the reactive effectiveness, η_R. In equation form:

$$\frac{V_{r_2}^2}{2g_c(J)} = \frac{(C_m V_{r_1})^2}{2g_c(J)} + \eta_R (A.E.)_R \qquad \textbf{(13-8)}$$

Finally from the figure:

$$V_2^2 = V_{r_2}^2 + V_b^2 - 2V_{r_2} V_b \cos \beta_2$$

The diagram stage work is found from Equation (13-7) by substitution of the several quantities determined above. The *combined blade efficiency*, which is similar to the nozzle-bucket efficiency of an impulse stage, is given by:

$$\eta_{cb} = \frac{(wk)_{st}}{(J)(A.E.)_{st}}$$

The actual stage efficiency as determined by test will be lower than the value given by the above equation because of additional internal losses due to leakage, rotation and radiation.

An approximation of the process through a single reaction stage is shown on the partial h-s diagram of Fig. 13-8. The initial state is at point 0 and the intermediate and final pressure lines are as indicated. Point 2 is the final state based upon a velocity vector diagram solution, and point x is the actual exit state as determined by test.

Referring to the figure, the diagram stage work for negligible entrance

STEAM TURBINES

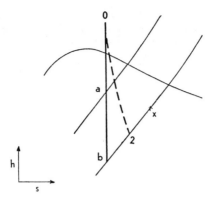

FIGURE 13-8. Partial h-s diagram for reaction stage.

velocity is given by:

$$\frac{(wk)_{st}}{(J)} = h_0 - h_2 - \frac{V_2^2}{2g_c(J)} \text{ Btu/lb (J/kg)}$$

Also: $(A.E.)_{st} = (h_0 - h_b)$ Btu/lb (J/kg)
Combined blade reheat $= (h_2 - h_b)$ Btu/lb (J/kg)
Reheat from other losses $= (h_x - h_2)$ Btu/lb (J/kg)

If several reaction stages are arranged in a continuous group on a single drum, the exit velocity from one stage is largely carried over to the next succeeding stage. In this situation, all stages other than the first in the group can utilize the kinetic energy associated with the *carryover velocity* (V_{co}), which is ideally equal to the absolute velocity (V_2) of the steam leaving the preceding stage. The work of such a stage is determined as before except that the kinetic energy of the jet leaving the fixed blades must be found from an energy summation involving the carryover velocity and the energy available for release in the fixed blades. The losses which occur are accounted for by use of a fixed blade velocity coefficient, C_f, and the fixed blade (nozzle) efficiency, η_n. Then,

$$\frac{V_1^2}{2g_c(J)} = \frac{(C_f V_{co})^2}{2g_c(J)} + \eta_n(A.E.)_n$$

The efficiency of the stages as a unit is determined by summing up the diagram work of each individual stage and dividing this total by the available energy as established by the conditions at entrance to the first stage and the exit pressure from the last stage.

STEAM TURBINES

Example 13-4: The first stage of a 50 percent reaction group receives steam with negligible approach velocity at a pressure of 300 psia with a temperature of 520°F. The available energy to the stage is 10 Btu/lb, the fixed blade efficiency is 96 percent, the velocity coefficient for the moving row is 0.88, the reactive effectiveness is 90 percent and the relative inlet and absolute exit velocities are 146 fps and 160 fps, respectively. Assume that the stage efficiency is the same as the combined blade efficiency and calculate:

(a) the steam velocity leaving the fixed blades, fps
(b) the relative exit velocity from the moving blades, fps
(c) the stage work, Btu/lb
(d) the combined blade efficiency, percent
(e) the enthalpy of the steam entering the succeeding stage, Btu/lb, for complete velocity carryover

Solution: Refer to Figs. 13-7 and 13-8:

$$p_0 = 300 \text{ psia}, \quad t_0 = 520°F, \quad (A.E.)_{st} = 10 \text{ Btu/lb}$$
$$\eta_n = 0.96, \quad C_m = 0.88, \quad \eta_R = 0.90$$
$$V_{r_1} = 146 \text{ fps}, \quad V_2 = 160 \text{ fps}$$

(a)
$$(A.E.)_n = (A.E.)_R = \frac{(A.E.)_{st}}{2} = \frac{10}{2} = 5 \text{ Btu/lb}$$
$$V_1 = [(2g_c J)\eta_n (A.E.)_n]^{\frac{1}{2}}$$
$$= [2 \times 32.2 \times 778 \times 0.96 \times 5]^{\frac{1}{2}} = \underline{490 \text{ ft/sec}}$$

(b)
$$\frac{V_{r_2}^2}{2g_c J} = \frac{(C_m V_{r_1})^2}{2g_c J} + \eta_R (A.E.)_R$$
$$V_{r_2} = [(C_m V_{r_1})^2 + (2g_c J)\eta_R (A.E.)_R]^{\frac{1}{2}}$$
$$= [(0.88 \times 146)^2 + 2 \times 32.2 \times 778 \times .90 \times 5]^{\frac{1}{2}} = \underline{492 \text{ ft/sec}}$$

(c)
$$\frac{(wk)_{st}}{J} = \frac{(V_1^2 - V_2^2) + (V_{r_2}^2 - V_{r_1}^2)}{2g_c J}$$
$$= \frac{[(490)^2 - (160)^2] + [(492)^2 - (146)^2]}{2 \times 32.2 \times 778} = \underline{8.69 \text{ Btu/lb}}$$

(d)
$$\eta_{cb} = \frac{(wk)_{st}}{J(A.E.)_{st}} = \frac{8.69}{10} = 0.869 = \underline{86.9\%}$$

(e)
$$h_2 = h_0 - \frac{(wk)_{st}}{J} - \frac{V_2^2}{2g_c J}$$
$$= 1269.4 - 8.7 - \frac{(160)^2}{2 \times 32.2 \times 778} = \underline{1260.2 \text{ Btu/lb}}$$

13-6. Comparison of Impulse and Reaction Stages. Theoretically, the efficiencies obtainable with an impulse stage and a corresponding reaction stage are exactly equal if each is operated at the optimum velocity ratio. However, to accommodate the same available energy, the optimum blade speed for a reaction stage will be greater than that for an impulse stage. In an ideal simple impulse stage with zero nozzle angle, maximum efficiency

occurs when $V_2 = 0$, and:

$$V_b = \tfrac{1}{2} V_1 = \tfrac{1}{2} \sqrt{2g_c(J)(A.E.)_{st}}$$

By analogy, an ideal 50 percent reaction stage with zero nozzle angle will also reach maximum efficiency when $V_2 = 0$. Analysis of the vector diagram for these conditions shows that V_{r_2} must equal V_b. Also $V_1 = V_b$ and $V_{r_1} = 0$. Then:

$$V_b = V_1 = \sqrt{2g_c(J)(A.E.)_n} = \sqrt{2g_c(J) \frac{(A.E.)_{st}}{2}}$$

or:
$$V_b = \frac{\sqrt{2}}{2} \sqrt{2g_c(J)(A.E.)_{st}}$$

Hence, for the same total *stage* available energy, the ideal blade speed for a reaction stage is greater than that for a simple impulse stage by a factor of 1.414 ($\sqrt{2}$). The higher blade speed demanded by the reaction turbine means higher rotational speeds or increased wheel diameters for the same available energy utilization. Usually the blade speeds used in both types are comparable, and for the same total enthalpy drop from throttle to exhaust, a reaction turbine will require more stages and be larger than an impulse machine.

On the basis of actual efficiencies, both types have their particular advantages, stemming generally from construction details. Eliminating leakage losses temporarily from the discussion, a group of reaction stages will show a higher efficiency than a corresponding group of impulse stages. The reasons for this are several and include: (a) reduced losses in the moving blades because of decreased relative entering velocities; (b) elimination of idle blade loss, since all reaction stages employ full peripheral admission; (c) relatively high utilization of carryover velocity from one stage to the next under usual conditions.

Unfortunately, the reaction stage is subject to a type of leakage loss not encountered in the impulse stage. *This leakage occurs across the clearance space between the moving blades and the casing and is motivated by the pressure drop that occurs across these blades. No such problem exists in the impulse stage, since the pressure is the same on both sides of the buckets.* The magnitude of the clearance is a mechanical design feature and remains substantially constant throughout a turbine. In the high pressure end, the blade area for a given flow rate is small because the steam is compressed. As the pressure drops through the turbine, the specific volume of the steam increases, the blades become longer and the leakage area relative to the blade area decreases. Consequently, the leakage losses are steadily decreasing in magnitude from inlet to exhaust. Many ingenious designs

for reducing the leakage losses are in use. However, the inherent efficiency of reaction blading is overcome by leakage losses in the high pressure end of a turbine. At the low pressure end, the leakage losses become relatively small, and the reaction stage here will outdo the pure impulse stage. At some intermediate pressure, the two types will give substantially equal efficiencies. This suggests the use of a turbine incorporating both impulse and reaction stages, and such practice is common.

Example 13-5: To illustrate the effect of tip leakage, it is assumed that 5 percent of the steam supplied to the stage of Ex. 13-4, Art. 13-5, flows past the stage without performing useful work. Assume there are no other stage losses and calculate:
(a) the stage work, Btu/lb
(b) the stage efficiency, percent
(c) the reheat due to leakage, Btu/lb
(d) the actual enthalpy entering the succeeding stage for complete velocity carryover, Btu/lb

Solution: Refer to Fig. 13-8.
From solution of Ex. 13-4,

$$h_0 = 1269.4 \text{ Btu/lb} \qquad (A.E.)_{st} = 10 \text{ Btu/lb}$$

$$\frac{(wk)_{st}}{J} = 8.69 \text{ Btu/lb} \qquad h_2 = 1260.2 \text{ Btu/lb}$$

(a) $$\frac{(wk)_{st}}{J} = 0.95(8.69) = \underline{8.26 \text{ Btu/lb}}$$

(b) $$\eta_{st} = \frac{(wk)_{st}}{J(A.E.)_{st}} = \frac{8.26}{10} = \underline{82.6\%}$$

(c) $$\text{leakage reheat} = 8.69 - 8.26 = \underline{0.43 \text{ Btu/lb}}$$

(d) $$h_x = 1260.2 + 0.43 = \underline{1260.6 \text{ Btu/lb}}$$

13-7. The Large Multi-stage Turbine. In modern high-capacity turbines built in this country, a velocity compounded impulse stage is almost universally used as the first stage after inlet. The Curtis stage is typical and consists of a set of nozzles followed in order by a row of buckets, a fixed row of blades to redirect the steam and a second row of buckets. The velocity of the nozzle jet is absorbed in two moving rows of buckets rather than in one. The optimum velocity ratio for such a stage is about 0.25, and a large nozzle exit velocity can be accepted for a given wheel speed. This in turn means an appreciable decrease in enthalpy and pressure across the fixed nozzles. Because of the high relative velocities, the losses in both fixed blades and moving buckets are rather high, and the efficiency of a Curtis stage is less than could be obtained with a series of simple impulse stages. However, the Curtis stage is of rugged construction, brings about an immediate and radical pressure reduction and has certain inherent

advantages for partial load operation; hence its considerable popularity. The velocity vector diagram for a Curtis stage is shown in Fig. 13-9.

The choice of the stages used in the remainder of the turbine is a matter of discretion. One large manufacturer uses a series of simple impulse

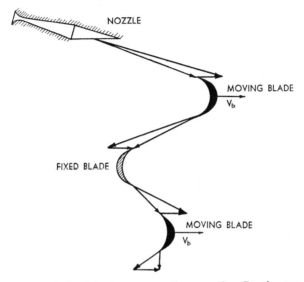

FIGURE 13-9. Velocity vector diagram for Curtis stage.

stages with gradually increasing amounts of reaction built into the moving blades of the later stages. Another employs 50 percent reaction in all stages after the Curtis wheel. There is little if any difference in the overall performance of the two types.

Up to this point, the discussion has generally assumed operation at designed conditions of load and rotational speed. Large turbines are widely used for electric power generation and marine propulsion. Because of frequency requirements, generator turbines must operate at constant speed under various conditions of load. Marine propulsion turbines, on the other hand, operate under extreme conditions of both load and speed. The two problems are considerably different in detail but alike in that *some decrease in efficiency can be expected whenever a turbine is operated at other than the designed conditions.* The loss in efficiency stems from a departure of the velocity ratio from the design value and from increased blade losses due to shock when the blade entrance angle and the relative entering jet angle do not coincide.

The design point of maximum efficiency may or may not occur at rated

load. In naval turbines, for instance, it may be desirable to have maximum efficiency at cruising speed with a secondary efficiency peak at full power. There are many methods, such as nozzle control and by-passing, for obtaining reasonable efficiencies over a wide range of operating conditions, but they cannot be detailed here.

13-8. Internal Work and Internal Engine Efficiency. *The engine efficiency of a turbine is defined as the ratio of the work output to the available energy.* The work output may be measured at any one of several points, and *a complete specification of the point of measurement is necessary if the engine efficiency figure is to have real comparative significance.* If other equipment is included between the turbine and the point of work determination, the engine efficiency will reflect the performance of the additional equipment as well as that of the turbine itself. Where the unit is offered as a single package by the manufacturer, such an efficiency is of primary interest to the prospective buyer. Consequently, the engine efficiency of turbo-generator sets is often quoted on the basis of the output at the generator terminals. Similarly, the engine efficiency of a marine propulsion unit, where the manufacturer furnishes both turbine and reduction gears, is frequently based on the work delivered at the gear output coupling. In evaluating the performance of a turbine alone, only two engine efficiencies are necessary; *internal engine efficiency,* based on the work output delivered internally to the rotor, and *brake engine efficiency,* based on the work output measured at the turbine stub shaft coupling. Both of these efficiencies are predicated on actual observed conditions.

The internal engine efficiency is subject to direct thermodynamic treatment based on the change in the properties of the steam from inlet to exhaust. The turbine is considered to begin with the throttle and end at the exhaust flange, since these are integral parts furnished by the manufacturer. By determining the condition of the steam entering the throttle and leaving through the exhaust trunk, one can find the internal work. In Fig. 13-10, line 1-2-3 is the approximate *state-line curve* or *condition curve* for a large turbine as it appears on *h-s* coordinates. Point 1 is the initial condition of the steam as determined at the boiler side flange of the turbine throttle or stop valve if no throttle is installed. Point 2 is the state after throttling. Point 3, known as the *state line end point,* represents the condition of the steam *at the turbine exhaust flange, and an appreciable velocity may exist here for all except light loads.* The kinetic energy associated with this velocity is known as the *leaving loss* and this is a true loss to the turbine, since it cannot possibly be utilized. Point 3′, known as the *exhaust point,* represents an equivalent exhaust state predicated on the conversion of the leaving loss to enthalpy. The slight pressure drop

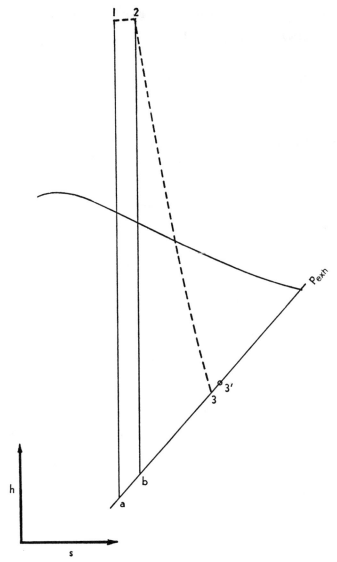

FIGURE 13-10. Approximate state-line curve for a complete turbine.

between the last stage and the exhaust flange is ignored. Points a, b, 3 and $3'$ all lie on a common *exhaust pressure line*.

Several important relationships can be written for the turbine as a whole, using the nomenclature of Fig. 13-10. From the flow energy

STEAM TURBINES

equation, the turbine internal work is given by:

$$\frac{(wk)_t}{(J)} = h_1 - h_3 - \frac{V_3^2}{2g_c(J)} = h_1 - h_3' \text{ Btu/lb (J/kg)} \quad (13\text{-}9)$$

The throttling process is characterized by equal enthalpy; so:

$$h_1 = h_2$$

The available energy to the turbine is evaluated by the isentropic change of enthalpy from the initial state to the exhaust pressure. Furthermore, *the throttle is part of the turbine*. Then:

$$(A.E.)_t = h_1 - h_a \text{ Btu/lb (J/kg)}$$

By definition, then, the *internal* engine efficiency is given by:

$$\eta_{ei} = \frac{(wk)_t}{(J)(A.E.)_t} = \frac{h_1 - h_{3'}}{h_1 - h_a} \quad (13\text{-}10)$$

Also $\quad (wk)_t/(J) = (A.E.)_t - $ internal losses

or $\qquad \dfrac{(wk)_t}{(J)} = (h_1 - h_a) - (h_{3'} - h_a)$

But $\qquad (h_{3'} - h_a) = (h_{3'} - h_3) + (h_3 - h_b) + (h_b - h_a)$

and $\qquad (h_{3'} - h_3) = \dfrac{V_3^2}{2g_c(J)} = $ leaving loss

$\qquad (h_b - h_a) = $ throttling loss

$\qquad (h_3 - h_b) = $ other internal losses

The leaving loss has already been identified. The throttling loss arises from the decrease in available energy, as referred to the exhaust pressure, in proceeding from 1 to 2 without the performance of useful work. The other internal losses are associated with all the individual stage losses previously mentioned.

Example 13-6: A large turbine receives steam at the throttle at 560 psia and 800°F at the rate of 100 lb/sec. The pressure after the throttle is 500 psia. At the exhaust flange the absolute pressure is 1.5 inches of mercury, the steam velocity is 1000 ft/sec and the moisture content is 10 percent. Find:
 (a) the available energy to the turbine, Btu/lb
 (b) the throttling loss, Btu/lb
 (c) the leaving loss, Btu/lb
 (d) the enthalpy at the exhaust point, Btu/lb
 (e) the internal turbine work, Btu/lb
 (f) the internal engine efficiency
 (g) other internal losses not accounted for by (b) and (c) above, Btu/lb
 (h) the internal power developed, hp

Solution: Refer to Fig. 13-10:

$$p_1 = 560 \text{ psia} \qquad p_2 = 500 \text{ psia}$$
$$t_1 = 800°F \qquad h_2 = h_1 = 1409.4$$
$$h_1 = 1409.4 \qquad s_b = s_2 = 1.6549$$
$$s_a = s_1 = 1.6430$$
$$p_a = p_b = p_3 = 1.5'' \text{ Hg} = 0.737 \text{ psia}$$
$$V_3 = 1000 \text{ ft/sec}, \quad m_3 = 0.10, \quad \dot{M} = 100 \text{ lb/sec}$$
$$h_a = 902.4, \quad h_b = 909.0, \quad h_3 = 997.2$$

(a) $(A.E.)_t = h_1 - h_a = 1409.4 - 902.4 = \underline{507.0 \text{ Btu/lb}}$

(b) throttling loss $= h_b - h_a = 909.0 - 902.4 = \underline{6.6 \text{ Btu/lb}}$

(c) leaving loss $= \dfrac{V_3^2}{2g_c J} = \dfrac{(1000)^2}{2 \times 32.2 \times 778} = \underline{20.0 \text{ Btu/lb}}$

(d) $h_{3'} = h_3 + \dfrac{V_3^2}{2g_c J} = 997.2 + 20.0 = \underline{1017.2 \text{ Btu/lb}}$

(e) $\dfrac{(wk)_t}{J} = h_1 - h_{3'} = 1409.4 - 1017.2 = \underline{392.2 \text{ Btu/lb}}$

(f) $\eta_{ei} = \dfrac{(wk)_t}{J(A.E.)_t} = \dfrac{392.2}{507.0} = 0.774 = \underline{77.4\%}$

(g) other losses $= h_3 - h_b = 997.2 - 909.0 = \underline{88.2 \text{ Btu/lb}}$

(h) $P = \dfrac{\dot{M}(wk)_t}{J} = 100 \times 392.2 = 39{,}220 \text{ Btu/sec}$

1 hp = 550 ft·lbf/sec

$$hp = \dfrac{39220 \times 778}{550} = \underline{55{,}478 \text{ hp}}$$

13-9. Governing. The throttling process is abhorrent thermodynamically, being accompanied by irreversibility, increased entropy and decreased available energy. It does, however, have the practical advantage of providing a simple method of governing. Another method of governing entails the use of nozzle valves to control the admission of steam to the first stage. Since reaction stages always employ full peripheral admission, nozzle governing can only be used if the first stage is of the impulse type. This is another good reason for using a Curtis stage at inlet. A third method of governing, applicable to any type of turbine, involves the admission of line steam directly to the lower stages, by-passing the high pressure stages. In large turbines, straight throttle governing is a thing of the past. In most cases, governing is accomplished more or less automatically by nozzle control or by-passing or both. However, many small turbines and an occasional large turbine incorporate both a main throttle valve and several manually operated nozzle control valves of various sizes. With this design the throttling losses can be kept very small over a wide

STEAM TURBINES

range of loads by intelligent operation. Every effort should be made to *keep the chest pressure after the throttle valve close to line pressure and to eliminate throttling in the nozzle valves by maintaining them either wide open or shut tight.* In practice, major load changes are accommodated by manipulation of the nozzle valves, and minor adjustments are made with the throttle. The secret of proper operation lies in the choice of the nozzle valve combination.

Example 13-7: A 60-kW auxiliary generator operates with dry saturated steam at 200 psia. It has three nozzles; one is always open, and the other two are fitted with manually operated nozzle control valves. With both nozzle control valves closed, the capacity of the machine is 30 kW. Speed control is obtained by use of a throttle valve actuated by a mechanical governor. At quarter load, the pressure in the steam chest is 115 psia when both hand valves are closed. At the same load with both hand valves open, the chest pressure becomes 62 psia. Referred to an exhaust pressure of 2.4 psia, find for this load condition:
 (a) the throttling loss which must be accepted for control purposes
 (b) the additional and unnecessary throttling loss if both hand nozzle valves are inadvertently opened

Solution:

$p_1 = 200$ psia (sat)
$h_1 = 1199.3$
$s_a = s_1 = 1.5464$
$p_a = 2.4$ psia
$h_a = 906.6$
$p_2 = 115$ psia
$h_2 = h_1 = 1199.3$
$s_b = s_2 = 1.6032$
$p_b = 2.4$ psia
$h_b = 940.3$
$p_{2'} = 62$ psia
$h_{2'} = h_1 = 1199.3$
$s_{b'} = s_{2'} = 1.6684$
$p_{b'} = 2.4$ psia
$h_{b'} = 978.9$

(a) $\quad h_b - h_a = 940.3 - 906.6 = \underline{33.7 \text{ Btu/lb}}$

(b) $\quad h_{b'} - h_b = 978.9 - 940.3 = \underline{38.6 \text{ Btu/lb}}$

13-10. Reheat and Reheat Factor. An interesting thermodynamic phenomenon occurs in multi-stage steam turbines in connection with stage reheat. As indicated in Fig. 13-11 in exaggerated form, the constant pressure lines on *h-s* coordinates diverge with increasing entropy. As a result, the sum of the available energies to the several individual stages is greater than the available energy to the turbine. For the simple case of the

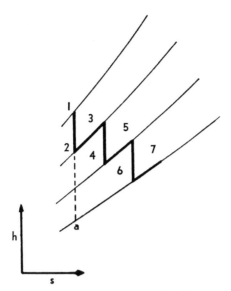

FIGURE 13-11. Effects of reheat.

three stages shown in the figure:

$$[(h_1 - h_2) + (h_3 - h_4) + (h_5 - h_6)] > (h_1 - h_a)$$

*The ratio of the sum of the stage available energies to the turbine available energy is known as the **reheat factor**.* As a corollary, the internal engine efficiency will always be greater than the average of the stage efficiencies, and their ratio will also be equal to the reheat factor. This will be shown below, introducing η_{st} to signify the average of the stage efficiencies as determined by actual test.

$$\eta_{st} = \frac{\Sigma(wk)_{st}}{(J)\Sigma(A.E.)_{st}} = \frac{(wk)_t}{(J)\Sigma(A.E.)_{st}}$$

$$\eta_{ei} = \frac{(wk)_t}{(J)(A.E.)_t}$$

Then,
$$\frac{\eta_{ei}}{\eta_{st}} = \frac{\Sigma(A.E.)_{st}}{(A.E.)_t} = RHF$$

The reheat factor will always be greater than unity and may be about 1.05 for large turbines. In general the value of the reheat factor varies directly with the initial temperature, the turbine available energy and the number of stages.

It should not be assumed that the irreversibilities which lead to stage reheats are desirable. Such processes are always accompanied by an increase in entropy and a final net increase in the heat rejected by the turbine. However, these losses are reduced by multi-staging and the subsequent use of some of the reheat energy in succeeding stages. The divergence of the pressure lines, as evidenced in the reheat factor, helps further to reduce the generally deleterious effects of reheat. Because of the partial reclamation of the stage losses, it is customary to speak of them as reheats when dealing with the turbine as a complete unit except in the last stage, where such losses are complete and irretrievable.

13-11. Brake Work and Brake Engine Efficiency. Turbines are frequently evaluated on the basis of **brake work** *defined as work delivered at the turbine shaft*. The brake work will be less than the internal work by an amount equal to the turbine *mechanical losses*. These losses include the work expended in overcoming journal, thrust bearing and other friction, and in driving parasitic equipment such as governors and integral oil pumps. These losses are a function of the turbine speed and remain substantially constant at all loads for constant speed turbines. For marine turbines employing high rotative speeds, the mechanical losses will amount to about 2 percent of the internal work at full power.

In dealing with turbine outputs, it is customary to speak of power rather than work. This introduces no particular complication, since they are related through the steam flow and the applicable conversion constants. In Engineers' units:

$$\text{hp} = \frac{wk}{J} \times \frac{\dot{M}}{2545}$$

and

$$\frac{wk}{J} = \frac{2545 \times hp}{\dot{M}} \text{ Btu/lb}$$

where

$$\frac{wk}{J} = \text{work, Btu/lb}$$

$$\dot{M} = \text{steam flow, lb/hr}$$

and \quad 1 hp = 2545 Btu/hr

From the first paragraph:

$$(wk)_{\text{brake}} = (wk)_t - (wk)_{ml}$$

where the subscript *ml* refers to mechanical losses.

STEAM TURBINES

Similarly:

$$BHP = IHP - MLHP \quad (13\text{-}11)$$

where
$BHP =$ brake horsepower
$IHP =$ internal horsepower
$MLHP =$ mechanical loss horsepower.

The *brake engine efficiency* is defined as the ratio of the brake work to the turbine available energy:

$$\eta_{eb} = \frac{(wk)_{\text{brake}}}{(J)(A.E.)_t} \quad (13\text{-}12)$$

The *mechanical efficiency* is defined as the ratio of the brake output to the internal output.

$$\eta_m = \frac{(wk)_{\text{brake}}}{(wk)_t} = \frac{BHP}{IHP} = 1 - \frac{MLHP}{IHP} \quad (13\text{-}13)$$

By dividing both numerator and denominator of the second term in the above equation by $(A.E.)_t$, we get:

$$\eta_m = \frac{\eta_{eb}}{\eta_{ei}} \quad \text{or} \quad \eta_{eb} = \eta_m \eta_{ei}$$

The brake engine efficiency provides a combined evaluation of the thermodynamic and the mechanical effectiveness of a turbine and is therefore important. It should be noted, however, that *the mechanical losses have no effect whatsoever on the condition of the working substance and can never be reflected in the condition curve.*

In marine practice, it is common to deal with output in terms of *shaft horsepower,* defined as the power output at the bull gear coupling in geared turbine units. The shaft horsepower (SHP) will be less than the brake horsepower by the mechanical loss horsepower occurring in the main reduction gears. Otherwise there is no difference in approach, and the equations developed above are directly applicable if *"shaft"* is interpolated mentally wherever *"brake"* appears. For good double reduction geared units, the total mechanical losses, including gear losses, will run about 4 to 5 percent of the turbine internal output at full power. At lower speeds the mechanical efficiency will be somewhat less than at full power, since the internal horsepower falls off more rapidly with decreased speed than does the mechanical loss horsepower.

STEAM TURBINES

Example 13-8: A geared turbine marine propulsion unit delivers 35,000 shaft horsepower at full power with a shaft speed of 240 rpm. The mechanical efficiency of the unit under these conditions is 95%. At 122 rpm the unit delivers 4750 shp, receiving steam at the throttle at 1250 psia and 940°F at the rate of 32,150 lb/hr and exhausting to a condenser at 0.7 psia. Assuming the mechanical losses vary as the square of the rotative speed, find for the 122-rpm condition:

(a) the estimated mechanical loss, hp
(b) the mechanical efficiency
(c) the shaft engine efficiency
(d) the internal engine efficiency

Solution:

(a)
$$N_A = 240 \text{ rpm} \qquad N_B = 122 \text{ rpm}$$
$$(SHP)_A = 35,000 \text{ hp} \qquad (SHP)_B = 4750 \text{ hp}$$
$$(\eta_m)_A = 0.95 \qquad \dot{M}_B = 32,150 \text{ lb/hr}$$

$$(IHP)_A = \frac{SHP}{\eta_m} = \frac{35,000}{0.95} = 36,840 \text{ hp}$$

$$(MLHP)_A = IHP - SHP = 36,840 - 35,000 = 1840 \text{ hp}$$

$$(MLHP)_B = (MLHP)_A \left(\frac{N_B}{N_A}\right)^2 = 1840 \left(\frac{122}{240}\right)^2 = \underline{475 \text{ hp}}$$

(b)
$$SHP = 4750, \quad MLHP = 475, \quad IHP = 4750 + 475 = 5225$$

$$\eta_m = \frac{SHP}{IHP} = \frac{4750}{5225} = 0.909 = \underline{90.9\%}$$

(c) Refer to Fig. 13-10:

$$p_1 = 1250 \text{ psia}, \quad t_1 = 940°F, \quad p_a = 0.7 \text{ psia}$$
$$h_1 = 1462.6 \qquad s_1 = s_a = 1.5994$$
$$h_a = 876.0$$

$$(A.E.)_t = h_1 - h_a = 1462.6 - 876.0 = 586.6 \text{ Btu/lb}$$

$$\frac{(wk)_{\text{shaft}}}{J} = \frac{2545 \times SHP}{\dot{M}} = \frac{2545 \times 4750}{32150} = 376.0 \text{ Btu/lb}$$

$$\eta_{es} = \frac{(wk)_{\text{shaft}}}{J(A.E.)_t} = \frac{376.0}{586.6} = 0.641 = \underline{64.1\%}$$

(d)
$$\frac{(wk)_t}{J} = \frac{2545 \times IHP}{\dot{M}} = \frac{2545 \times 5225}{32150} = 413.6 \text{ Btu/lb}$$

$$\eta_{ei} = \frac{(wk)_t}{J(A.E.)_t} = \frac{413.6}{586.6} = 0.705 = \underline{70.5\%}$$

or
$$\eta_{ei} = \frac{\eta_{es}}{\eta_m} = \frac{64.1}{90.9} = 0.705 = \underline{70.5\%}$$

Note that the efficiencies calculated above are relatively low, reflecting the fact that the plant is being operated well below the so-called design point.

13-12. Other Performance Criteria. Two other criteria are frequently quoted in connection with turbine performance. These are *heat rate* and

steam rate. Like engine efficiency they may be based on the output at any one of several points. *These quantities evaluate the cycle in which the turbine operates as well as the turbine itself.* Heat rate is defined as

$$HR = \frac{\dot{M}(h_1 - h_f)}{P} \text{ Btu/hp} \cdot \text{hr} \qquad (13\text{-}14)$$

where \dot{M} = steam flow, lb/hr
 h_1 = enthalpy of the steam at the throttle, Btu/lb
 h_f = enthalpy of the liquid at condenser pressure, Btu/lb
 P = power output, hp

The value of the heat rate naturally depends on where the power output is measured, and this point must be carefully specified. The value obtained indicates the amount of energy, including pump work and heat supplied in the boiler, that must be furnished to deliver unit work. By another interpretation of the units, *heat rate is the rate of energy supply required to produce one horsepower at the point selected for measuring output.*

Note that heat rate will depend on the cyclic effectiveness as well as the turbine engine efficiency. It is of primary interest in matching a turbine to particular inlet and exhaust conditions. Heat rate is of real economic significance since it reflects the fuel cost per unit of output.

Steam rate, also known as *water rate,* is defined as the throttle steam flow per horsepower output.

$$WR = SR = \frac{\dot{M}}{P} \text{ lb/hp} \cdot \text{hr} \qquad (13\text{-}15)$$

Note the differentiation between **steam flow** *(\dot{M}) pounds per hour and* **steam rate** *(SR) pounds per horsepower hour.* Steam rate indicates the amount of working substance that must be circulated per unit time to produce unit output. Again the point of power determination must be established and the value of the steam rate will depend upon the cyclic as well as the turbine characteristics. The steam rate is related to the heat rate by the following equation:

$$HR = \frac{\dot{M}}{P}(h_1 - h_f) = SR(h_1 - h_f)$$

In estimating the steam flow and steam rate to a turbine under varying load conditions, it is convenient to use the *Willans line,* which is simply a plot of the steam flow versus horsepower output. As a rule this line is substantially straight between no-load and full-load conditions. If data on these two points are available, the line can be constructed and steam rate

STEAM TURBINES

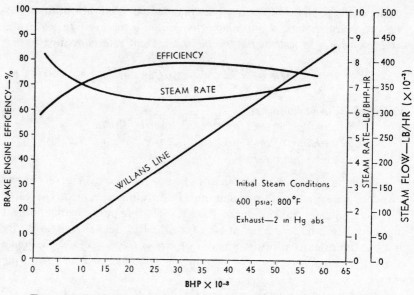

FIGURE 13-12. Performance curves of a large marine turbine.

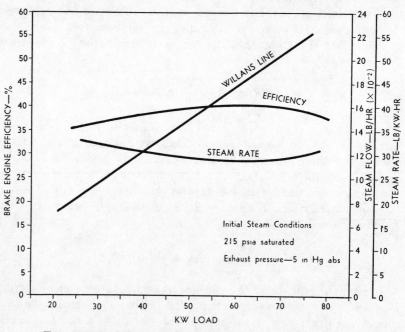

FIGURE 13-13. Performance curves of a small auxiliary turbine for generator drive.

STEAM TURBINES

and steam flow predicted for intermediate loads. Figure 13-12 shows the Willans line along with characteristic curves of brake engine efficiency and steam rate for a large multi-stage marine turbine at various loads. For purposes of comparison, Fig. 13-13, showing similar curves for a small single stage auxiliary turbine, is included.

Example 13-9: Turbine A receives steam at the throttle at 850 psia and 940°F and exhausts to the condenser at 0.7 psia; the throttle steam flow is 164,000 lb/hr. Turbine B receives steam at 620 psia and 900°F, exhausting at 1.0 psia, and the throttle steam flow is 187,000 lb/hr. Under the stated conditions, both turbines deliver 30,000 shaft horsepower with a mechanical efficiency of 96%. Find:
 (a) the water rate, the heat rate, the shaft engine efficiency and the internal engine efficiency for turbine A
 (b) the same quantities for turbine B
Solution: Refer to Fig. 13-10:

(a) $SHP = 30,000$ hp; $\eta_m = 0.96$; $\dot{M}_A = 164,000$ lb/hr
 $p_1 = 850$ psia $p_a = 0.7$ psia
 $t_1 = 940°F$ $h_a = 903.5$
 $h_1 = 1476.5$ $h_{fa} = 58.1$
 $s_1 = s_a = 1.6495$

$$WR = \frac{\dot{M}}{SHP} = \frac{164{,}000}{30{,}000} = 5.47 \text{ lb/shp·hr}$$

$$HR = WR(h_1 - h_{fa}) = 5.47(1476.5 - 58.1) = \underline{7759 \text{ Btu/shp·hr}}$$

$$\frac{(wk)_{shaft}}{J} = \frac{2545 \times SHP}{\dot{M}} = \frac{2545 \times 30{,}000}{164{,}000} = 465.5 \text{ Btu/lb}$$

$$(A.E.)_t = h_1 - h_a = 1476.5 - 903.5 = 573.0 \text{ Btu/lb}$$

$$\eta_{es} = \frac{(wk)_{shaft}}{J(A.E.)_t} = \frac{465.5}{573.0} = 0.812 = \underline{81.2\%}$$

$$\eta_{ei} = \frac{\eta_{es}}{\eta_m} = \frac{0.812}{0.96} = 0.846 = \underline{84.6\%}$$

(b) $p_1 = 620$ psia $p_a = 1$ psia
 $t_1 = 900°F$ $h_a = 934.4$
 $h_1 = 1462.2$ $h_{fa} = 69.7$
 $s_1 = s_a = 1.6726$ $\dot{M}_B = 187{,}000$ lb/hr

$$WR = \frac{\dot{M}}{SHP} = \frac{187{,}000}{30{,}000} = \underline{6.23 \text{ lb/shp·hr}}$$

$$HR = WR(h_1 - h_{fa}) = 6.23(1462.2 - 69.7) = \underline{8675 \text{ Btu/shp·hr}}$$

$$\frac{(wk)_{shaft}}{J} = \frac{2545 \times 30{,}000}{187{,}000} = 408.3 \text{ Btu/lb}$$

$$(A.E.)_t = h_1 - h_a = 1462.2 - 934.4 = 527.8 \text{ Btu/lb}$$

$$\eta_{es} = \frac{408.3}{527.8} = 0.774 = \underline{77.4\%}$$

$$\eta_{ei} = \frac{0.774}{0.96} = 0.806 = \underline{80.6\%}$$

STEAM TURBINES

Problems

13-1. In a simple impulse stage, the blade speed is 150 m/sec and the nozzle angle is 18 degrees. The velocity of the steam leaving the nozzles is 320 m/sec. The bucket entrance and exit angles are both 33 degrees. The bucket velocity coefficient is 0.89. Find (a) the relative velocity entering the buckets; (b) the relative velocity leaving the buckets; (c) the total change of velocity relative to and in the direction of motion of the buckets; (d) the bucket work, kJ/kg; (e) the absolute exit velocity.

Ans. (a) 184 m/sec; (b) 163.8 m/sec; (c) 291.7 m/sec; (d) 43.75 kJ/kg; (e) 90.1 m/sec

13-2. Steam enters the buckets of a simple impulse wheel at an absolute velocity of 1200 ft/sec and leaves with an absolute velocity of 245 ft/sec. The relative entering velocity is 640 ft/sec, and the relative exit velocity is 540 ft/sec. Determine (a) the available energy to the buckets, ft·lbf/lb; (b) the bucket loss, ft·lbf/lb; (c) the unused kinetic energy in the leaving jet, ft·lbf/lb; (d) the bucket work, ft·lbf/lb and Btu/lb; (e) the diagram efficiency; (f) the power developed by the buckets, hp, if steam is supplied at the rate of 5000 lb/hr.

Ans. (a) 22,360 ft·lbf/lb; (b) 1830 ft·lbf/lb; (c) 930 ft·lbf/lb; (d) 19,600 ft·lbf/lb or 25.2 Btu/lb; (e) 87.7%; (f) 49.5 hp

13-3. Steam enters the nozzles of a simple impulse stage with negligible velocity at 190 psia and 500°F. The velocity of the steam leaving the nozzles is 1175 ft/sec, and the stage pressure is 140 psia. The steam leaves the buckets with an absolute velocity of 300 ft/sec and an enthalpy of 1245.5 Btu/lb. Assume the stage efficiency is equal to the nozzle-bucket efficiency and that there is negligible velocity carryover to the next stage. Show the process through the stage on h-s coordinates and find (a) the available energy to the stage, Btu/lb; (b) the nozzle efficiency; (c) the nozzle reheat, Btu/lb; (d) the blade reheat, Btu/lb; (e) the exit reheat, Btu/lb; (f) the enthalpy of the steam entering the next stage, Btu/lb; (g) the stage work, Btu/lb; (h) the diagram efficiency of the buckets; (i) the nozzle-bucket efficiency.

Ans. (a) 29.8 Btu/lb; (b) 92.5%; (c) 2.2 Btu/lb; (d) 3.2 Btu/lb; (e) 1.8 Btu/lb; (f) 1247.3 Btu/lb; (g) 22.6 Btu/lb; (h) 82%; (i) 75.8%

13-4. In a 50 percent reaction stage, the carryover velocity from the preceding stage is 55 m/sec, the stage available energy is 50 kJ/kg, the fixed blade (nozzle) efficiency is 0.94 and the fixed blade velocity coefficient is 0.90. Find the velocity of the steam leaving the fixed blades, m/sec.

Ans. 222.4 m/sec

13-5. In a turbine stage with 30 percent reaction, the steam enters the moving blades with a relative velocity of 350 ft/sec, the stage available energy is 25 Btu/lb, the reactive effectiveness is 0.90 and the moving blade velocity coefficient is 0.92. Find the relative velocity of the steam leaving the moving blades, ft/sec.

Ans. 665 ft/sec

13-6. Steam enters a 50 percent reaction stage in a low pressure turbine with negligible carryover velocity at 12.0 psia and a moisture content of 0.06. The steam leaves the fixed blades with a velocity of 670 ft/sec. The relative velocities entering and leaving the moving blades are 230 ft/sec and 670 ft/sec, respectively. The steam at exit from the moving blades has an absolute velocity of 230 ft/sec and a pressure of 9.0 psia. Assume the combined blade efficiency and the stage efficiency are equal and find (a) the stage work, Btu/lb; (b) the available energy to the stage, Btu/lb; (c) the combined blade efficiency.

Ans. (a) 15.8 Btu/lb; (b) 19.2 Btu/lb; (c) 82.3%

STEAM TURBINES

13-7. The rotative speed of a high pressure turbine at full power is 6000 rpm. The first stage is of the simple impulse configuration, and the nozzles receive steam at 900 psia and 900°F with negligible velocity. The pitch diameter of the first stage buckets is 30 inches and the nozzle angle is 17 degrees. Find for these conditions (a) the blade speed, ft/sec; (b) the ideal blade speed–steam speed ratio; (c) the corresponding absolute ("spouting") velocity of the steam leaving the nozzles, ft/sec; (d) the enthalpy drop across the nozzles assuming an isentropic process, Btu/lb; (e) the entropy and enthalpy of the steam leaving the nozzles and its estimated pressure and temperature from the Mollier chart.

Ans. (a) 785 ft/sec; (b) 0.478; (c) 1642 ft/sec; (d) 53.8 Btu/lb; (e) 1.6257 Btu/lb·°R; 1398.1 Btu/lb; 620 psia, 790°F.

13-8. Steam is supplied to the throttle of a large turbine at 600 psia and 860°F. The pressure drop from the throttle entrance to the steam chest is negligible. At the exhaust flange the absolute pressure is 0.7 psia and the steam velocity is 400 ft/sec. The internal power developed is 20,000 hp and the steam flow is 135,000 lb/hr. Show the process on h-s coordinates and find (a) the available energy, Btu/lb; (b) the internal turbine work, Btu/lb; (c) the internal engine efficiency; (d) the leaving loss, Btu/lb; (e) the enthalpy at the state line end point, Btu/lb; (f) the required exhaust trunk cross-sectional area, ft².

Ans. (a) 531.5 Btu/lb; (b) 377.0 Btu/lb; (c) 70.9%; (d) 3.2 Btu/lb; (e) 1060.7 Btu/lb; (f) 42.1 ft²

13-9. Consider the operating conditions of the turbine described in the preceding problem to be unchanged except that the operator makes an improper nozzle setup such that the pressure after the throttle becomes 400 psia. Estimate the required steam flow, lb/hr, under these conditions, assuming the turbine losses other than throttling remain the same.

Ans. 144,070 lb/hr

13-10. An auxiliary turbo-generator set delivers 600 kW at the generator terminals. Steam is supplied to the throttle at 600 psia and 800°F. The exhaust pressure is 1.0 psia. The engine efficiency based on the generator output is 60 percent. Find the steam flow through the turbine, lb/hr.

Ans. 6900 lb/hr

13-11. A marine geared turbine unit delivers 18,500 kW at the bull gear coupling at full power with a propeller shaft speed of 400 rpm. Under these conditions, the mechanical losses total 925 kW. At 35 km/hr, the propeller makes 200 rpm and the shaft power is 2000 kW. Assume the mechanical losses vary as the square of the rpm and find: (a) the internal power developed at full power, kW; (b) the mechanical efficiency of the unit at full power; (c) the estimated mechanical loss power at 35 km/hr; (d) the estimated mechanical efficiency at this speed.

Ans. (a) 19,425 kW; (b) 95.2%; (c) 231 kW; (d) 89.6%

13-12. A large turbine receives steam at the throttle at 1000 psia and 900°F, exhausting to a condenser at 1.5 in. Hg absolute. The brake output is 40,000 hp, and the mechanical losses amount to 800 hp. The steam flow is 250,000 lb/hr. Find (a) the available energy, Btu/lb; (b) the brake work, Btu/lb; (c) the brake engine efficiency; (d) the brake steam rate, lb/bhp·hr; (e) the brake heat rate, Btu/bhp·hr; (f) the enthalpy of the steam at the exhaust point, Btu/lb.

Ans. (a) 562.8 Btu/lb; (b) 407.2 Btu/lb; (c) 72.4%; (d) 6.25 lb/bhp·hr; (e) 8678 Btu/bhp·hr; (f) 1032.8 Btu/lb

CHAPTER 14
HEAT TRANSMISSION

14-1. Introduction. In the preceding chapters it was assumed that the working substance of these discussions would receive and reject heat as required by the various processes. In a majority of these applications the working substance was confined within a system of one or more devices, while the heat which it received was furnished from an external source. In other applications, such as the regenerative cycles, the working fluid rejected heat to a storage device to be reabsorbed within a cycle, or there was a direct exchange of heat between the fluid at different temperature levels within the cycle. *In all cases there was a flow of heat from a fluid to a solid boundary and thence to a fluid.* Since heat flow is necessary for all thermodynamic cycles, a study of the laws governing heat transfer is essential to an understanding of the subject.

14-2. Modes of Heat Transfer. The two principal modes of heat transfer, energy transferred by a temperature difference, are conduction and radiation. Heat flow through a body (substance), or between bodies (substances) in direct contact, due to molecular motion is termed **conduction. Radiation** is the means by which heat is transferred between separated bodies through intervening space by electromagnetic wave motion. Both modes require a temperature difference within or between the body or bodies. Both result in a net loss of energy from a high temperature body or substance, and an equivalent net gain by the associated low temperature body or fluid. Time is necessary for the transfer of energy by either method. As a corollary, *heat flow is the inevitable result of any temperature difference which exists for a finite period of time.*

Convection is the transportation or flow of thermal energy within a fluid accompanied by an intermixing of one part of the fluid with another. In natural convection the fluid movement is caused by differences in density that are the result of unequal temperatures within the fluid. In forced convection the fluid motion is produced mechanically. In either case, convection influences the *conduction* of heat between a fluid and a solid surface. Hence the effects of convection are included in the empirical coefficients used with the conduction equations, and are not separately considered.

14-3. Conduction of Heat Through a Homogeneous Material. The basic equation for the conduction of heat is derived from Fourier's Law, which

HEAT TRANSMISSION

may be mathematically written for one dimension as

$$\frac{\delta Q}{d\tau} = -kA\frac{dt}{dL}$$

in which

$\frac{\delta Q}{d\tau}$ = amount of heat flow (δQ) in differential time ($d\tau$), Btu/hr (W)

A = area (normal to the path of flow) over which heat is transmitted, ft^2 (m^2)

$\frac{dt}{dL}$ = temperature gradient through material, or rate of change of temperature, t, with length of heat flow path, L, °F/ft (°C/m)

k = a dimensional coefficient which is characteristic of the material through which the heat flows.

As thus stated, Fourier's Law restricts the flow of heat in direction (along path L and in the direction of diminishing temperature) but permits a variation of temperature with time as well as with position along the path of flow. This unstable condition is eliminated from many engineering problems by the establishment of steady flow and thermal equilibrium. For this latter condition, *imposed temperatures remain constant with time,* an unvarying temperature gradient is established along the heat flow path *and the rate of heat flow is constant throughout path L.* For this

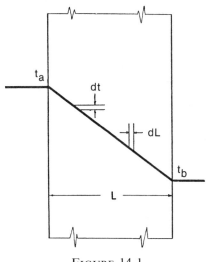

FIGURE 14-1

HEAT TRANSMISSION

steady flow of heat, Fourier's Law may be stated as

$$\dot{Q} = -kA\frac{dt}{dL} \text{ Btu/hr (W)} \tag{14-1}$$

in which \dot{Q} is the constant rate of heat flow through a particular material for the temperature gradient, dt/dL. (See Figure 14-1.)

The proportionality factor, k, is defined as the **coefficient of thermal conductivity** or, more simply, the **thermal conductivity** of a material or substance. As is necessary for various other physical properties, thermal conductivity is experimentally determined and is dimensionally expressed as

$$k = \frac{\text{Btu/hr}}{\text{ft}^2 \cdot {}^\circ\text{F/ft}} \left(\frac{\text{W}}{\text{m}^2 \cdot {}^\circ\text{C/m}}\right)$$

$$k = \frac{\text{Btu}}{\text{hr} \cdot {}^\circ\text{F} \cdot \text{ft}} \left(\frac{\text{W}}{\text{m} \cdot {}^\circ\text{C}}\right)$$

As pointed out in Art. 2-9, the degree intervals for the Fahrenheit and the Rankine temperature scales are equal. Hence

$$dt \ {}^\circ\text{F} \ ({}^\circ\text{C}) = dt \ {}^\circ\text{R} \ (\text{K})$$

and the thermal conductivity can also be expressed as

$$k = \frac{\text{Btu}}{\text{hr} \cdot {}^\circ\text{R} \cdot \text{ft}} \left(\frac{\text{W}}{\text{m} \cdot \text{K}}\right)$$

For a majority of engineering materials, k can be related to temperature according to the expression

$$k = k_0(1 + at)$$

in which k_0 is the thermal conductivity at a selected base temperature and a is a temperature coefficient.

Fourier's equation may now be developed for the case of steady heat flow in which the area along the path of flow is constant. Thus

$$\frac{\dot{Q}}{A}\int_0^L dL = -k_0 \int_{t_a}^{t_b} (1 + at)\, dt$$

$$\frac{\dot{Q}L}{A} = (t_a - t_b)k_0\left[1 + a\left(\frac{t_a + t_b}{2}\right)\right]$$

The quantity $k_0[1 + a(t_a + t_b)/2]$ determines an average value of thermal conductivity for the temperature, $(t_a + t_b)/2$, and the conduction equation of general utility is

$$\dot{Q} = \frac{kA(t_a - t_b)}{L} \text{ Btu/hr (W)}$$

HEAT TRANSMISSION

where \dot{Q} = rate of heat transfer, Btu/hr (W)
k = average thermal conductivity, Btu/hr · °F · ft (W/m · °C)
t_a = temperature of warmer surface of material, °F (°C)
t_b = temperature of cooler surface of material, °F (°C)
L = thickness of material between surfaces, ft (m)
A = area through which heat flows, sq ft (m²)

Example 14-1: A solid brick wall separates two rooms. The surface temperature on one side of the wall is 50°C, the surface temperature on the opposite side is 15°C, k for the brick is 1.32 W/m · °C and the wall is 20 cm thick. Calculate the heat transfer for a wall surface area of 10 m².
Solution:

$$t_a = 50°C \quad t_b = 15°C$$
$$A = 10 \text{ m}^2 \quad k = 1.32 \text{ W/m} \cdot °C$$
$$L = 0.20 \text{ m}$$

$$\dot{Q} = \frac{kA(t_a - t_b)}{L} = \frac{(1.32)(10)(50 - 15)}{(0.20)} = \underline{2,310 \text{ W}}$$

14.4. Fluid Films and the Effect of Convection Upon Heat Transfer. In all cases where fluids have contact with solids there is a surface layer of fluid which is substantially motionless, or stagnant. Such a contact layer is known as a **fluid film** and constitutes a layer of material through which heat must flow in passage to and through the confining solid. For any particular substance, the fluid film has the thermal properties of the main fluid body, including that of thermal conductivity. However, the film thickness is small and is variable, depending upon the viscosity of the fluid and the surface turbulence provided by the operating conditions. For free convection and for laminar flow the film thickness is maximum and is in the nature of 0.04 inch. Forced convection, produced by high fluid velocity or by mechanical means such as baffles, materially decreases the film thickness and is requisite for many devices. In either case the film heat transfer property must be separately investigated under a variety of fluid flow conditions. Since the film thickness in many cases cannot be measured, the facility with which it transfers heat is stated as a **surface** or **film coefficient.** A fluid film coefficient is therefore *the rate of heat transfer of a particular fluid for a particular condition of surface turbulence per unit area and unit temperature difference.* The fluid film coefficient is conventionally designated by the symbol **h** and dimensionally,

$$h = \frac{\text{Btu}}{\text{hr} \cdot \text{sq ft} \cdot °F} \left(\frac{W}{m^2 \cdot °C}\right)$$

Since a temperature drop accompanies the flow of heat through any material, the surface temperature, t_a, must be less than the fluid temperature, t_1; and on the low temperature side t_b must be greater than t_2, as

shown in Fig. 14-2. But because of the presence of the film, the surface temperatures t_a and t_b cannot be easily measured. Therefore, a new method for determining the conduction of heat through a solid and film(s) must be provided.

FIGURE 14-2

14-5. Overall Coefficient of Heat Transfer. The solution to the difficulties introduced by the fluid film phenomena is found in the employment of an **overall heat transfer coefficient** that may be used in conjunction with measurable fluid temperatures. Designated by the symbol U, *the overall coefficient of heat transfer is defined as the rate of heat flow between separated fluids per unit area and unit temperature difference between fluids.* As may be seen from the definition, U is an inclusive thermal *conductance* which can be determined for a specified combination of fluids and materials in a particular operating circumstance.

Again referring to Fig. 14-2, it is apparent that for steady flow between high and low temperature fluids, the heat which flows through the solid in unit time must have been passed by the high temperature surface film in

the preceding unit of time and will be transferred by the low temperature fluid film in the following unit of time. Or, mathematically,

$$\dot{Q}_{\text{film}_1} = \dot{Q}_{\text{solid}} = \dot{Q}_{\text{film}_2} = \dot{Q} \text{ Btu/hr (W)}$$

An equation for determining the rate of flow through a solid has already been developed; that is,

$$\dot{Q} = \dot{Q}_{\text{solid}} = \frac{k}{L} A_s(t_a - t_b)$$

A similar equation for the heat transferred by a fluid film is,

$$\dot{Q} = \dot{Q}_{\text{film}_1} = h_1 A_1 (t_1 - t_a),$$

and,

$$\dot{Q} = \dot{Q}_{\text{film}_2} = h_2 A_2 (t_b - t_2)$$

To obtain a mathematical statement for the overall heat transfer coefficient, U, the equations are re-arranged as follows:

$$t_1 - t_a = \frac{\dot{Q}}{A_1 h_1}$$

$$t_a - t_b = \frac{L\dot{Q}}{kA_s}$$

$$t_b - t_2 = \frac{\dot{Q}}{A_2 h_2}$$

and by addition,

$$t_1 - t_2 = \frac{\dot{Q}}{A_1 h_1} + \frac{L\dot{Q}}{kA_s} + \frac{\dot{Q}}{A_2 h_2}$$

From the definition of U,

$$\dot{Q} = UA(t_1 - t_2) \text{ Btu/hr (W)} \qquad (14\text{-}2)$$

from which

$$t_1 - t_2 = \frac{\dot{Q}}{UA} = \frac{\dot{Q}}{A_1 h_1} + \frac{L\dot{Q}}{kA_s} + \frac{\dot{Q}}{A_2 h_2}$$

and

$$\frac{1}{UA} = \frac{1}{A_1 h_1} + \frac{L}{kA_s} + \frac{1}{A_2 h_2}$$

HEAT TRANSMISSION

For plane surfaces, $A = A_1 = A_s = A_2$, thence,

$$\frac{1}{U} = \frac{1}{h_1} + \frac{L}{k} + \frac{1}{h_2}$$

$$U = \frac{1}{\dfrac{1}{h_1} + \dfrac{L}{k} + \dfrac{1}{h_2}}, \text{Btu/hr} \cdot \text{sq ft} \cdot {}^\circ\text{F} \ (\text{W/m}^2 \cdot {}^\circ\text{C}) \qquad (14\text{-}3)$$

An over-all heat transfer coefficient can be calculated for composite constructions of two or more solid materials with liquid, vapor or gaseous surface films or combinations thereof. Application of the general conduction equation is predicated primarily upon the determination of U, since in the usual case the operating temperatures can be measured and the heat flow area calculated with little difficulty.

Example 14-2: A typical furnace side wall is constructed of a 1-in. layer of diatomaceous earth insulating block, and $2\frac{1}{2}$ in. high temperature insulating brick faced with $4\frac{1}{2}$ in. firebrick within a $\frac{1}{8}$-in. steel casing. Average values of thermal conductivities are: diatomaceous earth insulating block, 0.063; insulating brick, 0.62; firebrick, 4.0; steel casing, 26 Btu/hr \cdot °F \cdot ft. Average film coefficients are 3.0 and 2.2 Btu/hr \cdot sq ft \cdot °F for inner and outer surface films, respectively. The wall area is 50 sq ft, the average gas temperature is 2100°F and the ambient air temperature is 100°F. (a) Calculate the heat transfer coefficient, U, for this furnace wall, and (b) calculate the heat transferred by the wall because of conduction.

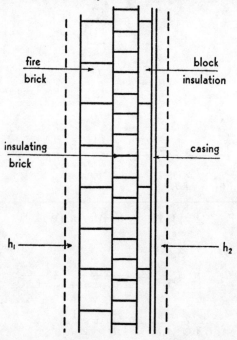

Solution:

(a) $k_{FB} = 4.0$, $k_{IB} = 0.62$, $k_{DE} = 0.063$
$k_S = 26$ Btu/hr · °F · ft
$h_1 = 3.0$, $h_2 = 2.2$ Btu/sq ft · hr · °F

$$\frac{1}{U} = \frac{1}{h_1} + \frac{L_{FB}}{k_{FB}} + \frac{L_{IB}}{k_{IB}} + \frac{L_{DE}}{k_{DE}} + \frac{L_S}{k_S} + \frac{1}{h_2}$$

$$= \frac{1}{3.0} + \frac{4.5}{12 \times 4.0} + \frac{2.5}{12 \times 0.62} + \frac{1}{12 \times 0.063} + \frac{0.125}{12 \times 26} + \frac{1}{2.2}$$

$$= 0.333 + 0.094 + 0.336 + 1.323 + 0.0004 + 0.454$$

$$U = \frac{1}{2.5404} = \underline{0.394 \text{ Btu/hr · sq ft · °F}}$$

(b) $\dot{Q} = UA(t_1 - t_2) = (0.394)(50)(2100 - 100)$
$\dot{Q} = \underline{39{,}400 \text{ Btu/hr}}$

14-6. The Resistance Concept. There is a marked similarity of analysis between heat flow by conduction and the flow of electricity. The temperature difference, Δt, is a thermal potential similar to voltage, ΔE; the rate of heat flow, \dot{Q}, is analogous to electrical current, I; and the overall heat transfer coefficient, U, is comparable to the conductance of several electrical resistances in series. In this sense, the thermal **resistivity** of a material is the reciprocal of its thermal conductivity, or

$$r = \frac{1}{k}$$

Similarly the **resistance** of a fluid film h_1, having area A_1, is

$$R_1 = \frac{1}{A_1 h_1}$$

the **resistance** of a solid of thickness L and parallel surfaces of area A_s is

$$R = \frac{L}{kA_s},$$

and the **total resistance** of a composite section is

$$R_t = \frac{1}{UA}$$

In many engineering problems, the temperatures of the material surfaces must be carefully considered. In the case of the furnace wall exemplified in the preceding article, insulating brick should not be used with temperatures greater than 2500°F, and the diatomaceous insulating material cannot withstand temperatures greatly in excess of 1500°F; and

HEAT TRANSMISSION

a refractory (firebrick) lining is required regardless of the operating temperature.

It has already been shown that

$$\dot{Q} = UA(t_1 - t_2) = h_1 A_1(t_1 - t_a) = \frac{k}{L} A_s(t_a - t_b) \cdots$$

and on the basis of resistance, Ohm's Law ($I = E/R$) has this thermal counterpart:

$$\dot{Q} = \frac{t_1 - t_a}{R_1} = \frac{t_a - t_b}{R_2} = \frac{t_1 - t_2}{R_t}$$

Thus *the temperature drop across a material varies directly with the resistance of the substance,* and the temperature drop across any material within a system can be calculated in terms of the resistance of the material, the total resistance and the temperature difference across the system. For the situation illustrated in Fig. 14-3,

FIGURE 14-3

HEAT TRANSMISSION

$$t_1 - t_a = \frac{R_1}{R_t}(t_1 - t_2)$$

and

$$t_1 - t_b = \frac{R_1 + R_2}{R_t}(t_1 - t_2)$$

Example 14-3: Calculate the temperature at the interface between the insulating brick and diatomaceous insulating block of Ex. 14-2.

Solution: For parallel surfaces, $A = A_1 = A_{FB} = A_{IB}$, and from Ex. 14-2,

$$R_1 = 0.333/A_1, \quad R_{FB} = 0.094/A_{FB}, \quad R_{IB} = 0.336/A_{IB}, \quad R_t = 2.54/A$$
$$t_1 = 2100°F, \quad t_2 = 100°F$$

$$\Delta t \text{ to interface} = \left(\frac{R_1 + R_{FB} + R_{IB}}{R_t}\right)(t_1 - t_2)$$

$$= \frac{A}{A}\left(\frac{0.333 + 0.094 + 0.336}{2.540}\right)(2100 - 100)$$

$$\Delta t = 600°F$$

temperature at interface $= 2100 - 600 = \underline{1500°F}$

14-7. Conduction Through Walls of Nonuniform Area. The overall conduction equation cannot be applied directly to problems in which heat flow occurs through tube walls, insulated pipe, and so on. Under equilibrium conditions, the heat flow path is always normal to the material surface. For curved walls, the path, L, is radial (as shown in Fig. 14-4) and the area of the outer surface is greater than that of the inner.

For *unit axial length* of tube having an average circumferential area, A_m, the heat flow is,

$$\dot{Q} = \frac{kA_m(t_a - t_b)}{L} = \frac{kA_m(t_a - t_b)}{r_2 - r_1}$$

At any intermediate radius, r, the circumferential area is $2\pi r$, and by Equation (14-1) the heat flow is,

$$\dot{Q} = -kA\left(\frac{dt}{dL}\right) = -k(2\pi r)\left(\frac{dt}{dr}\right)$$

Then,

$$\dot{Q}\left(\frac{dr}{r}\right) = -2\pi k(dt)$$

$$\dot{Q}\int_{r_1}^{r_2}\frac{dr}{r} = -2\pi k \int_{t_a}^{t_b} dt$$

$$\dot{Q}\left(\ln\frac{r_2}{r_1}\right) = 2\pi k(t_a - t_b)$$

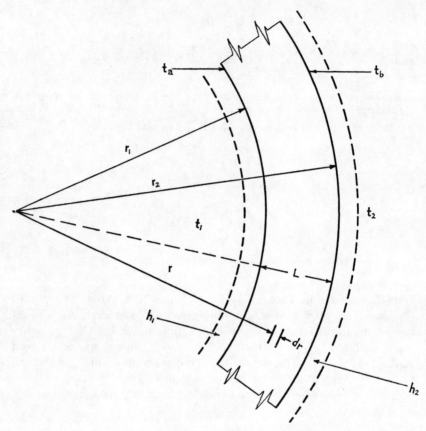

FIGURE 14-4

and combining equations,

$$\dot{Q} = \frac{2\pi k(t_a - t_b)}{\ln\dfrac{r_2}{r_1}} = \frac{kA_m(t_a - t_b)}{r_2 - r_1}$$

Thence, $A_m = 2\pi r_m = \dfrac{2\pi(r_2 - r_1)}{\ln\dfrac{r_2}{r_1}}$, ft²/ft length (m²/m length) (14-4)

By definition, r_m *is a logarithmic mean radius,* and A_m *is conventionally termed the log mean area* of the pipe wall or material.

The effect of variable area upon heat flow rate is usually included in the calculation of the overall heat transfer coefficient for the given situation. It

has previously been shown that

$$\frac{1}{UA} = \frac{1}{h_1 A_1} + \frac{L}{kA_s} + \frac{1}{h_2 A_2}$$

and for a curved wall

$$\frac{1}{UAx} = \frac{1}{h_1 A_1 x} + \frac{L}{kA_m x} + \frac{1}{h_2 A_2 x}$$

where, A_1 and A_2 = inner and outer surface areas per unit axial length respectively of the wall, ft²/ft (m²/m)

A_m = log mean area of section, ft²/ft (m²/m)
x = length of pipe or tube, ft (m)
A = a characteristic area of the section, ft²/ft (m²/m).

The area, A, associated with the overall coefficient is usually specified to be that of the external surface, since this area can be easily determined.

A simplification may be made in terms of radii, and the final equation for the overall coefficient of heat transfer for curved walls with $A = A_2$ is

$$\frac{1}{U} = \frac{1}{h_1 \dfrac{A_1 x}{Ax}} + \frac{L}{k \dfrac{A_m x}{Ax}} + \frac{1}{h_2 \dfrac{A_2 x}{Ax}}$$

$$U = \frac{1}{\dfrac{1}{h_1 \dfrac{r_1}{r_2}} + \dfrac{L}{k \dfrac{r_m}{r_2}} + \dfrac{1}{h_2}} \qquad \begin{array}{l} \text{Btu/hr} \cdot \text{ft}^2 \cdot {}^\circ\text{F} \\ \text{(based on external area)} \\ (\text{W/m}^2 \cdot {}^\circ\text{C}) \end{array}$$

As in the case of heat flow through a constant area, the above equation can be expanded to apply to composite systems, as will be shown in the accompanying typical problem. It is emphasized that A is arbitrarily selected, and that any statement of overall coefficient for curved sections must specify the particular area upon which it is based.

Example 14-4: A high temperature steam line is covered with two successive layers of insulation. The layer in contact with the pipe is a 1½-in. thickness of asbestos for which k is 0.08 Btu/hr · °F · ft. The asbestos is covered with a 1-in. thickness of magnesia insulation which has a value of k of 0.04. The internal pipe diameter is 2.90 in., the pipe wall thickness is 0.30 in. and k for the pipe is 26 Btu/hr · °F · ft. The steam temperature is 800°F, and the internal surface film coefficient is 40 Btu/hr · sq ft · °F, while the ambient outer temperature is 100°F and the outer surface film coefficient is 1.2. Calculate (a) the value of U based upon the external area of the magnesia covering and (b) the heat loss from the steam for a length of 180 feet of pipe, Btu/hr.

HEAT TRANSMISSION

Solution:

$h_1 = 40$, $h_2 = 1.2$ Btu/hr·ft²·°F
$k_a = 0.08$, $k_m = 0.04$, $k_p = 26$ Btu/hr·°F·ft
$t_1 = 800$°F, $t_2 = 100$°F, $x = 180$ ft length

(a)
$$r_p = \text{mean pipe radius} = \frac{r_2 - r_1}{\ln \dfrac{r_2}{r_1}} = \frac{1.75 - 1.45}{\ln \dfrac{1.75}{1.45}} = 1.60 \text{ in.}$$

$$r_a = \text{mean asbestos radius} = \frac{r_3 - r_2}{\ln \dfrac{r_3}{r_2}} = \frac{3.25 - 1.75}{\ln \dfrac{3.25}{1.75}} = 2.42 \text{ in.}$$

$$r_m = \text{mean magnesia radius} = \frac{r_4 - r_3}{\ln \dfrac{r_4}{r_3}} = \frac{4.25 - 3.25}{\ln \dfrac{4.25}{3.25}} = 3.73 \text{ in.}$$

$$\frac{1}{U} = \frac{1}{h_1 \dfrac{r_1}{r_4}} + \frac{L_p}{k_p \dfrac{r_p}{r_4}} + \frac{L_a}{k_a \dfrac{r_a}{r_4}} + \frac{L_m}{k_m \dfrac{r_m}{r_4}} + \frac{1}{h_2}$$

$$= \frac{1}{40 \left(\dfrac{1.45}{4.25}\right)} + \frac{0.3}{12(26)\left(\dfrac{1.60}{4.25}\right)} + \frac{1.5}{12(0.08)\left(\dfrac{2.42}{4.25}\right)}$$

HEAT TRANSMISSION

$$+ \frac{1.0}{12(0.04)\left(\frac{3.73}{4.25}\right)} + \frac{1}{1.2}$$

$$= 0.073 + 0.003 + 2.744 + 2.374 + 0.833$$

$$U = \frac{1}{6.027} = 0.1659 \text{ Btu/hr} \cdot \text{ft}^2 \cdot {}^\circ\text{F} \text{ (external area)}$$

(b) $\dot{Q} = UA(t_1 - t_2) = U\left(\frac{2\pi r_4}{12} x\right)(t_1 - t_2)$

$$\dot{Q} = (0.1659)\left[\frac{2\pi(4.25)}{12} \, 180\right](800 - 100) = \underline{46{,}516 \text{ Btu/hr}}$$

Frequently the difference between the log mean and the arithmetic mean radius is negligible. For ease in computation, it is convenient to use the arithmetic average when the ratio of the larger to the smaller radius is 1.5 or less.

Example 14-5: Calculate the overall heat transfer coefficient for a tubular heat exchanger wherein the liquid carried in the tubes is heated by steam which surrounds them. The tubes are $\frac{5}{8}$-in. Admiralty metal with a wall thickness of 0.049 in. The steam surface film coefficient is 1250 Btu/hr \cdot ft² \cdot °F, k for Admiralty metal is 70 Btu/hr \cdot °F \cdot ft and the liquid surface film coefficient is 20 Btu/hr \cdot ft² \cdot °F.
Solution:

$$h_1 = 20, \quad h_2 = 1250 \text{ Btu/hr} \cdot \text{ft}^2 \cdot {}^\circ\text{F}$$

$$r_2 = 0.3125 \text{ in.}, \quad r_1 = 0.2635 \text{ in.}, \quad k = 70 \text{ Btu/hr} \cdot {}^\circ\text{F} \cdot \text{ft}$$

$$r_t = \frac{r_1 + r_2}{2} = \frac{0.2635 + 0.3125}{2} = 0.288 \text{ in.}$$

$$\frac{1}{U} = \frac{1}{h_1 \frac{r_1}{r_2}} + \frac{L}{k_t \frac{r_t}{r_2}} + \frac{1}{h_2}$$

$$= \frac{1}{20\left(\frac{0.2635}{0.3125}\right)} + \frac{0.049}{12(70)\left(\frac{0.288}{0.3125}\right)} + \frac{1}{1250}$$

$$= 0.0593 + 0.00006 + 0.0008$$

$$U = \frac{1}{0.0602} = \underline{16.61 \text{ Btu/hr} \cdot \text{ft}^2 \cdot {}^\circ\text{F}} \text{ (external area)}$$

Example 14-6: The internal diameter of a boiler generating tube is 4 in. and the wall thickness is 0.375 in. In operation the external surface film coefficient is 26 Btu/hr \cdot ft² \cdot °F, the internal film coefficient is 1700 Btu/hr \cdot ft² \cdot °F and a 0.375-in. thickness of scale having a conductivity of 0.6 Btu/hr \cdot °F \cdot ft is deposited on the inner surface of the tube. The metal conductivity is 26 Btu/hr \cdot °F \cdot ft, the furnace gas temperature is 2100°F and the temperature of the water is 500°F. Calculate (a) the overall coefficient of heat transfer for the clean tube, Btu/hr \cdot ft² \cdot °F; (b) the overall heat transfer coefficient including the effect of the scale, Btu/hr \cdot ft² \cdot °F; (c) the external tube surface temperature before and after the scale has formed.

HEAT TRANSMISSION

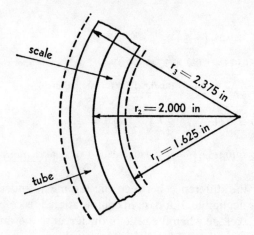

Solution:

$$r_1 = 1.625 \text{ in.}, \quad r_2 = 2.0 \text{ in.}, \quad r_3 = 2.375 \text{ in.}$$
gas temperature $= 2100°F$, water temperature $= 500°F$
$h_1 = 1700, \quad h_2 = 26 \text{ Btu/hr} \cdot \text{ft}^2 \cdot °F$
$k_{\text{tube}} = 26 \quad k_{\text{scale}} = 0.6 \text{ Btu/hr} \cdot °F \cdot \text{ft}$

$$r_t = r_2 + \frac{L}{2} = 2.0 + 0.1875 = 2.1875 \text{ in.}$$

$$r_s = r_1 + \frac{L}{2} = 1.625 + 0.1875 = 1.8125 \text{ in.}$$

(a) For clean tube:

$$\frac{1}{U} = \frac{1}{h_1 \left(\dfrac{r_2}{r_3}\right)} + \frac{r_3 - r_2}{k_t \left(\dfrac{r_t}{r_3}\right)} + \frac{1}{h_2}$$

$$= \frac{1}{(1700)\left(\dfrac{2.0}{2.375}\right)} + \frac{0.375}{12(26)\left(\dfrac{2.1875}{2.375}\right)} + \frac{1}{26}$$

$$= 0.0007 + 0.0013 + 0.0385 = 0.0405$$

$$U = \frac{1}{0.0405} = \underline{24.69 \text{ Btu/hr} \cdot \text{ft}^2 \cdot °F} \text{ (external area)}$$

(b) For tube and scale (R' = resistance per sq ft external area):

$$R'_{\text{scale}} + R'_1 = \frac{L}{k\left(\dfrac{r_s}{r_3}\right)} + \frac{1}{h_1\left(\dfrac{r_1}{r_3}\right)}$$

$$= \frac{0.375}{(12)(0.6)\left(\dfrac{1.8125}{2.375}\right)} + \frac{1}{1700\left(\dfrac{1.625}{2.375}\right)} = 0.0691$$

$$R'_{\text{total}} = 0.0691 + 0.0013 + 0.0385 = 0.1089$$

HEAT TRANSMISSION

$$U = \frac{1}{R'_{\text{total}}} = \frac{1}{0.1089} = 9.18 \text{ Btu/hr} \cdot \text{ft}^2 \cdot {}^{\circ}\text{F} \text{ (external area)}$$

(c) For clean tube:

$$\Delta t = \left(\frac{R'_2}{R'_{\text{total}}}\right)(t_1 - t_2) = \left(\frac{0.0385}{0.0405}\right)(2100 - 500) = 1521{}^{\circ}\text{F}$$

tube surface temperature $= 2100 - 1521 = \underline{579{}^{\circ}\text{F}}$

For tube with scale:

$$\Delta t = \left(\frac{R'_2}{R'_{\text{total}}}\right)(t_1 - t_2) = \left(\frac{0.0385}{0.1089}\right)(2100 - 500) = 566{}^{\circ}\text{F}$$

tube surface temperature $= 2100 - 566 = \underline{1534{}^{\circ}\text{F}}$

14-8. Fluid Films—Desirable and Undesirable. A full discussion of fluid films is beyond the scope of this text. The example problems of the preceding articles illustrate the dual nature of the heat flow problem in the engineering field. For the boiler wall, the high temperature steam line and a variety of similar applications, the problem is to reduce the loss of heat to a practical minimum. A low overall coefficient is essential to this end, and the fluid films may and frequently do contribute materially to the total heat flow resistance of the system. In the case of the boiler wall the resistance offered by the two surface films was second only to that of the diatomaceous insulating material, and in the example of the insulated pipe the combined film resistance was approximately one seventh of the total calculated for the system. Thus when heat flow is to be minimized, low values of film coefficient (high film resistances) are a tremendous advantage.

On the other hand, the function of all heat exchange devices is the transfer of heat from one fluid to another. In the interest of minimum surface area, these devices must have maximum values of U. In general, as illustrated by the example of the preceding article, heat exchanger tubes offer negligible resistance to heat flow, and the fluid films have predominant influence upon the overall coefficient. In the example cited the ratio of material resistance to film resistances is approximately $1/900$, and the tube effect may be dropped from the calculations with negligible error. It is thus seen that the performance of heat exchange devices is established by the fluid films encountered. For maximum U, fluid film effect must be minimized, and many ingenious arrangements are in use for this purpose.

Because of their low densities, gases have the lowest film coefficients of all fluids. Characteristic values for air range from 1.0 to 5.0 for free convection and to a maximum of approximately 30 with high velocity of

flow. Surface coefficients of liquids vary greatly. Fuel and lubricating oils inside tubes have characteristic values ranging from 30 to 70, whereas an average value for water under similar conditions is 800. For water flowing over tube banks, *h* may vary from 1000 to 2000, and for steam condensing on the outer tube surfaces *h* ranges from 1000 to 4000 Btu/hr · ft² · °F.

The fluid film coefficients here quoted are understood to represent operating rather than limiting values, and serve to illustrate the wide range of overall coefficient to be expected for heat transfer equipment. Thus a condenser might have a value of *U* of 600 Btu per hr · sq ft · °F, while an average value for a lubricating oil cooler might be 40 and that for an air preheater 5 Btu per hr · sq ft · °F.

14-9. Radiation. As previously stated, radiation is electromagnetic waves emitted from a substance by virtue of its absolute temperature. Thermal radiation is an electromagnetic wave phenomenon similar to light, but of a different wavelength. In general all substances at a temperature above absolute zero radiate heat waves to a greater or lesser degree. In turn the substance usually has radiant energy waves from other substances or bodies impinging upon its surface; and these may be absorbed, transmitted or reflected by the receiving substance.

The radiation from a solid or liquid surface may be evaluated by the following equation:

$$\dot{Q}_R = \sigma A e T^4 \text{ Btu/hr (W)}$$

where σ = the Stefan-Boltzmann constant
e = the emissivity of a body, a pure number having a maximum value of 1.0
A = the surface area of the body, ft² (m²)
T = the absolute temperature of the body °R (K).

The emissivity of a body depends upon the surface condition and color of the body. A so-called black body has an emissivity of 1.0. Further, a highly emissive body is also highly absorptive to radiant energy. Thus a *black body* has an absorptivity of 1.0. A *gray* body is one having an emissivity of less than 1.0, as is characteristic of most actual surfaces; and it can be shown that for any gray body at a given temperature the absorptivity and the emissivity are exactly equal. Thus if two gray bodies at the same temperature are placed in juxtaposition, each body is continually radiating to and receiving radiant energy from the other, but the net exchange of energy will be zero. If on the other hand a temperature difference exists between the two bodies, the body at the low temperature will receive a greater amount of energy than it in turn can radiate. In this

HEAT TRANSMISSION

case the net exchange of energy between the bodies can be evaluated by the Stefan-Boltzmann equation; thus:

$$\dot{Q}_{net} = \sigma A(T_1^4 - T_2^4)(F_a)(F_e) \text{ Btu/hr (W)} \qquad (14\text{-}5)$$

in which,
\dot{Q}_{net} = the net rate of heat transfer by radiation from the hotter to the colder body, Btu/hr (W)
$\sigma \approx 0.172 \times 10^{-8}$ Btu/hr · ft² · °R⁴ (5.70×10^{-8} W/m² · K⁴)
A = the surface area of one of the bodies as selected, usually the smaller of the two areas, ft² (m²)
T_1 = the absolute temperature of the hotter body, °R (K)
T_2 = the absolute temperature of the colder body, °R (K)
F_a = a dimensionless factor to allow for the average angle throughout which the larger surface *sees* the one chosen to evaluate the area, A, above. The numerical value of F_a will be one or less.
F_e = a dimensionless factor to allow for the departure of the two surfaces from *black body* conditions. F_e is a function of the individual emissivities of the bodies, and its value will be less than unity in all actual cases.

The evaluation of F_a and F_e is a complicated procedure and one that is highly specialized in application. The important things for the undergraduate student to remember are the nature of the various terms in the Stefan-Boltzmann equation, their general magnitudes and the fact that the rate of radiant heat exchange is a function of the difference between the fourth powers of the absolute temperatures of the surfaces of the bodies involved.

In addition to solids and liquids, certain gases are capable of absorbing and emitting radiant energy, while others are not. However, evaluation of the net energy exchange to or from a gaseous substance is even more complicated and more highly specialized in application than in the case of solids and liquids, and will not be developed in this text.

Problems

14-1. Calculate the rate of heat flow, in Btu/hr, through a 10-in. wall of solid concrete which is 20 ft long by 8 ft high. The thermal conductivity of concrete is 1.0 Btu/hr · °F · ft, the external surface temperature is 5°F and the internal surface temperature is 45°F.

Ans. 7680 Btu/hr

14-2. For the wall of problem 14-1 the outside film coefficient is 6.0 and the inside 1.5 Btu/hr · ft² · °F, the outside air temperature is -3°F and the inside air temperature is 77°F. Calculate (a) the overall heat transfer coefficient, Btu/hr · ft² · °F; (b) the rate of heat flow through the wall, Btu/hr.

Ans. (a) 0.60 Btu/hr · ft² · °F; (b) 7680 Btu/hr

HEAT TRANSMISSION

14-3. The wall of a cold storage room consists of an inside finish of $\frac{1}{2}$-in. cement plaster ($k = 0.67$), two layers of corkboard each 2 in. thick ($k = 0.03$) and an outside layer of building tile. The value of U for the entire wall is 0.058, the internal air film coefficient is 1.65, the inner temperature is 20°F and the outside temperature is 80°F. Calculate (a) the thermal resistance of the wall per ft² of surface area; (b) the corresponding resistance from the outside of the wall to the interface between the corkboard layers; (c) the temperature at the interface between the corkboard layers, °F; (d) the heat flow through unit wall area, Btu/hr · ft².

Ans. (a) 17.24; (b) 11.01; (c) 41.7°F; (d) 3.48 Btu/hr · ft²

14-4. A main steam line is covered with a $2\frac{1}{2}$-in. thickness of magnesia insulation. The actual internal diameter of the pipe is 7.625 in., the pipe wall thickness is 0.50 in., the respective values of k for the pipe and for the insulation are 25 and 0.04 Btu/hr · °F · ft, the steam and air film coefficients are 180 and 2.0 Btu/hr · ft² · °F respectively, the steam temperature is 800°F and the air temperature is 90°F. Calculate (a) the overall heat transfer coefficient based on the external area of the insulation, Btu/hr · ft² · °F; (b) the heat transferred per hundred linear feet of pipe, Btu/hr.

Ans. (a) 0.1428 Btu/hr · ft² · °F; (b) 36,165 Btu/hr

14-5. A refrigeration pipe line has an actual internal diameter of 1.38 in. and a wall thickness of 0.14 in. and is covered with a 2-in. layer of molded cork insulation. The refrigerant vapor film coefficient is 90 Btu/hr · ft² · °F, the thermal conductivities of the pipe and of the insulation are 25 and 0.025 Btu/hr · °F · ft respectively, the air film coefficient is 2.0 Btu/hr · ft² · °F, the refrigerant temperature is 20°F and the air temperature is 80°F. Calculate (a) the overall heat transfer coefficient, Btu/hr · ft² · °F, based on the external area of the insulation; (b) the heat transferred per linear foot of pipe, Btu/hr · ft; (c) the temperature of the external surface of the insulation, °F.

Ans. (a) 0.0825 Btu/hr · ft² · °F; (b) 7.34 Btu/hr · ft; (c) 77.5°F

14-6. Heat transfer by radiation is an important phenomenon at high temperatures in furnaces and internal combustion engines. Consider a one-square-foot black body plane surface at a temperature of 70°F. Calculate the heat transfer to the surface from another one-square-foot plane black body directly above and parallel to the first at temperatures of (a) 100°F, (b) 500°F, (c) 1500°F, and (d) 2000°F, Btu/hr. (Assume $F_a = F_e = 1.0$.)

Ans. (a) 33.4 Btu/hr; (b) 1325 Btu/hr; (c) 25,250 Btu/hr; (d) 62,850 Btu/hr

CHAPTER 15

HEAT TRANSFER EQUIPMENT

15-1. Foreword. In the preceding chapter the transfer of heat by conduction was discussed in some detail. It is the purpose of this chapter to develop the material of Chapter 14 and to present examples of typical heat transfer devices.

15-2. Mean Temperature Difference. All heat transfer devices have as their object the heating or cooling of a particular fluid. With the possible exceptions of evaporation or condensation, these heating or cooling processes involve temperature changes of two fluids. A general case, illustrated in Fig. 15-1, is that of a double pipe heat exchanger in which

FIGURE 15-1

cool water is circulated through the annular space while hot oil flows through the inner tube. In passage, as between sections A-A and B-B, heat flows from the oil to the water through the separating tube wall; and the

HEAT TRANSFER EQUIPMENT

temperature of the water increases while that of the oil decreases. As a result, the temperature difference between the fluids changes through the device, and the conduction equation of Chapter 14 must accordingly be modified.

Considering the differential section, dL, having an included heat transfer area, dA, we observe a decrement in oil temperature, and an increment in water temperature. Since a temperature difference, θ, is known to exist across the differential area, it follows that a rate of heat flow, $\delta\dot{Q}$, is established. Neglecting any effect of radiation, the rate of heat loss from the oil must be accompanied by an equivalent gain by the water. On this basis,

$$\delta\dot{Q} = UdA(t_0 - t_w) = -\dot{M}_0 c_0 dt_0 = \dot{M}_w c_w dt_w$$

where \dot{M}_0 = oil flow rate, lb/hr (kg/hr)
c_0 = specific heat of oil, Btu/lb · °F (J/kg · °C)
\dot{M}_w = water flow rate, lb/hr (kg/hr)
c_w = specific heat of water, Btu/lb · °F (J/kg · °C)
t_0 = temperature of oil at section dL, °F (°C)
t_w = temperature of water at section dL, °F (°C)
dt_0 = change in oil temperature, along dL, °F (°C)
dt_w = change in water temperature, along dL, °F (°C)

Rearranging terms,

$$\frac{UdA}{\dot{M}_0 c_0} = \frac{-dt_0}{t_0 - t_w}$$

and

$$\frac{UdA}{\dot{M}_w c_w} = \frac{dt_w}{t_0 - t_w}$$

By addition,

$$\frac{UdA}{\dot{M}_0 c_0} + \frac{UdA}{\dot{M}_w c_w} = \frac{dt_w - dt_0}{t_0 - t_w} = \frac{d\theta}{\theta}$$

where, $d\theta$ = differential change of fluid temperatures, °F (°C)
θ = difference between oil and water temperatures at section dL, °F (°C).

Then,

$$\left(\frac{U}{\dot{M}_0 c_0} + \frac{U}{\dot{M}_w c_w}\right)\int_0^A dA = -\int_{\theta_1}^{\theta_2} \frac{d\theta}{\theta}$$

and

$$\frac{UA}{\dot{M}_0 c_0} + \frac{UA}{\dot{M}_w c_w} = \ln\frac{\theta_1}{\theta_2}.$$

However, for the area included between A-A and B-B,

$$\dot{Q} = UA\theta_m = \dot{M}_o c_o(t_1 - t_2) = \dot{M}_w c_w(t_4 - t_3)$$

$$UA = \frac{\dot{M}_o c_o(t_1 - t_2)}{\theta_m} = \frac{\dot{M}_w c_w(t_4 - t_3)}{\theta_m}$$

and by substitution,

$$\frac{\dot{M}_o c_o(t_1 - t_2)}{\dot{M}_o c_o \theta_m} + \frac{\dot{M}_w c_w(t_4 - t_3)}{\dot{M}_w c_w \theta_m} = \ln\frac{\theta_1}{\theta_2}$$

$$\theta_m = \frac{(t_1 - t_2) + (t_4 - t_3)}{\ln\frac{\theta_1}{\theta_2}} = \frac{(t_1 - t_3) - (t_2 - t_4)}{\ln\frac{\theta_1}{\theta_2}}$$

Thence
$$\theta_m = \frac{\theta_1 - \theta_2}{\ln\frac{\theta_1}{\theta_2}} \text{ degrees Fahrenheit (°C)} \quad (15\text{-}1)$$

when t_1 and t_2 = entrance and exit temperatures, respectively, of the hot fluid, °F (°C)

t_3 and t_4 = entrance and exit temperatures, respectively, of the cool fluid, °F (°C)

θ_1 = **larger** difference between fluid temperatures, °F (°C)
θ_2 = **smaller** difference between fluid temperatures, °F (°C)
θ_m = **log mean difference between fluid temperatures**, °F (°C)

In this derivation oil and water are employed as typical fluids, but the use of the final equation is not restricted in terms of the hot and cold substances. However, it is assumed that there is no transfer of heat except between the hot and cold fluids; that there is a continuous, orderly change of fluid temperatures; and that the overall coefficient and the fluid specific heats are constant throughout the apparatus. These basic assumptions should be carefully noted, because in the strict sense they limit the direct use of log mean temperature difference to the types of fluid flow that will be described in the following article.

15-3. Parallel and Counter Flow. Generally speaking, heat exchangers are classified on the basis of design features, namely double pipe, shell and tube, submerged coil, finned tube, and so forth; and in terms of the direction of flow of the respective fluids. The derivation of the equation for log mean temperature difference employed **parallel flow** in a double pipe heat exchanger. That is, both fluids entered the double pipe from, and flowed through it in, the same direction. **Counter flow** is obtained by

reversing the direction of flow of either one of the fluids such that they enter the device from opposite ends and flow through it in opposite directions. Both types of flow and the resulting characteristic fluid temperature graphs are illustrated in Fig. 15-2. These temperature curves are

FIGURE 15-2

typical for an isolated section (as shown in Fig. 15-2) or for the entire flow length provided by a particular device, *provided only that neither fluid is a vapor which is changing phase.* In this latter case flow can be either parallel or counter, or directed across a tube bank as in a steam condenser; but the temperature distribution curves will in all cases be strikingly similar and of the general nature illustrated in Fig. 15-3.

FIGURE 15-3

HEAT TRANSFER EQUIPMENT

As in the preceding illustrations, Fig. 15-3 shows a section of double pipe with a condensing vapor in the inner pipe and a coolant flowing in the annular space. Since the vapor temperature remains constant during a phase change, the log mean temperature difference will be the same regardless of the relative directions of fluid flow.

15-4. Heat Transfer With Parallel and Counter Flow. The general equation for the flow of heat in heat transfer devices was stated with the derivation of the equation for log mean temperature difference. Thus for the transfer of heat between fluids in such equipment,

$$\dot{Q} = UA\theta_m \text{ Btu/hr (W)} \qquad (15\text{-}2)$$

in which \dot{Q} = rate of heat transfer between fluids, Btu/hr (W)
A = a selected representative area, such as the total internal or external tube area, ft² (m²)
U = overall coefficient of heat transfer based upon the selected area as indicated in Chapter 14, Btu/hr · ft² · °F temperature difference (W/m² · °C).

From a practical standpoint parallel flow has two major disadvantages, both of which may be deduced from a study of Fig. 15-4. With parallel

FIGURE 15-4

flow (solid lines) there is initially a large difference between fluid temperatures (t_1 and t_3), with an attendant high rate of heat transfer in this part of the apparatus. But as flow continues through the device, *the fluid temperatures approach a common limiting value,* and the difference between fluid temperatures decreases rapidly. Since the heat transfer rate is directly influenced by temperature difference, the latter part of the device (in the direction of flow) becomes progressively ineffective. With counter

HEAT TRANSFER EQUIPMENT

flow the lowest and the highest fluid temperatures are in juxtaposition at opposite ends of the apparatus, and the difference between fluid temperatures is uniform throughout the device. Consequently a counter-flow heat exchanger requires less area for a specified capacity than does its parallel-flow counterpart.

A second disadvantage of parallel flow is that the maximum capacity of any such device is established by the limiting common temperature of the fluids. Referring to Fig. 15-4, it is seen that t_2 decreases toward but can never be less than the maximum temperature of the coolant, t_4. However, with counter flow the exit temperature of the hot fluid can be made to approach the lowest temperature of the coolant; that is, t_2' can be made to approach t_3' as a limit, rather than t_4. Where the object of the device is the heating of the low temperature fluid, t_4' can be made to approach t_1.

Since the function of any heat exchanger is to provide maximum heating or cooling with the smallest possible area, parallel flow is seldom used, and most modern heat exchange devices, not involving condensation or evaporation, employ counter flow.

Example 15-1: A counter-flow lubricating oil cooler with a net heat transfer area of 258 sq ft cools 60,000 lb of oil per hour from a temperature of 145°F at inlet to 120°F at discharge. The temperatures of the cooling water are 75°F and 90°F, respectively, and the specific heat of the oil is 0.50 Btu/lb · °F. Calculate (a) the value of the overall heat transfer coefficient under these operating conditions, Btu/hr · ft² · °F, and (b) the required area for a parallel flow device having the same capacity under identical operating conditions.

Solution:

$\dot{M}_o = 60{,}000$ lb/hr
$c_o = 0.50$ Btu/lb · °F
$A = 258$ sq ft

(a) $\dot{Q} = \dot{M}_o c_o (t_1 - t_2) = (60{,}000)(0.50)(145 - 120) = 750{,}000$ Btu/hr

$$\theta_m = \frac{\theta_1 - \theta_2}{\ln \frac{\theta_1}{\theta_2}} = \frac{55 - 45}{\ln \frac{55}{45}} = 49.8°F$$

$$U = \frac{\dot{Q}}{A\theta_m} = \frac{750{,}000}{(258)(49.8)} = 58.4 \text{ Btu/hr} \cdot \text{ft}^2 \cdot °F$$

$t_1 = 145°F$
$\theta_1 = 55°F$
$t_2 = 120°F$
$\theta_2 = 45°F$
$t_4 = 90°F$
$t_3 = 75°F$

(b) for parallel flow:

$\dot{Q} = 750{,}000$ Btu/hr
$U = 58.4$ Btu/hr · ft² · °F

$$\theta_m = \frac{\theta_1 - \theta_2}{\ln \frac{\theta_1}{\theta_2}} = \frac{70 - 30}{\ln \frac{70}{30}}$$

$\theta_m = 47.2°F$

HEAT TRANSFER EQUIPMENT

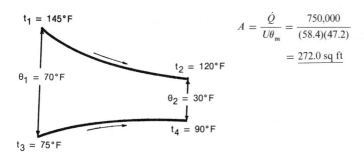

$$A = \frac{\dot{Q}}{U\theta_m} = \frac{750{,}000}{(58.4)(47.2)}$$

$$= \underline{272.0 \text{ sq ft}}$$

15-5. Actual Heat Transfer Equipment. In a great majority of cases true counter flow is not obtained in actual practice. As shown in Fig. 15-5, the baffles which produce high fluid turbulence (low surface resistance to heat flow) also introduce cross flow of the fluids. As a result the log mean

FIGURE 15-5

temperature difference based upon inlet and discharge fluid temperatures is no longer a true average and does not accurately represent the temperature difference between fluids at local sections within the apparatus. In addition, with varying fluid turbulence and for other reasons the overall coefficient may be subject to local variations; and even with true counter flow, U will change from inlet to exit with large changes of temperature of either fluid. Methods have been developed by which the effects of these variations are included in design calculations. In general such methods are based upon empirical formulae of a complicated nature and beyond the scope of this text.

Still another complication in the design of heat exchangers is the fact that in actual service various solid deposits accumulate upon the tube surfaces. Superheater tubes collect soot and fly ash, boiler generating

HEAT TRANSFER EQUIPMENT

tubes are subject to scaling due to mineral inclusions in the boiler feed water, and condenser tubes become coated and clogged with mud and vegetable material. These and other conditions that produce fouling of heat transfer surfaces result in a reduction of overall coefficient which is progressive with length of service, and which may be accompanied by a decrease in the rate of flow of one or both of the fluids.

Finally, it is possible to design a heat exchange device for only one set of operating conditions. In normal usage under variable load, fluid flow rates fluctuate over a wide range. These fluctuations are immediately reflected by unstable temperature conditions, which may be of a dangerous nature. Thus a sudden reduction of steam flow through a superheater may result in burning out the tubes unless there is a corresponding simultaneous reduction of the heat input to the apparatus. In any event, such changes of temperature and velocity will cause a change in the performance of heat transfer equipment and must be carefully considered in any estimate of the performance of these devices.

Despite the many variables involved, a knowledge of the design conditions and overall heat transfer coefficient is of great value to an operating engineer in estimating the performance of his equipment. The following problems will serve to illustrate the application of the basic material of this chapter to typical heat transfer equipment. The examples selected refer to actual apparatus, and the results are representative of conditions encountered in naval and marine service.

Example 15-2: A steam superheater has a net heat transfer area of 1620 sq ft and a design capacity of 221,000 lb of steam/hr when receiving saturated steam at 650 psia and discharging at 850°F with a pressure drop of not more than 25 psi through the heater. The design heat transfer coefficient is 30 Btu/hr · ft² · °F.

In operation the superheater receives 220,000 lb of saturated steam per hour at a pressure of 650 psia and discharges against a pressure of 630 psia with a temperature of 850°F when the flue gas temperatures at entrance and exit are 2100°F and 1430°F, respectively. Calculate the operating heat transfer coefficient for counter flow log mean temperature difference.

Solution:

entering superheater, $p = 650$ psia, $t = 495$ °F
leaving superheater, $p = 630$ psia, $t = 850$ °F
net heat transfer area $= 1620$ ft²
steam flow $= 220,000$ lb/hr
enthalpy entering $= 1203.1$, leaving $= 1434.2$ Btu/lb

$$\dot{Q} = (\dot{M}_s)(\Delta h_s) = (220,000)(1434.2 - 1203.1) = 51 \times 10^6 \text{ Btu/hr}$$

$$\theta_m = \frac{1250 - 935}{\ln \frac{1250}{935}} = 1085 °F$$

$$U = \frac{\dot{Q}}{A\theta_m} = \frac{51 \times 10^6}{(1620)(1085)}$$

$$\underline{U = 29 \text{ Btu/hr} \cdot \text{ft}^2 \cdot °F}$$

HEAT TRANSFER EQUIPMENT

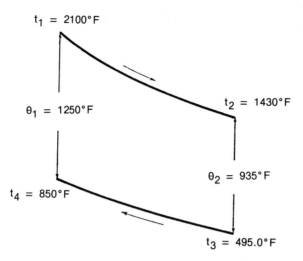

Example 15-3: A steam boiler economizer heats 160,000 lb of water per hr from 250°F to 335°F at an average pressure of 700 psia. Flue gas enters the economizer at a temperature of 1400°F and leaves at a temperature of 800°F. The net heat transfer area is 3,840 sq ft and the design value of U is 8 Btu/hr·ft²·°F for a capacity of 253,000 lb/hr at the temperatures given above. How does the heat transfer coefficient calculated for these operating conditions compare with the design value?

Solution:

$p = 700$ psia
water temperature entering $= 250°F$; leaving $= 335°F$
water flow $= 160,000$ lb/hr
net heat transfer area $= 3840$ ft²

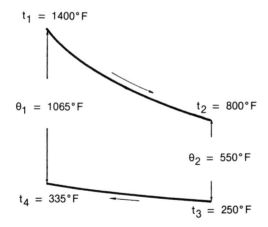

From steam table 4, entering water enthalpy $= 220.0$; leaving $= 307.2$ Btu/lb

$\dot{Q} = (\dot{M}_w)(\Delta h_w) = (160,000)(307.2 - 220.0) = 13.95 \times 10^6$ Btu/hr

HEAT TRANSFER EQUIPMENT

$$\theta_m = \frac{1065 - 550}{\ln \frac{1065}{550}} = 779°F$$

$$U = \frac{\dot{Q}}{A\theta_m} = \frac{13.95 \times 10^6}{(3840)(779)}$$

$$U = 4.66 \text{ Btu/hr} \cdot \text{ft}^2 \cdot °F$$

Assuming the same temperatures may be obtained if the flow rate is increased to 253,000 lb/hr, the value of U at full capacity is

$$U = (4.66)\left(\frac{253,000}{160,000}\right) = 7.37 \text{ Btu/hr} \cdot \text{ft}^2 \cdot °F$$

which compares favorably with the design value of 8.0 Btu/hr·ft²·°F.

Example 15-4: A steam condenser with a net heat transfer area of 23,500 sq ft has a design value of U of 486 Btu/hr·ft²·°F for a log mean temperature difference of 32.4 and a design operating pressure of 2.5 in. Hg abs. The water consumption at rated capacity is not to exceed 40,500 gpm with a temperature rise of 18.3°F. Measured temperatures of fresh water at this flow rate during an acceptance test were 71.5°F and 90°F at entrance and exit, respectively, and the average condenser pressure was 2.45 in. Hg abs. Assuming no subcooling of the condensate, calculate the overall heat transfer coefficient established by the acceptance test, and compare this with the manufacturer's design value given above.

Solution:

water temperature entering = 71.5°F; leaving = 90°F
steam pressure = 2.45 in. Hg abs
water quantity = 40,500 gpm
net heat transfer area = 23,500 ft²
water flow rate = (40,500)(8.34)(60) = 20.27 × 10⁶ lb/hr
\dot{Q} = (20.27 × 10⁶)(1)(90 − 71.5) = 375 × 10⁶ Btu/hr

$$\theta_m = \frac{36.5 - 18}{\ln \frac{36.5}{18}} = 26.2°F$$

$$U = \frac{\dot{Q}}{A\theta_m} = \frac{375 \times 10^6}{(23,500)(26.2)}$$

$$U = 609 \text{ Btu/hr} \cdot \text{ft}^2 \cdot °F$$

$t_1 = t_2 = 108°F$

$\theta_2 = 18°F$

$\theta_1 = 36.5°F$

$t_4 = 90°F$

$t_3 = 71.5°F$

Although the test value is much larger than the design coefficient, it provides a liberal

allowance for fouling under service conditions. Assuming a 20 percent fouling allowance, the predicted service value of U is 487.2 Btu/hr \cdot ft^2 \cdot °F, which is in close agreement with the manufacturer's value.

Example 15-5: An exhaust gas regenerator (counter-flow heat exchanger) for a marine gas turbine handles 1.8 kg/sec of air from its compressor and heats it by means of 1.88 kg/sec of hot exhaust gas. Exhaust gas enters the regenerator at 593°C and leaves at 310°C. Compressed air enters the regenerator at 266°C. For this temperature range a constant pressure specific heat for the exhaust gas may be estimated at 1090 J/kg \cdot °C. Assume no heat transfer other than between the regenerator fluids. Calculate:
(a) the energy exchanged as heat by the two fluids, kJ/sec
(b) the air temperature leaving the regenerator, °C
(c) the log mean temperature difference for the exchanger, °C

Solution:

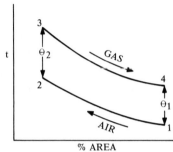

t_1 in $= 266$°C \quad t_2 out $= ?$ \quad $\dot{M}_a = 1.8$ kg/sec
t_3 in $= 593$°C \quad t_4 out $= 310$°C \quad $\dot{M}_g = 1.88$ kg/sec

(a) $\dot{Q} = \dot{M}_g c_{p_g} (\Delta t)_g$
$= 1.88$ kg/sec \cdot 1090 J/kg \cdot °C(593 − 310)°C
$= 579.92$ kJ/sec

(b) $|\dot{Q}_a| = |\dot{Q}_g|;\quad \dot{M}_a c_{p_a}(\Delta t)_a = \dot{M}_g c_{p_g}(\Delta t)_g;$
$\dot{M}_a c_{p_a}(t_2 - t_1) = \dot{M}_g c_{p_g}(t_3 - t_4)$

$t_2 = t_1 + \dfrac{\dot{Q}}{\dot{M}_a c_{p_a}} = 266 + \dfrac{579920}{1.8(1005)} = \underline{586.6\,°C}$

(c) L.M.T.D. $= \dfrac{\theta_1 - \theta_2}{\ln \dfrac{\theta_1}{\theta_2}} = \dfrac{(t_4 - t_1) - (t_3 - t_2)}{\ln \dfrac{(t_4 - t_1)}{(t_3 - t_2)}}$

$= \dfrac{(310 - 266) - (593 - 586.6)}{\ln \dfrac{44}{6.4}} = \underline{19.5\,°C}$

Problems

15-1. A heat exchanger receives oil having a specific heat of 0.45 Btu/lb \cdot °F and a temperature of 160°F at the rate of 40,000 lb/hr. Fresh water at an initial temperature of 60°F flows through the apparatus at the rate of 120,000 lb/hr. Assume unlimited heat

transfer area and calculate (a) the common temperature which the fluids will reach under parallel flow, °F; (b) the heat transferred by the exchanger, Btu/hr.

Ans. (a) 73°F; (b) 1,560,000 Btu/hr

15-2. A counter-flow lubricating oil cooler is presumed to operate under the conditions of problem 15-1 with both oil and water being discharged at 73°F. Calculate (a) the log mean temperature difference, °F; (b) the heat transfer area required if the overall heat transfer coefficient is 52 Btu/hr · ft² · °F.

Ans. (a) 38.9°F; (b) 771 ft²

15-3. Calculate the heat transfer area required by a parallel flow oil cooler which removes 1,524,000 Btu/hr from the oil while cooling it from 160°F at inlet to 75°F at discharge, when the cooling water temperatures are 60°F and 72.7°F respectively. U for the heater is 52 Btu/hr · ft² · °F.

Ans. 1132 ft²

15-4. A performance test of a finned-tube air heater having a heating surface area of 24.12 sq ft gave the following results: temperature of entering steam, 252°F, temperature of condensate, 252°F, temperature in calorimeter, 220°F; pressure in calorimeter, 14.696 psia; condensate, 10 lb; time to collect condensate, 8 min 35 sec; temperatures of air entering and leaving heater, 74.5°F and 86.5°F, respectively. Calculate the overall heat transfer coefficient established by the test, Btu/hr · ft² · °F.

Ans. 15.79 Btu/hr · ft² · °F

15-5. Steam enters a closed feedwater heater with a quality of 95% at a pressure of 50 psia and leaves through the drain valve as a saturated liquid at this pressure. Feedwater enters the heater at the rate of 160,000 lb/hr with a pressure of 700 psia and temperature of 100°F and leaves at a temperature of 250°F with negligible change of pressure. The heat transfer coefficient for the heater based on external tube area is 500 Btu/hr · ft² · °F, and the water velocity entering the tubes is not to exceed 4 ft/sec. Calculate (a) the quantity of steam required by the heater, lb/hr; (b) the heat transfer area required, ft²; (c) the number and length, in ft, of $\frac{5}{8}$-in. OD Admiralty metal tubes required (the tube wall thickness is 0.049 in).

Ans. (a) 27,353 lb/hr; (b) 564.7 ft²; (c) 118, 29.0 ft

15-6. A fuel oil heater delivers 30,000 lb of oil/hr at a temperature of 156°F. The initial temperature of the oil is 72°F, the specific heat is 0.50 Btu/lb · °F and U for the heater is 40 Btu/hr · ft² · °F. The heater receives saturated steam at 350°F, and the condensate leaves the heater as saturated liquid at this temperature. Calculate (a) the log mean temperature difference, °F; (b) the required surface area, ft²; (c) the quantity of steam required for the operation of the heater, lb/hr.

Ans. (a) 233.5 °F; (b) 134.9 ft²; (c) 1446 lb/hr

15-7. A superheater having a surface area of 1800 ft² receives saturated steam at 630 psia and discharges superheated steam at a pressure of 600 psia with a temperature of 740°F. It is estimated that the maximum flow rate of flue gas through the heater will be 3,000 lb/min with an entering temperature of 1200°F and an exit temperature of 760°F. For counter flow of steam and flue gas and a specific heat of 0.26 Btu/lb · °F for the gas, calculate, (a) the log mean temperature difference, °F; (b) the value of U for the heater, Btu/hr · ft² · °F; (c) the maximum quantity of superheated steam that can be produced by the heater, lb/hr.

Ans. (a) 355.6°F; (b) 32.17 Btu/hr · ft² · °F; (c) 121,000 lb/hr

HEAT TRANSFER EQUIPMENT

15-8. Steam enters the condenser of a marine main propulsion plant at 2 psia and a quality of 86 percent at the rate of 175,000 lb/hr and with a velocity of 800 ft/sec. It leaves the condenser as a saturated liquid without any change in pressure. The salt water inlet (injection) temperature is 75° F and the overboard temperature is 90°F. Sea water has a specific heat of 0.96 Btu/lb · °F and a specific weight of 64 lbf/ft^3. The injection and overboard velocities are substantially equal. Calculate (a) the energy rejection rate from the condensing steam, Btu/min; (b) the flow of sea water in gpm; (c) the log mean temperature difference; (d) the overall heat transfer coefficient for an assumed heat transfer area of 6035 ft^2.

Ans. (a) 2,601,000 Btu/min; (b) 21,111 gpm; (c) 43.1°F; (d) 600 Btu/hr · ft^2 · °F.

15-9. A superheater receives 100,000 lb/hr of saturated steam at 1250 psia. The conditions at the superheater outlet are 950°F and 1200 psia. The gases of combustion enter the superheater tube bank at 2500°F and leave at 1890°F. Determine (a) the log mean temperature difference assuming parallel flow; (b) the log mean temperature difference assuming counter flow; (c) the required heat transfer area assuming the log mean temperature difference to be 1400°F and the overall heat transfer coefficient to be 25 Btu/hr · ft^2 · °F.

Ans. (a) 1375°F; (b) 1430°F; (c) 825 ft^2

15-10. A gas turbine exhaust stack cooler drops the gas temperature from 1000°F to 600°F by means of a water cooling coil. Cooling water enters the coil at 60°F and leaves, after a counter-flow passage, with a temperature of 180°F. If c_p for the gas is 0.26 Btu/lb · °F, calculate (a) the quantity of water required to cool properly 100 lb/sec of exhaust gas; (b) log mean temperature difference; (c) the overall heat transfer coefficient for an exchanger area of 2000 ft^2.

Ans. (a) 86.7 lb/sec; (b) 670.3°F; (c) 27.93 Btu/hr · ft^2 · °F.

CHAPTER 16

PROPERTIES OF MIXTURES

16-1. Foreword. The occurrence of mixtures in engineering practice is very common. Air, a demonstratedly important working substance, is itself a mixture of gases. As a working substance, air is dependent upon a fuel for its source of thermal energy, and in many applications an intimate mixture of air and fuel is applied to a prime mover as a source of energy and leaves the engine (after delivering energy to it) as exhaust gas. In other applications, in which air is not the primary working substance, it is nevertheless necessary for the release of thermal energy from a fuel source. Thus in a boiler the air-fuel mixture delivers thermal energy to steam, and the products of combustion are exhausted through a stack.

In each of the above cases the exhaust products are mixtures of varying amounts of known or predictable gases, and these mixtures may be quantitatively analyzed to determine whether or not the release of thermal energy was efficiently accomplished.

In addition, atmospheric air always contains some water vapor (steam at very low absolute pressures), and this may be of primary concern to the engineer. The compression of air for pneumatic devices is accompanied by an accumulation of water, which must be eliminated from the system; the air-conditioning industry must be able to add or abstract moisture to or from air as desired; and in a steam condenser the air must be separated from the vapor and removed from the system if low condenser pressures are to be maintained.

It is the purpose of this chapter to study the properties and basic laws governing the behavior of these mixtures. For practical reasons, Engineers' units will be used exclusively in the following discussions.

16-2. Dalton's Law and The Universal Gas Constant. In a gaseous mixture the molecules of each constituent gas are evenly dispersed throughout the entire volume of the mixture, and, since they are in motion, are in continual collision with themselves and with the walls of their container, the latter absorbing a force proportional to the momentum of the molecule from each molecule that strikes it. The result of these boundary collisions is that the walls of the container are subjected to an internal pressure equal to the net intensity of all molecular impingement forces. If the various constituent substances are now segregated and each permitted to occupy the full volume of the container alone and at the

PROPERTIES OF MIXTURES

temperature of the mixture, the internal pressure upon the walls of the container will change with each substance so long as the constituents are present in unequal mol fractions. Under these circumstances Avogadro's Law (Art. 16-4) indicates a different number of molecules for each constituent substance, thus more or fewer boundary collisions per unit of time. Hence although the average molecular kinetic energy is the same for each substance, the net intensity of the impingement forces differs for each constituent. The individual pressures thus produced are designated as the **partial pressures** of the constituent substances, and the entire procedure may be summarized by Dalton's Law, which states that the total pressure of a mixture of gases is equal to the sum of the partial pressures of its constituents. Thus:

$$p_m = p_1 + p_2 + p_3 \qquad (16\text{-}1)$$

where p_m = the total pressure of a mixture of constituents 1, 2 and 3
p_1 = the partial pressure of constituent 1
p_2 = the partial pressure of constituent 2
p_3 = the partial pressure of constituent 3.

Since we are dealing primarily with gases, and with vapors which under certain conditions have characteristics similar to gases, we may apply the combined laws of Boyle and Charles and write for the mixture

$$p_m V_m = M_m R_m T_m$$

from which

$$p_m = \frac{M_m R_m T_m}{V_m}$$

and for any of the constituents at the temperature and volume of the mixture,

$$p_1 = \frac{M_1 R_1 T_m}{V_m}$$

To facilitate the application of these relationships, it is useful to recall the universal gas constant studied in physics, and to which reference has previously been made in this text. The value of the particular gas constant can be determined from the appropriate universal gas constant and the molecular weight of the specified substance. Thus, from the expressions for R in Art. 3-2:

$$R = \frac{1545}{m} \text{ ft} \cdot \text{lbf/lb} \cdot {}^\circ R \qquad (16\text{-}2)$$

where **m equals the molecular weight of the substance.** In this manner the

PROPERTIES OF MIXTURES

value of R for any given substance may be easily determined when its molecular weight is known.

It should be noted that while Dalton's Law is applicable to a large number of mixtures, there are important exceptions in the engineering field. One of these exceptions is gasoline vapor (which is actually a mixture of several different hydrocarbon vapors), while another consists of the mixtures of ammonia and water vapor commonly used in absorption refrigeration systems.

16-3. Volumetric Analysis, Partial Pressures and Mass Analysis. The quantitative analysis previously mentioned is commonly made on a volumetric basis, because of convenience, and subsequently converted to a mass basis (if desired) by methods that will presently be demonstrated. However, in addition to convenience, the volumetric analysis is immediately useful, since it provides an index of the partial pressures of the constituents of the mixture. For example, the cyclindrical container of Fig. 16-1 is filled with a mixture of gases (constituents 1, 2, and 3) in the volume percentages indicated, and the pressure and temperature of the mixture are known. If each constituent is now permitted to occupy the

FIGURE 16-1

PROPERTIES OF MIXTURES

container alone, at the temperature of the mixture, Boyle's law may be applied and, in each case,

$$p_m V_1 = p_1 V_m,$$

from which,

$$p_1 = p_m \frac{V_1}{V_m}.$$

Example 16-1: The mixture of gases shown in Fig. 16-1 has a pressure of one atmosphere. Using the volumetric analysis shown in the figure, calculate the partial pressures of the indivdual constituents

Solution:

$$1 \text{ atm} = 14.696 \text{ psia}$$

$$\frac{V_1}{V_m} = 0.7, \quad \frac{V_2}{V_m} = 0.1, \quad \frac{V_3}{V_m} = 0.2$$

$$p_1 = p_m \frac{V_1}{V_m} = (14.696)(0.7) = \underline{10.287 \text{ psia}}$$

$$p_2 = p_m \frac{V_2}{V_m} = (14.696)(0.1) = \underline{1.470 \text{ psia}}$$

$$p_3 = p_m \frac{V_3}{V_m} = (14.696)(0.2) = \underline{2.939 \text{ psia}}$$

The mass analysis of the mixture may also be determined from the volumetric analysis in the following manner. Referring to Fig. 16-1,

$$M_m = M_1 + M_2 + M_3$$

As indicated in Art. 16-2.

$$p_1 V_m = M_1 R_1 T_m$$

and,

$$M_1 = \frac{p_1 V_m}{R_1 T_m}$$

Then since,

$$p_1 = p_m \frac{V_1}{V_m} \quad \text{and} \quad R_1 = \frac{1545}{m_1},$$

$$M_1 = \frac{p_1 V_m m_1}{1545 \, T_m} = \frac{p_m (V_1/V_m) V_m m_1}{1545 \, T_m} = \left(\frac{p_m V_m}{1545 \, T_m}\right) m_1 \left(\frac{V_1}{V_m}\right)$$

When similar equations are developed for constituents 2 and 3, the proportion by mass may be determined. Hence,

$$\frac{M_1}{M_m} = \frac{M_1}{M_1 + M_2 + M_3}$$

PROPERTIES OF MIXTURES

$$\frac{M_1}{M_m} = \frac{\left(\frac{p_m V_m}{1545\, T_m}\right) m_1 \left(\frac{V_1}{V_m}\right)}{\left(\frac{p_m V_m}{1545\, T_m}\right)\left[m_1\left(\frac{V_1}{V_m}\right) + m_2\left(\frac{V_2}{V_m}\right) + m_3\left(\frac{V_3}{V_m}\right)\right]}$$

$$\frac{M_1}{M_m} = \frac{m_1\left(\frac{V_1}{V_m}\right)}{m_1\left(\frac{V_1}{V_m}\right) + m_2\left(\frac{V_2}{V_m}\right) + m_3\left(\frac{V_3}{V_m}\right)} \tag{16-3}$$

For any particular mixture the conversion from volumetric to mass analysis is completed by substituting the products $m_2(V_2/V_m), \ldots$, etc., in the numerator of Equation (16-3). However it is frequently convenient to arrange the entire conversion in tabular form.

Example 16-2: For the mixture shown in Fig. 16-1 constituent 1 is nitrogen, constituent 2 is carbon dioxide and constituent 3 is oxygen. Calculate the analysis (percentage) by mass of the mixture.
Solution:

$$\frac{V_1}{V_m} = 0.7 \qquad \frac{V_2}{V_m} = 0.1 \qquad \frac{V_3}{V_m} = 0.2$$

$$m_1 = 28 \qquad m_2 = 44 \qquad m_3 = 32$$

Volumetric Analysis	Mass Analysis
N_2, $0.70 \times 28 = 19.6$	$\frac{M_1}{M_m} = \frac{19.6}{30.4} = 64.47\%\ N_2$
CO_2, $0.10 \times 44 = 4.4$	$\frac{M_2}{M_m} = \frac{4.4}{30.4} = 14.47\%\ CO_2$
O_2, $0.20 \times 32 = \underline{6.4}$	$\frac{M_3}{M_m} = \frac{6.4}{30.4} = 21.06\%\ O_2$
30.4	

Several theoretical considerations necessitate a conversion from mass to volume. This is readily accomplished when it is noted from Fig. 16-1 that the volume of the mixture is the sum of the volumes of its constituents *when each is at the pressure and temperature of the mixture.*

$$V_m = V_1 + V_2 + V_3$$

and employing the Engineers' value for R,

$$pV = MRT \quad \text{and} \quad R = \frac{1545}{m}$$

$$V_1 = \frac{M_1 R_1 T_m}{p_m} = \frac{1545\, T_m}{p_m}\left(\frac{M_1}{m_1}\right)$$

PROPERTIES OF MIXTURES

Then,

$$\frac{V_1}{V_m} = \frac{V_1}{V_1 + V_2 + V_3}$$

$$= \frac{\dfrac{1545\, T_m}{p_m}\left(\dfrac{M_1}{m_1}\right)}{\dfrac{1545\, T_m}{p_m}\left[\left(\dfrac{M_1}{m_1}\right) + \left(\dfrac{M_2}{m_2}\right) + \left(\dfrac{M_3}{m_3}\right)\right]}$$

Clearing and introducing M_m into numerator and denominator puts the bracketed terms into mass analysis form, thus:

$$\frac{V_1}{V_m} = \frac{\dfrac{1}{m_1}\left(\dfrac{M_1}{M_m}\right)}{\dfrac{1}{m_1}\left(\dfrac{M_1}{M_m}\right) + \dfrac{1}{m_2}\left(\dfrac{M_2}{M_m}\right) + \dfrac{1}{m_3}\left(\dfrac{M_3}{M_m}\right)} \qquad (16\text{-}4)$$

The conversion from mass to volumetric analysis is also most conveniently accomplished in tabular form, as shown by the following problem.

Example 16-3: Determine the volumetric analysis of a mixture which consists of 56 percent nitrogen, 12 percent carbon dioxide and 32 percent oxygen as calculated on a mass basis.
Solution:

$$\frac{M_1}{M_m} = 0.56 \qquad \frac{M_2}{M_m} = 0.12 \qquad \frac{M_3}{M_m} = 0.32$$

$$m_1 = 28 \qquad m_2 = 44 \qquad m_3 = 32$$

Mass Analysis		Volumetric Analysis	
N_2, $0.56 \div 28 = 0.02000$		$\dfrac{V_1}{V_m} = \dfrac{0.02000}{0.03273} = 61.11\%\ N_2$	
CO_2, $0.12 \div 44 = 0.00273$		$\dfrac{V_2}{V_m} = \dfrac{0.00273}{0.03273} = 8.34\%\ CO_2$	
O_2, $0.32 \div 32 = \dfrac{0.01000}{0.03273}$		$\dfrac{V_3}{V_m} = \dfrac{0.01000}{0.03273} = 30.55\%\ O_2$	

16-4. Specific Heats and Thermal Properties. When little or no intermolecular attraction exists between the constituents of a gaseous mixture, the specific heat at constant pressure and at constant volume of the mixture is a function of both the relative quantity and the specific heat of the constituents. Therefore either specific heat may be calculated as follows:

$$c_m = \frac{M_1 c_1 + M_2 c_2 + M_3 c_3}{M_m} \qquad (16\text{-}5)$$

in which the numerical subscripts indicate the individual constituents of a specific mixture. Other thermal properties such as internal energy, enthalpy, entropy and the gas constant, R_m, may be calculated from similar equations if the specific values of each property, for each constituent, are known. However, the proper specific heat having been determined by Equation (16-5), the enthalpy and internal energy are easily calculated for any mixture temperature, and the property entropy is rarely used in connection with mixtures. As stated, the gas constant, R_m, may be calculated by an equation similar to (16-5) employing the proper values of R for each constituent. However, it may also be obtained by using the universal gas constant when the molecular weight of the mixture is known, and the latter is readily obtainable from the volumetric analysis of the mixture.

Avogadro's law states that, for any specified pressure and temperature, equal volumes of different gases contain the same number of molecules. The significance of this statement is that, for the same pressure and temperature conditions, the volume of one mol of any specified gas is exactly the same as the volume of one mol of any other gas. This statement is illustrated by the gas laws of Boyle and Charles as follows:

$$V = \frac{MRT}{p}$$

Substituting Engineers' values,

$$M = m \quad \text{and} \quad R = \frac{1545}{m}$$

$$V = \frac{1545\, mT}{mp} = \frac{1545\, T}{p} \text{ cubic feet/mol} \tag{16-6}$$

Based on Avogadro's law, one mol of the mixture of nitrogen, carbon dioxide and oxygen illustrated in Fig. 16-1 contains 0.7 mol of nitrogen, 0.1 mol of carbon dioxide and 0.2 mol of oxygen, and has a mass of 30.4 pounds (see Example 16-2). Since the amount of the mixture is one mol, and the mass is 30.4 pounds, this figure *is the molecular weight of the specified mixture.*

16-5. Atmospheric Air. Atmospheric air consists of a mixture of gases and water vapor. In this mixture, nitrogen and oxygen are in such predominant quantities that the small amounts of other gases are not considered. Furthermore, on a mass basis the amount of water vapor usually present has little effect upon calculations employing Boyle's and Charles' laws and is frequently disregarded. However, as indicated in Art. 16-1,

PROPERTIES OF MIXTURES

there are several applications in which the presence of vapor in the mixture *must* be considered, and equations must be developed to accommodate these situations.

In these developments, since the proportions of nitrogen and oxygen in air are constant, it is standard practice to consider atmospheric air to be composed of a mixture of *two* constituents only, these being the *dry air* and the *water vapor* that it contains.

16-6. Partial Pressures, Specific Humidity, Relative Humidity and Dewpoint. The water vapor present in atmospheric air is steam at very low pressures. As such, its characteristic properties may be found in the steam tables when the state of the vapor has been determined, and the graphical representation on *T-s* coordinates may be used to good advantage. Figure 16-2 represents a low pressure segment of the saturation line, and point *A* represents the state of the vapor in a given mixture. However, in the range

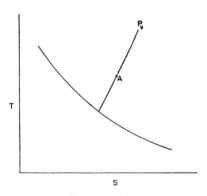

FIGURE 16-2

of temperatures of atmospheric mixtures, the vapor in the mixtures conforms to the gas laws and its properties may be calculated using the gas equations. Hence, in Engineers' units

$$M_v = \frac{p_v V_m}{R_v T_m} = \frac{p_v V_m}{85.8 T_m}$$

in which
 R_v = gas constant for water vapor = 85.8 ft · lbf/lb · °R
 M_v = mass of vapor in the mixture, pounds
 p_v = vapor pressure (partial pressure of the vapor in the mixture) psfa (pounds force per square foot absolute)
 V_m = total volume occupied by the mixture, cubic feet
 T_m = temperature of mixture, degrees R.

PROPERTIES OF MIXTURES

It should be noted that the gas laws apply to water vapor only in the saturated and superheated states, for vapor pressures up to approximately 5 psia.

Combination of the gas laws with the mixture theories previously developed affords a direct and convenient approach to the analysis of atmospheric mixtures. Dalton's law indicates that (for an atmosphere consisting of dry air and water vapor) the atmospheric pressure consists of the partial pressures of the two constituents, or,

$$p_B = p_a + p_v$$

in which

p_B = barometric pressure, units as below
p_a = partial pressure of dry air in the atmosphere, units as below
p_v = partial pressure of water vapor in the atmosphere, either psia or inches of mercury abs.

Both barometric and vapor pressures may be measured (the latter by methods that will presently be developed) and the dry air pressure determined by difference. And, after we have thus obtained the necessary partial pressures, the mass ratio of the mixture constituents may be determined as follows:

$$M_v = \frac{p_v V_m}{R_v T_m} \quad \text{and} \quad M_a = \frac{p_a V_m}{R_a T_m}$$

then, using values of R_a and R_v to four significant figures:

$$\frac{M_v}{M_a} = \frac{53.34 p_v}{85.77 p_a} = \frac{0.622 p_v}{p_a}$$

$$w_v = \frac{0.622 p_v}{p_B - p_v} \text{ lb/lb dry air} \qquad (16\text{-}7)$$

The above ratio is defined as the **specific humidity** of an atmospheric mixture and will hereafter be designated by the symbol w_v. In the past specific humidity was referred to as absolute humidity and frequently in modern usage is called humidity ratio. As above derived, specific humidity has the dimensions pounds of water vapor per pound of dry air; but, since the relative mass of vapor present in these mixtures is very small, a more convenient statement is

$$w_v = \frac{(7000)(0.622) p_v}{p_B - p_v}$$

$$= \frac{4354 p_v}{p_B - p_v} \text{ grains moisture per pound of dry air} \qquad (16\text{-}8)$$

where 7000 grains equals 1 pound.

PROPERTIES OF MIXTURES

Relative humidity is defined as the ratio of the vapor pressure of a given mixture to the pressure of saturated water vapor at the mixture temperature. It is customarily expressed as a percentage figure and may be illustrated by Fig. 16-3. Thus, on *T-s* coordinates, the state of the vapor in

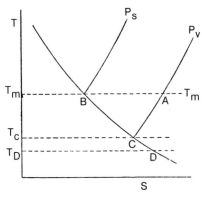

FIGURE 16-3

an atmospheric mixture is point *A* as fixed by the vapor pressure, p_v, and the mixture temperature, T_m. However, if the vapor existed in the given mixture in a saturated condition, as at point *B*, its pressure would have to be p_s, because the temperature of the mixture is fixed. *Relative humidity*, designated by the symbol ϕ, is *defined as the ratio of the vapor pressure, p_v, to the saturation pressure, p_s, at the same temperature.* Thus,

$$\phi = \frac{p_v}{p_s} \tag{16-9}$$

Relative humidity may also be stated in terms of either specific volumes or specific weights as

$$\phi = \frac{p_v}{p_s} = \frac{R_v T_m}{v_v} \times \frac{v_s}{R_v T_m} = \frac{v_s}{v_v} = \frac{\gamma_v}{\gamma_s}$$

If the vapor in the mixture does exist in its saturated state (as point *B*, Fig. 16-3), the mixture, too, is said to be saturated and contains the maximum mass of vapor possible for the given temperature. The mixture temperature, therefore, establishes this maximum.

Figure 16-3 also illustrates the dewpoint temperature of the mixture, or more simply, its **dewpoint**. The dewpoint of any mixture is the saturation temperature corresponding to the mixture vapor pressure which, in

PROPERTIES OF MIXTURES

Fig. 16-3, is point C. The significance of this point is best understood if the mixture containing vapor at state A is imagined to be cooled at constant vapor pressure. The relative mass of air and vapor and the barometric pressure remain unchanged during this imaginary process until T_m is reduced to T_c; but if the process continues until T_m becomes equal to T_D, some of the vapor will condense and drop out of the mixture, and the specific humidity will thus be reduced.

Example 16-4: Calculate the mass of moisture, in pounds, contained in 4000 cu ft of atmospheric air having a temperature of 90°F when the barometric pressure is 30.12 in. Hg (a) if the air is saturated, and (b) if the relative humidity is 50 percent. (c) What is the dewpoint of the atmosphere of part (b)? (d) What is the specific humidity of the atmosphere of part (b), grains/lb dry air?

Solution:

$$p_B = 30.12 \text{ in. Hg} \qquad t = 90°F$$

(a)
$$v_s = v_g = 467.7 \text{ ft}^3/\text{lb (steam tables at 90°F)}$$
$$M_v = \frac{4000}{467.7} = \underline{8.55 \text{ lb}}$$

(b)
$$p_s = 0.6988 \text{ psia (steam tables at 90°F)}$$
$$p_v = \phi p_s = (0.5)(0.6988) = 0.3494 \text{ psia}$$
$$M_v = \frac{p_v V_m}{R_v T_m} = \frac{(144)(0.3494)(4000)}{(85.8)(550)} = \underline{4.26 \text{ lb}}$$

(c) dewpoint $= \underline{68.8°F}$ (steam tables for 0.3494 psia)

(d)
$$p_B = (0.491)(30.12) = 14.79 \text{ psia}$$
$$w_v = \frac{4354 p_v}{p_B - pv} = \frac{(4354)(0.3494)}{14.79 - 0.3494} = \underline{105.3 \text{ grains/lb dry air}}$$

Although in the example problem the mass of vapor in the compartment when the relative humidity is 50 percent is approximately half of that under saturated conditions, it must be understood that relative humidity is not equal to the ratio of the specific humidity of a given mixture to the specific humidity of a corresponding saturated mixture. This may be illustrated by an inspection of the specific humidity equation

$$w_v = \frac{4354 p_v}{p_B - p_v}$$

It is seen at once that although p_v is a direct function of ϕ and p_s, the denominator of the expression also changes as p_v changes; therefore w_v cannot be a direct function of ϕ.

A majority of engineering problems deal with unconfined atmospheric mixtures, or with these mixtures in flow (as in a heating or an air conditioning system). The partial pressures, specific humidity and temperature of such mixtures may change, whereas the mass of dry air involved in a

given problem remains constant. For this reason it is standard practice to determine such factors as the specific humidity and enthalpy of these mixtures on the basis of one pound of dry air, and to base all calculations upon specific (rather than relative) humidity.

16-7. Psychrometry. As stated in the preceding article, the partial or vapor pressure of the water vapor in any given atmospheric mixture may be determined. Such a determination requires a basically simple instrument known as a psychrometer, which consists essentially of two similar thermometers, one of whose bulbs is surrounded by and in intimate contact with a wick soaked in water. In one application the instrument, known as a sling psychrometer, is rotated on a handle, and the thermometers are read when their fluid columns reach constant positions, the difference in the two readings being an indication of the pressure differential between the atmospheric vapor pressure and the saturation pressure corresponding to the lower, or wet bulb, psychrometer temperature.

In another application, the thermometers are stationary, and air is caused to move over the wet bulb by a fan or blower.

The accuracy of the psychrometer is dependent upon the relative velocity of the air past the wet bulb thermometer, the accepted optimum value being about 1200 feet per minute. The cleanliness of the wick is also important. If the wick is clogged with dirt or dust, the temperature attained by the wet bulb thermometer will be affected.

The reason for the temperature differential noted above is that the movement of air over the wet bulb causes water to be evaporated from the wick with a consequent cooling effect, and the wet bulb temperature is lowered. This effect is progressive up to a certain point, at which evaporation continues from the wick at a constant rate, the wet bulb temperature thus becoming constant.

Under these final equilibrium conditions, the vapor is escaping from the wick to the atmosphere through a film of air and vapor which blankets the wick, by a process known as diffusion, the rate of diffusion being a function of the vapor pressure of the atmosphere and the saturation pressure of the water contained in the wick. Thus:

$$\dot{M} = \frac{dM}{dt} = kA(p_s' - p_v)$$

in which
 \dot{M} = the mass rate of diffusion
 k = the film coefficient of diffusion
 A = the surface area of the wick
 p_v = atmospheric vapor pressure
 p_s' = saturation pressure at wick temperature.

PROPERTIES OF MIXTURES

Further, under the final equilibrium conditions in which the wick temperature is less than the atmospheric temperature, heat will flow from the atmosphere to the wick, through the film, being absorbed as the latent heat required for the evaporation of the diffused moisture. Hence:

$$\frac{dq}{dt} = \frac{dM}{dt}(h_{f_g}) = fA(t_D - t_W)$$

in which
h_{f_g} = the latent heat at temperature t_w
f = the coefficient of thermal conductance of the air-vapor surrounding the wick
A = the surface area of the wick
t_D = the dry bulb, or atmospheric, temperature, °F
t_W = the wet bulb, or wick, temperature, °F.

Combining equations results in

$$kAh_{f_g}(p_s' - p_v) = fA(t_D - t_W)$$

Hence,
$$p_s' - p_v = \frac{f}{kh_{f_g}}(t_D - t_W)$$

With the psychrometer in motion, the film thickness is reduced to a minimum, and the value of f/k approaches a constant. Also the value of h_{f_g} has a relatively small variation over the range of atmospheric temperature. However, the pressure at which diffusion occurs is a variable, because there is no constant atmospheric pressure.

Various psychrometric equations have been advanced, of which the Carrier and the Apjohn equations are typical. The latter, although slightly less accurate than the Carrier equation, has sufficient accuracy for most engineering purposes and will be used.

Carrier's equation is usually stated as:

$$p_v = p_s' - \frac{(p_B - p_s')(t_D - t_W)}{2755 - 1.28t_W}$$

The Apjohn equation, which will be used in this course, is

$$p_v = p_s' - \frac{p_B(t_D - t_W)}{2700} \qquad (16\text{-}10)$$

In both of these equations, the symbols are as follows:
p_v = partial pressure of water vapor in air, psia or in. Hg abs
p_s' = pressure of saturated vapor at wet bulb temperature, psia or in. Hg abs

PROPERTIES OF MIXTURES

p_B = total pressure of mixture (usually barometric pressure of atmospheric air), psia or in. Hg abs
t_D = dry bulb temperature, °F
t_W = wet bulb temperature, °F

The establishment of a psychrometric equation provides the means for calculating the vapor pressures of atmospheric mixtures, and opens the way to a direct calculation of many necessary factors.

Example 16-5: Atmospheric air has a temperature (dry bulb) of 80°F and a wet bulb temperature of 60°F when the barometric pressure is 14.696 psia. Determine (a) the dewpoint, °F; (b) the relative humidity, percent; (c) the specific humidity, grains/lb dry air.
Solution:

$$p_B = 14.696 \text{ psia} \qquad t_D = 80°F \qquad t_W = 60°F$$

(a) $\quad p_s' = 0.2563$ psia (Steam Tables for 60°F)

$$p_v = p_s' - \frac{p_B(t_D - t_W)}{2700} = 0.2563 - \frac{14.696(80 - 60)}{2700} = 0.1474 \text{ psia}$$

dewpoint = 45°F (Steam Tables)

(b) $\quad p_s = 0.5073$ psia (Steam Tables for 80°F)

$$\phi = \frac{p_v}{p_s} = \frac{0.1474}{0.5073} \times 100 = 29.1\%$$

(c) $$w_v = \frac{4354 p_v}{p_B - p_v} = \frac{(4354)(0.1474)}{14.696 - 0.1474} = 44.1 \text{ grains/lb dry air}$$

16-8. Enthalpy of Air-Vapor Mixtures. Enthalpy of atmospheric mixtures is customarily expressed as the enthalpy of the mixture per pound of dry air contained in the mixture. In a given quantity of mixture the mass of dry air remains fixed, whereas the vapor content may be decreased by condensation or increased by evaporation. It is therefore convenient to express those properties involving quantities of energy or matter in terms of unit mass of dry air rather than unit mass of atmospheric mixture. The enthalpy of atmospheric air is therefore the sum of the enthalpies of one pound of dry air and of that quantity of vapor which accompanies it in the given mixture.

In engineering practice temperatures are generally measured by thermometers graduated in degrees Fahrenheit, and the enthalpy datum for dry air is established as zero degrees on this scale. This is merely a matter of convenience which results in the elimination of one arithmetical step, because we are concerned with changes of enthalpy and not with its absolute value. The enthalpy of dry air is therefore

$$h_a = 0.24t \text{ Btu/lb} \qquad (16\text{-}11)$$

where t = temperature of the air, °F.

PROPERTIES OF MIXTURES

The enthalpy of water vapor in the range of pressures and temperatures of atmospheric mixtures is a linear function of temperature. This may be demonstrated by plotting tabulated values from the steam tables against temperature and striking a line through the points thus established. If this is done, the equation of the line produced will be

$$h_v = 1061 + 0.45t$$

where h_v = enthalpy of vapor, Btu/lb
t = temperature of vapor, °F

However, since vapor enthalpies are tabulated against temperature in the steam tables, this equation is usually not required for the solution of problems. The enthalpy of the vapor in one pound of atmospheric air is quickly determined (using the steam tables) and is

$$H_v = w_v h_g$$

where H_v = total enthalpy of vapor, Btu/lb dry air
h_g = enthalpy of saturated vapor at the atmospheric temperature.

The sum of the two enthalpies is the enthalpy of the atmospheric mixture and is

$$h_A = 0.24t + w_v h_g \text{ Btu/lb dry air} \tag{16-12}$$

Example 16-6: Calculate the enthalpy, Btu/lb dry air, for an atmosphere having a temperature (dry bulb) of 80°F and a specific humidity of 42.1 grains/lb dry air.
Solution:

$$t = 80°F \quad w_v = 42.1$$
$$h_g = 1096.4 \text{ Btu/lb (steam tables at 80°F)}$$
$$h_A = 0.24t + w_v h_g$$
$$= 0.24(80) + \frac{(42.1)(1096.4)}{7000} = \underline{25.8 \text{ Btu/lb dry air}}$$

16-9. Adiabatic Saturation, Psychrometry and Psychrometric Charts.

In the development of the basic psychrometric equation in Art. 16-7 it was seen that the wet bulb temperature obtained by a psychrometer depended upon a flow of heat from the atmosphere to the wick. However, no heat was supplied to the atmosphere, and that which was absorbed by the wick was returned to the atmosphere as latent heat of the moisture evaporated. The process was therefore adiabatic, with no heat either gained or lost by the atmosphere.

If, now, we imagine an apparatus such as in Fig. 16-4, consisting of an insulated chamber containing a supply of water at the wet bulb temperature of the air which enters the chamber at 1 and leaves at 2, we may

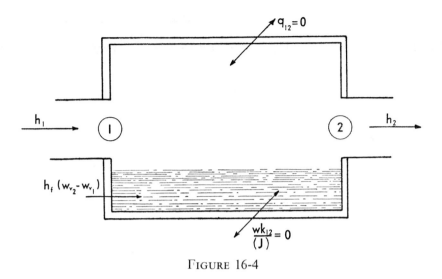

FIGURE 16-4

investigate another essential characteristic of atmospheric mixtures. If the air entering at 1 is not saturated and if adequate water surface is provided (frequently by the use of spray nozzles), experiments have established the fact that the air leaving the chamber *will be saturated* and that its temperature will be the wet bulb temperature of the entering air. The additional water vapor necessary to effect saturation of the entering air is supplied as *liquid* at the wet bulb temperature, t_W. The energy required to evaporate this liquid is supplied by the bulk air, which is cooled from t_D at entrance to t_W at exit.

Since neither heat nor work is involved and the potential and kinetic energy changes are negligible, the steady flow energy equation written between points 1 and 2 in the terms shown in Fig. 16-4, gives

$$h_1 + (w_{v_2} - w_{v_1})h_f = h_2$$

where: h_1 = enthalpy of the mixture entering the apparatus
h_2 = enthalpy of the mixture leaving the apparatus
$(w_{v_2} - w_{v_1})h_f$ = enthalpy of *water* absorbed by the air.

Again there has been no gain or loss of heat by the air; hence the saturation process has been accomplished adiabatically and at constant wet bulb temperature. The relative magnitude of the term $(w_{v_2} - w_{v_1})h_f$ is very small and may be neglected in most calculations; thus for practical purposes the enthalpy of air vapor mixtures is constant for any given wet bulb temperature. This approximation forms the basis for most psychrometric charts.

PROPERTIES OF MIXTURES

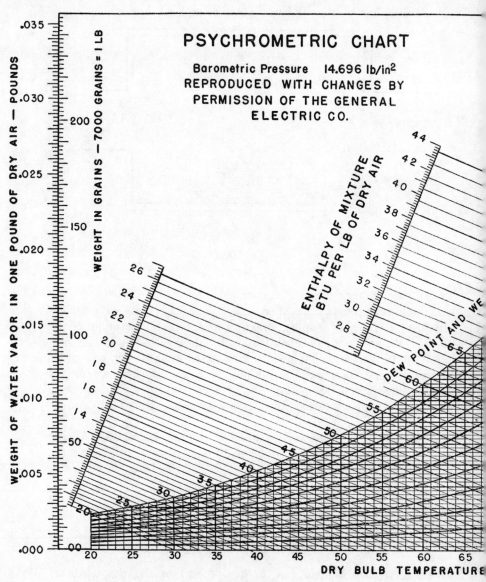

FIGURE 16-5
(This diagram appears as a foldout following the index)

The construction of a psychrometric chart similar to Fig. 16-5 is easily accomplished. After suitable scales are established for the ordinate (which is specific humidity) and the abscissa (which is dry bulb or atmospheric temperature), a barometric pressure for the chart is assumed (usually

PROPERTIES OF MIXTURES

29.92 in. Hg) and the saturation, or 100 percent relative humidity, line is established. This is done by taking the saturation pressures for a series of temperatures from the steam tables and calculating the specific humidity for each temperature of the series by Equation (16-8). The points thus obtained are plotted on the chart form, and the saturation line is drawn in.

PROPERTIES OF MIXTURES

Other relative humidity lines are obtained in a similar fashion for the same temperature series by first establishing the vapor pressures using Equation (16-9) and then repeating the specific humidity calculations using these vapor pressures. When the desired number of relative humidity lines have been drawn on the chart form, the lines of constant enthalpy (and therefore of constant wet bulb temperature) are established as follows: The enthalpy for a given point on the saturation line is calculated using Equation (16-12). Using this value of enthalpy and a temperature approximately 30 degrees higher, the same equation is used to determine the new value of w_v that will satisfy the equation. Two points are thus established which have the same enthalpy, and a straight line may be drawn through them, since the enthalpy of atmospheric mixtures varies directly with temperature. The wet bulb temperature for this line is the temperature of the first point, since at saturation, wet bulb, dry bulb and dew point temperatures coincide.

It should be remembered that psychrometric charts are accurate for one barometric pressure only. However, for the normal range of barometric pressures in any one locality the inaccuracies are negligible. For maritime or naval service where practically all operations are at sea level, one such chart is sufficient; and for submarine work a series of correction curves can be prepared. For any elevation at which the departure from standard barometer is considerable, a new chart should be prepared.

Problems

16-1. It is desired to produce a mixture of helium and hydrogen which shall have a specific heat of 1.0 Btu/lb · °F at constant volume. What must be the volumetric percentage of helium?

Gas	c_v
Helium	0.750
Hydrogen	2.440

Ans. 74.2 percent

16-2. A gaseous mixture has the following mass analysis: H_2, 10 percent. CO_2, 5 percent. N_2, 85 percent.

c_p for common gases at room temperature: O_2, 0.217; H_2, 3.42; N_2, 0.247; CO, 0.243; CO_2, 0.205.

Find:
(a) the specific heat at constant pressure
(b) the volumetric analysis
(c) the partial pressure of the N_2 in psia if barometer is standard and the mixture is at barometric pressure

Ans. (a) 0.562 Btu/lb · °F; (b) H_2 = 61.4%, CO_2 = 1.4%, N_2 = 37.2%; (c) 5.47 psia

16-3. A tank contains air at 50 psia. Air may be assumed to consist of 79.1 percent N_2 and 20.9 percent O_2 by volume. Calculate: (a) the partial pressure due to the oxygen, psia; (b) the partial pressure due to the nitrogen, psia.

Ans. (a) 10.45 psia; (b) 39.55 psia

PROPERTIES OF MIXTURES

16-4. A mass analysis of gases in a compartment shows the following: O_2, 20 lb; N_2, 140 lb; CO_2, 15 lb; H_2O, 4 lb. Find the volumetric analysis of the gases, percent.

Ans. $O_2 = 10.10\%$; $N_2 = 80.81\%$; $CO_2 = 5.50\%$; $H_2O = 3.59\%$

16-5. The volumetric analysis of a gaseous mixture is as follows: CO_2, 12 percent; O_2, 2 percent; N_2, 86 percent. Find the mass analysis, percent.

Ans. $CO_2 = 17.60\%$; $O_2 = 2.13\%$; $N_2 = 80.27\%$

16-6. Given, for atmospheric air: temperature, 82°F; barometric pressure 29.95 in. Hg; partial pressure of water vapor, 0.3632 psia. What is the dew point temperature?

Ans. 70°F

16-7. A room 14 ft × 16 ft × 10 ft contains atmospheric air at 72°F. The partial pressure of the water vapor in the air is 0.2140 psia. Barometer is standard. Calculate:
(a) mass of dry air in the room
(b) mass of water vapor in the room
(c) the dew point temperature

Ans. (a) 164.7 lb; (b) 1.51 lb; (c) 55°F

16-8. For atmospheric air, given: barometer, 30.50 in. Hg; temperature, 80°F; relative humidity, 40 percent.
(a) What is the partial pressure of the water vapor in the air?
(b) What is the approximate dew point temperature?

Ans. (a) 0.20292 psia; (b) 53.5°F

16-9. For atmospheric air, given: temperature, 78°F; relative humidity, 50 percent; barometric pressure, 29.92 in. Hg. Calculate, using steam tables, the mass of water vapor in one pound of dry air.

Ans. 0.01022 lb

16-10. A compartment contains 3000 ft³ of air at 75°F, 50 percent relative humidity, and at standard atmospheric pressure. Calculate:
(a) the mass of water vapor in the compartment, lb
(b) the mass of nitrogen in the compartment, lb
(c) the mass of oxygen in the dry air in the compartment, lb

Ans. (a) 2.023 lb; (b) 167.6 lb; (c) 50.6 lb

16-11. The observed dry bulb temperature of atmospheric air is 80°F and the wet bulb temperature 72°F. Barometer is standard.

$$p_v = p_s' - \frac{p_B(t_D - t_W)}{2700}$$

Find:
(a) partial pressure of water vapor
(b) pounds of water vapor per pound of dry air
(c) enthalpy of atmospheric air per pound of dry air
(d) relative humidity
(e) dew point

Ans. (a) 0.3452 psia; (b) 0.01496 lb/lb D.A.; (c) 35.6 Btu/lb D.A.; (d) 68%; (e) 68.5°F

16-12. For atmospheric air, given: dry bulb temperature—84°F; relative humidity—54 percent. Using psychrometric chart, determine the:
(a) wet bulb temperature, °F

(b) dew point temperature, °F
(c) enthalpy, Btu/lb of dry air
(d) water vapor, lb/lb of dry air

Ans. (a) 71.2°F; (b) 66.1°F; (c) 35 Btu/lb D.A.; (d) 0.0134 lb

16-13. For atmospheric air, given: dry bulb temperature, 85°F; wet bulb temperature, 70°F; barometer, standard. The air is cooled at constant total pressure to 50°F. Using psychrometric chart, calculate:
(a) water vapor condensed, grains/lb of dry air
(b) heat rejected, Btu/lb of dry air

Ans. (a) 32 grains; (b) 13.7 Btu/lb D.A.

16-14. The data below refer to a two-stage air compressor with interstage cooler. The compressor takes in 100 cfm of atmospheric air.

	Entering L.P. Cylinder 1	Entering Intercooler 2	Entering H.P. Cylinder 3
t_a °F	90	500	100
t_W °F	80	xxxx	xxxx
P psia	14.7	114.7	114.7

Find:
(a) dew point of atmospheric air, °F
(b) partial pressure of water vapor in atmospheric air, psia
(c) rate of flow of dry air, lb/hr
(d) moisture content at (1), lb vapor/lb D.A.
(e) moisture content at (3), lb vapor/lb D.A.
(f) condensate removed by intercooler, lb/hr.

Ans. (a) 76.8°F; (b) 0.4529 psia; (c) 419.9 lb/hr; (d) 0.01977 lb vapor/lb D.A.; (e) 0.00521 lb vapor/lb D.A.; (f) 6.11 lb/hr

CHAPTER 17

AIR CONDITIONING

17-1. Introduction. Air conditioning in general pertains to the control of the cleanliness, temperature, motion and humidity of air, with a view to improving the health and comfort of individuals occupying the spaces to which it is supplied. This is distinguished from "revitalization" of the air in such spaces as submarines, where a maintenance of the proper carbon dioxide and oxygen content is necessary in order to sustain life.

The importance of some form of air conditioning is appreciated when the effect of adverse conditions on the capacity of the human body for performing physical work is realized. If a relative humidity of 50 percent is taken as representative of average humidity conditions, the capacity of an individual drops to about 84 percent of its usual value at 95°F, to about 68 percent at 105°F and to about 45 percent at 115°F.

17-2. Conditioned Air. The human body maintains itself at a constant temperature by establishing an equilibrium between the heat supplied and the heat lost. The heat supplied has two main sources, that generated within the body as a result of the oxidation of food, and that received as radiation from the sun. Heat is lost from the body by three methods: radiation, convection and evaporation. The rate at which heat is lost by each of these methods depends on the condition of the surrounding air; the first on the temperature; the second on the temperature and the velocity; and the third on the temperature, velocity and the humidity. Thus, for various conditions of temperature, humidity, and air velocity, changes must take place in the body to re-establish equilibrium. It has been found that this readjustment takes from one and a half to three hours.

Ordinarily an individual feels more comfortable when conditions are such that the body easily adjusts itself. This is true for temperatures of 65° to 75°F.

Stripping to the waist has been found to increase the cooling power of the body about 50 percent. In situations where there is no air conditioning and where the temperature is not over 99°F, comfort may be considerably improved by setting the air in motion by means of fans. Ordinarily, however, a person at rest cannot stand a velocity of over 120 feet per minute without discomfort.

Tests were made on a number of individuals in order to determine just

AIR CONDITIONING

what conditions are required for comfort. These tests were made in "still air" (15 to 25 feet per minute), and sufficient time was allowed for each individual to become adjusted. It was noted that for a large variety of conditions the same sensation of warmth was experienced. Hence these states were all said to have the same effective temperature, the datum

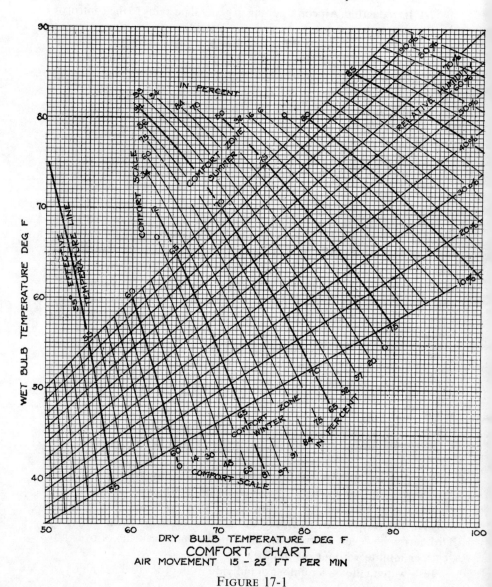

COMFORT CHART
AIR MOVEMENT 15 - 25 FT PER MIN

FIGURE 17-1

AIR CONDITIONING

being taken as saturated air at the corresponding dry bulb temperature. For example, air producing the same sensation of warmth as saturated air at 50°F is said to have an effective temperature of 50 degrees.

Using the data from these experiments a chart was plotted with wet and dry bulb temperatures as coordinates. Lines of constant humidity and constant effective temperature were drawn. Also zones were marked off showing the percentage of individuals that experienced a sensation of comfort for that condition. Such a chart is known as a comfort chart (Fig. 17-1). Within the range of comfort for winter and summer, relative humidities between 30 percent and 70 percent only are recommended.

As a rule it will be found more satisfactory for hot weather operation to maintain a comparatively small difference in temperature between the outside and inside air, and a humidity such that evaporation takes place easily on the skin, making it easier for the body to control its loss of heat. The loss of body heat by radiation and convection is controlled mainly by the air temperature and the amount of clothing, and is beyond the control of the body itself.

17-3. Psychrometric Processes for Air Conditioning. The subject of air conditioning is generally studied in two distinct phases. During winter or heating seasons it is necessary to heat and to humidify to "condition" air, whereas during summer seasons air must be cooled and dehumidified. In both cases the basic processes involved may be illustrated on the psychrometric chart, Fig. 16-5, which should be consulted for study with this chapter. A thorough understanding of these psychrometric processes is necessary before a detailed study of either phase of air conditioning is attempted. In particular it must be remembered that both the enthalpy

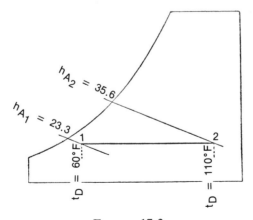

FIGURE 17-2

AIR CONDITIONING

and the specific humidity of an atmospheric mixture are determined on the basis of unit mass of dry air. In this discussion and throughout this chapter, "air" is used synonymously with "atmosphere." Hence "one pound of air" designates an atmospheric mixture consisting of one pound of dry air plus w_v pounds of associated water vapor.

In Fig. 17-2 the process line 1-2 represents sensible heating of one pound of air; that is, heating which is evidenced by a *change of temperature* of the air. For such a process the heat required can be determined by the methods developed in Chapter 16. However, use of the psychrometric chart eliminates these tedious calculations.

Example 17-1: One hundred pounds of air per minute are to be heated from 60°F dry and 55°F wet bulb temperatures to a final temperature of 110°F. There is no change of total moisture during the process. Determine the heat required for the process, Btu/min, (a) by the analytical methods developed in Chapter 16, and (b) by use of the psychrometric chart. Barometric pressure is 29.92 in. Hg.

Solution: Refer to Fig. 17-2:

$p_B = 29.92$ in. Hg $t_{D_1} = 60°F$ $t_{w_1} = 55°F$

$t_{D_2} = 110°F$ $w_{v_1} = w_{v_2}$ $\dot{M} = 100$ lb air/min

(a) $$p_v = p_s' - \frac{p_B(t_D - t_w)}{2700} = 0.4359 - \frac{29.92(60-55)}{2700}$$

$$= 0.3805 \text{ in. Hg abs}$$

$$w_{v_1} = \frac{4354 p_v}{p_B - p_v} = \frac{(4354)(0.3805)}{29.92 - 0.3805} = 56.1 \text{ grains/lb dry air}$$

From Equation (16-12),

$$h_A = 0.24 t_D + w_v h_g = 0.24 t_D + w_v(1061 + 0.45 t_D)$$
$$h_{A_1} = 0.24 t_{D_1} + w_{v_1}(1061 + 0.45 t_{D_1})$$
$$h_{A_2} = 0.24 t_{D_2} + w_{v_2}(1061 + 0.45 t_{D_2})$$
$$h_{A_2} - h_{A_1} = (0.24 + 0.45 w_{v_1})(t_{D_2} - t_{D_1})$$

$$= \left(0.24 + \frac{0.45 \times 56.1}{7000}\right)(110 - 60) = 12.2 \text{ Btu/lb dry air}$$

$$\dot{Q}_{12} = \dot{M}(h_{A_2} - h_{A_1}) = 100(12.2) = \underline{1220 \text{ Btu/min}}$$

(b) From psychrometric chart, $h_{A_1} = 23.3$, $h_{A_2} = 35.6$

$$\dot{Q}_{12} = \dot{M}(h_{A_2} - h_{A_1}) = 100(35.6 - 23.3) = \underline{1230 \text{ Btu/min}}$$

Figure 17-3 illustrates both sensible cooling and dehumidification. The air is first sensibly cooled to its dewpoint, process 1-*a* and then dehumidified and sensibly cooled to the final required temperature. In this case the psychrometric chart may be used to determine both the amount of heat removed and the dehumidification accomplished.

AIR CONDITIONING

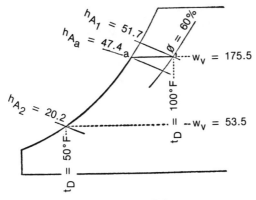

FIGURE 17-3

Example 17-2: One hundred pounds of air per minute at a temperature of 100°F with a relative humidity of 60 percent are cooled and dehumidified to a final temperature of 50°F. Using the psychrometric chart determine (a) the heat abstracted by the process, Btu/min, and (b) the moisture removed, lb/min.

Solution: Refer to Fig. 17-3:

From psychrometric chart,

$$h_{A_1} = 51.7 \quad w_{v_1} = 175.5$$
$$h_{A_a} = 47.4 \quad w_{v_a} = 175.5$$
$$h_{A_2} = 20.2 \quad w_{v_2} = 53.5$$
$$\dot{M} = 100 \text{ lb air per min}$$

(a)
$$\dot{Q}_{1a} = \dot{M}(h_{A_1} - h_{A_a}) = 100(51.7 - 47.4) = 430 \text{ Btu/min}$$
$$\dot{Q}_{a2} = \dot{M}(h_{A_a} - h_{A_2}) = 100(47.4 - 20.2) = 2720 \text{ Btu/min}$$
$$\dot{Q}_{12} = \dot{Q}_{1a} + \dot{Q}_{a2} = 430 + 2720 = \underline{3150 \text{ Btu/min}}$$

(b)
$$\text{dehumidification} = \dot{M}(w_{v_1} - w_{v_2})/7000$$
$$= \frac{100(175.5 - 53.5)}{7000} = \underline{1.743 \text{ lb/min}}$$

In addition to illustrating the use of the psychrometric chart, a comparison of the above example problems clearly indicates the effect of atmospheric moisture upon air conditioning problems. Although the change of temperature is the same in both cases, the heat removed in Example 17-2 is approximately 2½ times greater than the heat supplied in Example 17-1. However the reason for this considerable difference becomes clear when it is understood that the total heat removed in Example 17-2 includes the latent heat of vaporization of the condensed water vapor.

Chemical dehumidification, illustrated in Fig. 17-4, is accomplished by passing the moist air through an adsorbing agent such as silica gel,

FIGURE 17-4

calcium chloride, activated alumina, and so forth. The resulting processes are almost always exothermic and productive of a high final dry bulb temperature. This heat of reaction must be removed from the air prior to its introduction into the air-conditioned spaces. The psychrometric chart is of little value in an analysis of the actual dehumidification process.

Humidification, or the addition of moisture to air, is usually accomplished by spraying water or steam directly into the air stream which is being treated. If a small steam jet is used, the resulting process will be similar to 1–2″ of Fig. 17-5, in which there is a simultaneous increase in

FIGURE 17-5

specific humidity and in temperature. If a water spray is used, the process may be similar to 1–2′, in which the temperature decreases as the specific humidity increases. If the air is passed through a spray chamber (or air washer) in which the water is recirculated but neither heated nor cooled

AIR CONDITIONING

externally, the water will reach the wet bulb temperature of the entering air, and the humidification process 1–2 will be adiabatic at that temperature. For purposes of simplicity the last process will be employed in this text.

Usually it is unnecessary to provide air-conditioned spaces with completely fresh air, but a certain amount of fresh air is required for ventilation. This ventilation air may be cooled and dehumidified so that when combined with recirculated air from the conditioned space the mixture will be adequate for the purpose of the system. A flow diagram for such a system is illustrated in Fig. 17-6, and the properties of the air supplied to the compartment can be calculated as indicated in Example 17-3.

FIGURE 17-6

Example 17-3: Ten pounds of air at a dry bulb temperature of 50°F with a specific humidity of 40 grains/lb dry air are mixed with 25 lb of air having a temperature of 85°F and a specific humidity of 90 grains/lb dry air. By the methods developed in Chapter 16, calculate (a) the specific humidity of the mixture, grains/lb dry air; (b) the dry bulb temperature, °F; (c) the enthalpy of the mixture, Btu/lb dry air.

Solution: Refer to Fig. 17-6:

$$M_4 = 10 \text{ lb} \quad t_{D_4} = 50°F \quad w_{v_4} = 40 \text{ grains/lb dry air}$$
$$M_2 = 25 \text{ lb} \quad t_{D_2} = 85°F \quad w_{v_2} = 90 \text{ grains/lb dry air}$$

(a)
$$w_{v_1} = \frac{M_4(w_{v_4}) + M_2(w_{v_2})}{M_1} = \frac{M_4(w_{v_4}) + M_2(w_{v_2})}{M_4 + M_2}$$

$$w_{v_1} = \frac{10(40) + 25(90)}{10 + 25} = 75.7 \text{ grains/lb dry air}$$

(b) Since the process is adiabatic, and no moisture is gained or lost, the energy balance for the process is as follows:

$$H_{A_1} = H_{A_2} + H_{A_4}$$
$$M_1(h_{A_1}) = M_2(h_{A_2}) + M_4(h_{A_4})$$

By substitution and the elimination of the quantities M_1 and w_{v_1}, the energy equation can be

solved for t_{D_1}. Thus,

$$M_4(0.24 + 0.45w_{v_4})(t_{D_1} - t_{D_4}) = M_2(0.24 + 0.45w_{v_2})(t_{D_2} - t_{D_1})$$

$$10\left(0.24 + \frac{0.45 \times 40}{7000}\right)(t_{D_1} - 50) = 25\left(0.24 + \frac{0.45 \times 90}{7000}\right)(85 - t_{D_1})$$

$$t_{D_1} = \underline{75.1°F}$$

(c) $$h_{A_1} = 0.24t_{D_1} + w_{v_1}(1061 + 0.45t_{D_1})$$

$$= 0.24(75.1) + \frac{75.7(1061 + 0.45 \times 75.1)}{7000}$$

$$= \underline{29.9 \text{ Btu/lb dry air}}$$

Application of the psychrometric chart to the practical solution of this type of problem is illustrated in Fig. 17-7. To this end it is assumed that

FIGURE 17-7

the involved moisture has no effect upon the dry bulb temperature of the mixture, hence that a straight line connecting the state points of the components will contain the state point of the resultant mixture. Although mathematically incorrect, the inaccuracy introduced is of a similar order of magnitude as that accruing from the assumption of constant specific heats for gases in this field.

Example 17-4: Using the psychrometric chart, determine (a) the enthalpy, Btu/lb of dry air, and (b) the specific humidity, grains/lb dry air, for the mixture described in Example 17-3.

Solution: Refer to Fig. 17-7:

$$M_4 = 10 \text{ lb} \qquad t_{D_4} = 50°F$$
$$M_2 = 25 \text{ lb} \qquad t_{D_2} = 85°F$$

$$t_{D_1} = \frac{M_4 t_{D_4} + M_2 t_{D_2}}{M_4 + M_2} = \frac{10(50) + 25(85)}{10 + 25} = 75°F$$

AIR CONDITIONING

Connecting points 4 and 2 with a straight line crossing the 75°F ordinate at point 1, the state point of the mixture, then from the chart,

(a) $\quad\quad\quad\quad\quad\quad h_{A_1} = \underline{29.9 \text{ Btu/lb dry air}}$

(b) $\quad\quad\quad\quad\quad\quad w_{v_1} = \underline{75.7 \text{ grains/lb dry air}}$

17-4. Winter Air Conditioning. In a winter air conditioning system warm air is supplied to the rooms or spaces to be heated in such quantities that the decrease in sensible heat of the air in passage through the spaces offsets the loss of heat from the exterior surfaces. Because of low outdoor temperatures during this season, the air introduced for ventilation purposes is relatively dry, and the system is required to humidify as well as heat the air handled. Copper tube finned coils placed in the air stream and supplied with steam usually provide the source of heat, and this analysis will assume that humidification is accomplished in an air washer.

For a preliminary study it will be assumed that fresh air only is being treated, that outside temperatures are known, and that the final condition of the air is specified, as shown by points A and D, respectively, of Fig. 17-8. Establishment of point D also fixes the condition of the air leaving

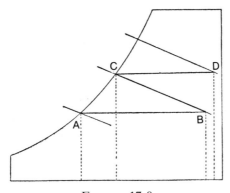

FIGURE 17-8

the washer at point C, because no moisture is added outside of the washer and because the air leaving the washer is saturated. The required condition of the air entering the washer can now be determined because the entering wet bulb temperature must be the same as that of point C and the specific humidity is that of the fresh air. To achieve process A-B, that is to bring the cold air to the required wet bulb temperature prior to entry into the washer, heat must be supplied to the air ahead of the washer. In general this amount of heat does not provide the dry bulb temperature specified at point D; and even if this temperature was maintained at B, it must be remembered that in passing through the washer all of the sensible

AIR CONDITIONING

heat above the wet bulb temperature is absorbed in evaporating the moisture required for saturation. Additional heat must therefore be supplied along the process line *C-D* to bring the air leaving the washer to the specified dry bulb temperature. This sequence of processes establishes the basic equipment required for a winter air conditioning system, and the arrangement of this equipment within the system is illustrated by Fig. 17-9. The heater supplied for process *A-B* is designated the preheating

FIGURE 17-9

coil, the heater for process *C-D* the reheating coil, and the fan for this system may be ahead of or after the reheating coil, depending upon convenience and desirability.

Example 17-5: The heat losses from a group of compartments have been determined to be 420,000 Btu/hr. Air is furnished to the compartments at a temperature of 100°F and leaves the spaces with a temperature of 70°F and a relative humidity of 50 percent. Assuming the system to use 100 percent outdoor air at a temperature of 20°F with 100 percent relative humidity, determine (a) the mass of air which must be circulated, lb/hr; (b) the capacity of the preheating coil, Btu/hr; (c) the capacity of the reheating coil, Btu/hr; (d) the water vapor absorbed from the washer, lb/hr.

Solution: Refer to Fig. 17-10.

$$\dot{Q}_{DE} = 420{,}000 \text{ Btu/hr} \qquad t_{D_A} = 20°\text{F} \qquad \phi_A = 100\%$$
$$t_{D_E} = 70°\text{F} \qquad \phi_E = 50\% \qquad t_{D_D} = 100°\text{F}$$

From psychrometric chart:

$$h_{A_D} = 33.0 \qquad h_{A_E} = 25.4$$
$$h_{A_C} = h_{A_B} = 20.5 \qquad h_{A_A} = 7.2 \text{ Btu/lb dry air}$$
$$w_{v_C} = w_{v_E} = w_{v_D} = 55 \qquad w_{v_A} = w_{v_B} = 16 \text{ grains/lb dry air}$$

(a) $$\dot{M} = \frac{\dot{Q}_{DE}}{h_{A_D} - h_{A_E}} = \frac{420{,}000}{33.0 - 25.4} = \underline{55{,}260 \text{ lb/hr}}$$

(b) $$\dot{Q}_{AB} = \dot{M}(h_{A_B} - h_{A_A}) = 55{,}260(20.5 - 7.2) = \underline{735{,}000 \text{ Btu/hr}}$$

(c) $\dot{Q}_{CD} = \dot{M}(h_{A_D} - h_{A_C}) = 55,260(33.0 - 20.5) = \underline{690,750 \text{ Btu/hr}}$

(d) humidification $= \dot{M}(w_{v_C} - w_{v_B})$

$$= \frac{55,260(55 - 16)}{7000} = \underline{308 \text{ lb/hr}}$$

In general the fresh air requirements of a winter air conditioning system do not exceed 25 percent of the total air circulated, this amount being sufficient for good ventilation of the various spaces. Under these circumstances 75 percent of the air supplied is being recirculated after being mixed with the required fresh air in a plenum chamber. Cursory consideration of this process (illustrated by connecting points A and E of Fig. 17-10 by a straight line) would indicate that no preheating coil is

FIGURE 17-10

needed since point F (the condition produced by such a mixing) falls on the line connecting B and C, and it would appear that the mixture could be taken directly into the washer. However, it must be noted that the mixture line (A, F, E) is tangent to the saturation line between the temperatures of 20°F and 31°F, and it is highly probable that some vapor in the mixture will condense and freeze in the plenum chamber. To prevent such an occurrence a preheating coil should always be included in the system, although its capacity may be greatly reduced in terms of the relative amounts of fresh and recirculated air. In many cases the preheating coil is provided with an automatic control which will maintain the fresh air temperature at from 40°F to 60°F, and the desired mixture

conditions are achieved by varying the proportions of fresh and recirculated air.

It is possible to supply all of the heat before the humidification process, and various means of humidifying the air may be used; but such systems become increasingly complicated in the matter of automatic control of the various processes, and their performance is usually unsatisfactory for this reason. This difficulty of control is minimized in the preheat-reheat system, which is therefore generally preferred.

17-5. Summer Air Conditioning. The objective of summer air conditioning is to provide comfortable indoor conditions while at the same time avoiding large differences between inside and outside temperatures, since such differences have undesirable physiological effects upon many people.

Summer air conditioning is specifically concerned with cooling and dehumidifying the air within a given system. Usually direct refrigeration employing a finned tube coil provides both cooling and dehumidification, and quite frequently supplementary heating is necessary. Cooling and dehumidification may also be accomplished by a washer, and occasionally the need for reheat can be eliminated by by-passing.

The most economical summer conditioning cycle is illustrated on the psychrometric chart in Fig. 17-11. This system employs refrigeration and by-passing, but requires no reheat. Air leaving the conditioning equipment at state D is supplied to the spaces served by the system, in which it absorbs both heat and moisture such that it leaves these spaces at state 2. After being exhausted from the conditioned spaces, a small part of the air at state 2 is discharged from the system and is replaced with an equal amount of fresh air at state 1 to produce the mixture at state A. This freshened air at state A is returned to the conditioning equipment, where a fraction of it is by-passed around the cooling coil while the remainder is cooled and dehumidified as represented by the process A-B-C of Fig. 17-11. Finally, the by-passed fraction, at state A, is remixed with the cooled air, at state C, to produce state D and complete the flow cycle.

When by-passing is employed the special problem which arises is the determination of the relative amounts of air that must be heated or cooled. This is essentially a mixing problem in which the properties of the components and of the final mixture are known or specified, and the psychrometric chart can be used to facilitate the determination of the required relative weights of each component.

Referring to Fig. 17-11,

$$w_{v_D} = \frac{M_A(w_{v_A}) + M_C(w_{v_C})}{M_A + M_C}$$

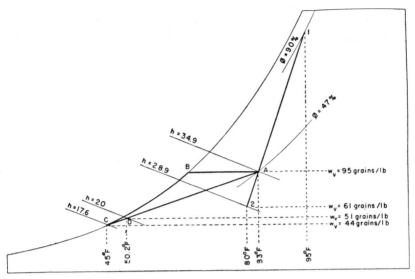

FIGURE 17-11

and for unit mass of mixture at state D,

$$M_A + M_C = 1$$

Let,

$$F = M_A = \text{fraction by-passed, lb/lb of mixture}$$

then,

$$M_C = 1 - F$$

$$w_{v_D} = \frac{(F)(w_{v_A}) + (1 - F)(w_{v_C})}{1}$$

and,

$$F = \frac{w_{v_D} - w_{v_C}}{w_{v_A} - w_{v_C}} \text{ lb/lb of mixture}$$

Similar equations for the by-pass fraction can be derived in terms of the enthalpies and in terms of the dry bulb temperatures of states A, C and D.

Example 17-6: For an air conditioning system such as that described in this article and illustrated in Fig. 17-11, the freshened air (state A) has dry and wet bulb temperatures of 83°F and 71°F respectively, the temperature leaving the cooling coils (state C) is 45°F, the specific humidity after remixing is 51 grains/lb dry air and the temperature and relative humidity leaving the conditioned spaces (state 2) are 80°F and 40 percent respectively. If the total sensible and latent heat gain from the spaces is 240,000 Btu/hr, calculate (a) the mass of air that must be circulated, lb/hr; (b) the amount of moisture that can be absorbed from the conditioned spaces, lb/hr; (c) the fraction of air that by-passes the cooling coil, lb/lb; (d) the

air that flows through the coil, lb/hr; (e) the capacity of the refrigerating plant required for this system, tons.

Solution: Refer to Fig. 17-11:

$$\dot{Q}_{D2} = 240{,}000 \text{ Btu/hr} \qquad t_{D_A} = 83°F \qquad t_{W_A} = 71°F$$
$$t_{D_C} = t_{W_C} = 45°F \qquad t_{D_2} = 80°F \qquad \phi_2 = 40\% \qquad w_{v_D} = 51$$

From psychrometric chart:

$$w_{v_2} = 61 \text{ grains/lb dry air}$$
$$w_{v_A} = 95 \qquad w_{v_C} = 44 \text{ grains/lb dry air}$$
$$h_{A_2} = 28.9 \qquad h_{A_D} = 20 \qquad h_{A_A} = 34.9$$
$$h_{A_C} = 17.6 \text{ Btu/lb dry air}$$

(a) $$\dot{M} = \frac{\dot{Q}_{D2}}{h_{A_2} - h_{A_D}} = \frac{240{,}000}{28.9 - 20} = \underline{27{,}000 \text{ lb/hr}}$$

(b) $$\text{moisture absorbed} = \frac{\dot{M}(w_{v_2} - w_{v_D})}{7000}$$

$$= \frac{27{,}000(61 - 51)}{7000} = \underline{38.6 \text{ lb/hr}}$$

(c) $$F = \frac{w_{v_D} - w_{v_C}}{w_{v_A} - w_{v_C}} = \frac{51 - 44}{95 - 44} = \underline{0.1372 \text{ lb/lb}}$$

(d) $$\text{air through coil} = \dot{M}_C = \dot{M}_D - F(\dot{M}_D)$$
$$\dot{M}_C = 27{,}000 - (27{,}000)(0.1372) = \underline{23{,}300 \text{ lb/hr}}$$

(e) $$\text{refrigeration} = \frac{\dot{M}_C(h_{A_A} - h_{A_C})}{60 \times 200}$$

$$= \frac{(23{,}300)(34.9 - 17.6)}{12{,}000} = \underline{33.6 \text{ tons}}$$

Although the above processes provide maximum economy, the system described may be unsatisfactory because of the large temperature difference between the conditioned air and that which is exhausted from the compartments. The reduction of this temperature difference entails considerations beyond the scope of this course, and only the two principal design variations will be presented.

The first of these variations consists of reheating the conditioned air prior to mixing it and is illustrated in Fig. 17-12. In this arrangement the cooling and dehumidification process is *ABC* as before, but the air leaving the cooling coil at *C* is reheated along the process line *CD* before being mixed with untreated air to produce the conditions at point *E*. Should the condition at *E* still be unsatisfactory, the final recourse is the elimination of by-passing. Thus in Fig. 17-12 when there is no by-passing, the air at *A* is cooled to point *F* and then reheated to point *G*. This last method is the least economical but has many practical advantages.

AIR CONDITIONING

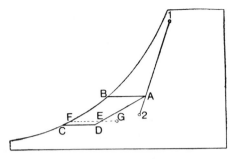

FIGURE 17-12

Problems

17-1. In a winter air conditioning unit consisting of a preheater, humidifier and reheater, similar to Fig. 17-9, the following conditions are maintained:

Point	t_D	ϕ
1. Air to preheater	51°F	80 percent
2. Air to humidifier		
3. Air to reheater		100 percent
4. Air from reheater	77°F	50 percent

(a) What is the wet bulb temperature at point 2?
(b) What is the dry bulb temperature at point 2?
(c) If the rate of flow is 350 lb per min of dry air, how many pounds per min of water are evaporated in the humidifier?

Ans. (a) 57°F; (b) 73°F; (c) 1.3 lb/min.

17-2. Air at 45°F (DB) and 41°F (WB) is heated and humidified to 72°F (DB) and 59°F (WB). To what temperature should the air be heated before humidification?

Ans. 64°F

17-3. Air is to be conditioned from $t_D = 39°F$ and $\phi = 80$ percent to $t_D = 74°F$ and $\phi = 70$ percent. (a) To what temperature should the air be heated before humidifying? (b) How much moisture in grains is added during humidification?

Ans. (a) 102.5°F; (b) 61 grains/lb D.A.

17-4. Air is conditioned from $t_D = 46°F$ and $\phi = 60$ percent to $t_D = 70°F$ and $t_W = 61°F$. (a) To what temperature should the air be heated before humidification? (b) How much moisture in pounds per pound of dry air conditioned must be added in the humidifier? (c) How many Btu must be supplied per pound of dry air conditioned?

Ans. (a) 80°F; (b) 0.0055 lb/lb D.A.; (c) 11.8 Btu/lb D.A.

17-5. Outside air at $t_D = 35°F$ and $\phi = 50$ percent is to be conditioned to $t_D = 70°F$ and $\phi = 60$ percent. (a) To what temperature must the air be heated before humidifying? (b) How much heat must be supplied per pound of dry air in the heaters? (c) How much heat must be supplied if the air is heated to the final temperature without humidifying? (d) What would be the final relative humidity under the conditions of part (c)?

Ans. (a) 88°F; (b) 16.4 Btu/lb D.A.; (c) 8.4 Btu/lb D.A.; (d) 13%.

AIR CONDITIONING

17-6. In a summer air conditioning unit, part of the air is cooled in a refrigerated chamber and then mixed with the remaining untreated air to obtain the desired conditions of temperature and humidity. The following conditions are maintained:

Point	t_D	ϕ
1. Untreated air	100°F	50 percent
2. Air from cooler		100 percent
3. Air after mixing	80°F	80 percent

(a) What is the temperature at point 2?
(b) What percentage of the total air must pass through the cooler?
(c) If the total flow is 1000 lb/min of dry air, how much refrigeration in standard tons is required?

Ans. (a) 71°F; (b) 69%; (c) 42.3 tons

17-7. The outside air is at $t_D = 84°F$ and $\phi = 60$ percent. A large supply of water is available at 55°F for an air washer. The air through the washer is mixed with outside air to obtain air at $t_D = 68°F$ for compartment cooling. (a) Using the enthalpy method, find the percentage of air by-passed. (b) What is the dew point of the air entering the compartment?

Ans. (a) 45%; (b) 62°F

17-8. Outside air is conditioned from $t_D = 92°F$ and $t_w = 80°F$ to $t_D = 75°F$ and $\phi = 80$ percent. Heating is accomplished by mixing outdoor air with cooled air. (a) What percentage of air must pass through the cooler? (b) To what temperature must the air passing through the cooler be cooled?

Ans. (a) 52%; (b) 59°F

CHAPTER 18

BOILERS AND THE CHEMISTRY OF COMBUSTION

18-1. Introduction. The term *boiler* was originally applied to a furnace and a generating section used for production of saturated, or nearly saturated, steam. As refinements were devised and additional sections for heat transfer such as superheaters, economizers and air preheaters were added to boilers, they soon became known as steam generators. However, the name *boiler* has persisted, and, because of its brevity, will be used in this chapter to denote the combination of furnace, saturated steam generating section, superheater section, economizer and air preheater, or any of these pieces of apparatus used in combination to generate steam.

The purpose of a boiler is to transmit the greatest possible quantity of heat to the feedwater in order to generate steam. Substantial heat release rates can be obtained by combining certain compounds, called fuels, with oxidizing mediums such as air, in chemical processes classified as *combustion*. To utilize this energy-liberating process effectively, the properties of the fuel to be burned, the conditions affecting combustion and the limitations on the size of the boiler furnace must be taken into account in any boiler design.

The capacity of a boiler in pounds of steam per hour at a *specified pressure and temperature* determines the amount of heat that must be supplied to the furnace, when losses are properly accounted for. For a given fuel, the quantity of heat liberated in the furnace will depend on the quality of combustion, which, in turn, is dependent on the atomization of the fuel, the air supplied, and the manner in which the air and fuel are mixed. The proper number and size of burners to accommodate this heat requirement of the furnace may then be determined. The quantity of air supplied and the products of combustion formed are directly proportional to the quantity of fuel fired. Therefore, furnace size must be sufficient to accommodate the fuel and air required for combustion at maximum capacity, as well as to permit proper mixing of constituents and maximum heat liberation from the combustion process.

The following articles include definitions of standard boiler terms and a consideration of the various factors affecting energy liberation by the boiler fuel and absorption by the working substance.

18-2. Boiler Terminology. As a result of the large number of boiler types and the lack of uniformity concerning boiler terminology, certain basic

definitions will be given in Engineers' Units, based on the American Society of Mechanical Engineers' Power Test Code and the Naval Ships' Technical Manual.

A *boiler horsepower* at one time was defined as the generation of 34.5 pounds per hour of dry and saturated steam at 212°F from feedwater at the same temperature. This is equivalent to a heat transfer of 33,475 Btu per hour. At one time there was a significant relationship between boiler steam and the power output in horsepower of the engines supplied. Today that relationship no longer exists, and the term is meaningless, although it may still be observed on the name plates of some small boilers.

Boiler capacity is defined as the heat absorbed by the water and steam in Btu per hour. In large boilers it is usually more convenient to express this heat absorption rate in multiples of the Btu such as the $kB = 1000$ Btu and the $mB = 1,000,000$ Btu.

The *total heating surface* of a boiler is that portion of the heat transfer apparatus exposed on one side to the gas being cooled and on the other to the fluid or steam being heated. In other words, the total heating surface is the sum of the generating surface, the superheater surface and the economizer surface.

The *generating surface* is that portion of the total heating surface in which the fluid being heated forms a part of the circulating system; this surface is measured on the flue-gas side. This includes the boiler tube bank, water walls, water screens and water floor when installed.

The *superheater surface* is that portion of the total heating surface in which the steam is heated after leaving the saturated steam drum. This surface is also measured on the flue-gas side.

The *economizer surface* is that portion of the total heating surface outside the boiler circulating system which has the hot gases on one side and the liquid being heated on the other. Economizer surface is measured on the flue-gas side of the tubes.

Total furnace volume is the cubical content of the combustion space up to the plane of first entry of the gases into or between the tubes.

Actual evaporation in pounds of steam per hour is a term frequently used to indicate the performance of a boiler, but it is essential that the steam conditions be known before it has any real meaning.

Factor of evaporation is defined as the ratio of the heat actually absorbed per pound of steam to the heat involved in converting one pound of saturated water to one pound of saturated steam at 212°F. Multiplication of the actual evaporation by the factor of evaporation results in an *equivalent evaporation* based on standard conditions normally designated as evaporation of one pound of water "from and at 212°F." The

BOILERS AND THE CHEMISTRY OF COMBUSTION

factor of evaporation is expressed as

$$f = \frac{h_{out} - h_{in}}{970.3}$$

where h_{out} is the enthalpy of wet, saturated or superheated steam at the boiler outlet, h_{in} is the enthalpy of the compressed liquid entering the boiler, and 970.3 Btu per pound is the latent heat added to one pound of saturated water at 212°F to produce one pound of saturated steam at 212°F.

Furnace rating is defined as the heat developed per cubic foot of furnace volume per hour, also called the furnace heat-release rate. It may be calculated by dividing the product of pounds of fuel per hour and the higher heating value of the fuel by the total furnace volume in cubic feet. The heating value of a fuel is the ideal energy that could be released if complete combustion occurred at standard conditions.

Boiler efficiency is defined as the ratio of heat absorbed by the water and steam to the higher heating value of the fuel as fired during the same unit of time. Thus,

$$\eta_b = \frac{\text{output}}{\text{input}} = \frac{M_s(h_{out} - h_{in})}{M_F(\text{H.H.V.})}$$

where M_s represents pounds of steam generated per hour, M_F the pounds of fuel fired per hour, h_{out} the enthalpy of steam leaving the superheater, Btu per pound, h_{in} the enthalpy of compressed liquid at entrance to the boiler, Btu per pound, and H. H. V. the higher heating value of the fuel, Btu per pound.

Example 18-1: A boiler produces 250,000 pounds of steam per hour at 1200 psia and 1050°F from feedwater entering the boiler at 1500 psia and 300°F. Fuel oil having a higher heating value of 18,000 Btu/lb is supplied to the burners at a rate of 20,500 lb/hr. Furnace volume is 1500 cubic feet. Calculate:

(a) boiler capacity in mB/hr
(b) factor of evaporation
(c) equivalent evaporation in lb/hr
(d) furnace heat-release rate in Btu/hr·cu ft
(e) boiler efficiency in percent

Solution:

h for steam = 1528.9 Btu/lb
h for water = h_f(300°F and 1500 psia) (table 4, Steam Tables)
= 272.39 Btu/lb

(a) capacity = 250,000(1528.9 − 272.4)
= 314.1 mB/hr

(b) $$f = \frac{h_{out} - h_{in}}{970.3} = \frac{1528.9 - 272.4}{970.3} = \underline{1.295}$$

(c) $$\text{equivalent evaporation} = 1.295(250,000)$$
$$= \underline{323,750 \text{ lb/hr}}$$

(d) $$\text{furnace heat-release rate} = \frac{20,500(18,000)}{1500}$$
$$= \underline{246,000 \text{ Btu/hr} \cdot \text{ft}^3}$$

(e) $$\eta_b = \frac{250,000(1528.9 - 272.4)}{20,500(18,000)}$$
$$\eta_b = \underline{85\%}$$

18-3. Types of Fossil Fuel. Fuels may be classified by phase as being solid, gaseous or liquid, and by their availability as either natural or manufactured. The natural solid fuels are numerous, including wood, coal, lignite and peat. Wood is composed chiefly of cellulose, various resin compounds and ash. Coal consists of varying percentages of carbon, hydrogen, oxygen, nitrogen, sulfur, ash and moisture. Lignite and peat are similar to coal, but are less desirable as fuels because of their high percentage of moisture and low percentage of carbon. Coke is an example of a manufactured solid fuel and is an important fuel in the steel industry. Gaseous fuels can be categorized either as natural gas from petroleum wells, or from gas wells usually located in a territory having an abundance of petroleum, or as manufactured gas from any of a number of gas-producing processes. Natural gas consists of varying percentages of the hydrocarbons methane (CH_4) and ethane (C_2H_6), accompanied by smaller percentages of hydrogen, carbon monoxide and inert gases. The major natural liquid fuel source is crude petroleum, while the liquid processed fuels are primarily products of the petroleum refining industry. Prominent among these refinery products as liquid fuels are gasoline, kerosene and fuel oil.

18-4. Properties of Fuel Oil. Crude petroleum, as obtained from wells in various parts of the world, consists of carbon, hydrogen, oxygen, nitrogen, sulfur and moisture in varying proportions. These proportions are given by an *ultimate analysis,* which is a mass analysis of the constituents of a fuel on the *dry* or *moisture-free* basis. A simpler *proximate* analysis determines moisture, volatile matter, fixed carbon and ash.

The several fractions or cuts of which crude petroleum consists may be separated by a process known as fractional distillation, in which the crude is divided into components determined by the volatility or boiling-point range of each. Another means of refining crude petroleum is thermal

cracking. In this method, some of the heavier fractions are exposed to controlled pressure and temperature conditions in such a way that the lighter, more volatile hydrocarbons are removed, leaving a relatively heavier and less volatile residue.

Fuel oil for naval and marine consumption generally has been a blend of the residue from the cracking process, residue from the fractional distillation process and general refinery residue. The word residue, as used here, refers to the heavier, less volatile products remaining in the still after a distillation or cracking process. More recently a trend toward using the same or similar lighter fuel oils for boilers and some combustion engines has been started in order to reduce logistic problems faced by the Navy. The name fuel oil has been generally applied to a large number of commercial products, including distillates known as Diesel oil and household heating oil.

A fuel oil for naval or marine use must have certain characteristics to be burned efficiently. It must be free of moisture and sediment, relatively low in sulfur content and capable of being atomized properly with reasonable heating, and should neither deteriorate in storage nor foul fuel oil heaters.

Specific gravity of fuel oils is ordinarily determined by a direct reading on the A.P.I. (American Petroleum Institute) hydrometer scale. The modulus of the A.P.I. scale is such that ten degrees A.P.I. equals a specific gravity of 1.0. An A.P.I. reading more than ten indicates a specific gravity less than 1.0, and lower than ten, a specific gravity greater than 1.0. The relation of A.P.I. degrees to specific gravity is expressed by the following equation:

$$\text{A.P.I. degrees} = \frac{141.5}{\text{specific gravity } 60/60} - 131.5$$

In the above equation specific gravity 60/60 refers to the specific gravity of the oil measured at 60°F referred to that of water at 60°F.

According to the A.S.M.E. Power Test Code, the calorific or heating value of a liquid fuel is preferably expressed in Btu per pound, although it is often stated in Btu per gallon. The *higher heating value* of a fuel is defined as the quantity of heat that may be absorbed from complete combustion when the products of combustion are cooled back to the original temperature of the fuel with complete condensation of all moisture formed from the combustion of hydrogen. This definition does not limit the process to one of constant pressure or constant volume. It is recognized that a discrepancy will exist between the higher heating values of a given fuel determined at constant volume and at constant pressure, but this discrepancy will be negligible for engineering applications. A

lower heating value, use of which is standard practice in European countries, is based on the heat that may be absorbed from complete combustion of a fuel when the products of combustion are cooled back to the original fuel temperature but the moisture formed by the combustion of hydrogen is not condensed. A standard of pressure and temperature commonly employed for expression of heating values is 29.92 inches of mercury and 68°F.

When comparing thermal efficiencies of either internal combustion or external combustion devices, one should remember that use of the lower heating value will result in an increase in efficiency of several percent over that obtained by use of the higher heating value.

The flash point of a fuel oil is that temperature at which the vapors emanating from the oil will flash momentarily in the presence of an open flame brought near the oil surface. Fire point is that temperature at which the emanating vapors will flash and continue to burn. Flash and fire points are of primary concern in safety of storage and transfer.

Viscosity of fuel oils is measured either in Saybolt Seconds Universal or Saybolt Seconds Furol. Transfer and atomization of fuel oils are directly affected by viscosity.

18-5. Combustion Equations. Although the actual combustion reactions are very complex, for purposes of analysis the simple combining equations of the combustible elements of fuels will be used. When carbon is burned in the presence of pure oxygen, the ideal combustion reaction can be written chemically as

$$C + O_2 \rightarrow CO_2 + 14{,}095 \text{ Btu/lb C} \qquad (18\text{-}1)$$

Similarly,

$$2H_2 + O_2 \rightarrow 2H_2O + 121{,}994 \text{ Btu/2 lb H}_2 \qquad (18\text{-}2)$$

and

$$S + O_2 \rightarrow SO_2 + 3984 \text{ Btu/lb S} \qquad (18\text{-}3)$$

The conservation of mass for combustion requires the number of atoms of all elements entering the chemical process of combustion to be equal to the number appearing in the products of the process. Thus in Equation (18-2) above, 4 atoms of hydrogen initially present in H_2 equals the 2×2 or 4 atoms of hydrogen in the water after reaction. Further, the law of combining weights indicates that $2(2) = 4$ pounds of hydrogen combine with $2(16) = 32$ pounds of oxygen to form $2(2 + 16) = 36$ pounds of water. Finally, if all substances in the equation are assumed to be in the gaseous state, 2 moles (hereafter written mols) of hydrogen combine with 1 mol of oxygen to form 2 mols of water; or 2 cubic feet of hydrogen combine with 1 cubic foot of oxygen to produce 2 cubic feet of water

BOILERS AND THE CHEMISTRY OF COMBUSTION

(vapor) when all quantities are measured at the pressure and temperature of the fuel. From the foregoing statements it becomes clear that the mass of substances entering into combustion must equal the mass of the products of combustion; but the volumes occupied by the entering constituents may be greater than, equal to, or less than that of the products of combustion.

In Table 18-I are listed the chemical elements and compounds entering into combustion reactions with their approximate molecular weights.

TABLE 18-I

Substance	Molecular symbol	Molecular weight (approximate)
Oxygen	O_2	32
Hydrogen	H_2	2
Carbon (solid)	C	12
Sulfur	S	32
Nitrogen	N_2	28
Carbon monoxide	CO	28
Carbon dioxide	CO_2	44
Sulfur dioxide	SO_2	64
Water vapor	H_2O	18

In Equations (18-1), (18-2) and (18-3) it was assumed that oxygen was present in just the proper quantity for complete combination with the fuel to produce what is commonly called stoichiometric or "theoretical" combustion. Actually, in almost all cases of combustion, the oxygen is supplied from the atmosphere, which is a mixture of oxygen, nitrogen and traces of inert gases. For purposes of this discussion, the inert gases will be included with the nitrogen, giving the composition for air shown in Table 18-II.

TABLE 18-II

	Percent Mass	Percent Volume
Oxygen	23.2	20.9
Nitrogen	76.8	79.1

BOILERS AND THE CHEMISTRY OF COMBUSTION

The nitrogen associated with the oxygen in air is inert as far as combustion processes are concerned, the only result being a lower final temperature of the products of combustion because of the diluting effect of the added mass of products for a fixed mass of fuel.

It was shown that with pure oxygen

$$C + O_2 \rightarrow CO_2 + 14{,}095 \text{ Btu/lb C} \qquad (18\text{-}4)$$
$$12 \text{ lb C} + 32 \text{ lb } O_2 = 44 \text{ lb } CO_2$$

The same constituent, carbon, burned with the theoretical quantity of air needed to provide the required oxygen would give

$$C + O_2 + 3.78 \text{ } N_2 \rightarrow CO_2 + 3.78 \text{ } N_2$$

since air contains $0.791/0.209 = 3.78$ parts of nitrogen by volume to one part of oxygen. Thus, accompanying each part of oxygen by volume in atmospheric air are 3.78 parts by volume of nitrogen. Calculated on a mass basis, air contains $0.768/0.232 = 3.31$ parts of nitrogen to one part of oxygen. The combustion equation then gives for carbon

$$\mathbf{C + O_2 + 3.78N_2 \rightarrow CO_2 + 3.78N_2} \qquad (18\text{-}5)$$
$$12 + 32 + (3.78)(28) = 44 + (3.78)(28)$$
$$12 \text{ lb C} + 32 \text{ lb } O_2 + 105.8 \text{ lb } N_2 = 44 \text{ lb } CO_2 + 105.8 \text{ lb } N_2$$

Dividing this last equation by the number of pounds of carbon indicated gives

$$1 \text{ lb C} + 2.67 \text{ lb } O_2 + 8.82 \text{ lb } N_2 = 3.67 \text{ lb } CO_2 + 8.82 \text{ lb } N_2$$

Finally,

$$12.49 \text{ lb fuel and air} = 12.49 \text{ lb combustion products}$$

Analyzing this same equation on the basis of volumes entering into combustion as mentioned previously,

$$1 \text{ mol C} + 1 \text{ mol } O_2 + 3.78 \text{ mols } N_2 = 1 \text{ mol } CO_2 + 3.78 \text{ mols } N_2$$

or,

$$0 \text{ ft}^3 \text{ C} + 1 \text{ ft}^3 \text{ } O_2 + 3.78 \text{ ft}^3 \text{ } N_2 = 1 \text{ ft}^3 \text{ } CO_2 + 3.78 \text{ ft}^3 \text{ } N_2 \quad (18\text{-}6)$$

Since carbon is rarely found uncombined in any but the solid form, the volume of the solid carbon is assumed to be negligible with respect to the volumes of the gaseous constituents in the combustion process. It should be noted here that the total volumes on the left side of combustion equations such as (18-5) above will not always equal the total volumes on

BOILERS AND THE CHEMISTRY OF COMBUSTION

the right side. Evidence of this is the equation

$$2H_2 + O_2 \rightarrow 2H_2O$$

$$2 \text{ mols } H_2 + 1 \text{ mol } O_2 \rightarrow 2 \text{ mols } H_2O$$

$$2 \text{ ft}^3 \ H_2 + 1 \text{ ft}^3 \ O_2 \rightarrow 2 \text{ ft}^3 \ H_2O \text{ (as a vapor)} \tag{18-7}$$

The same equation in terms of masses gives

$$4 \text{ lb } H_2 + 32 \text{ lb } O_2 = 36 \text{ lb } H_2O$$

$$\mathbf{1 \text{ lb } H_2 + 8 \text{ lb } O_2 = 9 \text{ lb } H_2O} \tag{18-8}$$

From Equation (18-5) above, written on the basis of one pound of carbon, it is shown that for theoretical combustion 2.67 pounds of oxygen and 8.82 pounds of nitrogen are required, and their sum is equal to 11.49 pounds of air. It is also possible to multiply the required oxygen, 2.67 pounds, by the pounds of air per pound of oxygen in the atmosphere, 4.31, and arrive at 11.50 pounds of air required per pound of carbon. The small discrepancy arises from rounding in the intermediate calculations. Because carbon is the predominant fuel component, there is merit in carrying the oxygen/carbon ratio (32/12) to one more significant figure, i.e., 2.667 vice 2.67.

In a similar manner, the combustion equation for hydrogen shows that the air required to burn one pound of fuel will be made up of 8 pounds of oxygen and 26.48 pounds of nitrogen for a total of 34.48 pounds. Sulfur, which requires one pound of oxygen per pound of fuel, requires 4.31 pounds of air.

Example 18-2: The following is an ultimate analysis of a typical naval fuel oil, ash and moisture free:

$$
\begin{aligned}
\text{carbon} &= 0.8663 \text{ lb} \\
\text{hydrogen} &= 0.1127 \text{ lb} \\
\text{oxygen} &= 0.0019 \text{ lb} \\
\text{nitrogen} &= 0.0028 \text{ lb} \\
\text{sulfur} &= \underline{0.0163 \text{ lb}} \\
&\ 1.0000 \text{ lb}
\end{aligned}
$$

Calculate the following for "complete" or "theoretical" combustion with air, in pounds per pound of fuel:
(a) oxygen required from air
(b) nitrogen required from air
(c) air required
(d) water formed from combustion
(e) carbon dioxide formed
(f) total mass of flue gases

BOILERS AND THE CHEMISTRY OF COMBUSTION

Solution:

Fuel constituents pounds per pound	Pounds of O_2 required from air per pound of fuel constituent
C = 0.8663	0.8663(2.667) = 2.3104
H = 0.1127	0.1127(8.000) = 0.9016
O = 0.0019	reduces external O_2 = −0.0019
N = 0.0028	inert = 0.0000
S = 0.0163	0.0163(1.00) = 0.0163
	lb O_2 required from air = 3.2264

(a) oxygen required = 3.2264 lb/lb fuel

(b) nitrogen from air = $3.2264 \left(\dfrac{0.768}{0.232} \right)$ = 10.6805 lb N_2/lb fuel

(c) air required = 3.2264 + 10.6805 = 13.9069 lb air/lb fuel

Element + Oxygen =	Combustion product
C 0.8663 + 2.3104 =	3.1767 lb CO_2/lb fuel
H 0.1127 + 0.9016 =	1.0143 lb H_2O/lb fuel
O 0.0019 + ——— =	———————————————
N 0.0028 + 0.0000 =	0.0028 lb N_2/lb fuel
S 0.0163 + 0.0163 =	0.0326 lb SO_2/lb fuel
	4.2264 lb combustion products per lb fuel

(d) water from combustion = 1.0143 lb/lb fuel

(e) carbon dioxide formed = 3.1767 lb/lb fuel

(f) total mass of flue gases = 4.2264 + 10.6805
 = 14.9069 lb/lb fuel

It should be noted in the above example that the total mass of flue gases per pound of fuel equals the mass of air required plus the mass of fuel fired when no losses as unburned fuel occur.

18-6. Combustion With Excess Air. The combustion reactions developed up to this point have been concerned only with theoretical or stoichiometric combustion. Actual conditions in the combustion chamber are such that proper mixing of the fuel and air does not always occur. Consequently, an excess of air is usually supplied in a furnace in order that all the fuel elements have an opportunity to combine with their required quantity of oxygen. As a result, that portion of oxygen supplied by the air finding no fuel elements with which to combine appears in the flue gases as uncombined oxygen. Combustion equations such as (18-5) can be altered for use with excess air by multiplying the O_2 and N_2 by (1 + fraction excess air) and including an excess O_2 term along with the products of combustion on the right side of the equation. This excess oxygen carries along its usual complement of nitrogen from the air; and

BOILERS AND THE CHEMISTRY OF COMBUSTION

since neither the excess oxygen nor the nitrogen assists the combustion, the net result is a transfer of heat to these excess agents, a large part of which is carried out the stack with consequent reduction of overall efficiency of the furnace and heat exchanger unit.

The effect of excess air on fuel-lean combustion calculations can best be shown by calculating the same items as required in the example of the previous article for 50 percent excess air.

Example 18-3: Calculate the following quantities in pounds per pound of fuel for combustion with 50 percent excess air using the same fuel analysis as in the example of the preceding article:
 (a) oxygen supplied from air
 (b) nitrogen accompanying oxygen
 (c) air supplied
 (d) water formed from combustion
 (e) carbon dioxide formed
 (f) total mass of flue gases

Solution:

Fuel constituents	Pounds of O_2 required from air per pound of fuel
C = 0.8663	0.8663(2.667) = 2.3104
H = 0.1127	0.1127(8.000) = 0.9016
O = 0.0019	reduces external O_2 = −0.0019
N = 0.0028	
S = 0.0163	0.0163(1.000) = 0.0163
1.0000	3.2264

(a) oxygen supplied with 50% excess air = 1.5(3.2264)

 = 4.8396 lb/lb fuel

(b) nitrogen from air = $4.8396 \left(\dfrac{0.768}{0.232}\right)$ = 16.0207 lb/lb fuel

(c) air supplied = 4.8396 + 16.0207 = 20.8603 lb/lb fuel

Element + Oxygen	= Combustion product
C 0.8663 + 2.3104	= 3.1767 lb CO_2
H 0.1127 + 0.9016	= 1.0143 lb H_2O
O 0.0019 (appears with other constituents)	Excess O_2 1.6132 lb O_2 (= 0.5 × 3.2264)
N 0.0028 + 0.0000	= 0.0028 lb N_2
S 0.0163 + 0.0163	= 0.0326 lb SO_2
	5.8396 lb products

(d) water formed = 1.0143 lb/lb fuel

(e) carbon dioxide formed = 3.1767 lb/lb fuel

(f) total mass of flue gases = 5.8396 + 16.0207

 = 21.8603 lb flue gases/lb fuel

BOILERS AND THE CHEMISTRY OF COMBUSTION

A comparison of the quantities calculated in the example above and that of the preceding article shows that for the same heating value, increasing quantities of flue gases are formed as the percent excess air is increased. Two important results of increasing excess air are the lowering of furnace temperature by dilution and the increase in energy carried out the stack by the greater mass of flue gases per pound of fuel fired. Other aspects of this case will be investigated later with the study of boiler heat balances.

18-7. Effects of Air Deficiency on Combustion. The possibility of having insufficient air for proper combustion is just as important from an operational standpoint as having too much air. With fuel oil consisting mainly of hydrogen and carbon as combustible constituents, the hydrogen is so much more active, chemically, than the carbon, that in the case of insufficient air supplied, the hydrogen is assumed to combine with its required quantity of oxygen to form water, leaving the carbon to combine with the remaining oxygen. In the absence of sufficient oxygen to form carbon dioxide, carbon will combine with such oxygen as is present to form carbon monoxide. Thus, the combustion of carbon for the fuel-rich case becomes

$$C + \tfrac{1}{2}O_2 \rightarrow CO + 3960 \text{ Btu/lb C} \tag{18-9}$$

$$12 \text{ lb C} + 16 \text{ lb } O_2 = 28 \text{ lb CO} + 47{,}520 \text{ Btu}$$

$$1 \text{ lb C} + 1.333 \text{ lb } O_2 = 2.333 \text{ lb CO} + 3960 \text{ Btu}$$

It follows, then, that for each pound of carbon burned to CO rather than to CO_2, a loss in heating value of $14{,}095 - 3{,}960 = 10{,}135$ Btu is incurred. Carbon monoxide will further combine with oxygen to form carbon dioxide:

$$CO + \tfrac{1}{2}O_2 \rightarrow CO_2 + 4344 \text{ Btu/lb CO} \tag{18-10}$$

$$1 \text{ lb CO} + 0.571 \text{ lb } O_2 = 1.571 \text{ lb } CO_2 + 4344 \text{ Btu}$$

Therefore, burning one pound of carbon to 2.333 pounds of carbon monoxide produces 3,960 Btu, and completing the combustion of the 2.333 pounds of carbon monoxide to carbon dioxide produces $(2.333)(4344) = 10{,}135$ Btu, the sum of these two heat liberations equaling the heating value of one pound of carbon burning directly to carbon dioxide.

18-8. Analysis of Flue Gases. The term *flue gases* includes all gases and vapors emanating from a combustion process, whether active in the process or inert. A typical qualitative analysis of boiler flue gases for a fuel oil includes CO_2, CO, O_2, N_2 and H_2O. A practical method of determining

BOILERS AND THE CHEMISTRY OF COMBUSTION

the percent by volume of each constituent gas is to use an analyzer of the Orsat type, which absorbs CO_2, CO and O_2, in various chemical solutions, the N_2 being calculated by difference. The gas sample is collected and analyzed over water, with the result that the analysis is always that of a dry gas sample. It should be remembered, also, with apparatus of this type, that any SO_2 formed by combustion of sulfur in the fuel will be absorbed with the CO_2. Most naval fuel oils contain very little sulfur, making this method of determination satisfactory for practical purposes. Similar results can be obtained using gas chromatographic or spectroscopic techniques to yield dry volumetric product analysis.

Orsat and other volumetric analyses may be converted to the mass basis by using Equation (16-3) in the suggested tabular format.

Example 18-4: An analysis of the flue gases of a combustion process, percent by volume is as follows:

$$CO_2 = 10.0\%, \quad CO = 2.0\%, \quad O_2 = 8.0\%, \quad N_2 = 80.0\%$$

Calculate the mass fraction of each constituent and the apparent molecular weight of the mixture. Calculate also an approximate value of the gas constant, R, for the gas mixture.

Solution:

Constituent	Volume fraction	Mol wt.			Percent mass of const.	
CO_2	0.10	$\dfrac{\text{mols } CO_2}{\text{mol gas}} \times 44 \dfrac{\text{lb}}{\text{mol}}$	$=$	$4.40 \dfrac{\text{lb } CO_2}{\text{mol gas}}$	$\dfrac{4.40}{29.92} =$	14.71
CO	0.02	$\dfrac{\text{mols CO}}{\text{mol gas}} \times 28 \dfrac{\text{lb}}{\text{mol}}$	$=$	$0.56 \dfrac{\text{lb CO}}{\text{mol gas}}$	$\dfrac{0.56}{29.92} =$	1.87
O_2	0.08	$\dfrac{\text{mols } O_2}{\text{mol gas}} \times 32 \dfrac{\text{lb}}{\text{mol}}$	$=$	$2.56 \dfrac{\text{lb } O_2}{\text{mol gas}}$	$\dfrac{2.56}{29.92} =$	8.56
N_2	0.80	$\dfrac{\text{mols } N_2}{\text{mol gas}} \times 28 \dfrac{\text{lb}}{\text{mol}}$	$=$	$22.40 \dfrac{\text{lb } N_2}{\text{mol gas}}$	$\dfrac{22.40}{29.92} =$	74.86
	1.00			$29.92 \dfrac{\text{lb gas}}{\text{mol gas}}$		100.00

$$\text{Apparent molecular weight of gas mixture} = 29.92 \dfrac{\text{lb gas}}{\text{mol gas}}$$

$$R \text{ for the mixture} = \dfrac{1545}{29.92} = 51.64 \dfrac{\text{ft} \cdot \text{lbf}}{\text{lb} \cdot {}^\circ R}$$

Expressing the constituent gas quantities in mols has the advantage of easy conversion from volume to mass analysis as well as allowing the determination of constituent volumes at any time for a stated pressure and temperature.

The presence of both O_2 and CO in the gas analysis results from incomplete mixing of the fuel with air.

BOILERS AND THE CHEMISTRY OF COMBUSTION

The mass of dry flue gas from a combustion process in a boiler furnace may now be calculated from a knowledge of the flue gas analysis and the quantity of carbon in the fuel, assuming all carbon is burned.

$$M_G(\text{dry}) = \frac{\dfrac{\text{pounds carbon}}{\text{pound fuel}}}{\dfrac{\text{pounds carbon}}{\text{pound dry gas}}} = \frac{\text{pounds dry flue gas}}{\text{pound fuel}} \quad (18\text{-}11)$$

The carbon content per pound of fuel is available from the ultimate analysis of the fuel. All the carbon in the fuel must appear either as CO_2 or CO in the flue gases. Since each *pound* of CO_2 contains $12/44$ *pound* of carbon, and each *pound* of CO contains $12/28$ *pound* of carbon, the total pounds of carbon per pound of flue gas is

$$\frac{12}{44} CO_2 + \frac{12}{28} CO \; \frac{\text{pounds of carbon}}{\text{pound flue gas}}$$

Substituting the last relation in Equation (18-11) yields

$$M_G(\text{dry}) = \frac{C_F}{\dfrac{12}{44} CO_2 + \dfrac{12}{28} CO} = \frac{\text{pounds dry flue gas}}{\text{pound fuel}} \quad \underline{(18\text{-}12)}$$

where C_F is the mass fraction of carbon in the fuel, and CO_2 and CO are the mass fractions of those constituents in the flue gas.

The actual mass of air supplied to a furnace may be calculated very closely from an analysis of the flue gases. In this case, M_A, the pounds of air supplied per pound of fuel fired is required. When the fuel contains little or no nitrogen, it may be reasonably assumed that all nitrogen appearing in the flue gases comes from the air supplied. Originally associated with the nitrogen is a quantity of air determined by $N_2/0.768$, where N_2 is the mass of nitrogen in the flue gases per pound of flue gas and 0.768 is the mass fraction of nitrogen in atmospheric air. Then

$$M_A = \frac{\dfrac{N_2}{0.768}(C_F)}{\dfrac{12}{44} CO_2 + \dfrac{12}{28} CO} = \frac{\text{lb air supplied}}{\text{lb fuel}} \quad \underline{(18\text{-}13)}$$

and

$$M_A = \frac{\left(\dfrac{\text{lb air}}{\text{lb gas}}\right)\left(\dfrac{\text{lb carbon}}{\text{lb fuel}}\right)}{\left(\dfrac{\text{lb carbon}}{\text{lb gas}}\right)} = \frac{\text{lb air supplied}}{\text{lb fuel}}$$

BOILERS AND THE CHEMISTRY OF COMBUSTION

Occasionally it may be desirable to calculate the mass of wet gases from a combustion process. In that case it is necessary to remember only that the moisture in the flue gases comes from (1) the air supplied for combustion, (2) the moisture in the fuel and (3) the moisture formed from the combustion of hydrogen. Most frequently the moisture in the air and the moisture in the fuel will be negligible with respect to the moisture from burning hydrogen. Consequently, combustion calculations are usually completed on the basis of one pound of fuel, ash and moisture free, with atmospheric air containing only nitrogen and oxygen.

As pointed out previously, the quantity of excess air determines, to a great extent, the completeness of the combustion process and the overall efficiency of the combustion and heat transfer apparatus. This most important item, the percent excess air, may be determined directly from the flue gas analysis and the fuel analysis, and is expressed as

Percent excess air

$$= \left[\frac{\text{lb air actually supplied}}{\text{lb air theoretically required}} - 1.0 \right] 100 \qquad (18\text{-}14)$$

where both the masses of air supplied and required are in terms of one pound of fuel. Equation (18-13) yields the mass of air actually supplied, while equations similar to (18-5) may be used to determine the mass of air theoretically required.

Example 18-5: A fuel oil has the following analysis on an ash- and moisture-free basis and yields, on burning, the following Orsat analysis:

Fuel analysis, lb/lb fuel	Orsat analysis
C = 0.8700	CO_2 = 11.0 percent
H = 0.1200	CO = 2.0 percent
N = 0.0020	O_2 = 4.5 percent
S = 0.0020	N_2 = 82.5 percent
O = 0.0060	

Calculate:

(a) the mass of dry flue gas, lb/lb fuel
(b) the mass of air actually supplied, lb/lb fuel
(c) the air required for perfect combustion, lb/lb fuel
(d) the percent excess air supplied

BOILERS AND THE CHEMISTRY OF COMBUSTION

Solution:

Mols of constituent per mol of flue gas			Lb per mol		Lb per mol flue gas	Percent mass	
CO_2	0.11	×	44	=	4.84	$\dfrac{4.84}{29.94} =$	16.17
CO	0.02	×	28	=	0.56	$\dfrac{0.56}{29.94} =$	1.87
O_2	0.045	×	32	=	1.44	$\dfrac{1.44}{29.94} =$	4.81
N_2	0.825	×	28	=	$\dfrac{23.10}{29.94}$	$\dfrac{23.10}{29.94} =$	$\dfrac{77.15}{100.00}$

(a) mass of dry flue gas per lb of fuel:

$$M_G = \frac{C_F}{\frac{12}{44}CO_2 + \frac{12}{28}CO} = \frac{0.8700}{\frac{12}{44}(0.1617) + \frac{12}{28}(0.0187)} = 16.69 \; \frac{\text{lb gas}}{\text{lb fuel}}$$

(b) mass of air actually supplied:

$$M_A = \frac{\left(\dfrac{N_2}{0.768}\right)(C_F)}{\dfrac{12}{44}CO_2 + \dfrac{12}{28}CO} = \frac{\left(\dfrac{0.7715}{0.768}\right)(0.8700)}{\dfrac{12}{44}(0.1617) + \dfrac{12}{28}(0.0187)}$$

$$M_A = 16.77 \; \frac{\text{lb air}}{\text{lb fuel}}$$

(c) air required for perfect combustion of one pound of fuel:

$$\begin{aligned}
\text{carbon } 0.8700(11.49) &= 9.9963 \text{ lb air for carbon} \\
\text{hydrogen } 0.1200(34.48) &= 4.1376 \text{ lb air for hydrogen} \\
\text{sulfur } 0.0020(4.31) &= \underline{0.0086} \text{ lb air for sulfur} \\
& \; 14.1425 \text{ total air required, lb}
\end{aligned}$$

Since the fuel contains 0.006 pound of oxygen, assume the hydrogen combines with it, and reduces the external oxygen required:

$$4.1376 - \left[\left(\frac{\text{lb } O_2}{\text{lb fuel}}\right)\left(\frac{\text{lb air}}{\text{lb } O_2}\right)\right] = 4.1376 - (0.006)(4.31)$$

$$= \underline{4.1117 \text{ lb air for hydrogen}}$$

Then, perfect combustion of one pound of fuel requires

$$14.1425 - 0.0259 = \underline{14.1166 \text{ lb air/lb fuel}}$$

(d) % excess: $\dfrac{16.77 - 14.12}{14.12} = \underline{18.8\%}$

18-9. The Boiler Energy Balance. In order that the efficiency of a boiler be kept at its highest possible value at all times, a reliable method of

checking the many variables involved is required. A boiler energy balance serves that purpose. The following information must be available from test data to calculate an energy balance on an oil-fired boiler:
1. Total mass of steam delivered by the boiler for a specified test period
2. Average steam temperature and pressure maintained by the boiler; average feedwater temperature and pressure
3. Mass of fuel fired during test
4. An ultimate analysis of the fuel with its higher heating value
5. An average analysis of the flue gases, sampled after leaving the last heat exchanger passage
6. Flue gas temperature leaving the last heat exchanger passage
7. The relative humidity, barometric pressure and temperature of the air supplied for combustion
8. Fuel temperature entering the burners

The items constituting a heat or energy balance for a boiler are as follows:
1. Energy absorbed by water and steam in boiler
2. Energy loss to dry flue gases
3. Energy loss to moisture formed from burning hydrogen
4. Energy loss to moisture in fuel
5. Energy loss to moisture in air supplied for combustion
6. Energy loss to incomplete combustion
7. Energy loss to radiation and unaccounted-for losses
8. Higher heating value of fuel

All items in an energy balance of this type are usually calculated on a basis of Btu per pound of dry fuel fired and expressed also as a percentage of the higher heating value of the fuel. The dry fuel basis is used because heating values are usually expressed in Btu per pound of dry fuel. An explanation of each of the items appearing in the above heat balance is given below:

Item 1. Energy absorbed by water and steam in the boiler per pound of fuel fired:

$$Q_1 = M_s(h_2 - h_1) \text{ Btu/lb fuel}$$

M_s = pounds of steam generated per pound of fuel fired
h_1 = enthalpy of feed water, Btu/lb
h_2 = enthalpy of steam at boiler or superheater outlet, Btu/lb

Item 2. Energy loss to dry flue gases:

$$Q_2 = M_G c_p(t_2 - t_1) \text{ Btu/lb fuel}$$

BOILERS AND THE CHEMISTRY OF COMBUSTION

M_G = pounds of dry flue gas per pound fuel fired
c_p = the mean specific heat of the dry flue gases at constant pressure for the temperature range from t_1, the temperature of the air supply, to t_2, the temperature of the flue gases taken after the last heat exchange passage (An acceptable value over the usual flue to boiler room temperature range is 0.24 Btu/lb · °F.)
t_1 = temperature of air supplied for combustion, °F
t_2 = flue gas temperature leaving last heat exchanger passage, °F

Item 3. Energy loss to moisture formed from burning hydrogen:

$$Q_3 = \frac{9H}{100}(h_4 - h_3) \text{ Btu/lb fuel}$$

H = percent hydrogen in fuel
h_4 = enthalpy of superheated steam at 1 psia and flue gas temperature at last heat exchanger passage
h_3 = enthalpy of saturated liquid at fuel supply temperature

Item 4. Energy loss to moisture associated with fuel:

$$Q_4 = \frac{M_M}{1 - M_M}(h_4 - h_3) \text{ Btu/lb fuel (dry)}$$

M_M = lb of moisture per lb of fuel as received
$\dfrac{M_M}{1 - M_M}$ = lb of moisture in fuel per lb of *dry* fuel
h_4 and h_3 same as in Item 3

Item 5. Energy loss to moisture in air supplied for combustion:

$$Q_5 = M_A M_W (h_4 - h_5) \text{ Btu/lb fuel}$$

M_A = lb of air actually supplied per lb of fuel fired
M_W = lb of moisture per lb of dry air supplied for combustion
h_4 = same as in item 3
h_5 = enthalpy of superheated steam at air supply temperature and vapor partial pressure

Item 6. Energy loss to incomplete combustion:

$$Q_6 = C_F \left(\frac{CO}{CO_2 + CO}\right) 10{,}100 \text{ Btu/lb fuel}$$

C_F = lb carbon per lb fuel
CO_2 and CO = **volume** fractions of CO and CO_2 from Orsat analysis of flue gases

BOILERS AND THE CHEMISTRY OF COMBUSTION

Item 7. Energy loss to radiation and unaccounted-for losses:

$$Q_7 = \text{H.H.V.} - (Q_1 + Q_2 + Q_3 + Q_4 + Q_5 + Q_6)$$

H.H.V. = higher heating value of fuel, Btu/lb of dry fuel

The most important point in calculation of an energy balance of this type is keeping all quantities on the same basis. A convenient basis for the higher heating value of the fuel and the ultimate analysis must be chosen and then adhered to strictly.

Example 18-6: The following data were observed during an oil-fired boiler test:

1. Duration of test — 1 hr
2. Steam delivered by boiler — 200,000 lb
3. Average steam temperature at superheater outlet — 760°F
4. Average steam pressure at superheater outlet — 600 psia
5. Feedwater temperature — 240°F
6. Feedwater pressure — 700 psia
7. Fuel fired (dry basis) — 15,385 lb
8. Flue gas temperature leaving last heat transfer passage — 450°F
9. Dry bulb temperature of air supplied for combustion — 80°F
10. Wet bulb temperature of air supplied for combustion — 70°F
11. Barometric pressure at test location — 29.92 in. Hg
12. Temperature of fuel supplied to burners — 80°F
13. Ultimate analysis of fuel on an as-fired basis:

Carbon	0.8095 lb/lb fuel	
Hydrogen	0.1143 lb/lb fuel	
Nitrogen	0.0048 lb/lb fuel	
Sulfur	0.0143 lb/lb fuel	
Oxygen	0.0095 lb/lb fuel	
Moisture	0.0476 lb/lb fuel	
Ash	0.0000 lb/lb fuel	
	1.0000	

14. Volume analysis of flue gases in percent (Orsat):

CO_2 =	11.34 percent	
CO =	00.71 percent	
O_2 =	5.06 percent	
N_2 =	82.89 percent	
	100.00 percent	

15. Higher heating value of fuel is 19,500 Btu/lb "dry" fuel.

Calculate an energy balance for the tested boiler.

Solution:

Calculation of this energy balance is based on one pound of dry fuel. Since the ultimate analysis of the fuel is on the as-received basis, the first step in the solution is conversion of the fuel analysis to the dry basis. One pound of fuel as received contains 0.0476 pound of

BOILERS AND THE CHEMISTRY OF COMBUSTION

moisture, so dividing each term in the analysis except the moisture by $(1 - 0.0476)$ yields the dry-basis analysis.

As-fired basis		Dry basis
Carbon	0.8095	0.8500 lb/lb
Hydrogen	0.1143	0.1200 lb/lb
Nitrogen	0.0048	0.0050 lb/lb
Sulfur	0.0143	0.0150 lb/lb
Oxygen	0.0095	0.0100 lb/lb
Moisture	0.0476	1.0000

$\{$ Carbon through Oxygen $\} \div (1 - 0.0476) = \{$ dry basis values $\}$

Inspection of the energy balance calculations involving flue gas quantities reveals that, in all but one instance, mass fractions rather than volume fractions are required. Consequently, the next step is conversion of the flue gas analysis by volume to its equivalent by mass.

Flue Gas Analysis:

	Percent volume dry		Mol wt.		Pounds per 100 mols gas	Percent mass	
CO_2 =	11.34	×	44	=	498.96	$\dfrac{498.96}{3001.68}$ =	16.62
CO =	0.71	×	28	=	19.88	$\dfrac{19.98}{3001.68}$ =	0.66
O_2 =	5.06	×	32	=	161.92	$\dfrac{161.92}{3001.68}$ =	5.40
N_2 =	82.89	×	28	=	2320.92	$\dfrac{2320.92}{3001.68}$ =	77.32
					3001.68		100.00

Energy Balance:

(1) Energy absorbed by water and steam in boiler, Btu per pound of dry fuel:

$$Q_1 = M_s(h_2 - h_1) = \frac{200,000}{15,385}(1385.1 - 209.9)$$

$= 13(1175.2) = \underline{15,277.6 \text{ Btu/lb fuel (dry)}}$
h_2 at 600 psia and 760°F = 1385.1 Btu/lb
h_1 at 700 psia and 240°F = 209.9 Btu/lb

(2) Energy loss to dry flue gases, Btu per pound dry fuel:

$Q_2 = M_G c_p (t_2 - t_1) = 17.65(0.24)(450 - 80)$
$= \underline{1567.32 \text{ Btu/lb dry fuel}}$

$$M_G = \frac{C_F}{\dfrac{12}{44}CO_2 + \dfrac{12}{28}CO} = \frac{0.8500}{\dfrac{12}{44}(0.1662) + \dfrac{12}{28}(0.0066)}$$

$= 17.65$ lb dry gas/lb fuel
$c_p = 0.24$ (average specific heat of flue gases)
$t_1 =$ entering air temperature = 80°F
$t_2 =$ leaving flue gas temperature = 450°F

BOILERS AND THE CHEMISTRY OF COMBUSTION

(3) Energy loss due to moisture from burning hydrogen, Btu per pound of dry fuel:

$$Q_3 = \frac{9H}{100}(h_4 - h_3) = \frac{9(12)}{100}(1265.0 - 48.1)$$

$$= 1314.25 \text{ Btu/lb dry fuel}$$

H = percent hydrogen in 1 lb dry fuel
h_4 = 1265.0 (table 3, steam tables at 1 psia and 450°F)
$h_3 = h_f$ at 80°F = 48.1 Btu/lb

(4) Energy loss to moisture accompanying one pound dry fuel:

$$Q_4 = \frac{M_M}{1 - M_M}(h_4 - h_3) = \frac{0.0476}{1 - 0.0476}(1265.0 - 48.1)$$

$$= 0.05(1216.9) = 60.85 \text{ Btu/lb dry fuel}$$

In spite of calculating on the dry fuel basis, account must be taken of the moisture accompanying the fuel as fired:

M_M = mass of moisture in 1 lb of fuel as received = 0.0476 lb

$$\frac{M_M}{1 - M_M} = 0.05 \text{ lb of moisture/lb dry fuel}$$

(5) Energy loss to moisture in air supplied for combustion:

$$Q_5 = M_A M_W(h_4 - h_5) = 17.77(0.0134)(1265.0 - 1096.6)$$

$$= 40.1 \text{ Btu/lb dry fuel}$$

$$M_A = \frac{\frac{N_2}{0.768}(C_F)}{\frac{12}{44}CO_2 + \frac{12}{28}CO} = \frac{\left(\frac{0.7732}{0.768}\right)(0.8500)}{\frac{12}{44}(0.1662) + \frac{12}{28}(0.0066)}$$

$$= 17.77 \text{ lb air supplied/lb dry fuel}$$

N_2, CO_2, CO = percents by mass in flue gases
M_W = 0.0134 lb moisture/lb dry air (from psychrometric chart at 80°F dry bulb and 70°F wet bulb)
h_5 = 1096.6 Btu/lb (enthalpy of superheated steam at 80°F, read most easily from Mollier chart; partial pressure has relatively small effect on enthalpy at this temperature evidenced by practically horizontal 80°F line on chart)

(6) Energy loss to incomplete combustion, Btu per pound of dry fuel:

$$Q_6 = C_F\left(\frac{CO}{CO_2 + CO}\right)(10,100)$$

$$= 0.85 \frac{(0.71)}{(11.34 + 0.71)}(10,100) = 505.84 \text{ Btu/lb dry fuel}$$

CO_2, CO = *volume* percentages from Orsat analysis of flue gases
C_F = 0.8500 lb carbon/lb dry fuel
10,100 = Btu loss/lb of carbon burned to CO rather than to CO_2

(7) Energy loss to radiation and unaccounted-for losses:

$$Q_7 = \text{H.H.V.} - (Q_1 + Q_2 + Q_3 + Q_4 + Q_5 + Q_6)$$
$$= 19{,}500 - (15{,}277.6 + 1567.3 + 1314.3 + 60.8 + 40.1 + 505.8)$$
$$= 19{,}500 - 18{,}765.9 = \underline{734.1 \text{ Btu/lb dry fuel}}$$

A convenient representation of the energy balance from the example above is shown in the table following:

Boiler Energy Balance
Basis—One Pound of Dry Fuel

No.	Item	Btu/lb	Percent
1.	Gain by water and steam	15,277.6	78.35
2.	Loss to dry flue gases	1567.3	8.04
3.	Loss to moisture from burning hydrogen	1314.3	6.74
4.	Loss to moisture in fuel	60.8	0.31
5.	Loss to moisture in air	40.1	0.21
6.	Loss to incomplete combustion	505.8	2.59
7.	Radiation and unaccounted-for losses	734.1	3.76
8.	Higher heating value of fuel	19,500.0	100.00

It might be well to point out that the method used for calculation of the mass of air actually supplied for combustion and the mass of dry flue gases per pound of fuel in the foregoing energy balance is only one of several available methods. An alternate method based on the mols of constituent gases in the Orsat analysis is interesting and is calculated as a check on these two items.

The 15,385 pounds of fuel burned in one hour contained

$$15{,}385(0.85) = 13{,}077.25 \frac{\text{lb carbon}}{\text{hr}}$$

In the flue gases there were 11.34 mols of carbon in the CO_2 and 0.71 mol of carbon in the CO, totaling 12.05 mols of carbon per 100 mols of stack gas. Since one mol of carbon contains 12 pounds, there are $12.05(12) = 144.60$ pounds of carbon per 100 mols of flue gas. Now,

$$\frac{\frac{\text{lb carbon}}{\text{hr}}}{\frac{\text{lb carbon}}{\text{mol gas}}} = \frac{\text{mols gas}}{\text{hr}}$$

and

$$\frac{\text{mols gas}}{\text{hr}} = \frac{13{,}077.25}{1.4460} = 9{,}043.74$$

BOILERS AND THE CHEMISTRY OF COMBUSTION

Then,

$$\frac{\frac{\text{mols gas}}{\text{hr}}}{\frac{\text{lb fuel}}{\text{hr}}} = \frac{9{,}043.74}{15{,}385.0} = 0.5878 \; \frac{\text{mol gas}}{\text{lb fuel}}$$

In the conversion of flue gas analysis from volumes to masses in the energy balance, the apparent molecular weight of the flue gases was found to be 30.02 pounds per mol. Therefore,

$$(30.02)(0.5878) = 17.65 \text{ lb flue gas/lb fuel}$$

which is an exact check on the previous calculation.

A similar calculation on the molar basis may be made for the actual air supplied for combustion. Assuming, as before, that the nitrogen in the flue gases all originated in the air supplied, the 0.8289 mol of nitrogen indicated by the volume analysis of the flue gases is employed. The previous calculation showed 0.5878 mol of flue gas per pound of fuel. Therefore,

$$\left(\frac{\text{mols } N_2}{\text{mols gas}}\right)\left(\frac{\text{mols gas}}{\text{lb fuel}}\right) = \frac{\text{mols } N_2}{\text{lb fuel}}$$

and

$$(0.8289)(0.5878) = 0.4872 \text{ mol } N_2/\text{lb fuel}$$

Then,

$$\frac{\left(\dfrac{\text{mols } N_2}{\text{lb fuel}}\right)\left(\dfrac{\text{lb } N_2}{\text{mol } N_2}\right)}{\left(\dfrac{\text{lb } N_2}{\text{lb air}}\right)} = \frac{(0.4872)(28)}{(0.768)} = 17.76 \text{ lb air/lb fuel}$$

This result substantially checks that obtained from the mass equation employed in the energy balance.

An energy balance for a boiler can be no more accurate than the data from which the balance is calculated. It becomes imperative, then, to exercise the utmost care during a boiler test to obtain the most accurate information practical considerations will allow. The flue gas analysis is probably the most difficult part of the data to obtain accurately, and since it plays such an important part in the energy balance calculation, extreme care must be taken to assure correct results.

BOILERS AND THE CHEMISTRY OF COMBUSTION

Problems

18-1. A ship's boiler consumes 16,800 pounds of fuel/hour when producing 210,000 pounds steam/hour at 620 psia and 800°F from feedwater at 650 psia and 300°F. The heating value of the fuel is 18,500 Btu/pound and the furnace volume is 1250 cubic feet. Calculate (a) the boiler capacity in mB/hr; (b) the factor of evaporation; (c) the equivalent evaporation in lb/hr; (d) the furnace heat-release rate in Btu/hr · ft^3; (e) the boiler efficiency in percent.

Ans. (a) 238.6 mB/hr; (b) 1.17; (c) 245,700 lb/hr; (d) 248,600 Btu/hr · ft^3; (e) 76.8%

18-2. A fuel oil has the following ultimate analysis, ash and moisture free:

$$\begin{array}{ll} \text{Carbon} & = 0.85 \text{ lb} \\ \text{Hydrogen} & = 0.13 \text{ lb} \\ \text{Oxygen} & = 0.00 \text{ lb} \\ \text{Nitrogen} & = 0.01 \text{ lb} \\ \text{Sulfur} & = \underline{0.01 \text{ lb}} \\ & 1.00 \text{ lb} \end{array}$$

For complete combustion with air, calculate the following in pounds per pound of fuel:
(a) oxygen required from air
(b) nitrogen accompanying oxygen from air
(c) air required
(d) water formed from combustion of hydrogen
(e) carbon dioxide formed
(f) total mass of flue gases

Ans. (a) 3.32 lb; (b) 10.98 lb; (c) 14.30 lb; (d) 1.17 lb; (e) 3.12 lb; (f) 15.30 lb

18-3. For the fuel analysis of problem 18-2 calculate the following items for combustion with 25 percent excess air in pounds per pound of fuel:
(a) oxygen supplied from air
(b) nitrogen accompanying oxygen from air
(c) air supplied
(d) water formed from combustion of hydrogen
(e) carbon dioxide formed
(f) mass of flue gases

Ans. (a) 4.15 lb; (b) 13.72 lb; (c) 17.87 lb; (d) 1.17 lb; (e) 3.12 lb; (f) 18.87 lb

18-4. A volumetric analysis of the flue gases of a combustion process is as follows: $CO_2 = 12.0\%$, $CO = 2.0\%$, $O_2 = 6.0\%$, $N_2 = 80.0\%$. Calculate the percent by mass of each constituent and the apparent molecular weight of the mixture. Then determine an approximate value of the gas constant, R, for the mixture.

Ans. $CO_2 = 17.50\%$, $CO = 1.86\%$, $O_2 = 6.37\%$, $N_2 = 74.27\%$, apparent molecular weight = 30.16 lb/mol; $R_{mix} = 51.23$ ft · lbf/lb · °R

18-5. A fuel oil has the following analysis on the ash- and moisture-free basis:

$$\begin{array}{ll} \text{Carbon} & = 0.8700 \text{ lb/lb fuel} \\ \text{Hydrogen} & = 0.1200 \text{ lb/lb fuel} \\ \text{Nitrogen} & = 0.0020 \text{ lb/lb fuel} \\ \text{Sulfur} & = 0.0020 \text{ lb/lb fuel} \\ \text{Oxygen} & = \underline{0.0060 \text{ lb/lb fuel}} \\ & 1.0000 \text{ lb fuel} \end{array}$$

BOILERS AND THE CHEMISTRY OF COMBUSTION

The dry flue gas analysis by volume is as follows:

$$CO_2 = 14.0\%$$
$$O_2 = 6.0\%$$
$$N_2 = 80.0\%$$

Calculate (a) the mass of dry flue gas per pound of fuel; (b) the air actually supplied for combustion per pound of fuel; (c) the air required for theoretically perfect combustion of the fuel, pounds of air per pound of fuel; (d) the percent excess air.

Ans. (a) 15.78 lb gas/lb fuel; (b) 15.10 lb air/lb fuel; (c) 14.12 lb air/lb fuel; (d) 6.94%

18-6. A certain fuel oil has the following ultimate analysis:

Carbon = 0.8500 lb/lb fuel
Hydrogen = 0.1100 lb/lb fuel
Nitrogen = 0.0030 lb/lb fuel
Sulfur = 0.0090 lb/lb fuel
Oxygen = 0.0100 lb/lb fuel
Water = 0.0175 lb/lb fuel
Ash = 0.0005 lb/lb fuel

Calculate (a) the fuel analysis on a moisture- and ash-free basis; (b) the oxygen required for each fuel constituent from part (a) for perfect combustion, lb/lb of constituent; (c) the total mass of flue gas, lb/lb of moisture- and ash-free fuel; (d) the mass analysis of dry flue gas constituents, lb/lb of dry and ash-free fuel; (e) the percent by mass of each dry flue gas constituent; (f) the total mass of dry flue gas constituents, lb/lb of dry fuel; (g) the lb of air required for theoretically perfect combustion of one pound of the dry and moisture-free fuel.

Ans. (a)
Carbon = 0.8656 lb
Hydrogen = 0.1120 lb
Nitrogen = 0.0030 lb
Sulfur = 0.0092 lb
Oxygen = 0.0102 lb
 ———
 1.0000 lb

(b) oxygen required from air for each constituent, lb/lb of dry and ash-free fuel:

Carbon = 2.3086
Hydrogen = 0.8960
Nitrogen = 0.0000
Sulfur = 0.0092
Oxygen = −0.0102 (reduces external oxygen required)

(c) 14.81 lb total products
(d) mass analysis of dry gas constituents, lb/lb of dry and ash-free fuel:

$$CO_2 = 3.17$$
$$N_2 = 10.61$$
$$SO_2 = 0.02$$
 13.80 lb total

(e) percent of each dry gas constituent:

$$CO_2 = 23.00\%$$
$$N_2 = 76.87\%$$
$$SO_2 = 00.13\%$$
 100.00%

BOILERS AND THE CHEMISTRY OF COMBUSTION

(f) 13.80 lb
(g) 13.81 lb

18-7. The following data were observed during an oil-fired steam boiler test:

1. Duration of test — 24 hr
2. Steam delivered by boiler — 6,072,000 lb
3. Average steam temperature at superheater outlet — 850°F
4. Average steam pressure at superheater outlet — 600 psia
5. Feedwater temperature — 300°F
6. Feedwater pressure — 700 psia
7. Fuel fired (dry and ash free) — 477,600 lb
8. Flue gas temperature leaving last heating transfer passage — 450°F
9. Dry bulb temperature of air supplied for combustion — 90°F
10. Wet bulb temperature of air supplied for combustion — 80°F
11. Barometric pressure at test location — 29.92″ Hg
12. Temperature of fuel supplied to burners — 90°F
13. Ultimate analysis of fuel on an "as fired" basis, lb/lb of fuel:

Carbon	0.8500
Hydrogen	0.1100
Nitrogen	0.0030
Oxygen	0.0100
Sulfur	0.0090
Moisture	0.0175
Ash	0.0005
	1.0000

14. Volume analysis of dry flue gases, percent:

CO_2	12.29
CO	00.84
O_2	04.07
N_2	82.80

15. Higher heating value of the fuel is 18,500 Btu/pound *dry* and *ash free*.

Calculate an energy balance for the tested boiler with results based on one pound of dry and ash-free fuel.

Ans.

Item	Btu/lb	Percent
Q_1	14,804	80.02
Q_2	1,430	7.73
Q_3	1,217	6.58
Q_4	21	0.12
Q_5	54	0.29
Q_6	559	3.02
Q_7	415	2.24
	18,500	100.00

APPENDIX A

All thermodynamic calculations require the proper use of certain physical laws. To this end, the proper mathematical formulation of any problem must be correct both qualitatively, i.e., algebraically, and quantitatively, i.e. numerically.

Thermodynamic relationships must be written using a consistent set of terms, or *dimensions*. A list of principal dimensions includes:

force	F
mass	M
length	L
time	T
temperature	t

Using these quantities we can express the fundamental laws of thermodynamics algebraically. For example, velocity has the dimensions of length divided by time, LT^{-1}.

Now the magnitude of any dimension is measured in terms of *units*. Consider, for example, familiar unit choices for the basic dimension of length:

$$L = \begin{cases} \text{inches} \\ \text{miles} \\ \text{angstroms} \\ \text{kilometres} \end{cases}$$

There are standards for mass, length, time and temperature established and maintained under the supervision of the General Conference on Weights and Measures, an international body.

Four dimension and unit systems have been developed for engineering and scientific work: The English Gravitational, the English Engineers', the Metric (centimetre-gram-second) and the new International System of Units (SI). Length (L) and time (T) are primary dimensions in all of these measurement systems (see Table A-I). In the SI and Metric systems, mass (M) is also taken as a primary dimension, and force (F) is a derived, or secondary, dimension. Conversely, in the British Gravitational System, force (F) is taken as a primary dimension, and mass (M) then becomes a secondary or derived quantity.

The Engineers' System differs from the systems above in that mass (M) and force (F) are *both* taken as primary dimensions together with length (L) and time (T). Table A-I summarizes the primary dimensions and corresponding units for these four systems.

Table A-I

BASIC UNITS

Dimensions	System Internationalé (SI)	Metric (CGS)	English Engineers'	British Gravitational
PRIMARY				
force $[F]$	—	—	pound force (lbf)	pound force (lbf)
mass $[M]$	kilogram (kg)	gram (gm)	pound mass (lb)	—
length $[L]$	metre (m)	centimetre (cm)	feet (ft)	feet (ft)
time $[T]$	second (sec)	second (sec)	second (sec)	second (sec)
SECONDARY				
force $[M][L][T^{-2}]$	newton (kg·m/sec²)	dyne (gm·cm/sec²)	—	—
mass $[M]$	—	—	—	slug (lbf·sec²/ft)
pressure $[F][L^{-2}]$	pascal (N/m²)	bar (dyne/cm²)	psf (lbf/ft²)	psf (lbf/ft²)
energy $[F][L]$	joule (N·m)	erg (dyne·cm)	ft·lbf	ft·lbf
power $[F][L][T^{-1}]$	watt (J/sec)	—	horsepower (ft·lbf/sec)	horsepower (ft·lbf/sec)
g_c	$1.0\ \dfrac{\text{kg}\cdot\text{m}}{\text{N}\cdot\text{sec}^2} = 1.0$	$1.0\ \dfrac{\text{gm}\cdot\text{cm}}{\text{dyne}\cdot\text{sec}^2} = 1.0$	$32.2\ \dfrac{\text{lb}\cdot\text{ft}}{\text{lbf}\cdot\text{sec}^2}$	$1.0\ \dfrac{\text{slug}\cdot\text{ft}}{\text{lbf}\cdot\text{sec}^2} = 1.0$

APPENDIX A

Newton's Second Law expresses the interrelationship between force, mass and acceleration as

$$F \sim Ma \qquad (A\text{-}1)$$

or in terms of the dimensions discussed

$$F \sim MLT^{-2}$$

Since force, mass, length and time are distinct dimensions, Newton's Second Law is more properly written in general form as

$$F = \frac{Ma}{g_c} \qquad (A\text{-}2)$$

where g_c is an arbitrary constant necessary to provide both dimensional and unit homogeneity. Thus

$$g_c = \frac{Ma}{F} \qquad (A\text{-}3)$$

or

$$g_c = MLT^{-2}F^{-1}$$

Using g_c, Newton's Second Law yields, dimensionally,

$$F = \frac{MLT^{-2}}{MLT^{-2}F^{-1}} = F$$

In the SI System the secondary dimension of force specified as the newton is defined as that force which corresponds to the acceleration of one unit of mass, the kilogram, at a rate of 1 metre/sec². For the SI System, then

$$g_c = \frac{Ma}{F} = \frac{(1\ \text{kg})(1\ \text{m/sec}^2)}{1\ \text{N}}$$

$$= 1\ \frac{\text{kg} \cdot \text{m}}{\text{N} \cdot \text{sec}^2}$$

Note, though, the newton is defined as

$$1\ \text{N} \equiv 1\ \text{kg} \cdot \text{m/sec}^2$$

Thus:

$$g_c = 1\ \frac{\text{kg} \cdot \text{m}}{\left(\dfrac{\text{kg} \cdot \text{m}}{\text{sec}^2}\right)\text{sec}^2} = 1.0$$

Thus the SI System results in a *magnitude of unity and dimensionless character* for g_c.

APPENDIX A

In the Metric System a similar result yields for g_c a value of 1 gm·cm/dyne sec². But since from physics 1 dyne \equiv gm·cm/sec², g_c in the metric system is also unity and dimensionless.

In the English Engineers' System force, mass, length and time are all primary quantities. Also from physics the unit of force, the pound force (lbf), is the force that will accelerate a unit of mass, the pound mass (lb), at 32.1740 ft/sec². Using the above values, the value of g_c in the English Engineers' System is

$$g_c = \frac{(1 \text{ lb})(32.1740 \text{ ft/sec}^2)}{(1 \text{ lbf})}$$

$$= 32.1740 \frac{\text{lb} \cdot \text{ft}}{\text{lbf} \cdot \text{sec}^2}$$

It is important to emphasize that g_c in Engineers' units is a *dimensional* constant and with the choice of primary dimensions in Table A-I has both magnitude and dimensions.

Using the primary units in the British Gravitational System of pound force, feet and second, a unit of mass becomes a secondary quantity termed a slug. The slug is the unit of mass that will accelerate at 1 ft/sec² under the force of 1 lbf. Then in the Gravitational System:

$$g_c = \frac{(1 \text{ slug})(1 \text{ ft/sec}^2)}{(1 \text{ lbf})}$$

$$= 1.0 \frac{\text{slug} \cdot \text{ft}}{\text{lbf} \cdot \text{sec}^2}$$

But since the slug = 1 lbf·sec²/ft,

$$g_c = 1.0$$

In this text we will confine ourselves to the Engineers' and the SI systems of measurement. Difficulties in thermodynamic calculations are compounded by the fact that these two systems of dimensions and units are presently in use.

Some calculations will involve more than one system of dimensions and units. Conversion factors allow an easy interchange between systems of units. An equation can be multiplied by a conversion factor without its affecting the dimensional characteristics because conversion factors are unities (dimensionless). Two examples:

$$\frac{2.54 \text{ cm}}{\text{in.}} \equiv 1 \quad \frac{60 \text{ sec}}{\text{min}} \equiv 1$$

Table A-II provides a list of some useful conversion factors between the English Engineers' and SI systems.

TABLE A-II

CONVERSION TABLE FROM ENGLISH ENGINEERS' UNITS TO SI UNITS FOR SOME PHYSICAL QUANTITIES OF THERMODYNAMIC INTEREST[1]

To convert from (English Engineers' units)	To (SI units)	Multiply by
Acceleration (ft/sec^2)	metre/second2 (m/s^2)	3.048 000 E − 01
Btu	joules (J)	1.055 056 E + 03
$\dfrac{\text{Btu}}{\text{lb}}$	joules/kilogram (J/kg)	2.326 000 E + 03
$\dfrac{\text{Btu} \cdot \text{ft}}{\text{hr} \cdot \text{ft}^2 \cdot \text{F}}$ (thermal conductivity)	watts/metre kelvin $\dfrac{\text{W}}{\text{m} \cdot \text{K}}$	1.730 735 E + 00
$\dfrac{\text{Btu}}{\text{hr}}$	watt (W)	2.930 711 E − 01
degree Celsius	kelvin (K)	t K = t°C + 273.15
degree Fahrenheit	degree Celsius	t°C = (t°F − 32)/1.8
degree Fahrenheit	kelvin (K)	t K = (t°F + 459.67)/1.8
degree Rankine	kelvin (K)	t K = t°R/1.8
density (lb/ft^3)	kg/m^3	1.601 846 E + 01
fathom	metre (m)	1.828 800 E + 00
foot	metre (m)	3.048 000 E − 01
foot of H$_2$O (39.2 F)	pascal (Pa)	2.988 980 E + 03
ft^2	metre2 (m^2)	9.290 304 E − 02
ft^3 (volume)	metre3 (m^3)	2.831 685 E − 02
ft^3/lb	metre3/kg (m^3/kg)	6.242 797 E − 02
ft · lbf	joule (J)	1.355 818 E + 00
ft · lbf/hr	watt (W)	3.766 161 E − 04
ft · lbf/lb	joules/kilogram (J/kg)	2.989 067 E + 00
horsepower (550 ft · lbf/s)	watt (W)	7.456 999 E + 02
inch	metre (m)	2.540 000 E − 02
inch of Hg (32 F)	pascal (Pa)	3.386 380 E + 03
inch2	metre2 (m^2)	6.451 600 E − 04
kelvin	degree Celsius	t°C = t K − 273.15
kilowatt · hour	joule (J)	3.600 000 E + 06
knot (international)	metre/second (m/s)	5.144 444 E − 01
litre	metre3 (m^3)	1.000 000 E − 03
mile (statute)	metre (m)	1.609 300 E + 03
mile (U. S. nautical)	metre (m)	1.852 000 E + 03
pound (lb avoirdupois)	kilogram (kg)	4.535 924 E − 01
pound force (lbf)	newton (N)	4.448 222 E + 00
lbf/in^2 (psi)	pascal (Pa)	6.894 757 E + 03
ton (long, 2240 lb)	kilogram (kg)	1.016 047 E + 03
ton (refrigeration)	watt (W)	3.516 800 E + 03
ton (short, 2000 lb)	kilogram (kg)	9.071 847 E + 02
velocity (ft/sec)	metre/sec (m/sec)	3.048 000 E − 01
volume flow rate (ft^3/sec)	metre3/sec (m^3/sec)	2.831 685 E − 02

[1] ASTM Standard for Metric Practice E380-76, 1976 (Printed by Permission)

APPENDIX A

For use with the SI units there is a set of 16 prefixes (see Table A-III) to form multiples and submultiples of these units.

TABLE A-III SI Prefixes

Factor	Prefix	Symbol
10^{18}	exa	E
10^{15}	peta	P
10^{12}	tera	T
10^{9}	giga	G
10^{6}	mega	M
10^{3}	kilo	k
10^{2}	hecto	h
10^{1}	deka	da
10^{-1}	deci	d
10^{-2}	centi	c
10^{-3}	milli	m
10^{-6}	micro	μ
10^{-9}	nano	n
10^{-12}	pico	p
10^{-15}	femto	f
10^{-18}	atto	a

The following examples will illustrate the use of Table A-II.

Example A-1: A mass of 1.0 kg is moving at a velocity of 5 metres per second. Determine the kinetic energy on a unit mass basis:
(a) in SI units, J/kg
(b) in Engineers' units, ft·lbf/lb
Solution:

(a) Kinetic energy (SI):

$$K.E. = \frac{M}{2g_c} V^2$$

or on a unit mass basis:

$$K.E. = \frac{V^2}{2g_c}$$

Then:

$$K.E. = \left(\frac{1}{2}\right)\left(1.0 \, \frac{N \cdot sec^2}{kg \cdot m}\right)\left(5 \, \frac{m}{sec}\right)^2$$

$$= 12.5 \, \frac{N \cdot m}{kg} = 12.5 \, \frac{J}{kg}$$

If g_c is omitted, one obtains:

$$K.E. = \left(\frac{1}{2}\right)\left(5 \, \frac{m}{sec}\right)^2$$

$$= 12.5 \, \frac{m^2}{sec^2}$$

APPENDIX A

and since $1 \text{ N} = 1 \frac{\text{kg} \cdot \text{m}}{\text{sec}^2}$, multiplying and dividing by a newton yields

$$K.E. = 12.5 \left(\frac{\text{m}^2}{\text{sec}^2}\right) \times \frac{1 \text{ N}}{\left(1 \frac{\text{kg} \cdot \text{m}}{\text{sec}^2}\right)}$$

$$= 12.5 \frac{\text{N} \cdot \text{m}}{\text{kg}} = 12.5 \frac{\text{J}}{\text{kg}}$$

Note that by carrying g_c through the calculation although the quantity is unity and dimensionless, the answer comes out in correct units with ease.

(b) Kinetic energy (Engineers'):

Now,

$$K.E. = \left(12.5 \frac{\text{J}}{\text{kg}}\right)$$

Using Table A-II,

$$K.E. = \frac{\left(12.5 \frac{\text{J}}{\text{kg}}\right)}{\left(2.989067 \frac{\text{J/kg}}{\text{ft} \cdot \text{lbf/lb}}\right)}$$

$$= 4.182 \frac{\text{ft} \cdot \text{lbf}}{\text{lb}}$$

Example A-2: Steam has a value of internal energy $u = 1171.9$ Btu/lb at a temperature of 500°F and pressure of 150 psia. Determine the values of the above in their SI equivalences.
Solution:

(a) $\quad u = 1171.9$ Btu/lb. From Table A-II

$$u = (1171.9 \text{ Btu/lb})\left(2{,}326.0 \frac{\text{J/kg}}{\text{Btu/lb}}\right)$$

$$2.726 \times 10^6 \text{ J/kg} = \underline{2.726 \times 10^3 \text{ kJ/kg}}$$

(b) $\quad p = 150$ psia, again from Table A-II

$$p = (150 \text{ psi})\left(6.894757 \times 10^3 \frac{\text{pascals}}{\text{psi}}\right)$$

$$= 1.034 \times 10^6 \text{ pascals} = \underline{1.034 \times 10^3 \text{ kP}}$$

(c) $\quad t = 500°$F. From Table A-II
$\quad t = (500 - 32)/1.8 = \underline{260°\text{C}}$

Example A-3: The Universal Gas Constant has the precise value of $R = 1545.32$ ft·lbf/lb·mole°R. Calculate the value of R in SI units.
Solution:

$$R = \frac{\left(1545.32 \frac{\text{ft} \cdot \text{lbf}}{\text{lb} \cdot \text{mole } °\text{R}}\right) \times \left(2.989067 \frac{\text{J/kg}}{\text{ft} \cdot \text{lbf/lb}}\right)}{\left(\frac{1}{1.8} \frac{\text{K}}{°\text{R}}\right)}$$

$$= \underline{8314.32 \text{ J/kg} \cdot \text{mole K}}$$

APPENDIX B

FREON-12

Properties of Saturated Vapor

Temp. °F t	Pressure Abs lb/in.² p	Volume Liquid ft³/lb v_f	Volume Vapor ft³/lb v_g	Enthalpy from −40° Liquid Btu/lb h_f	Enthalpy from −40° Evap. Btu/lb h_{fg}	Enthalpy from −40° Vapor Btu/lb h_g	Entropy from −40° Liquid Btu/lb °F s_f	Entropy from −40° Vapor Btu/lb °F s_g
−40	9.32	0.0106	3.911	0	73.50	73.50	0	0.17517
−38	9.82	.0106	3.727	0.40	73.74	73.74	0.00094	.17490
−36	10.34	.0106	3.553	0.81	73.17	73.98	.00188	.17463
−34	10.87	.0106	3.389	1.21	73.01	74.22	.00282	.17438
−32	11.43	.0107	3.234	1.62	72.84	74.46	.00376	.17412
−30	12.02	0.0107	3.088	2.03	72.67	74.70	0.00471	0.17387
−28	12.62	.0107	2.950	2.44	72.50	74.94	.00565	.17364
−26	13.26	.0107	2.820	2.85	72.33	75.18	.00659	.17340
−24	13.90	.0108	2.698	3.25	72.16	75.41	.00753	.17317
−22	14.58	.0108	2.583	3.66	71.98	75.64	.00846	.17296
−20	15.28	0.0108	2.474	4.07	71.80	75.87	0.00940	0.17275
−18	16.01	.0108	2.370	4.48	71.63	76.11	.01033	.17253
−16	16.77	.0108	2.271	4.89	71.45	76.34	.01126	.17232
−14	17.55	.0109	2.177	5.30	71.27	76.57	.01218	.17212
−12	18.37	.0109	2.088	5.72	71.09	76.81	.01310	.17194
−10	19.20	0.0109	2.003	6.14	70.91	77.05	0.01403	0.17175
−8	20.08	.0109	1.922	6.57	70.72	77.29	.01496	.17158
−6	20.98	.0110	1.845	6.99	70.53	77.52	.01589	.17140
−4	21.91	.0110	1.772	7.41	70.34	77.75	.01682	.17123
−2	22.87	.0110	1.703	7.83	70.15	77.98	.01775	.17107
0	23.87	0.0110	1.637	8.25	69.96	78.21	0.01869	0.17091
2	24.89	.0110	1.574	8.67	69.77	78.44	.01961	.17075
4	25.96	.0111	1.514	9.10	69.57	78.67	.02052	.17060
5	26.51	.0111	1.485	9.32	69.47	78.79	.02097	.17052
6	27.05	.0111	1.457	9.53	69.37	78.90	.02143	.17045
8	28.18	.0111	1.403	9.96	69.17	79.13	.02235	.17030
10	29.35	0.0112	1.351	10.39	68.97	79.36	0.02328	0.17015
12	30.56	.0112	1.301	10.82	68.77	79.59	.02419	.17001
14	31.80	.0112	1.253	11.26	68.56	79.82	.02510	.16987
16	33.08	.0112	1.207	11.70	68.35	80.05	.02601	.16974
18	34.40	.0113	1.163	12.12	68.15	80.27	.02692	.16961
20	35.75	0.0113	1.121	12.55	67.94	80.49	0.02783	0.16949
22	37.15	.0113	1.081	13.00	67.72	80.72	.02873	.16938
24	38.58	.0113	1.043	13.44	67.51	80.95	.02963	.16926
26	40.07	.0114	1.007	13.88	67.29	81.17	.03053	.16913
28	41.59	.0114	0.973	14.32	67.07	81.39	.03143	.16900
30	43.16	0.0115	0.939	14.76	66.85	81.61	0.03233	0.16887
32	44.77	.0115	.908	15.21	66.62	81.83	.03323	.16876
34	46.42	.0115	.877	15.65	66.40	82.05	.03413	.16865
36	48.13	.0116	.848	16.10	66.17	82.27	.03502	.16854
38	49.88	.0116	.819	16.55	65.94	82.49	.03591	.16843
40	51.68	0.0116	0.792	17.00	65.71	82.71	0.03680	0.16833
42	53.51	.0116	.767	17.46	65.47	82.93	.03770	.16823
44	55.40	.0117	.742	17.91	65.24	83.15	.03859	.16813
46	57.35	.0117	.718	18.36	65.00	83.36	.03948	.16803
48	59.35	.0117	.695	18.82	64.74	83.57	.04037	.16794
50	61.39	0.0118	0.673	19.27	64.51	83.78	0.04126	0.16785
52	63.49	.0118	.652	19.72	64.27	83.99	.04215	.16776
54	65.63	.0118	.632	20.18	64.02	84.20	.04304	.16767
56	67.84	.0119	.612	20.64	63.77	84.41	.04392	.16758
58	70.10	.0119	.593	21.11	63.51	84.62	.04480	.16749
60	72.41	0.0119	0.575	21.57	63.25	84.82	0.04568	0.16741
62	74.77	.0120	.557	22.03	62.99	85.02	.04657	.16733
64	77.20	.0120	.540	22.49	62.73	85.22	.04745	.16725
66	79.67	.0120	.524	22.95	62.47	85.42	.04833	.16717
68	82.24	.0121	.508	23.42	62.20	85.62	.04921	.16709
70	84.82	0.0121	0.493	23.90	61.92	85.82	0.05009	0.16701
72	87.50	.0121	.479	24.37	61.65	86.02	.05097	.16693
74	90.20	.0122	.464	24.84	61.38	86.22	.05185	.16685
76	93.00	.0122	.451	25.32	61.10	86.42	.05272	.16677
78	95.85	.0123	.438	25.80	60.81	86.61	.05359	.16669
80	98.76	0.0123	0.425	26.28	60.52	86.80	0.05446	0.16662

APPENDIX B

FREON-12

Properties of Saturated Vapor—*Concluded*

Temp. °F t	Pressure Abs lb/in.² p	Volume		Enthalpy from −40°			Entropy from −40°	
		Liquid ft³/lb v_f	Vapor ft³/lb v_g	Liquid Btu/lb h_f	Evap. Btu/lb h_{fg}	Vapor Btu/lb h_g	Liquid Btu/lb °F s_f	Vapor Btu/lb °F s_g
80	98.76	0.0123	0.425	26.28	60.52	86.80	0.05446	0.16662
82	101.7	.0123	.413	26.76	60.23	86.99	.05534	.16655
84	104.8	.0124	.401	27.24	59.94	87.18	.05621	.16648
86	107.9	.0124	.389	27.72	59.65	87.37	.05708	.16640
88	111.1	.0124	.378	28.21	59.35	87.56	.05795	.16632
90	114.3	0.0125	0.368	28.70	59.04	87.74	0.05882	0.16624
92	117.7	.0125	.357	29.19	58.73	87.92	.05969	.16616
94	121.0	.0126	.347	29.68	58.42	88.10	.06056	.16608
96	124.5	.0126	.338	30.18	58.10	88.28	.06143	.16600
98	128.0	.0126	.328	30.67	57.78	88.45	.06230	.16592
100	131.6	0.0127	0.319	31.16	57.46	88.62	0.06316	0.16584
102	135.3	.0127	.310	31.65	57.14	88.79	.06403	.16576
104	139.0	.0128	.302	32.15	56.80	88.95	.06490	.16568
106	142.8	.0128	.293	32.65	56.46	89.11	.06577	.16560
108	146.8	.0129	.285	33.15	56.12	89.27	.06663	.16551
110	150.7	0.0129	0.277	33.65	55.78	89.43	0.06749	0.16542
112	154.8	.0130	.269	34.15	55.43	89.58	.06836	.16533
114	158.9	.0130	.262	34.65	55.08	89.73	.06922	.16524
116	163.1	.0131	.254	35.15	54.72	89.87	.07008	.16515
118	167.4	.0131	.247	35.65	54.36	90.01	.07094	.16505
120	171.8	0.0132	0.240	36.16	53.99	90.15	0.07180	0.16495
122	176.2	.0132	.233	36.66	53.62	90.28	.07266	.16484
124	180.8	.0133	.227	37.16	53.24	90.40	.07352	.16473
126	185.4	.0133	.220	37.67	52.85	90.52	.07437	.16462
128	190.1	.0134	.214	38.18	52.46	90.64	.07522	.16450
130	194.9	0.0134	0.208	38.69	52.07	90.76	0.07607	0.16438
132	199.8	.0135	.202	39.19	51.67	90.86	.07691	.16425
134	204.8	.0135	.196	39.70	51.26	90.96	.07775	.16411
136	209.9	.0136	.191	40.21	50.85	91.06	.07858	.16396
138	215.0	.0137	.185	40.72	50.43	91.15	.07941	.16380
140	220.2	0.0138	0.180	41.24	50.00	91.24	0.08024	0.16363

APPENDIX B

FREON-12

Properties of Superheated Vapor

Temp. °F	Abs Pressure 8 lb/in.² (Sat'n. Temp. −45.8° F)			Abs Pressure 9 lb/in.² (Sat'n. Temp. −41.4° F)			Abs Pressure 10 lb/in.² (Sat'n. Temp. −37.3° F)			Abs Pressure 11 lb/in.² (Sat'n. Temp. −33.5° F)		
t	v	h	s	v	h	s	v	h	s	v	h	s
(at sat'n.)	(4.502)	(72.80)	(0.17596)	(4.036)	(73.32)	(0.17533)	(3.652)	(73.80)	(0.17480)	(3.356)	(74.27)	(0.17432)
−40	4.569	73.56	0.17777	4.050	73.51	0.17576						
−30	4.684	74.87	.18085	4.152	74.83	.17884	3.728	74.77	0.17704	3.383	74.73	0.17540
−20	4.799	76.20	.18390	4.255	76.15	.18188	3.821	76.11	.18008	3.467	76.06	.17845
−10	4.914	77.54	.18691	4.357	77.49	.18490	3.913	77.46	.18310	3.551	77.40	.18146
0	5.028	78.89	0.18991	4.460	78.84	0.18791	4.006	78.81	0.18611	3.635	78.75	0.18448
10	5.142	80.26	.19284	4.562	80.20	.19084	4.098	80.18	.18905	3.719	80.12	.18742
20	5.257	81.64	.19574	4.663	81.58	.19374	4.189	81.56	.19194	3.802	81.50	.19032
30	5.370	83.02	.19860	4.766	82.98	.19661	4.280	82.94	.19482	3.887	82.90	.19320
40	5.484	84.43	.20143	4.867	84.39	.19945	4.371	84.35	.19766	3.971	84.31	.19605
50	5.598	85.85	0.20425	4.969	85.80	0.20227	4.463	85.77	0.20047	4.055	85.73	0.19887
60	5.711	87.27	.20703	5.071	87.24	.20505	4.556	87.19	.20326	4.138	87.16	.20165
70	5.824	88.72	.20977	5.171	88.68	.20779	4.648	88.64	.20601	4.221	88.61	.20440
80	5.938	90.18	.21250	5.272	90.13	.21051	4.740	90.11	.20874	4.304	90.07	.20713
90	6.051	91.64	.21519	5.374	91.60	.21321	4.832	91.58	.21144	4.388	91.54	.20984
100	6.165	93.13	0.21786	5.475	93.09	0.21588	4.923	93.05	0.21411	4.471	93.03	0.21251
110	6.278	94.63	.22051	5.576	94.59	.21853	5.015	94.56	.21676	4.553	94.52	.21515
120	6.391	96.13	.22314	5.677	96.10	.22116	5.107	96.07	.21940	4.636	96.03	.21778
130	6.504	97.64	.22573	5.778	97.63	.22375	5.198	97.59	.22199	4.718	97.56	.22037
140	6.617	99.18	.22831	5.879	99.16	.22634	5.289	99.14	.22458	4.800	99.09	.22296
150	6.730	100.73	0.23087	5.979	100.70	0.22889	5.379	100.66	0.22713	4.882	100.63	0.22551
160	6.843	102.29	.23340	6.080	102.26	.23143	5.470	102.24	.22967	4.965	102.20	.22805
170	6.955	103.87	.23591	6.180	103.84	.23394	5.560	103.81	.23218	5.047	103.78	.23057
180	7.068	105.44	.23842	6.280	105.43	.23645	5.650	105.40	.23469	5.130	105.37	.23308
190	7.181	107.05	.24090	6.380	107.03	.23893	5.740	107.00	.23717	5.214	106.97	.23557
200	7.294	108.67	0.24337	6.481	108.64	0.24140	5.831	108.63	0.23963	5.297	108.58	0.23804
210	7.407	110.28	.24581	6.581	110.26	.24384	5.921	110.25	.24208	5.379	110.21	.24049
220	7.520	111.93	.24825	6.682	111.90	.24628	6.011	111.88	.24451	5.462	111.85	.24291
230	7.633	113.57	.25066	6.782	113.55	.24868	6.191	113.53	.24692	5.544	113.50	.24532
240										5.626	115.18	.24773

Temp. °F	Abs Pressure 12 lb/in.² (Sat'n. Temp. −30.0° F)			Abs Pressure 13 lb/in.² (Sat'n. Temp. −26.8° F)			Abs Pressure 14 lb/in.² (Sat'n. Temp. −23.7° F)			Abs Pressure 15 lb/in.² (Sat'n. Temp. −20.8° F)		
(at sat'n.)	(3.093)	(74.69)	(0.17389)	(2.875)	(75.08)	(0.17350)	(2.677)	(75.45)	(0.17314)	(2.518)	(75.78)	(0.17282)
−30	3.093	74.69	0.17389									
−20	3.172	76.02	.17695	2.920	75.98	0.17556	2.706	75.94	0.17427	2.521	75.89	0.17307
−10	3.250	77.37	.17998	2.992	77.32	.17859	2.773	77.28	.17731	2.583	77.23	.17611
0	3.328	78.73	0.18299	3.064	78.69	0.18160	2.841	78.64	0.18032	2.646	78.59	0.17913
10	3.405	80.10	.18594	3.136	80.05	.18455	2.908	80.01	.18328	2.708	79.97	.18208
20	3.483	81.48	.18884	3.207	81.43	.18746	2.974	81.40	.18618	2.771	81.37	.18499
30	3.560	82.87	.19173	3.278	82.83	.19034	3.041	82.80	.18907	2.833	82.77	.18788
40	3.637	84.28	.19458	3.349	84.23	.19319	3.107	84.21	.19192	2.895	84.18	.19074
50	3.714	85.71	0.19739	3.420	85.66	0.19601	3.173	85.63	0.19475	2.957	85.60	0.19357
60	3.790	87.14	.20018	3.491	87.10	.19880	3.239	87.06	.19753	3.019	87.03	.19635
70	3.867	88.59	.20293	3.562	88.54	.20156	3.303	88.51	.20028	3.081	88.48	.19911
80	3.943	90.05	.20566	3.632	90.01	.20428	3.369	89.97	.20302	3.143	89.94	.20185
90	4.019	91.52	.20836	3.703	91.48	.20699	3.435	91.44	.20572	3.204	91.41	.20455
100	4.095	93.00	0.21104	3.774	92.96	0.20967	3.501	92.93	0.20841	3.266	92.91	0.20723
110	4.170	94.50	.21367	3.844	94.47	.21231	3.567	94.43	.21106	3.327	94.41	.20989
120	4.246	96.01	.21631	3.915	95.98	.21495	3.633	95.95	.21369	3.388	95.91	.21252
130	4.323	97.53	.21891	3.986	97.50	.21755	3.699	97.48	.21630	3.450	97.44	.21513
140	4.400	99.07	.22151	4.056	99.04	.22014	3.763	99.01	.21889	3.510	98.98	.21772
150	4.474	100.62	0.22406	4.126	100.59	0.22270	3.828	100.56	0.22144	3.571	100.53	0.22028
160	4.549	102.18	.22659	4.196	102.16	.22524	3.894	102.13	.22400	3.632	102.10	.22282
170	4.624	103.75	.22911	4.266	103.73	.22775	3.960	103.71	.22651	3.694	103.68	.22535
180	4.700	105.34	.23162	4.337	105.32	.23027	4.025	105.30	.22902	3.755	105.27	.22786
190	4.774	106.94	.23409	4.408	106.92	.23276	4.090	106.90	.23150	3.816	106.87	.23034
200	4.850	108.55	0.23656	4.478	108.53	0.23523	4.155	108.52	0.23398	3.877	108.49	0.23282
210	4.926	110.18	.23901	4.547	110.16	.23767	4.220	110.14	.23643	3.938	110.12	.23527
220	5.000	111.82	.24144	4.617	111.81	.24011	4.285	111.78	.23886	3.998	111.77	.23770
230	5.076	113.47	.24385	4.686	113.45	.24252	4.350	113.44	.24128	4.059	113.42	.24011
240	5.152	115.15	.24626	4.755	115.13	.24492	4.414	115.11	.24368	4.120	115.10	.24253
250							4.479	116.79	0.24607	4.181	116.78	0.24491

APPENDIX B

FREON-12

PROPERTIES OF SUPERHEATED VAPOR—*Continued*

Temp. °F	Abs Pressure 16 lb/in.² (Sat'n. Temp. −18.0° F)			Abs Pressure 17 lb/in.² (Sat'n. Temp. −15.4° F)			Abs Pressure 18 lb/in.² (Sat'n. Temp. −12.9° F)			Abs Pressure 19 lb/in.² (Sat'n. Temp. −10.5° F)		
t	v	h	s	v	h	s	v	h	s	v	h	s
(at sat'n.)	*(2.370)*	*(76.11)*	*(0.17254)*	*(2.240)*	*(76.41)*	*(0.17226)*	*(2.123)*	*(76.70)*	*(0.17202)*	*(2.019)*	*(76.99)*	*(0.17180)*
−10	2.417	77.20	0.17498	2.268	77.15	0.17390	2.137	77.09	0.17289	2.022	77.05	0.17195
0	2.476	78.56	0.17800	2.325	78.52	0.17693	2.191	78.47	0.17592	2.072	78.42	0.17497
10	2.535	79.94	.18095	2.382	79.90	.17990	2.244	79.85	.17889	2.122	79.80	.17794
20	2.594	81.33	.18387	2.438	81.29	.18283	2.297	81.24	.18181	2.172	81.19	.18087
30	2.652	82.73	.18676	2.493	82.69	.18571	2.350	82.65	.18471	2.222	82.60	.18376
40	2.710	84.13	.18962	2.548	84.10	.18857	2.402	84.06	.18757	2.272	84.02	.18663
50	2.768	85.55	0.19244	2.603	85.52	0.19140	2.454	85.49	0.19041	2.321	85.45	0.18946
60	2.827	86.99	.19524	2.658	86.96	.19420	2.506	86.93	.19321	2.371	86.90	.19226
70	2.886	88.45	.19800	2.712	88.41	.19696	2.558	88.38	.19597	2.420	88.35	.19503
80	2.944	89.91	.20075	2.767	89.88	.19969	2.610	89.84	.19872	2.469	89.81	.19777
90	3.002	91.39	.20345	2.822	91.36	.20241	2.662	91.32	.20142	2.518	91.29	.20047
100	3.059	92.88	0.20613	2.876	92.85	0.20510	2.713	92.81	0.20411	2.567	92.78	0.20317
110	3.117	94.38	.20879	2.930	94.35	.20774	2.764	94.32	.20676	2.616	94.28	.20583
120	3.175	95.89	.21143	2.984	95.86	.21038	2.815	95.83	.20941	2.665	95.80	.20847
130	3.232	97.42	.21403	3.038	97.39	.21298	2.867	97.36	.21201	2.714	97.33	.21108
140	3.290	98.95	.21662	3.092	98.93	.21558	2.918	98.90	.21462	2.763	98.88	.21368
150	3.347	100.50	0.21917	3.147	100.48	0.21814	2.969	100.45	0.21717	2.812	100.43	0.21625
160	3.404	102.07	.22172	3.201	102.05	.22070	3.020	102.02	.21971	2.861	102.00	.21880
170	3.461	103.65	.22425	3.256	103.63	.22323	3.072	103.60	.22223	2.910	103.59	.22133
180	3.519	105.24	.22677	3.310	105.22	.22574	3.124	105.19	.22477	2.958	105.18	.22384
190	3.576	106.85	.22925	3.364	106.82	.22822	3.175	106.80	.22725	3.006	106.78	.22633
200	3.633	108.47	0.23173	3.417	108.44	0.23070	3.226	108.42	0.22973	3.054	108.39	0.22880
210	3.690	110.10	.23418	3.470	110.07	.23315	3.277	110.05	.23219	3.102	110.02	.23126
220	3.747	111.74	.23662	3.524	111.71	.23558	3.327	111.69	.23461	3.150	111.66	.23369
230	3.803	113.40	.23903	3.578	113.37	.23800	3.378	113.34	.23703	3.198	113.31	.23612
240	3.860	115.07	.24143	3.632	115.04	.24041	3.428	115.00	.23944	3.246	114.99	.23852
250	3.917	116.76	0.24382	3.686	116.72	0.24279	3.478	116.68	0.24182	3.294	116.68	0.24091
260				3.739	118.41	.24517	3.528	118.36	.24420	3.342	118.37	.24327

Temp. °F	Abs Pressure 20 lb/in.² (Sat'n. Temp. −8.2° F)			Abs Pressure 22 lb/in.² (Sat'n. Temp. −3.8° F)			Abs Pressure 24 lb/in.² (Sat'n. Temp. 0.3° F)			Abs Pressure 26 lb/in.² (Sat'n. Temp. 4.1° F)		
(at sat'n.)	*(1.925)*	*(77.27)*	*(0.17160)*	*(1.762)*	*(77.77)*	*(0.17121)*	*(1.623)*	*(78.24)*	*(0.17089)*	*(1.510)*	*(78.68)*	*(0.17059)*
0	1.965	78.39	0.17407	1.779	78.30	0.17237						
10	2.013	79.76	.17704	1.823	79.68	.17534	1.664	79.59	0.17378	1.530	79.50	0.17233
20	2.060	81.14	.17996	1.866	81.07	.17827	1.705	80.99	.17673	1.568	80.90	.17529
30	2.107	82.55	.18286	1.909	82.48	.18118	1.745	82.40	.17964	1.606	82.31	.17821
40	2.155	83.97	.18573	1.953	83.90	.18405	1.785	83.82	.18251	1.643	83.74	.18109
50	2.203	85.40		1.997	85.33	0.18691	1.825	85.26	0.18536	1.680	85.19	.018395
60	2.250	86.85	.19138	2.040	86.78	.18971	1.865	86.71	.18817	1.716	86.64	.18675
70	2.297	88.31	.19415	2.083	88.25	.19248	1.905	88.17	.19096	1.754	88.10	.18954
80	2.343	89.78	.19688	2.125	89.71	.19522	1.944	89.64	.19371	1.790	89.57	.19229
90	2.390	91.26	.19959	2.168	91.19	.19794	1.984	91.13	.19643	1.826	91.05	.19501
100	2.437	92.75	0.20229	2.211	92.68	0.20064	2.023	92.63	0.19912	1.863	92.55	0.19772
110	2.483	94.26	.20494	2.253	94.19	.20330	2.062	94.13	.20179	1.900	94.06	.20039
120	2.530	95.78	.20759	2.296	95.71	.20595	2.101	95.64	.20444	1.936	95.58	.20305
130	2.577	97.31	.21020	2.339	97.25	.20856	2.140	97.17	.20704	1.972	97.12	.20565
140	2.623	98.85	.21280	2.381	98.80	.21116	2.179	98.72	.20966	2.008	98.67	.20827
150	2.669	100.40	0.21537	2.423	100.35	0.21373	2.218	100.28	0.21223	2.045	100.23	0.21085
160	2.716	101.97	.21792	2.465	101.92	.21628	2.257	101.86	.21479	2.082	101.80	.21341
170	2.762	103.56	.22045	2.507	103.50	.21881	2.296	103.45	.21732	2.117	103.39	.21594
180	2.808	105.15	.22297	2.550	105.10	.22133	2.335	105.05	.21984	2.153	104.99	.21847
190	2.854	106.76	.22545	2.592	106.71	.22382	2.373	106.66	.22233	2.189	106.60	.22096
200	2.901	108.38	0.22794	2.634	108.33	0.22630	2.412	108.28	0.22481	2.224	108.23	0.22344
210	2.947	110.01	.23039	2.676	109.96	.22876	2.450	109.92	.22727	2.259	109.86	.22589
220	2.992	111.65	.23283	2.717	111.60	.23120	2.489	111.57	.22972	2.295	111.51	.22834
230	3.038	113.31	.23524	2.758	113.26	.23361	2.528	113.23	.23214	2.331	113.17	.23076
240	3.084	114.98	.23766	2.800	114.94	.23603	2.566	114.90	.23455	2.366	114.85	.23318
250	3.130	116.67	0.24005	2.842	116.62	0.23843	2.605	116.58	0.23695	2.402	116.54	0.23557
260	3.177	118.36	.24242	2.884	118.32	.24080	2.644	118.28	.23933	2.438	118.24	.23796
270	3.223	120.07	.24477	2.926	120.04	.24316	2.682	120.00	.24169	2.474	119.95	.24033
280										2.510	121.68	.24267

APPENDIX B

FREON-12

Properties of Superheated Vapor—Continued

Temp. °F	Abs Pressure 28 lb/in.² (Sat'n. Temp. 7.7° F)			Abs Pressure 30 lb/in.² (Sat'n. Temp. 11.1° F)			Abs Pressure 32 lb/in.² (Sat'n. Temp. 14.3° F)			Abs Pressure 34 lb/in.² (Sat'n. Temp. 17.4° F)		
t	v	h	s	v	h	s	v	h	s	v	h	s
(at sat'n.)	*(1.409)*	*(79.10)*	*(0.17032)*	*(1.323)*	*(79.47)*	*(0.17008)*	*(1.245)*	*(79.84)*	*(0.16985)*	*(1.175)*	*(80.20)*	*(0.16965)*
10	1.415	79.41	0.17099
20	1.450	80.81	.17393	1.350	80.73	0.17269	1.262	80.67	0.17152	1.183	80.58	0.17040
30	1.485	82.23	.17685	1.383	82.15	.17562	1.293	82.08	.17445	1.212	82.00	.17333
40	1.520	83.66	.17975	1.415	83.58	.17851	1.323	83.51	.17734	1.241	83.44	.17623
50	1.555	85.11	0.18261	1.448	85.03	0.18138	1.354	84.96	0.18022	1.270	84.88	0.17910
60	1.590	86.56	.18544	1.480	86.48	.18420	1.384	86.41	.18304	1.299	86.34	.18194
70	1.625	88.03	.18823	1.512	87.95	.18699	1.414	87.88	.18583	1.328	87.81	.18474
80	1.659	89.51	.19097	1.544	89.43	.18974	1.444	89.36	.18860	1.356	89.29	.18750
90	1.693	90.99	.19371	1.576	90.91	.19249	1.474	90.85	.19133	1.385	90.78	.19025
100	1.727	92.49	0.19642	1.608	92.41	0.19519	1.504	92.35	0.19404	1.413	92.29	0.19295
110	1.761	94.01	.19909	1.640	93.93	.19787	1.535	93.87	.19673	1.441	93.81	.19563
120	1.795	95.53	.20174	1.672	95.46	.20053	1.565	95.40	.19940	1.470	95.34	.19831
130	1.828	97.07	.20436	1.703	97.00	.20315	1.595	96.94	.20202	1.498	96.88	.20094
140	1.862	98.62	.20698	1.735	98.54	.20577	1.624	98.50	.20463	1.526	98.43	.20356
150	1.896	100.18	0.20956	1.767	100.11	0.20836	1.654	100.06	0.20721	1.554	100.00	0.20614
160	1.930	101.75	.21212	1.799	101.69	.21092	1.683	101.64	.20977	1.582	101.58	.20871
170	1.963	103.33	.21466	1.829	103.28	.21344	1.713	103.23	.21232	1.610	103.17	.21125
180	1.997	104.93	.21719	1.860	104.88	.21597	1.743	104.83	.21486	1.638	104.78	.21379
190	2.030	106.55	.21967	1.891	106.49	.21846	1.772	106.45	.21735	1.666	106.40	.21629
200	2.063	108.17	0.22216	1.923	108.12	0.22096	1.802	108.08	0.21985	1.693	108.03	0.21878
210	2.096	109.81	.22462	1.954	109.76	.22342	1.831	109.72	.22231	1.721	109.67	.22125
220	2.129	111.46	.22706	1.986	111.41	.22588	1.860	111.36	.22476	1.749	111.32	.22370
230	2.163	113.12	.22949	2.017	113.08	.22830	1.889	113.03	.22718	1.776	112.98	.22613
240	2.196	114.80	.23191	2.048	114.75	.23072	1.918	114.72	.22960	1.804	114.66	.22856
250	2.229	116.49	0.23430	2.079	116.44	0.23312	1.948	116.41	0.23200	1.833	116.35	0.23097
260	2.262	118.19	.23669	2.110	118.15	.23550	1.977	118.11	.23439	1.860	118.06	.23335
270	2.295	119.91	.23905	2.141	119.87	.23787	2.006	119.82	.23676	1.888	119.79	.23573
280	2.329	121.65	.24141	2.172	121.60	.24023	2.035	121.55	.23912	1.916	121.54	.23809
290	2.065	123.30	.24146	1.944	123.31	.24043

Temp. °F	Abs Pressure 36 lb/in.² (Sat'n. Temp. 20.4° F)			Abs Pressure 38 lb/in.² (Sat'n. Temp. 23.2° F)			Abs Pressure 40 lb/in.² (Sat'n. Temp. 25.9° F)			Abs Pressure 42 lb/in.² (Sat'n. Temp. 28.5° F)		
(at sat'n.)	*(1.113)*	*(80.54)*	*(0.16947)*	*(1.058)*	*(80.86)*	*(0.16931)*	*(1.009)*	*(81.16)*	*(0.16914)*	*(0.963)*	*(81.44)*	*(0.16897)*
30	1.140	81.90	0.17227	1.076	81.82	0.17126	1.019	81.76	0.17030	0.967	81.65	0.16939
40	1.168	83.35	.17518	1.103	83.27	.17418	1.044	83.20	.17322	0.991	83.10	.17231
50	1.196	84.81	0.17806	1.129	84.72	0.17706	1.070	84.65	0.17612	1.016	84.56	0.17521
60	1.223	86.27	.18089	1.156	86.19	.17991	1.095	86.11	.17896	1.040	86.03	.17806
70	1.250	87.74	.18369	1.182	87.67	.18272	1.120	87.60	.18178	1.063	87.51	.18086
80	1.278	89.22	.18647	1.208	89.16	.18551	1.144	89.09	.18455	1.087	89.00	.18365
90	1.305	90.71	.18921	1.234	90.66	.18826	1.169	90.58	.18731	1.110	90.50	.18640
100	1.332	92.22	0.19193	1.260	92.17	0.19096	1.194	92.09	0.19004	1.134	92.01	0.18913
110	1.359	93.75	.19462	1.285	93.69	.19365	1.218	93.62	.19272	1.158	93.54	.19184
120	1.386	95.28	.19729	1.310	95.22	.19631	1.242	95.15	.19538	1.181	95.09	.19451
130	1.412	96.82	.19991	1.336	96.76	.19895	1.267	96.70	.19803	1.204	96.64	.19714
140	1.439	98.37	.20254	1.361	98.32	.20157	1.291	98.26	.20066	1.227	98.20	.19979
150	1.465	99.93	0.20512	1.387	99.89	0.20416	1.315	99.83	0.20325	1.250	99.77	0.20237
160	1.492	101.51	.20770	1.412	101.47	.20673	1.340	101.42	.20583	1.274	101.36	.20496
170	1.518	103.11	.21024	1.437	103.07	.20929	1.364	103.02	.20838	1.297	102.96	.20751
180	1.545	104.72	.21278	1.462	104.67	.21183	1.388	104.63	.21092	1.320	104.57	.21005
190	1.571	106.34	.21528	1.487	106.29	.21433	1.412	106.25	.21343	1.343	106.19	.21256
200	1.597	107.97	0 21778	1.512	107.93	0.21681	1.435	107.88	0.21592	1.365	107.82	0.21505
210	1.623	109.61	.22024	1.537	109.57	.21928	1.459	109.52	.21840	1.388	109.47	.21754
220	1.650	111.27	.22270	1.562	111.22	.22176	1.482	111.17	.22085	1.411	111.12	.22000
230	1.676	112.94	.22513	1.587	112.89	.22419	1.506	112.84	.22329	1.434	112.80	.22244
240	1.702	114.62	.22756	1.612	114.58	.22662	1.530	114.52	.22572	1.457	114.49	.22486
250	1.728	116.31	0.22996	1.637	116.28	0.22903	1.554	116.21	0.22813	1.480	116.19	0.22728
260	1.754	118.02	.23235	1.662	117.99	.23142	1.577	117.92	.23052	1.502	117.90	.22967
270	1.780	119.74	.23472	1.687	119.71	.23379	1.601	119.65	.23289	1.524	119.62	.23204
280	1.807	121.47	.23708	1.712	121.45	.23616	1.625	121.40	.23526	1.547	121.36	.23441
290	1.833	123.22	.23942	1.737	123.20	.23850	1.649	123.15	.23760	1.570	123.11	.23675
300	1.762	124.95	0.24083	1.673	124.92	0.23994	1.592	124.87	0.23909

APPENDIX B

FREON-12

Properties of Superheated Vapor—Continued

Temp. °F	Abs Pressure 44 lb/in.² (Sat'n. Temp. 31.0 °F)			Abs Pressure 46 lb/in.² (Sat'n. Temp. 33.5° F)			Abs Pressure 48 lb/in.² (Sat'n. Temp. 35.8° F)			Abs Pressure 50 lb/in.² (Sat'n. Temp. 38.3° F)		
t	v	h	s	v	h	s	v	h	s	v	h	s
(at satn.)	(0.922)	(81.72)	(0.16882)	(0.885)	(82.00)	(0.16867)	(0.849)	(82.25)	(0.16855)	(0.817)	(82.52)	(0.16841)
40	0.943	83.03	0.17142	0.899	82.94	0.17057	0.858	82.85).16974	0.821	82.76	0.16895
50	0.966	84.48	0.17432	0.921	84.40	0.17347	0.880	84.32	0.17266	0.842	84.24	0.17187
60	0.989	85.96	.17717	0.943	85.88	.17633	0.092	85.80	.17554	0.863	85.72	.17475
70	1.012	87.45	.18000	0.965	87.37	.17916	0.923	87.29	.17837	0.884	87.22	.17760
80	1.035	88.94	.18279	0.988	88.88	.18198	0.944	88.79	.18117	0.904	88.72	.18040
90	1.058	90.44	.18556	1.010	90.39	.18474	0.965	90.30	.18394	0.924	90.23	.18317
100	1.080	91.95	0.18828	1.031	91.90	0.18746	0.986	91.82	0.18668	0.944	91.75	0.18591
110	1.103	93.48	.19099	1.053	93.43	.19016	1.007	93.35	.18939	0.964	93.29	.18862
120	1.125	95.02	.19367	1.074	94.96	.19285	1.028	94.89	.19208	0.984	94.83	.19132
130	1.147	96.57	.19630	1.096	96.51	.19551	1.048	96.44	.19472	1.004	96.39	.19397
140	1.170	98.14	.19895	1.117	98.08	.19814	1.069	98.01	.19737	1.024	97.96	.19662
150	1.192	99.72	0.20154	1.139	99.66	0.20075	1.089	99.59	0.19997	1.044	99.54	0.19923
160	1.214	101.31	.20412	1.160	101.25	.20333	1.110	101.18	.20256	1.064	101.14	.20182
170	1.236	102.91	.20667	1.181	102.85	.20588	1.130	102.79	.20513	1.084	102.75	.20439
180	1.258	104.52	.20922	1.202	104.46	.20843	1.150	104.40	.20766	1.103	104.36	.20694
190	1.280	106.14	.21173	1.223	106.09	.21094	1.170	106.02	.21017	1.123	105.98	.20946
200	1.302	107.78	0.21424	1.244	107.73	0.21344	1.191	107.00	0.21260	1.142	107.62	0.21196
210	1.324	109.42	.21672	1.265	109.38	.21592	1.211	109.31	.21517	1.162	109.28	.21444
220	1.346	111.08	.21918	1.286	111.04	.21839	1.231	110.98	.21763	1.181	110.95	.21691
230	1.367	112.75	.22161	1.307	112.71	.22083	1.251	112.66	.22007	1.200	112.62	.21935
240	1.389	114.44	.22405	1.327	114.39	.22326	1.271	114.35	.22251	1.220	114.31	.22179
250	1.411	116.14	0.22646	1.348	116.09	0.22567	1.291	116.05	0.22492	1.239	116.00	0.22419
260	1.432	117.85	.22885	1.369	117.81	.22806	1.311	117.77	.22731	1.258	117.71	.22660
270	1.454	119.57	.23123	1.390	119.54	.23044	1.331	119.49	.22970	1.277	119.44	.22898
280	1.475	121.31	.23359	1.410	121.27	.23281	1.351	121.23	.23207	1.296	121.18	.23134
290	1.496	123.06	.23592	1.431	123.02	.23515	1.370	122.98	.23440	1.314	122.93	.23367
300	1.518	124.82	0.23826	1.452	124.79	0.23749	1.390	124.75	0.23674	1.332	124.69	0.23600
310	1.539	126.59	.24058	1.472	126.57	.23981	1.410	126.53	.23907	1.350	126.45	.23831

Temp. °F	Abs Pressure 52 lb/in.² (Sat'n. Temp. 40.4° F)			Abs Pressure 54 lb/in.² (Sat'n. Temp. 42.5° F)			Abs Pressure 56 lb/in.² (Sat'n. Temp. 44.6° F)			Abs Pressure 58 lb/in.² (Sat'n. Temp. 46.7° F)		
(at sat'n.)	(0.788)	(82.75)	(0.16831)	(0.759)	(82.98)	(0.16820)	(0.734)	(83.22)	(0.16810)	(0.710)	(83.44)	(0.16800)
50	0.808	84.17	0.17114	0.774	84.07	0.17036	0.744	83.99	0.16965	0.716	83.91	0.16896
60	0.827	85.65	.17400	0.794	85.56	.17326	0.763	85.48	.17255	0.734	85.40	.17185
70	0.847	87.14	.17684	0.814	87.06	.17612	0.782	86.98	.17541	0.752	86.90	.17471
80	0.867	88.64	.17966	0.833	88.57	.17894	0.801	88.49	.17824	0.770	88.41	.17753
90	0.886	90.15	.18244	0.852	90.08	.18172	0.819	90.01	.18102	0.788	89.93	.18033
100	0.906	91.68	0.18518	0.871	91.61	0.18446	0.837	91.54	0.18377	0.806	91.46	0.18309
110	0.925	93.22	.18789	0.890	93.16	.18718	0.856	93.08	.18651	0.824	93.00	.18583
120	0.945	94.77	.19059	0.908	94.71	.18989	0.874	94.63	.18921	0.842	94.55	.18854
130	0.964	96.33	.19325	0.927	96.27	.19255	0.892	96.19	.19188	0.860	96.11	.19122
140	0.983	97.90	.19590	0.945	97.84	.19520	0.910	97.77	.19453	0.877	97.68	.19387
150	1.002	99.48	0.19850	0.964	99.43	0.19782	0.928	99.36	0.19715	0.894	99.26	0.19648
160	1.021	101.07	.20109	0.982	101.03	.20043	0.946	100.96	.19975	0.912	100.86	.19908
170	1.040	102.68	.20365	1.001	102.64	.20299	0.964	102.57	.20232	0.929	102.47	.20166
180	1.059	104.30	.20621	1.019	104.25	.20554	0.981	104.19	.20487	0.946	104.09	.20423
190	1.078	105.93	.20873	1.037	105.88	.20806	0.999	105.83	.20739	0.963	105.72	.20676
200	1.097	107.58	0.21125	1.055	107.52	0.21057	1.016	107.48	0.20991	0.980	107.36	0.20927
210	1.116	109.23	.21374	1.073	109.17	.21305	1.034	109.14	.21241	0.997	109.02	.21177
220	1.134	110.89	.21620	1.091	110.84	.21553	1.051	110.81	.21487	1.014	110.70	.21424
230	1.153	112.56	.21865	1.109	112.51	.21797	1.068	112.49	.21731	1.031	112.39	.21669
240	1.172	114.26	.22110	1.127	114.20	.22042	1.086	114.18	.21977	1.048	114.10	.21914
250	1.190	115.96	0.22352	1.145	115.91	0.22283	1.103	115.88	0.22218	1.064	115.82	0.22155
260	1.208	117.67	.22591	1.163	117.63	.22523	1.120	117.59	.22458	1.081	117.55	.22396
270	1.227	119.40	.22829	1.181	119.37	.22762	1.138	119.31	.22698	1.098	119.29	.22635
280	1.247	121.14	.23065	1.199	121.11	.22999	1.155	121.05	.22934	1.114	121.04	.22871
290	1.263	122.90	.23299	1.216	122.86	.23233	1.172	122.80	.23169	1.130	122.80	.23105
300	1.281	124.66	0.23532	1.234	124.63	0.23467	1.189	124.57	0.23403	1.147	124.57	0.23340
310	1.298	126.42	.23763	1.251	126.40	.23699	1.206	126.36	.23635	1.164	126.35	.23574
320							1.223	128.17	.23867	1.180	128.14	.23806

499

APPENDIX B

FREON-12

Properties of Superheated Vapor—Continued

Temp. °F	Abs Pressure 60 lb/in.² (Sat'n. Temp 48.7° F)			Abs Pressure 70 lb/in.² (Sat'n. Temp. 57.9° F)			Abs Pressure 80 lb/in. (Sat'n. Temp. 66.3° F)			Abs Pressure 90 lb/in.² (Sat'n. Temp. 73.9° F)		
t	v	h	s	v	h	s	v	h	s	v	h	s
(at sat'n.)	(0.688)	(83.65)	(0.16791)	(0.594)	(84.61)	(0.16749)	(0.521)	(85.45)	(0.16716)	(0.465)	(86.21)	(0.16685)
50	0.690	83.83	0.16829									
60	0.708	85.33	.17120									
70	0.726	86.84	.17407	0.597	84.94	0.16810						
80	0.743	88.35	.17689	0.612	86.44	.17097	0.526	86.01	0.16819			
90	0.760	89.87	.17963	0.628	87.96	.17382	0.540	87.56	.17108	0.473	87.18	0.16862
				0.643	89.49	.17665	0.554	89.12	.17394	0.486	88.74	.17149
100	0.778	91.41	0.18246	0.658	91.03	0.17943	0.568	90.68	0.17675	0.499	90.31	0.17433
110	0.795	92.96	.18519	0.673	92.59	.18219	0.582	92.26	.17954	0.511	91.89	.17713
120	0.812	94.51	.18789	0.689	94.16	.18493	0.596	93.84	.18229	0.523	93.48	.17990
130	0.829	96.07	.19056	0.704	95.75	.18763	0.609	95.43	.18500	0.535	95.08	.18262
140	0.846	97.65	.19323	0.719	97.34	.19030	0.623	97.03	.18771	0.547	96.69	.18533
150	0.863	99.24	0.19585	0.733	98.94	0.19293	0.636	98.64	0.19035	0.559	98.31	0.18799
160	0.880	100.84	.19846	0.748	100.54	.19555	0.649	100.26	.19298	0.571	99.94	.19065
170	0.897	102.45	.20104	0.763	102.16	.19814	0.662	101.88	.19558	0.584	101.58	.19327
180	0.913	104.07	.20360	0.777	103.80	.20071	0.675	103.52	.19817	0.596	103.23	.19588
190	0.930	105.71	.20613	0.792	105.45	.20325	0.688	105.18	.20073	0.607	104.89	.19845
200	0.946	107.36	0.20865	0.806	107.10	0.20579	0.701	106.84	0.20328	0.619	106.56	0.20101
210	0.962	109.02	.21113	0.820	108.76	.20829	0.714	108.51	.20580	0.630	108.24	.20353
220	0.979	110.69	.21361	0.835	110.43	.21079	0.726	110.19	.20828	0.642	109.93	.20603
230	0.995	112.37	.21607	0.849	112.13	.21325	0.739	111.88	.21076	0.653	111.63	.20852
240	1.012	114.06	.21853	0.863	113.83	.21570	0.751	113.58	.21321	0.665	113.35	.21100
250	1.028	115.77	0.22094	0.878	115.55	0.21815	0.764	115.30	0.21566	0.676	115.08	0.21345
260	1.044	117.49	.22334	0.892	117.28	.22057	0.777	117.03	.21809	0.688	116.82	.21589
270	1.060	119.23	.22573	0.906	119.02	.22296	0.789	118.78	.22049	0.699	118.57	.21831
280	1.076	120.97	.22810	0.920	120.76	.22534	0.802	120.54	.22289	0.710	120.33	.22070
290	1.092	122.73	.23045	0.934	122.52	.22770	0.814	122.30	.22525	0.721	122.10	.22306
300	1.108	124.50	0.23280	0.948	124.29	0.23006	0.826	124.08	0.22760	0.732	123.88	0.22542
310	1.124	126.28	.23513	0.961	126.07	.23239	0.839	125.88	.22995	0.743	125.67	.22776
320	1.140	128.07	.23745	0.975	127.88	.23471	0.851	127.70	.23229	0.754	127.48	.23008
330				0.989	129.70	.23702	0.864	129.52	.23461	0.765	129.31	.23240

Temp. °F	Abs Pressure 100 lb/in.² (Sat'n. Temp. 80.9° F)			Abs Pressure 110 lb/in.² (Sat'n. Temp. 87.3° F)			Abs Pressure 120 lb/in.² (Sat'n. Temp. 93.4° F)			Abs Pressure 130 lb/in.² (Sat'n. Temp. 99.1° F)		
(at sat'n.)	(0.419)	(86.89)	(0.16659)	(0.382)	(87.50)	(0.16635)	(0.350)	(88.05)	(0.16610)	(0.323)	(88.547)	(0.16588)
90	0.430	88.32	0.16926	0.385	87.91	0.16711						
100	0.442	89.93	0.17210	0.396	89.51	0.17001	0.357	89.13	0.16803	0.324	88.69	0.16615
110	0.454	91.54	.17493	0.407	91.12	.17287	0.367	90.75	.17090	0.333	90.33	.16905
120	0.465	93.15	.17773	0.417	92.74	.17568	0.377	92.38	.17374	0.343	91.98	.17193
130	0.477	94.76	.18049	0.428	94.37	.17845	0.387	94.01	.17654	0.353	93.64	.17476
140	0.488	96.37	.18321	0.438	96.01	.18122	0.397	95.65	.17932	0.362	95.30	.17756
150	0.499	97.99	0.18590	0.449	97.66	0.18394	0.407	97.30	0.18207	0.371	96.97	0.18030
160	0.510	99.63	.18856	0.459	99.31	.18660	0.417	98.96	.18474	0.380	98.65	.18302
170	0.521	101.28	.19120	0.469	100.97	.18924	0.426	100.63	.18743	0.389	100.34	.18571
180	0.531	102.94	.19381	0.479	102.64	.19187	0.436	102.31	.19011	0.398	102.04	.18839
190	0.542	104.61	.19638	0.489	104.32	.19447	0.445	104.00	.19271	0.407	103.74	.19102
200	0.553	106.29	0.19894	0.499	106.01	0.19706	0.454	105.70	0.19529	0.416	105.45	0.19362
210	0.563	107.98	.20148	0.509	107.71	.19962	0.463	107.41	.19785	0.424	107.16	.19620
220	0.574	109.68	.20401	0.519	109.42	.20216	0.472	109.13	.20041	0.433	108.89	.19877
230	0.585	111.39	.20650	0.528	111.14	.20464	0.482	110.86	.20294	0.442	110.62	.20130
240	0.595	113.11	.20899	0.538	112.87	.20712	0.491	112.60	.20545	0.450	112.36	.20382
250	0.606	114.84	0.21145	0.548	114.61	0.20959	0.500	114.35	0.20792	0.458	114.11	0.20629
260	0.616	116.58	.21389	0.557	116.36	.21205	0.508	116.11	.21035	0.467	115.87	.20876
270	0.626	118.33	.21631	0.567	118.12	.21448	0.517	117.88	.21279	0.475	117.64	.21122
280	0.636	120.10	.21870	0.576	119.89	.21690	0.526	119.66	.21521	0.483	119.42	.21364
290	0.646	121.88	.22108	0.586	121.68	.21930	0.534	121.45	.21760	0.492	121.21	.21605
300	0.657	123.67	0.22347	0.595	123.48	0.22167	0.543	123.25	0.22000	0.500	123.01	0.21846
310	0.667	125.47	.22583	0.605	125.29	.22405	0.552	125.07	.22238	0.508	124.82	.22084
320	0.677	127.28	.22817	0.614	127.11	.22639	0.560	126.90	.22472	0.516	126.64	.22320
330	0.687	129.10	.23050	0.623	128.94	.22872	0.569	128.74	.22707	0.524	128.47	.22554
340	0.697	130.94	.23281	0.632	130.78	.23103	0.578	130.59	.22940	0.531	130.31	.22786
350	0.707	132.80	0.23510	0.641	132.63	0.23333	0.586	132.45	0.23171	0.539	132.17	0.23016
360	0.718	134.68	.23738	0.651	134.50	.23562	0.595	134.32	.23400	0.547	134.05	.23246
370										0.555	135.94	.23475

FREON-12

Properties of Superheated Vapor—*Concluded*

Temp. °F	Abs Press. 140 lb/in.² (Sat'n. Temp. 104.5° F)			Abs Press. 160 lb/in.² (Sat'n. Temp. 114.5° F)			Abs Press. 180 lb/in.² (Sat'n. Temp. 123.7° F)			Abs Press. 200 lb/in.² (Sat'n. Temp. 132.1° F)			Abs Press. 220 lb/in.² (Sat'n. Temp. 139.9° F)		
t	*v*	*h*	*s*	*v*	*h*	*s*	*v*	*h*	*s*	*v*	*h*	*s*	*v*	*h*	*s*
(at sat'n.)	(0.298)	(88.99)	(0.16566)	(0.259)	(89.77)	(0.16522)	(0.228)	(90.58)	(0.16476)	(0.202)	(90.86)	(0.16424)	(0.181)	(91.50)	(0.16375)
100															
110	0.304	89.92	0.16725												
120	0.314	91.60	.17021	0.264	90.68	0.16682									
130	0.323	93.28	.17306	0.273	92.40	.16977	0.233	91.47	0.16665						
140	0.332	94.96	.17590	0.282	94.12	.17269	0.241	93.23	.16964	0.208	92.30	0.16661	0.181	91.52	0.16378
150	0.341	96.65	0.17868	0.290	95.84	0.17553	0.249	94.99	0.17254	0.216	94.10	0.16966	0.188	93.32	0.16685
160	0.350	98.34	.18142	0.298	97.57	.17832	0.257	96.75	.17541	0.224	95.90	.17262	0.195	95.13	.16986
170	0.358	100.03	.18412	0.306	99.31	.18106	0.265	98.52	.17823	0.231	97.70	.17551	0.202	96.94	.17282
180	0.366	101.72	.18678	0.313	101.05	.18377	0.272	100.29	.18102	0.238	99.51	.17838	0.209	98.75	.17576
190	0.374	103.42	.18941	0.321	102.80	.18646	0.280	102.07	.18377	0.245	101.32	.18115	0.216	100.57	.17861
200	0.383	105.14	0.19205	0.329	104.55	0.18913	0.287	103.85	0.18648	0.252	103.13	0.18388	0.223	102.39	0.18142
210	0.391	106.86	.19466	0.336	106.31	.19175	0.294	105.63	.18912	0.258	104.94	.18659	0.229	104.22	.18420
220	0.399	108.59	.19724	0.344	108.07	.19435	0.301	107.42	.19174	0.265	106.76	.18927	0.236	106.05	.18694
230	0.407	110.33	.19976	0.351	109.83	.19693	0.307	109.21	.19433	0.272	108.58	.19192	0.242	107.89	.18962
240	0.415	112.09	.20229	0.358	111.60	.19949	0.314	111.01	.19693	0.278	110.40	.19455	0.248	109.74	.19226
250	0.423	113.85	0.20479	0.366	113.38	0.20203	0.321	112.81	0.19947	0.284	112.23	0.19713	0.254	111.59	0.19487
260	0.431	115.63	.20728	0.373	115.17	.20453	0.327	114.62	.20199	0.290	114.06	.19967	0.259	113.44	.19745
270	0.439	117.42	.20974	0.380	116.97	.20700	0.334	116.44	.20449	0.296	115.89	.20217	0.265	115.30	.20001
280	0.447	119.22	.21219	0.387	118.78	.20946	0.340	118.26	.20698	0.302	117.73	.20467	0.271	117.16	.20255
290	0.455	121.03	.21461	0.394	120.60	.21189	0.347	120.09	.20944	0.308	119.58	.20715	0.277	119.03	.20504
300	0.462	122.85	0.21701	0.401	122.43	0.21432	0.353	121.92	0.21187	0.314	121.44	0.20961	0.282	120.91	0.20753
310	0.470	124.67	.21939	0.408	124.27	.21672	0.359	123.76	.21428	0.320	123.31	.21204	0.288	122.80	.21002
320	0.477	126.50	.22174	0.414	126.12	.21909	0.365	125.19	.21665	0.326	125.19	.21445	0.293	124.70	.21244
330	0.485	128.33	.22411	0.421	127.98	.22145	0.371	127.47	.21904	0.331	127.08	.21685	0.299	126.60	.21485
340	0.492	130.17	.22646	0.428	129.85	.22381	0.377	129.34	.22140	0.337	128.98	.21923	0.304	128.51	.21724
350	0.500	132.02	0.22880	0.435	131.73	0.22616	0.383	131.23	0.22374	0.343	130.89	0.22159	0.309	130.42	0.21960
360	0.507	133.89	.23109	0.442	133.63	.22849	0.390	133.13	.22608	0.348	132.81	.22392	0.314	132.34	.22195
370	0.515	135.78	.23336	0.448	135.55	.23079	0.396	135.05	.22840	0.354	134.74	.22624	0.320	134.27	.22430
380							0.402	136.98	.23072	0.360	136.68	.22856	0.325	136.21	.22665
390							0.408	138.91	.23301	0.365	138.63	.23085	0.330	138.16	.22895
400							0.414	140.85	0.23529	0.370	140.59	0.23314	0.335	140.12	0.23124

APPENDIX B

FREON-12
Pressure-Enthalpy Diagram

APPENDIX B

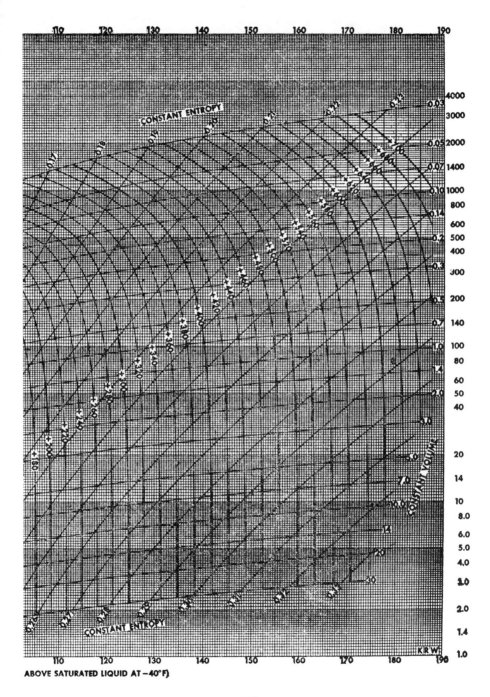

ABOVE SATURATED LIQUID AT −40°F)

APPENDIX C

Equations (8-10) and (8-15), which were introduced in Chapter 8, are completely developed here, starting with the thermal efficiency expression for the Diesel and dual combustion cycles as developed in Equations (8-9) and (8-14) respectively.

Diesel Cycle

$$\eta_t = 1 - \frac{T_4 - T_1}{k(T_3 - T_2)}$$

Using the appropriate process relations to express all temperatures in terms of T_1, we have:

$$T_2 = T_1 \left(\frac{v_1}{v_2}\right)^{k-1}$$

$$T_3 = T_2 \left(\frac{v_3}{v_2}\right) = T_1 \left(\frac{v_3}{v_2}\right)\left(\frac{v_1}{v_2}\right)^{k-1}$$

$$T_4 = T_3 \left(\frac{v_3}{v_4}\right)^{k-1} = T_1 \left(\frac{v_3}{v_2}\right)\left(\frac{v_1}{v_2}\right)^{k-1}\left(\frac{v_3}{v_4}\right)^{k-1}$$

Now, v_1/v_2 is the compression ratio and v_3/v_2 is the fuel cutoff ratio. Expressing these respectively as r_v and r_L, we obtain:

$$T_2 = T_1 r_v^{k-1}$$
$$T_3 = T_1 r_v^{k-1} r_L$$
$$T_4 = T_1 r_v^{k-1} r_L \left(\frac{r_L}{r_v}\right)^{k-1} = T_1 r_L^k$$

APPENDIX C

Finally, the thermal efficiency of the ideal Diesel cycle becomes:

$$\eta_t = 1 - \frac{T_1 r_L^k - T_1}{k(T_1 r_v^{k-1} r_L - T_1 r_v^{k-1})}$$

$$\eta_t = 1 - \frac{T_1 (r_L^k - 1)}{k T_1 r_v^{k-1} (r_L - 1)}$$

$$\eta_t = 1 - \frac{1}{r_v^{k-1}} \left[\frac{(r_L^k - 1)}{k(r_L - 1)} \right]$$

The Dual Combustion Cycle. The dual combustion cycle thermal efficiency may be expressed in terms of compression ratio, fuel cutoff ratio, v_4/v_3, and pressure ratio, p_3/p_2.

The dual combustion cycle thermal efficiency is

$$\eta_t = 1 - \frac{(T_5 - T_1)}{(T_3 - T_2) + k(T_4 - T_3)}$$

Expressing each temperature in terms of the initial temperature and the ratios mentioned above:

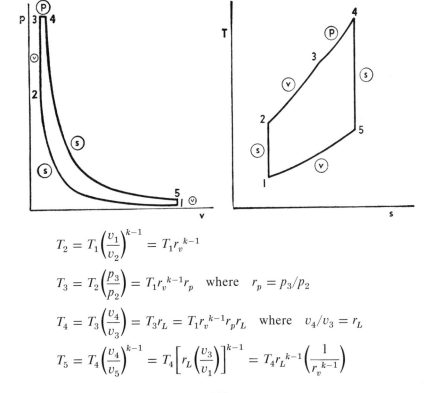

$$T_2 = T_1 \left(\frac{v_1}{v_2}\right)^{k-1} = T_1 r_v^{k-1}$$

$$T_3 = T_2 \left(\frac{p_3}{p_2}\right) = T_1 r_v^{k-1} r_p \quad \text{where} \quad r_p = p_3/p_2$$

$$T_4 = T_3 \left(\frac{v_4}{v_3}\right) = T_3 r_L = T_1 r_v^{k-1} r_p r_L \quad \text{where} \quad v_4/v_3 = r_L$$

$$T_5 = T_4 \left(\frac{v_4}{v_5}\right)^{k-1} = T_4 \left[r_L \left(\frac{v_3}{v_1}\right)\right]^{k-1} = T_4 r_L^{k-1} \left(\frac{1}{r_v^{k-1}}\right)$$

APPENDIX C

$$T_5 = T_1 r_v^{k-1} r_p r_L r_L^{k-1}\left(\frac{1}{r_v^{k-1}}\right) = T_1 r_L^k r_p$$

Finally,

$$\eta_t = 1 - \frac{T_1 r_L^k r_p - T_1}{T_1 r_v^{k-1} r_p - T_1 r_v^{k-1} + k(T_1 r_v^{k-1} r_p r_L - T_1 r_v^{k-1} r_p)}$$

$$\eta_t = 1 - \frac{T_1(r_L^k r_p - 1)}{T_1 r_v^{k-1}[(r_p - 1) + k r_p(r_L - 1)]}$$

$$\eta_t = 1 - \frac{1}{r_v^{k-1}}\left[\frac{(r_L^k r_p - 1)}{(r_p - 1) + k r_p(r_L - 1)}\right]$$

Development of Optimum Pressure Ratio for Maximum Net Work of Brayton Cycle

$$\frac{wk}{(J)} = q_S - q_R = (h_3 - h_2) - (h_j - h_a)$$

$$\frac{wk}{(J)} = h_a\left(\frac{h_3}{h_a} - \frac{h_2}{h_a} - \frac{h_3}{h_2} + 1\right)$$

$$\frac{d[wk/(J)]}{dh_2} = h_a\left(0 - \frac{1}{h_a} + \frac{h_3}{h_2^2} + 0\right) = 0$$

Since h_a is not equal to zero, $h_3/h_2^2 = 1/h_a$. Multiplying through by h_a, rearranging and then expressing in terms of temperatures, we get:

$$\frac{T_a}{T_3} = \left(\frac{T_a}{T_2}\right)^2$$

For the isentropic process a-2, (Fig. 8-12),

$$\frac{T_a}{T_2} = \left(\frac{p_a}{p_2}\right)^{(k-1)/k}$$

Combining these last two equations and solving for p_a/p_2 gives:

$$\frac{p_a}{p_2} = \left(\frac{T_a}{T_3}\right)^{k/2(k-1)} \quad ; \quad \frac{p_2}{p_a} = \left(\frac{T_3}{T_a}\right)^{k/2(k-1)}$$

Regenerative Brayton Cycle Thermal Efficiency

$$\eta_t = 1 - \frac{T_{4'} - T_1}{T_3 - T_{2'}}$$

$$= 1 - \frac{T_1}{T_3}\left[\frac{\frac{T_{4'}}{T_1} - 1}{1 - \frac{T_{2'}}{T_3}}\right]$$

APPENDIX C

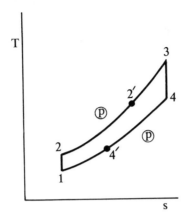

For the ideal regenerator:

$$T_{2'} = T_4$$
$$T_2 = T_{4'}$$

$$\eta_t = 1 - \frac{T_1}{T_3} \left[\frac{\dfrac{T_2}{T_1} - 1}{1 - \dfrac{T_4}{T_3}} \right]$$

Now, the isentropic processes 1-2 and 3-4 give:

$$\frac{T_2}{T_1} = \left(\frac{p_2}{p_1}\right)^{\frac{k-1}{k}} \quad \text{and} \quad \frac{T_4}{T_3} = \left(\frac{p_4}{p_3}\right)^{\frac{k-1}{k}}$$

Then:

$$\eta_t = 1 - \frac{T_1}{T_3} \left[\frac{\left(\dfrac{p_2}{p_1}\right)^{\frac{k-1}{k}} - 1}{1 - \left(\dfrac{p_4}{p_3}\right)^{\frac{k-1}{k}}} \right]$$

Since

$$p_4 = p_1$$
$$p_3 = p_2$$

substituting gives:

$$\eta_t = 1 - \frac{T_1}{T_3} \left[\frac{\left(\dfrac{p_2}{p_1}\right)^{\frac{k-1}{k}} - 1}{1 - \left(\dfrac{p_1}{p_2}\right)^{\frac{k-1}{k}}} \right]$$

APPENDIX C

Let $a = (p_2/p_1)^{(k-1)/k}$; then:

$$\eta_t = 1 - \frac{T_1}{T_3}\left[\frac{a-1}{1-\dfrac{1}{a}}\right]$$

$$\eta_t = 1 - \frac{T_1}{T_3}\left[\frac{a-1}{\dfrac{a-1}{a}}\right]$$

$$= 1 - \frac{T_1}{T_3}a = 1 - \frac{T_1}{T_3}\left(\frac{p_2}{p_1}\right)^{\frac{k-1}{k}}$$

APPENDIX D

The general shape of a gas nozzle (typical for all compressible fluids) will be developed from the basic process and flow equations of Art. 12-4. The continuity of flow equation can be stated as:

$$\dot{M} = \rho A V = \text{constant}$$

which, upon differentiating, yields:

$$\rho A dV + A V d\rho + \rho V dA = 0$$

or

$$\frac{dA}{A} = -\frac{dV}{V} - \frac{d\rho}{\rho} \qquad \text{(D-1)}$$

The nozzle flow energy equation expressed in terms of stagnation conditions is:

$$\frac{V^2}{2g_c(J)} + h = h_0 \qquad \text{(12-1)}$$

which, after differentiation, becomes:

$$\frac{VdV}{g_c(J)} + c_p dT = 0$$

Since

$$(J)c_p = \frac{Rk}{k-1}$$

then

$$\frac{VdV}{g_c} = \frac{-RkdT}{(k-1)} \quad \text{and} \quad dT = \frac{-(k-1)VdV}{Rkg_c} \qquad \text{(D-2)}$$

The isentropic process relationship for an ideal gas can be written as:

$$\left(\frac{\rho}{\rho_0}\right)^{k-1} = \left(\frac{T}{T_0}\right)$$

or

$$\frac{\rho^{k-1}}{T} = \frac{\rho_0^{k-1}}{T_0}$$

Differentiating the isentropic process relationship gives:

$$\frac{1}{T}(k-1)\rho^{k-2} d\rho + \rho^{k-1}(-1)\frac{dT}{T^2} = 0$$

APPENDIX D

and
$$\frac{d\rho}{\rho} = \frac{1}{(k-1)}\frac{dT}{T}$$

Substitution of the differential nozzle flow energy Equation (D-2) yields

$$\frac{d\rho}{\rho} = -\frac{1}{(k-1)}\frac{1}{T}(k-1)\frac{VdV}{Rkg_c}$$

$$\frac{d\rho}{\rho} = -\frac{VdV}{RkTg_c}$$

The mach number and sonic velocity are given by:

$$N_m = \frac{V}{c} \tag{12-9}$$

and

$$c = \sqrt{kRTg_c} \tag{12-8}$$

Using these relationships, the derivative of density becomes:

$$\frac{d\rho}{\rho} = -\frac{N_m^2 dV}{V} \tag{D-3}$$

Substitution of Equation (D-3) into Equation (D-1) gives the required gas nozzle shape relationship:

$$\frac{dA}{A} = -\frac{dV}{V} + \frac{N_m^2 dV}{V}$$

$$\frac{dA}{A} = -[1 - N_m^2]\frac{dV}{V} \tag{12-11}$$

APPENDIX E

Thermodynamic Properties of Air at Low Pressures[a]
Engineers'

T, °R	h, Btu/lb	p_r	u, Btu/lb	v_r	\emptyset Btu/lb · °R
200	47.67	0.04320	33.96	1714.9	0.36303
220	52.46	0.06026	37.38	1352.5	0.38584
240	57.25	0.08165	40.80	1088.8	0.40666
260	62.03	0.10797	44.21	892.0	0.42582
280	66.82	0.13986	47.63	741.6	0.44356
300	71.61	0.17795	51.04	624.5	0.46007
320	76.40	0.22290	54.46	531.8	0.47550
340	81.18	0.27545	57.87	457.2	0.49002
360	85.97	0.3363	61.29	396.6	0.50369
380	90.75	0.4061	64.70	346.6	0.51663
400	95.53	0.4858	68.11	305.0	0.52890
420	100.32	0.5760	71.52	270.1	0.54058
440	105.11	0.6776	74.93	240.6	0.55172
460	109.90	0.7913	78.36	215.33	0.56235
480	114.69	0.9182	81.77	193.65	0.57255
500	119.48	1.0590	85.20	174.90	0.58233
520	124.27	1.2147	88.62	158.58	0.59173
540	129.06	1.3860	92.04	144.32	0.60078
560	133.86	1.5742	95.47	131.78	0.60950
580	138.66	1.7800	98.90	120.70	0.61793
600	143.47	2.005	102.34	110.88	0.62607
620	148.28	2.249	105.78	102.12	0.63395
640	153.09	2.514	109.21	94.30	0.64159
660	157.92	2.801	112.67	87.27	0.64902
680	162.73	3.111	116.12	80.96	0.65621
700	167.56	3.446	119.58	75.25	0.66321
720	172.39	3.806	123.04	70.07	0.67002
740	177.23	4.193	126.51	65.38	0.67665
760	182.08	4.607	129.99	61.10	0.68312
780	186.94	5.051	133.47	57.20	0.68942
800	191.81	5.526	136.97	53.63	0.69558
820	196.69	6.033	140.47	50.35	0.70160
840	201.56	6.573	143.98	47.34	0.70747
860	206.46	7.149	147.50	44.57	0.71323
880	211.35	7.761	151.02	42.01	0.71886
900	216.26	8.411	154.57	39.64	0.72438
920	221.18	9.102	158.12	37.44	0.72979
940	226.11	9.834	161.68	35.41	0.73509
960	231.06	10.610	165.26	33.52	0.74030
980	236.02	11.430	168.83	31.76	0.74540
1000	240.98	12.298	172.43	30.12	0.75042
1020	245.97	13.215	176.04	28.59	0.75536
1040	250.95	14.182	179.66	27.17	0.76019

[a] Abridged from Table 1 in *Gas Tables*, by Joseph H. Keenan and Joseph Kaye. Copyright 1948, by Joseph H. Keenan and Joseph Kaye. Published by John Wiley & Sons, Inc., New York.

APPENDIX E

Engineers'

T, °R	h, Btu/lb	p_r	u, Btu/lb	v_r	ϕ Btu/lb·°R
1060	255.96	15.203	183.29	25.82	0.76496
1080	260.97	16.278	186.93	24.58	0.76964
1100	265.99	17.413	190.58	23.40	0.77426
1120	271.03	18.604	194.25	22.30	0.77880
1140	276.08	19.858	197.94	21.27	0.78326
1160	281.14	21.18	201.63	20.293	0.78767
1180	286.21	22.56	205.33	19.377	0.79201
1200	291.30	24.01	209.05	18.514	0.79628
1220	296.41	25.53	212.78	17.700	0.80050
1240	301.52	27.13	216.53	16.932	0.80466
1260	306.65	28.80	220.28	16.205	0.80876
1280	311.79	30.55	224.05	15.518	0.81280
1300	316.94	32.39	227.83	14.868	0.81680
1320	322.11	34.31	231.63	14.253	0.82075
1340	327.29	36.31	235.43	13.670	0.82464
1360	332.48	38.41	239.25	13.118	0.82848
1380	337.68	40.59	243.08	12.593	0.83229
1400	342.90	42.88	246.93	12.095	0.83604
1420	348.14	45.26	250.79	11.622	0.83975
1440	353.37	47.75	254.66	11.172	0.84341
1460	358.63	50.34	258.54	10.743	0.84704
1480	363.89	53.04	262.44	10.336	0.85062
1500	369.17	55.86	266.34	9.948	0.85416
1520	374.47	58.78	270.26	9.578	0.85767
1540	379.77	61.83	274.20	9.226	0.86113
1560	385.08	65.00	278.13	8.890	0.86456
1580	390.40	68.30	282.09	8.569	0.86794
1600	395.74	71.73	286.06	8.263	0.87130
1620	401.09	75.29	290.04	7.971	0.87462
1640	406.45	78.99	294.03	7.691	0.87791
1660	411.82	82.83	298.02	7.424	0.88116
1680	417.20	86.82	302.04	7.168	0.88439
1700	422.59	90.95	306.06	6.924	0.88758
1720	428.00	95.24	310.09	6.690	0.89074
1740	433.41	99.69	314.13	6.465	0.89387
1760	438.83	104.30	318.18	6.251	0.89697
1780	444.26	109.08	322.24	6.045	0.90003
1800	449.71	114.03	326.32	5.847	0.90308
1820	455.17	119.16	330.40	5.658	0.90609
1840	460.63	124.47	334.50	5.476	0.90908
1860	466.12	129.95	338.61	5.302	0.91203
1880	471.60	135.64	342.73	5.134	0.91497
1900	477.09	141.51	346.85	4.974	0.91788
1920	482.60	147.59	350.98	4.819	0.92076
1940	488.12	153.87	355.12	4.670	0.92362

APPENDIX E

Engineers'

T, °R	h, Btu/lb	p_r	u, Btu/lb	v_r	∅ Btu/lb · °R
1960	493.64	160.37	359.28	4.527	0.92645
1980	499.17	167.07	363.43	4.390	0.92926
2000	504.71	174.00	367.61	4.258	0.93205
2020	510.26	181.16	371.79	4.130	0.93481
2040	515.82	188.54	375.98	4.008	0.93756
2060	521.39	196.16	380.18	3.890	0.94026
2080	526.97	204.02	384.39	3.777	0.94296
2100	532.55	212.1	388.60	3.667	0.94564
2120	538.15	220.5	392.83	3.561	0.94829
2140	543.74	229.1	397.05	3.460	0.95092
2160	549.35	238.0	401.29	3.362	0.95352
2180	554.97	247.2	405.53	3.267	0.95611
2200	560.59	256.6	409.78	3.176	0.95868
2220	566.23	266.3	414.05	3.088	0.96123
2240	571.86	276.3	418.31	3.003	0.96376
2260	577.51	286.6	422.59	2.921	0.96626
2280	583.16	297.2	426.87	2.841	0.96876
2300	588.82	308.1	431.16	2.765	0.97123
2320	594.49	319.4	435.46	2.691	0.97369
2340	600.16	330.9	439.76	2.619	0.97611
2360	605.84	342.8	444.07	2.550	0.97853
2380	611.53	355.0	448.38	2.483	0.98092
2400	617.22	367.6	452.70	2.419	0.98331
2450	631.48	400.5	463.54	2.266	0.98919
2500	645.78	435.7	474.40	2.125	0.99497
2550	660.12	473.3	485.31	1.9956	1.00064
2600	674.49	513.5	496.26	1.8756	1.00623
2650	688.90	556.3	507.25	1.7646	1.01172
2700	703.35	601.9	518.26	1.6617	1.01712
2750	717.83	650.4	529.31	1.5662	1.02244
2800	732.33	702.0	540.40	1.4775	1.02767
2850	746.88	756.7	551.52	1.3951	1.03282
2900	761.45	814.8	562.66	1.3184	1.03788
2950	776.05	876.4	573.84	1.2469	1.04288
3000	790.68	941.4	585.04	1.1803	1.04779
3500	938.40	1829.3	698.48	0.7087	1.09332
4000	1088.26	3280	814.06	0.4518	1.13334
4500	1239.86	5521	931.39	0.3019	1.16905
5000	1392.87	8837	1050.12	0.20959	1.20129
6000	1702.29	20120	1291.00	0.11047	1.25769
6500	1858.44	28974	1412.87	0.08310	1.28268

APPENDIX E

SI

T, K	h, kJ/kg	p_r	u, kJ/kg	v_r	∅ kJ/kg·K
100	99.76	0.029 90	71.06	2230	1.4143
110	109.77	0.041 71	78.20	1758.4	1.5098
120	119.79	0.056 52	85.34	1415.7	1.5971
130	129.81	0.074 74	92.51	1159.8	1.6773
140	139.84	0.096 81	99.67	964.2	1.7515
150	149.86	0.123 18	106.81	812.0	1.8206
160	159.87	0.154 31	113.95	691.4	1.8853
170	169.89	0.190 68	121.11	594.5	1.9461
180	179.92	0.232 79	128.28	515.6	2.0033
190	189.94	0.281 14	135.40	450.6	2.0575
200	199.96	0.3363	142.56	396.6	2.1088
210	209.97	0.3987	149.70	351.2	2.1577
220	219.99	0.4690	156.84	312.8	2.2043
230	230.01	0.5477	163.98	280.0	2.2489
240	240.03	0.6355	171.15	251.8	2.2915
250	250.05	0.7329	178.29	227.45	2.3325
260	260.09	0.8405	185.45	206.26	2.3717
270	270.12	0.9590	192.59	187.74	2.4096
280	280.14	1.0889	199.78	171.45	2.4461
290	290.17	1.2311	206.92	157.07	2.4813
300	300.19	1.3860	214.09	144.32	2.5153
310	310.24	1.5546	221.27	132.96	2.5483
320	320.29	1.7375	228.45	122.81	2.5802
330	330.34	1.9352	235.65	113.70	2.6111
340	340.43	2.149	242.86	105.51	2.6412
350	350.48	2.379	250.05	98.11	2.6704
360	360.58	2.626	257.23	91.40	2.6987
370	370.67	2.892	264.47	85.31	2.7264
380	380.77	3.176	271.72	79.77	2.7534
390	390.88	3.481	278.96	74.71	2.7796
400	400.98	3.806	286.19	70.07	2.8052
410	411.12	4.153	293.45	65.83	2.8302
420	421.26	4.522	300.73	61.93	2.8547
430	431.43	4.915	308.03	58.34	2.8786
440	441.61	5.332	315.34	55.02	2.9020
450	451.83	5.775	322.66	51.96	2.9249
460	462.01	6.245	329.99	49.11	2.9473
470	472.25	6.742	337.34	46.48	2.9693
480	482.48	7.268	344.74	44.04	2.9909
490	492.74	7.824	352.11	41.76	3.0120
500	503.02	8.411	359.53	39.64	3.0328
510	513.32	9.031	366.97	37.65	3.0532
520	523.63	9.684	374.39	35.80	3.0733

APPENDIX E

SI

T, K	h, kJ/kg	p_r	u, kJ/kg	v_r	Ø kJ/kg · K
530	533.98	10.372	381.88	34.07	3.0930
540	544.35	11.097	389.40	32.45	3.1124
550	554.75	11.858	396.89	30.92	3.1314
560	565.17	12.659	404.44	29.50	3.1502
570	575.57	13.500	411.98	28.15	3.1686
580	586.04	14.382	419.56	26.89	3.1868
590	596.53	15.309	427.17	25.70	3.2047
600	607.02	16.278	434.80	24.58	3.2223
610	617.53	17.297	442.43	23.51	3.2397
620	628.07	18.360	450.13	22.52	3.2569
630	638.65	19.475	457.83	21.57	3.2738
640	649.21	20.64	465.55	20.674	3.2905
650	659.84	21.86	473.32	19.828	3.3069
660	670.47	23.13	481.06	19.026	3.3232
670	681.15	24.46	488.88	18.266	3.3392
680	691.82	25.85	496.65	17.543	3.3551
690	702.52	27.29	504.51	16.857	3.3707
700	713.27	28.80	512.37	16.205	3.3861
710	724.01	30.38	520.26	15.585	3.4014
720	734.20	31.92	527.72	15.027	3.4156
730	745.62	33.72	536.12	14.434	3.4314
740	756.44	35.50	544.05	13.900	3.4461
750	767.30	37.35	552.05	13.391	3.4607
760	778.21	39.27	560.08	12.905	3.4751
770	789.10	41.27	568.10	12.440	3.4894
780	800.03	43.35	576.15	11.998	3.5035
790	810.98	45.51	584.22	11.575	3.5174
800	821.94	47.75	592.34	11.172	3.5312
810	832.96	50.08	600.46	10.785	3.5449
820	843.97	52.49	608.62	10.416	3.5584
830	855.01	55.00	616.79	10.062	3.5718
840	866.09	57.60	624.97	9.724	3.5850
850	877.16	60.29	633.21	9.400	3.5981
860	888.28	63.09	641.44	9.090	3.6111
870	899.42	65.98	649.70	8.792	3.6240
880	910.56	68.98	658.00	8.507	3.6367
890	921.75	72.08	666.31	8.233	3.6493
900	932.94	75.29	674.63	7.971	3.6619
910	944.15	78.61	682.98	7.718	3.6743
920	955.38	82.05	691.33	7.476	3.6865
930	966.64	85.60	699.73	7.244	3.6987
940	977.92	89.28	708.13	7.020	3.7108
950	989.22	93.08	716.57	6.805	3.7227
960	1000.53	97.00	725.01	6.599	3.7346
970	1011.88	101.06	733.48	6.400	3.7463

APPENDIX E

SI

T, K	h, kJ/kg	p_r	u, kJ/kg	v_r	∅ kJ/kg·K
980	1023.25	105.24	741.99	6.209	3.7580
990	1034.63	109.57	750.48	6.025	3.7695
1000	1046.03	114.03	759.02	5.847	3.7810
1020	1068.89	123.12	775.67	5.521	3.8030
1040	1091.85	133.34	793.35	5.201	3.8259
1060	1114.85	143.91	810.61	4.911	3.8478
1080	1137.93	155.15	827.94	4.641	3.8694
1100	1161.07	167.07	845.34	4.390	3.8906
1120	1184.28	179.71	862.85	4.156	3.9116
1140	1207.54	193.07	880.37	3.937	3.9322
1160	1230.90	207.24	897.98	3.732	3.9525
1180	1254.34	222.2	915.68	3.541	3.9725
1200	1277.79	238.0	933.40	3.362	3.9922
1220	1301.33	254.7	951.19	3.194	4.0117
1240	1324.89	272.3	969.01	3.037	4.0308
1260	1348.55	290.8	986.92	2.889	4.0497
1280	1372.25	310.4	1004.88	2.750	4.0684
1300	1395.97	330.9	1022.88	2.619	4.0868
1320	1419.77	352.5	1040.93	2.497	4.1049
1340	1443.61	375.3	1059.03	2.381	4.1229
1360	1467.50	399.1	1077.17	2.272	4.1406
1380	1491.43	424.2	1095.36	2.169	4.1580
1400	1515.41	450.5	1113.62	2.072	4.1753
1420	1539.44	478.0	1131.90	1.9808	4.1923
1440	1563.49	506.9	1150.23	1.8942	4.2092
1460	1587.61	537.1	1168.61	1.8124	4.2258
1480	1611.80	568.8	1187.03	1.7350	4.2422
1500	1635.99	601.9	1205.47	1.6617	4.2585
1550	1696.63	691.4	1251.78	1.4948	4.2983
1600	1757.55	791.2	1298.35	1.3485	4.3369
1650	1818.70	902.0	1345.17	1.2197	4.3745
1700	1907.39	1025.1	1392.18	1.1065	4.4112
1750	1941.63	1160.5	1439.38	1.0056	4.4469
1800	2003.36	1310.3	1486.76	0.9164	4.4817
1850	2065.27	1475.0	1534.33	0.8370	4.5156
1900	2127.37	1655.4	1582.09	0.7659	4.5487
1950	2182.72	1830.4	1624.91	0.7084	4.5777
2000	2252.06	2067.9	1678.07	0.6449	4.6127
2500	2883.91	5521	2166.41	0.3019	4.8946
3000	3526.54	12490	2665.55	0.16015	5.1288
3500	4177.22	25123	3172.71	0.09289	5.3294

INDEX

Abbreviations: table of, xvi
Absolute pressure, 7
Absolute temperature, Kelvin, Rankine, 12
Absorbent, 317
Absorption referigeration system, 317-18
Acceleration of gravity, 3
Adiabatic, defined, 71
 compression of air, 169
 flow in nozzles, 336, 349-50
 process, 71 72
 saturation, 438-39
Admission, steam engine, 327
Air, atmosphere, 448
Air: composition of, 467
Air compression; compressors
 aftercooling, 167
 axial flow, 153, 168, 175-78
 capacity, 166-67
 centrifugal, 153, 168, 175-78
 clearance, 160-63
 compression exponent, n, 159-60
 cylinder cooling, 159-60, 169-70, 172
 efficiency: compression, 172, 177; mechanical, 172; overall, 173; volumetric, 162-63
 horsepower: brake, 171; friction, 172; indicated, 172
 indicator card, 153-55
 intercooling, 164-66, 169-70
 multi-stage, 164-66
 pressure: ideal interstage, 165-66
 receiver, 153, 167, 169
 reciprocating, 153-74
 single stage, 153, 155-59
 staging and cooling, 169-70
 steady flow analysis, 138, 166-68, 175-77
 volumetric efficiency, 162-63
 work: minimum, 165-66
Air conditioning
 air washer, 450, 454
 by-passing, 456-58
 comfort, 445-47
 cooling coil, 456
 dehumidification, 448-50

 heating coils, 453-54
 humidification, 450-51, 453
 mixing, 451, 455, 456-58
 plenum chamber, 455
 preheat-reheat system, 455-56
 psychrometric processes, 447-52
 recirculation, 451, 455
 refrigeration, 456
 summer air conditioning, 456-58
 winter air conditioning, 453-56
 ventilation, fresh air, 451, 453, 455-56
Aircraft gas turbine, 254-62
Air refrigeration cycle, 313-16
Air required for combustion, 469-70
Air standard analyses, 223-24
Air Tables, Appendix E
 analysis of gas cycles, 263-66
 formulation and use, 84-91
Air washer, 450, 453
American Petroleum Institute, 465
Ammonia, 311-13, 317-18
Angle of divergence, nozzle, 348, 355
A.P.I. gravity, 465
Apjohn's equation, 436
Apparent volumetric efficiency, 163
Aqua ammonia, 317-18
Atmospheric air, mixture, 430-31, 448, 467
Atmospheric pressure, 7
Available energy, 204-10
 to nozzle, 349, 355, 367
 to Rankine cycle, 274
 to turbine, 380, 383, 385
 to turbine buckets, 362
 to turbine stage, 367, 370-72
Avogadro's law, 430

Barometer, 7, 9
Bernoulli's equation, 133
Black body, 408
Blades (turbine), 359, 370
 efficiency, 362-64, 372-73
 force on, 360
 loss, 362, 367-69, 372-73
 reheat, 367, 382-84
 speed, 360, 364

INDEX

Blades (turbine) (*cont.*)
 vector diagrams for, 360, 371, 377
 velocity coefficient, 362, 372
 velocity ratio, 363–64
Boiler (steam generator), 130, 461
 efficiency, 463
 energy balance (heat losses), 476–79
 fuels, 464
 terminology, 461–63
Boyle's law, 56, 425
Brake engine efficiency, turbine, 378, 385
Brake horsepower
 air compressor, 171
 steam engine, 333
 turbine, 385
Brayton cycle, 238–62
 aircraft gas turbine, 254–62
 basic, 238–43
 intercooled, 244–45
 open, 249–52
 regenerative, 246–48
 reheat, 245–46
 reversed, 313–16
 split shaft, 252
Buckets, turbine. *See* Blades
By-pass governing, turbine, 381
By-passing (air conditioning), 456–58

Calorific value, of fuel, 465
Calorimeter, throttling, 103, 142–44
Capacity
 of air compressor, 166–67
 of boiler, 462
 of refrigeration system, 302
Carbon dioxide
 produced by combustion, 466–68, 472–74
Carnot, cycle, 192–94
 reversed, 198–99, 299–300
Carnot, N. L. S., 199
Carnot, thermodynamic propositions, 200
Carrier's equation, 436
Celsius temperature, 12
Charles' law, 56, 425
Chemical dehumidification, 449
Clausius, Rudolph, 200
Clearance in compressors, 160–63
Clearance volume
 air compressor, 161
 steam engine, 328
Closed cycle, 183
Coefficient, of heat transfer
 fluid film, 395
 overall, 396–98
 thermal conductivity, 394
Coefficient of performance, 303
Cold air standard, 223–24
Combined blade efficiency, 372
Combustion engines, 223
Combustion equations, 466–69, 472
Combustion products, 466–69, 472
 analysis of, 472–75
Comfort chart, 446
Composite walls (heat flow), 398, 403
Compound (multi-cylinder) engines, 332
Compressed air, 153
 refrigeration system, 313–15
Compressed liquid, 106–7
Compressible fluid, 342, 345
Compression efficiency, air compressor, 172, 177
Compression ratio, 229
Compression process, 64
Compression, steam engine, 327
Compression work
 in air compressor, 156
 of cycle, 198, 245, 250
Compressor. *See* Air compression
Condensation and re-evaporation
 steam engine, 331
Condensers
 refrigeration, 306, 317
 steam, 146, 272, 276, 280
 transfer of heat in, 414, 420
Condition curve, turbine, 378–79
Conditioned air, 445
Conduction, 20–21, 392–95
 through nonuniform area, 401–5
Conductivity, thermal, 394
Conservation of energy, 23
 of mass, 466
Constant entropy non-flow process
 gas, 62, 65, 71, 80
 steam, 118–19
Constant pressure non-flow process
 gas, 61, 65, 71, 78
 steam, 116–17
Constant temperature non-flow process
 gas, 62, 66, 71, 79
 steam, 117–18
Constant volume non-flow process
 gas, 61, 65, 71, 77
 steam, 116
Continuity equation, 124–25, 335–36, 341, 353
Convection, 21, 392
 effect upon heat transfer, 395
Conversion table, Engineers' and SI

INDEX

units, 491
Corliss valve, 332
Counter flow, 413-16
Critical flow, 345-47
Critical point, steam, 97
Critical pressure, 345
 ratios, table of, 345
Critical temperature, steam, 98
Curtis impulse stage, 376-77
Cutoff, steam engine, 327
Cycles, defined, 182-83
 four stroke, two stroke, 227
 power, gas (see under name of cycle)
 power, steam (see under name of cycle)
Cylinder condensation, 331

Dalton's Law, 424-25, 432
Dehumidification, 448-50
Dewpoint, 433-34
Diagram efficiency, 362
Diaphragm leakage, 368
Dichlorodifluoromethane (Freon-12), 312
 thermodynamic properties, Appendix B
Diesel, Rudolph, 223
 cycle, 230-33
Diffuser, 335, 343
Discharge work, compressors, 156
Disk losses (turbine), 368
Displacement energy, 22
Double-acting steam engine, 329
Double-pipe heat exchanger, 411-12
Dry bulb temperature, 435-36
Dry compression (refrigeration), 305
Dry vapor, 96
Dual combustion cycle, 235-38

Economizer, heat transfer in, 419
Efficiency
 of air compressors
 compression, 172, 177, 251
 overall, 173
 volumetric, 162-63
 boiler, 463
 engine, 213, 219, 378
 brake, 330, 385
 indicated, 330
 internal, 380
 isentropic turbine, 252
 mechanical
 air compressor, 172

steam engine, 333
steam turbine, 385
propulsive (aircraft), 256
of steam turbine components
 combined blade, 372
 diagram, 362
 nozzle, 349
 nozzle-bucket, 368
 stage, 383
thermal, 187, 213, 219
 Carnot cycle, 194
 (gas and steam power cycles—see under name of cycle)
Ellison calorimeter, 143
Emissivity, 408
Energy, available, 204-10
Energy
 defined, 13
 displacement, 22
 equation for non-flow, 34-36
 equation for steady flow, 27, 125
 heat, 20-21
 internal, 16-20
 kinetic, 15-16
 mechanical, 13-22
 potential, 13-14
 stored, 13
 thermal, 13, 16, 20
 in transition, 13, 20, 22
Energy (heat) balance
 boiler, 476-82
 multi-fluid, 147-49
 two-fluid, 144-46
Energy thermometric scale, 12
Engine efficiency, 213
 gas turbine, 252
 steam engine, 330
 steam turbine, 380, 385
Engine indicator, 154, 328
Engineers' system (of units), 2-4
Enthalpy, 37-38
 of air-vapor mixture, 437-38
 of perfect gas, 68
 of steam, 98
 tables of for Freon-12, Appendix B
Entropy, 43
 change of and unavailability, 213-16
 equations for perfect gas, 70-71
 function, ϕ (Air Tables), 87
 of steam, 98
Equilibrium, thermodynamic, 124-25, 393
Ericsson cycle, 226
Evaporation, boiler
 actual, 462

519

INDEX

Evaporation, boiler (*cont.*)
 equivalent, 462
 factor of, 462
Excess air, 470–72, 475
Exhaust point, turbine, 378
Exhaust pressure line, turbine, 379
Expansion process, 64
Expansion ratio
 gas power cycles, 229
 steam engine, 328
Expansion valve, 306, 317
External irreversibility, 187–90

Factor of evaporation, 462
Fahrenheit temperature, 12
Film coefficient, heat transfer, 395
 effect of, 407
 range of values, 407–8
Fire point, 466
First Law (of Thermodynamics), 25
Fixed blades, 370, 377
Flash point, 466
Flow energy equation, 126
 for air compressor, 138, 167
 for boiler, 121
 for gas turbine, 129, 249–51
 for impulse stage, 368
 for nozzles, 127, 336
 for pumps, 133–36
 for steam engine, 129, 322
 for steam turbine, 129, 380
 for throttles, 140
Flow, gas through nozzle, 337–38
 vapor through nozzle, 351–53
 maximum, 347
Flow, of heat, 392
Flow work, 22, 23
Flue gas analysis, 472–73
Fluid films, heat transfer, 395, 407
 coefficient, 395
 range of values, 407–8
 resistance of, 399
Fluid flow
 counter, parallel, 413–16
 compressible, 342, 345–47
 incompressible, 132–36, 341
Fluid friction, internal, 132–33, 188
Forced convection, 395
Fourier's law, 392
 equation for heat transfer, 394
Four-stroke cycle, 227
Free convection, 395
Freon-12, 312
 table of properties and *p-h* diagram, Appendix B

Friction
 and irreversibility, 159, 188
 horsepower, 172, 333, 385
 in nozzles, 348
 in piping and pumps, 132–34
 in steam engines, 331
 in turbines, 362, 384
Fuel cut-off ratio, 232, 237
Fuel analysis, 464–66
Fuel, types of, 464
Furnace, boiler, 461
 rating, 463
 volume, 462

Gage pressure, 7
Gas, 4
Gas constant(s), 57, 425
 of mixtures, 430
 of water vapor, 431–32
Gas film conductance, 407
Gas laws, 56–57, 425, 430
Gas turbine cycle (open), 249–52
Generating surface, 462
Governing, turbine, 381–82
Gravitational acceleration, 3

Heat, 20–21, 43–45
Heat balance. *See* Energy balance
Heat and entropy, 43–45, 70–72
Heat exchangers
 classification of, 413
 counter flow, 413–16
 cross flow, 414, 417
 log mean temperature difference, 413
 parallel flow, 413–16
Heat pump, 199, 300, 318
Heat rate, turbine, 387
Heat sink, 183
Heat source, 182
Heat transfer equipment, 417
 economizer, 419
 gas regenerator, 246, 421
 steam superheater, 418
 surface condenser, 420
Heat transfer modes, 392
Heating (air conditioning), 453–54
Heating, with refrigeration system, 318
Heating value, of fuel, 465–66
Hot air engine, 225
Hot air standard (cycle), 224
h-s diagram for steam, 112–14
Humidification, 450–51, 453
Humidity

INDEX

relative, 433
specific, 432
Hydrogen, combustion, 466

Ideal air compression process, 155–58
Ideal cycle analysis, 190–96
Idle blade losses, 368–69
Impulse stage, 359, 366–69
Impulse turbine, 359
Incomplete combustion, 472, 481
Incomplete expansion cycle (steam), 292–93
Incomplete expansion steam engine, 323–24
Incompressible flow, 132–36, 341
Indicated horsepower
 air compressor, 172
 steam engine, 329
Indicator diagram (card)
 air compressor, 154, 172
 Otto cycle, 227
 steam engine, 327–28
Intercooler, 164, 244
 pressure for equal work, 165–66, 244–45
Internal combustion cycles
 Diesel, 230
 Ericsson, 226
 Joule, 238
 Otto, 226
 Sabathé (dual combustion), 235
 Stirling, 224
Internal combustion engine, 223
Internal energy, 16–20
 of a gas, 67–68
 of steam, 98
Internal engine efficiency
 gas turbine, 252
 steam turbine, 380
Internal irreversibility, 188
Internal losses, of turbines, 380
 impulse stage, 368–69
 reaction stage, 372–73
Irreversible process, 187–90
 on T-s system, 141, 216
 on h-s system, 142, 215–16
Irreversibility, 141–42, 187–90, 213–16
Irreversibility and change of entropy, 213–16
Isentropic, defined, 72
Isentropic exponent, k, 62
Isentropic non-flow process
 gas, 62, 65, 71, 80
 steam, 118–19

Isothermal air compression, 159–60, 172–73
Isothermal non-flow process
 gas, 62, 66, 71, 79
 steam, 117–18

Joule cycle, 238
 reversed, 313–15
Joule's equivalent, 24
Joule, Sir James, 24

Keenan and Kaye, 85
Keenan, Keyes et al., 94
Kelvin, Lord, 12
Kelvin (heat pump), 318
Kelvin temperature, 12
kB (boiler capacity), 462
Kinetic energy, 15–16
 nozzles, 335, 349
 turbine buckets, 362

Leaving loss, turbine, 378, 380
Liquid film, 395
 conductance of, 395, 408
 resistance of, 399
Log mean area, radius, 402, 405
Log mean temperature difference, 413
Losses
 in boilers, 477–79
 internal turbine, 368–69, 375, 380
 mechanical turbine, 384–85
 reduction gearing, 385
 steam engine, 330–33

Mach number, 340
Manometers, 9
Mass, 1–4
Mass analysis
 of flue gas, 473
 of mixtures, 426–29
Mean effective pressure
 air compressor, 172
 steam engine, 328–29
Mean area, heat transfer, 403
Mean temperature difference, 411–13
Mechanical efficiency
 of air compressor, 172
 of steam engines, 333
 of steam turbine, 385
Mechanical energy, 13, 22
Mechanical equivalent of heat, 24

INDEX

Mechanical kinetic energy, 15–16
Mechanical irreversibility, 187–89
Mechanical work, 22, 38–39
Minimum work (air compression)
　single stage, 160
　multi-stage, 165–66
Mixtures, 424–42
Moisture content, of steam, 102
Mol, 430, 473
Molecular weight(s), 57, 430, 467
Mollier chart, for steam, 112–114

Net work
　of ideal compressor, 155–58, 162
　of cycle, 186, 198
Newcomen, James, 321
Non-expansion cycle (steam), 293–95
Non-expansion steam engine, 325–26
Non-flow energy equation, 34–36
Non-flow reversible process, gas
　applications, 76–84
　constant entropy, 62, 65, 71, 80
　constant pressure, 61, 65, 71, 78
　constant temperature, 62, 66, 71, 79
　constant volume, 61, 65, 71, 77
　general polytropic, 58–61, 62, 64, 70, 81
　on p-v diagram, 63, 75
　on T-s diagram, 75
Non-flow reversible process, steam, 115
　constant entropy, 118, 120
　constant pressure, 116
　constant temperature, 117
　constant volume, 116
Nozzles, 126, 335–55
　convergent, 341
　convergent-divergent, 346
　critical flow in, 345–47
　stagnation state, 337, 343–44
　steam flow in, 352–53
　throat, 339, 343, 347
Nozzle angle, 360, 363–64
Nozzle-bucket efficiency, 368
Nozzle efficiency, 349, 351
Nozzle reheat, 350
Nozzle shape, 341–43
Nozzle valve governing, turbine, 381–82
Nozzle velocity coefficient, 349, 351
Nuclear reactor steam plants, 286–90

Oil, fuel, 464–66
Open cycle, 183, 249
Orsat, analysis, 473, 475

Otto cycle, 226–29
Overall coefficient, heat transfer, 396–98
Overall efficiency, air compressor, 173
Overexpansion, nozzles, 355

Parallel flow, 413–16
Partial pressure, 425, 426–27
　of air vapor mixture, 432
Performance, coefficient of, 303, 306
p-h diagram, 306–7
Phase
　of working substance, 32
　of vapor, 94
Polytropic exponent, 59
Polytropic process (gas), 58–60, 62, 64, 70
Polytropic process in air compressors, 81, 156–58
Polytropic specific heat, 70
Potential energy, 13–14
Power, 28
Power cycles (gas), 223–67
Power cycles (steam), 271–95
Pressure, 7–8
Process line, 33
Process, thermodynamic, 32
Products of combustion, 466–69, 472
　analysis of, 472–75
Properties, thermodynamics, 5, 31–33
Psychrometer, 435
Psychrometric chart, 440–41
　(Also see chart in pocket at back)
Psychrometric equations, 436
Psychrometric processes, 447–52
p-v diagram, 39–40
　for steam, 108–10
　non-flow gas process, 63–64
Pump work, 136, 198, 272–73

Quality, of steam, 102

Radiation, 21, 392, 408–9
Rankine, W. J. M., 13
Rankine cycle, 272–74
　incomplete/non-expansion, 292–95
　with regeneration, 278–84
　with reheat, 275–78
　with reheat and regeneration, 284–86
Rankine cycle, reversed refrigeration, 304
Rankine temperature, absolute, 13
Rated capacity, air compressor, 167

INDEX

Rate of heat transfer
 through composite constructions, 398, 403
 through fluid films, 395–96
 through homogeneous solid, 392–95
Rating cycle, refrigerants, 308
Reaction stage, 359, 370–74
Reaction turbine, 359
Reactive effectiveness, 371, 372
Real Mixture Standard, 224
Reciprocating
 compressor, 153
 engine (steam), 321–34
Refrigerants,
 absorption systems, 317–18
 vapor compression, 311–13
Refrigerating effect, 302
Refrigeration (air conditioning), 456
Refrigeration
 defined, 299
 terms and definitions, 301–3
Refrigeration cycles
 reversed Brayton, 313, 316
 reversed Carnot, 299–301
 reversed Joule, 313–15
 reversed Rankine, 304
Refrigeration systems
 compressed air, 313–16
 vapor absorption, 317–18
 vapor compression, 303–7
 dual expansion, 311, 312
 practical aspects, 309–11
Regenerative power cycles
 Brayton, 246–49
 Ericsson, 226
 Rankine, 278–84
 Stirling, 224–26
Regenerative reheat cycles
 Brayton, 249
 Rankine, 284–86
Regenerator effectiveness, 247
Reheat
 in nozzles, 350, 367
 in steam turbine, 367, 382–84
Reheat cycle
 Brayton, 245–46
 Rankine, 275–78
Reheat factor, turbine, 383
Relative humidity, 433
Relative pressure and volume, 88–89
Release, steam engine, 327
Resistance, to heat transfer, 399–400
Resistivity, heat transfer, 399
Reversed Carnot cycle, 198–99, 299–301

Reversible cycle, 192
Reversible process, 34

Sabathé cycle, 235
Saturated air, 433, 439
Saturated liquid line, 96, 97
Saturated steam, 96
Saturated vapor line, 96, 97
Saturated water, 96
Saturation lines, steam, 97
Saturation pressure, temperature, 96
Savery, Thomas, 321
Saybolt viscosity, 466
Second Law of Thermodynamics, 187, 200
Sensible heating, cooling, 448
Shaft horsepower, 385
SI measurement system, 2
Single stage air compression, 155
Sink, heat, 183
Sonic velocity, 338–40
Source, heat, 182
Spark-ignition engine, 226
Specific gravity, of fuel oil, 465
Specific heat, general, 46–47
 mean values for air, 62
 of mixture, 429
 of perfect gases, 47–49
 of real gases, 49–51
 polytropic, 70
 variation with temperature, 50, 84–85
Specific humidity, 432
Specific volume, 5–6
Specific weight, 5–6
Stage (turbine), defined, 359
 impulse, 366–69
 reaction, 370–73
 comparison, 374–76
Standard rating cycle (for refrigerants), 308
State, of working substance(s), 31–32
State-line curve, turbine, 378–79
State-line end point, 378
State path, 32–33
Steady flow system, 25–26, 124
Steady flow energy equation, 27
Steady flow
 in boilers, 130
 in compressors, 138, 167
 in gas turbine, 129, 249–51
 in liquid pumps, 133–36
 in nozzles, 127, 336
 in steam engine, 129, 322
 in throttles, 140
 in turbines, steam, 129, 380

INDEX

Steady flow (*cont.*)
 of compressible fluids, 342, 345–47
 of incompressible fluids, 132–36, 341
Steam engine(s), 321–34
 characteristics, 321
 cyclic events, 327–28
 engine efficiency, 330
 incomplete expansion, 323–24
 indicated horsepower, 329
 indicator diagram (actual), 327–28
 losses, 331–32
 mechanical efficiency, 333
 multi-cylinder, 332
 non-expansion, 325–26
 steam rate, 330
 superheat and back pressure, 333
 uniflow, 332
Steam flow in nozzles, 352–53
Steam flow, turbine, 387–88
Steam generator (boiler), 130, 461
Steam rate, steam engine, 330
Steam rate, turbine, 387–89
Steam tables, 94 et seq.
Steam turbines, 358–89
 Curtis stage, 377
 governing, 381–82
 efficiencies: brake engine, 378, 385; engine, 378; internal engine, 378, 380; combined blade, 372; diagram, 362; nozzle, 349; nozzle-bucket, 368; stage, 383; mechanical, 385
 impulse blading, 359–66
 impulse stage, 366–69
 mechanical losses, 385
 reaction blading, 370
 reaction stage, 370–74
 reaction/impulse stages compared, 374–78
 reactive effectiveness, 372
 reheat, 367, 380, 384
 reheat factor, 383
 state line, 379
 steam flow, 387
 steam (water) rate, 387
 types, 359
 velocity coefficient
 bucket, 362
 moving blade, 372
 nozzle, 349
 velocity (blade speed–steam speed) ratio, 363
 Willans line, 387–89
Stefan-Boltzmann equation, 408–9
Stirling cycle, 224–26

Stored energy, 13
Subcooled liquid, 107
Suction work, compressors, 157
Summer air conditioning, 456–59
Supersaturation, 105, 351–52
Superheated steam, 96, 105
Superheater, heat transfer in, 418–19
Superheater surface, 462
Surface coefficient, heat transfer, 395
 range of values, 407–8
Surface condenser, heat transfer in, 420
Symbols (and Basic Units), xiii

Tables
 abbreviations, xvi
 air, composition of, 467
 Air Tables, Appendix E
 conversion table, Engineers' and SI units, 491
 critical pressure ratios for common gases and vapors, 345
 dimensions and basic units, 4
 energy forms and types, 25
 Freon-12, thermodynamic properties, Appendix B
 gas constant, R, values, 57
 isentropic exponent, k, for selected gases and vapors, 345
 measurement systems and units, comparison, 488
 molecular weights of:
 combustion constituents, 467
 common gases, 57
 nozzle and diffuser shapes, 343
 refrigerant comparison, 313
 SI prefixes, 492
 symbols with basic units, xiii
Temperature, 12–13
 dewpoint, 433–34
 dry bulb, 435–36
 saturation, 96
 wet bulb, 435–36
Temperature difference, 13, 399
 log mean, 413
Temperature drop, through materials, 393, 400
Temperature-entropy diagram. *See T-s diagram*
Temperature scale intervals, 46
Thermal conductivity, 394
Thermal efficiency, 184–87, 212–13
 of cycle(s). *See cycle by name*
Thermal energy, 13, 16, 20
Thermal equilibrium, 124–25, 393
Thermal irreversibility, 187–190

INDEX

Thermal potential, 399
Thermal resistance, resistivity, 399
Thermodynamic cycle, 182
Thermodynamic equilibrium, 124–25
Thermodynamic process, 32
Thermodynamic properties, 5, 31–33
 of air at low pressure, Appendix E
 of steam, 94–107
 of Freon-12, Appendix B
Thermodynamics, defined, 1
 First Law, 25
 Second Law, 187, 200
Throat, nozzle, 339, 343, 347
Throttle governing, turbine, 381
Throttling calorimeter, 142–44
Throttling loss, steam engines, 331–32
Throttling loss, turbine, 380
Throttling process, 139–41
Ton, of refrigeration, 302
Triple point, steam, 98
T-s diagram, 43–45
T-s diagram, of non-flow gas processes, 74–75
T-s diagram, for steam, 110–11
Turbines
 aircraft gas, 254–62
 gas, 249–52
 steam. *See* Steam turbines
Two-stage air compressor, 164–66
 ideal work, 166
 interstage pressure, 165
Two-stroke cycle, 227

Unavailable energy, 205–6
Underexpansion, nozzles, 355
Uniflow steam engine, 332
Units, 1–4; Appendix A
Units and symbols, xiii
Universal gas constant, 425

Vacuum, 8
Vapor, phase, 32, 94, 97

Vapor refrigerants, 311–13
Variable specific heats, 49, 84, 224
 curves of for gases, 50
Velocity
 bucket, 360
 of gas in nozzle, 337–38
 of steam in nozzle, 352–55
 sonic, 338–40
Velocity coefficient
 nozzle, 349, 351
 turbine buckets, 362
 turbine reaction blading, 372
Velocity ratio, 363
Velocity vector diagram
 impulse blading, 359–63
 reaction blading, 371
 Curtis stage, 377
Volumetric analysis of mixtures, 426–29, 473
Volumetric efficiency, air compressor, 162–63
Volume, specific, 5

Water rate, turbine, 387
Watt, James, 321
Weight, specific, 5
Wet-bulb temperature, 435–36
Wet compression (refrigeration), 305
Wet region, steam, 98
Wet vapor, steam, 101–4
Wheel speed, 360, 364
Willans line, 387–89
Winter air conditioning, 453–56
Work
 and p-v diagram, 38–40, 63–65
 flow, 22–23
 mechanical, defined, 22
 net of cycle, 186, 198, 273
 pump (compression), 132–36, 198, 272–73
Working substance(s), 4–5, 182
 properties, 5, 31–33
 states, 31–33

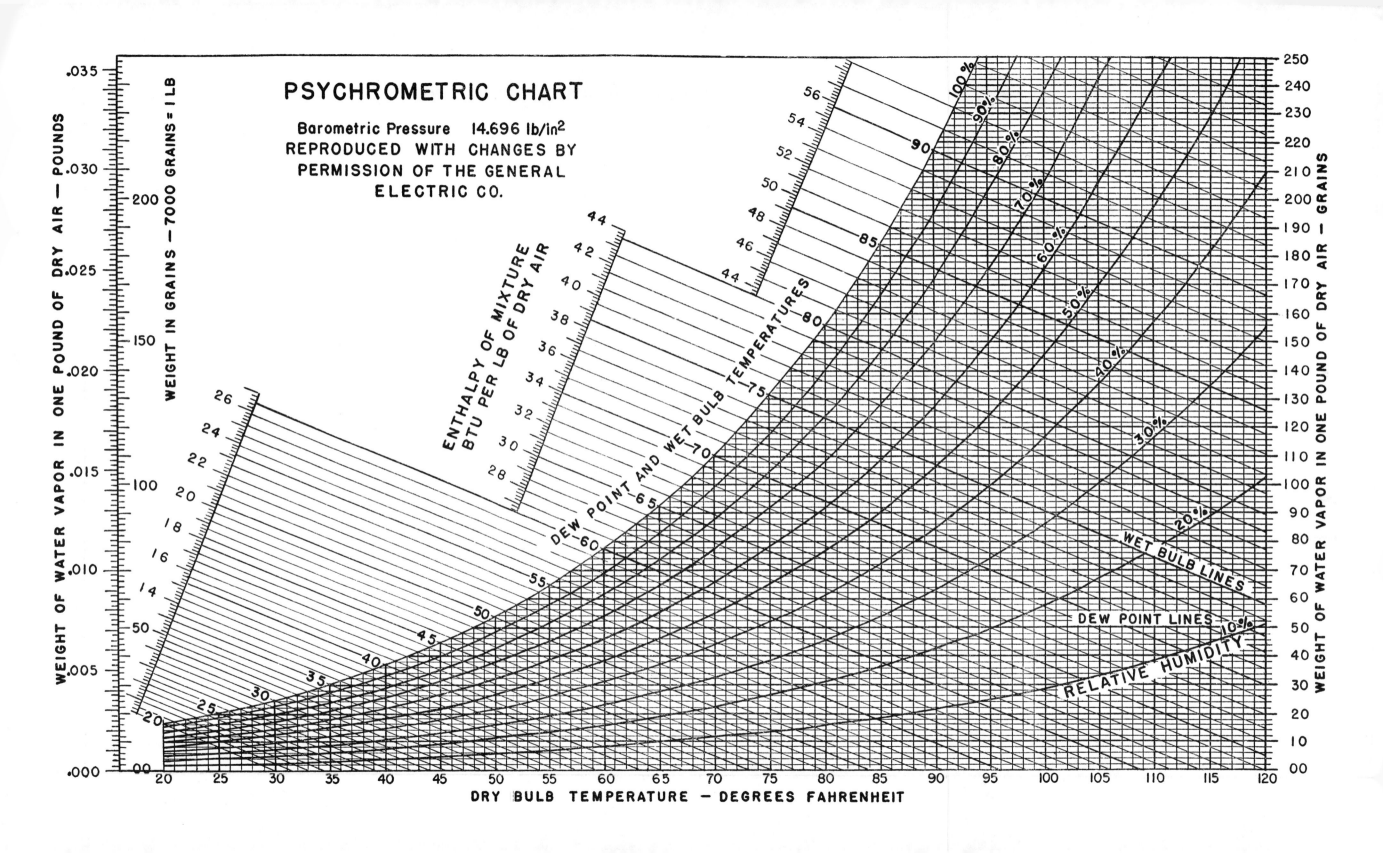